Topological Algebras

Other Books of Interest...

Algebra: *V. Sahai and V. Bist*

Algebra (Vol. 1. Groups): *I.B.S. Passi and I.S. Luthar*

Algebra (Vol. 2. Rings): *I.B.S. Passi and I.S. Luthar*

Current Trends in Mathematics—A Tribute to Harishchandra: *S.D. Adhikari (Ed)*

A First Course in Mathematical Analysis: *D. Somasundaram and B. Choudhary*

A First Course on Parametric Inference: *B.B. Kale*

Foundation of Complex Analysis: *S. Ponnusamy*

Frontiers in Probability and Statistics: *S.K. Mukherjee et. al (Eds)*

Functional Analysis—Selected Topics: *P.K. Jain (Ed)*

An Introduction to Measure and Integration: *I.K. Rana*

Introduction to Rings and Modules (2/e): *C. Musili*

Inverse Methods—General Principles and Applications to Earth Systems Science: *N.K. Indira and P.K. Gupta (Eds)*

Lie Groups and Ergodic Theory: *S.G. Dani (Ed)*

Mathematical Theory of Continuum Mechanics: *R. Chatterjee*

Mathematics and Its Applications in Engineering and Industry: *B. Singh et al. (Eds)*

Metric Spaces (Rev. Ed.): *P.K. Jain and K. Ahmad*

Nonlinear Diffusion Phenomenon: *P.L. Sachdev and R.E. Grundy (Eds)*

Nonlinear Functional Analysis: *R. Akerkar*

Sample Survey Theory: *Des Raj and P.K. Chandhok*

Sequence Spaces and Applications: *P.K. Jain and E. Malkowsky (Eds)*

Small Area Estimation in Survey Sampling: *P. Mukhopadhyay*

Stochastic Process and their Applications: *A. Vijaykumar and M. Srinivasan (Eds)*

Forthcoming

A First Course in Asymptotic Theory: *T.K. Chandra*

A First Course in Probability: *D. Chatterjee and T.K. Chandra*

Fundamentals of Approximation Theory: *H.N. Mahaskar and D.V. Pai*

An Introduction to Commutative Algebra and Number Theory: *S.D. Adhikari*

Mathematical Analysis and Applications: *A.P. Dwivedi (Ed)*

Mathematics and Statistics in Engineering and Technology: *A. Chattopadhyay (Ed)*

TOPOLOGICAL ALGEBRAS

V.K. Balachandran

Narosa Publishing House
New Delhi Madras Bombay Calcutta
London

V.K. Balachandran
Ramanujan Institute for Advanced Studies in Mathematics
Chennai, India

NAROSA PUBLISHING HOUSE

6 Community Centre, Panchsheel Park, New Delhi 110 017
22 Daryaganj, Prakash Deep, Delhi Medical Association Road, New Delhi 110 002
35–36 Greams Road, Thousand Lights, Madras 600 006
306 Shiv Centre, D.B.C. Sector 17, K.U. Bazar P.O., New Bombay 400 705
2F–2G Shivam Chambers, 53 Syed Amir Ali Avenue, Calcutta 700 019
3 Henrietta Street, Covent Garden, London WC2E 8LU, UK

ISBN 81-7319-282-0

Published by N.K. Mehra for Narosa Publishing House, 6 Community Centre, Panchsheel Park, New Delhi 110 017 and printed at Replika Press Pvt Ltd, Delhi 110 040 (India).

To my grand-children

Shobhana and Vivek

for delaying the completion
of the writing of the book

PREFACE

There are very few books devoted to general topological algebras. This book is the outcome of an attempt to present a fairly self-contained and systematic exposition of a number of basic topics concerning such algebras. For the sake of completeness and to increase the usefulness of the book as a reference source (for the material treated) I have not hesitated in stating explicitly (and proving) herein several corollaries and deductions emerging from the main results. I hope to be able to follow this volume with another treating algebras with involution and other topics.

In this book I have considered both complex and real algebras, with and without unity. There is found in the literature on Banach algebras two types of real algebras: the strictly real algebras and the formally real algebras. Both these types of algebras are treated here in more general settings and it turns out that for such strictly real algebras most of the results available for corresponding complex algebras can be extended while such formally real algebras share a few properties with the corresponding complex algebras.

In the treatment of the spectrum I have made some conventional changes. The usage of the term "spectrum of an element" is limited to unital algebras. To take care of the non-unital case the following procedure is adopted. If A is any algebra (non-unital or not) it has its formal unitization A_1 a unital algebra. The spectrum with respect to A_1 is denoted by σ_{A_1}, and set for x in A, $\sigma'(x) = \sigma_A'(x) = \sigma_{A_1}(x)$ (the spectrum of x in A_1), and call $\sigma'(x)$ the quasi-spectrum of x. If A is itself unital the spectrum $\sigma(x) = \sigma_A(x)$ make sense and we have the simple relation $\sigma'(x) = \sigma(x) \bigcup \{0\}$. It is to be noted that always $0 \in \sigma'(x)$ so that $\sigma'(x) \neq \emptyset$. For a real algebra A we have, for x in A, besides $\sigma'(x), \sigma(x)$ also the extended quasi-spectrum $\tilde{\sigma}'(x)$ and extended spectrum $\tilde{\sigma}(x)$ (when A is unital) which are defined in the following way. Every real algebra A has a complexification, a complex algebra $\tilde{A}$. Set $\tilde{\sigma}'(x) = \sigma_{\tilde{A}}'(x)$ and $\tilde{\sigma}(x) = \sigma_{\tilde{A}}(x)$ (note that when A is unital $\tilde{A}$ is also unital). The above changes or extensions in the terminology for the spectrum I feel are ideologically justified and technically useful.

The book consists of nine chapters. Chapter 1 is devoted to algebraic preliminaries. Here I have included more material than strictly needed in the following chapters; for instance it contains an interesting result generalizing a well-known property of the Heisenberg commutation equation but this is not used anywhere in the book. The chapter can be profitably read independently for the topics treated. For the definition of the circle operation in a ring I have followed Perlis-Kaplansky rather than Hille-Jacobson; thus the definition adopted here is at variance with the one employed by Ricart or Bonsall-Duncan in their books (this has to be bourne in mind when comparing results) but agrees with that in Neumark's book.

Chapter 2 deals with some of the basic definition and results concerning topological groups, topological linear spaces and topological algebras. For dealing with continuity questions I have used largely the approach via net convergence since I find this particularly suitable. I have included in this chapter a construction for the completion of a topological algebra using nets, imitating the classical construction of Hausdorff for the completion of a metric space. In connection with the construction I have isolated a property which I have called "essentially bounded" applicable to a net in a topological linear space. This property is weaker than boundeness of the net (as a set). In fact there are convergent nets which are not bounded but every convergent - or even Cauchy - net is essentially bounded.

In Chapter 3 I have considered some generalizations of the norm: quarter-norm, (F) norm, ρ-seminorm, ρ-norm. These give rise to different types of topological algebras: quarter-normed algebras, (F) algebras, ρ-seminormed algebras, ρ-normed algebras and ρ-Banach algebras. Some properties of these algebras are studied here. Based on special properties of quasi-inversion or inversion, topologival algebras come under the following categories: C algebras, Q algebras, I algebras, CQ algebras and CI algebras. Various results concerning there categories are obtained. Finally the chapter contains a large number of results pertaining to topological zero divisors.

Chapter 4 is concerned with a generalization of the notion of convexity called ρ-convexity. This concept leads to some generalizations of locally convex spaces and algebras giving rise to; locally ρ-convex, locally pseudo-convex spaces and algebras; locally sm. ρ-convex, locally sm. pseudo-convex algebras; pseudo Frèchet algebras. After giving a basic treatment of these spaces and algebras the projective limit decomposition, discovered by Michael, is obtained for certain locally sm. pseudo-convex algebras. This decomposition enables the extension to these algebras of some results available for ρ-Banach algebras. Also contained in this chapter is a section on ample algebras and another on topological spectral radius. The latter contains a proof of an important theorem, due to Żelazko, for generating seminorm from ρ-seminorm.

In chapter 5 some differential and integral analysis involving vector valued functions is developed. The extensions to ρ-Banach algebras of the Banach algebra theorems of Nagumo (on the range of the exponential function) and of Gleason-Kahane-Żelazko (on characters) are obtained. There is a section devoted to square roots and quasi-square roots which ends up with a useful result regarding existence of idempotents. The final section concerns power series operation in topological algebras and includes the proof of a theorem of Mitjagin-Rolewvicz-Żelazko affirming local submultiplicativity property for Frechet algebras on which all entire functions can operate.

Chapter 6 is concerned with spectral analysis and applications. Besides quasi-spectrum and resolvent function, I also consider the pseudo-resolvent function which is a useful tool in the study of algebras without unity. As applications of the spectral analysis is obtained a number of Gelfand-Mazur type theorems which include those due to Arens, Shilov, Żelazko. The last section of the chapter is taken up with the proof of an interesting theorem due to Turpin, affirming the local submultiplicativity property for all commutative Frechet algebras which are Q algebras.

The Gelfand representation theory is the subject-matter of chapter 7. For properly understanding and appreciating Gelfand's results I have found it convenient to introduce two classes of topological algebras called Gelfand algebras and spectrally Gelfand algebras. Besides I have also introduced the class called GB algebras i.e. algebras in which the spectral radius formulae of Gelfand and Beurling (in a modified form) hold. The proof of the Beurling-

Gelfand-Żelazko theorem that every complex ρ-Banach algebra is a GB algebra is presented. Other topics considered in this chapter are holomorphic functional calculus for a single element and, automorphirms and derivations. The functional calculus in the strong form has been developed for an element of a ρ-Banach algebra and using the projective limit decomposition theorem the final result extended in the weak form to pseudo-Michael algebras. The Singer-Wermer theorem on derivation has been obtained for ρ-Banach algebras and an extended version for pseudo-Michael algebras.

Chapter 8 deals with function algebras, Shilov boundary, hull-kernel topology, completely regular algebras, holomorphic functional calculus for several elements of a commutative ρ-Banach algebra and Shilov idempotent theorem. Finally, in chapter 9 an exposition of the norm uniqueness theorems of Gelfand and Johnson (extended to ρ-Banach algebras) is given.

For writing this book I have drawn material and ideas from the Banach algebra books of Neumark, Rickart, and Bonsall-Duncan, from the memoir of Michael on locally convex algebra and lecture notes of Żelazko on topological algebras. Besides, I have also been influenced by the topological algebra book of Guichardet and the treatment of Banach algebras in Rudin's book on functional analysis. I wish to acknowledge here my indebtedness to these authors.

I wish to thank P.S. Rema (a former Director) and S.S. Sribala (the present Director) of the Ramanujan Institute for their unstinted help in connection with the publication of the book. I thank N. Vijayarangan (UGC project assistant) for his help in proof reading.

I wish to record my thanks to G. Narayanan, (Assistant Technical Officer (Computer)) of the Ramanujan Institute for his help in the preparation of the well-executed laser print copy of the book. Finally, I wish to express my appreciation to N.K. Mehra of Narosa Publishing House for readily agreeing to publish the book and for his understanding role in the production of the book.

V.K. Balachandran

CONTENTS

CHAPTER I

ALGEBRAIC PRELIMINARIES

§ 1. Some Basic Concepts and Results

1.1.1. Recall that in a ring R an element e_l (respy.[†] e_r) is called a *left* (respy. *right*) *unity* if $e_l a = a$ (respy. $ae_r = a$) for all a in R. If e is both a l. (=left) unity and a r. (=right) unity then e is called a *unity*. If R has a unity, R is called *unital*. We always assume that the unity $e \neq 0$.

The meaning of a positive power a^m of an element a is clear. Also, in a ring R with unity e, we define for any element a, $a^0 = e$. This is well-defined since unity is unique (see 1.1.2).

1.1.2. LEMMA. *If R has a l. unity e_l and a r. unity e_r then necessarily $e_l = e_r = e$ (say) and e is a unity of R which is moreover unique.*

PROOF. $e_l = (e_l)e_r = e_l(e_r) = e_r$. The uniqueness of e is clear.

1.1.3. Let R be unital with unity e. An element a_l^{-1} (respy. a_r^{-1}) is called a *l.* (respy. *r.*) *inverse* or *l.i.* (respy. *r.i.*) of an element a if $a_l^{-1}a = e$ (respy. $aa_r^{-1} = e$). An element a^{-1} is called an *inverse* of a if it is both a l.i. and a r.i. If the inverse a^{-1} of a exits we call a *invertible* or regular.

1.1.4. LEMMA. (a) *If a has a l.i. a_l^{-1} and a r.i. a_r^{-1} then $a_l^{-1} = a_r^{-1} = a^{-1}$ (say) and a^{-1} is the unique inverse of a. In particular, the inverse of an element, whenever it exists, is unique.*

(b) *The invertible elements of a unital ring R form a group $G_i = G_i(R)$ under multiplication.*

(c) *If $a \in G_i$ then $-a \in G_i$ and $(-a)^{-1} = -a^{-1}$.*

PROOF. (a) $a_l^{-1} = a_l^{-1}e = a_l^{-1}(aa_r^{-1}) = (a_l^{-1}a)a_r^{-1} = ea_r^{-1} = a_r^{-1}$.

[†] respy. = respectively.

(b) It suffices to observe that if a, b are invertible then $b^{-1}a^{-1}$ is the inverse of ab and a is the inverse of a^{-1}.

(c) This is an immediate consequence of the identity $xy = (-x)(-y)$ $(x, y \in R)$.

1.1.5. LEMMA. (Kaplansky). *In a unital ring R (with unity e) if an element a has a unique l. (or r.) inverse then a is invertible.*

PROOF. Suppose that a has a unique l. inverse b, so that $ba = e$. Then

$$(ab - e + b)a = aba - a + ba = a - a + e = e.$$

By uniqueness of l. inverse, $ab - e + b = b$ or $ab = e$. Therefore, by 1.1.4 (a), a is ivertible.

1.1.6. Let F be a field and A an (associative, i.e., linear associative) algebra over F. Given a subset S of A, there exists a smallest subalgebra $A(S)$ containing S, called the subalgebra of A *generated by* S; $A(S)$ is the intersection of all subalgebras of A containing S. Explicitly, $A(S)$ is the set of all finite sums of the form $\sum \lambda_k x_k$, where $\lambda_k \in F$ and (each) x_k a finite product of elements from S (clearly this set is a subalgebra $A(S)$ containing S and every subalgebra containing S contains $A(S)$). If A has a unity e then we have also the smallest subalgebra $A_1(S)$ containing S and e. We have

$$A_1(S) = Fe + A(S) = \{\lambda e + x : \lambda \in F, x \in A(S)\}.$$

If A is only a *unital ring* then we have the analogous subrings $A(S)$ and $A_1(S)$. Here $A_1(S) = \mathbb{Z}e + A(S)$.

A subalgebra (respy. subring) A_0 of a unital algebra (respy. ring) A is called a *subunital algebra* (respy. subunital ring) of A if the unity e(of A) $\in A_0$; then A_0 is automatically unital with e as its unity. Note, however, that a subalgebra (respy. subring) of A can be unital without being subunital (i.e. A_0 can have a unity $e_0 \neq e$). For example in the algebra A of 2×2 diagonal matrices over a field F the subalgebra A_0 comprising diagonal matrices with second entry 0 is unital but not subunital ($e =$ diag $(1, 1)$,

$e_0 =$ diag $(1,0)$).

1.1.7. Let A be an algebra (or a ring). For $a, b \in A$ we write $a \leftrightarrow b$ if $ab = ba$ and say that a and b *commute*. For two commuting elements a, b we have the Binomial Theorem.

$$(a+b)^n = \sum_{k=0}^{n} \binom{n}{k} a^{n-k} b^k, \text{ where } \binom{n}{k} = \frac{n!}{k!(n-k)!}, n \qquad (*)$$

a positive integer. (This can be proved by induction as in the classical case of the theorem, see [13, p.52].)

For two subsets S_1, S_2 of A we write $S_1 \leftrightarrow S_2$ if for every $a \in S_1, b \in S_2$ we have $a \leftrightarrow b$. We also write for an element $a \in A$ and a subset S of A, $a \leftrightarrow S$ if $a \leftrightarrow b$, for every $b \in S$.

If $S \leftrightarrow S$ we say that the (subset) S is *commutative*. For a subset S of A we set

$S' = \{x \in A : x \leftrightarrow S\}$ and call S' the *commutant* (or centralizer) of S in A. $S'' = (S')'$ is called the double commutant of S.

1.1.8. LEMMA. *Let S be a subset of an algebra (or ring) A. Then*

(i) *S' is a subalgebra (subring) of A.*

(ii) *$S \subseteq S''$ and consequently $A(S) \subseteq S''$; also when A is unital, $A_1(S) \subseteq S''$.*

(iii) *If $S \subseteq T \subseteq A$, where T is a subset, then $T' \subseteq S'$.*

(iv) *$S' = S'''$.*

(v) *S is commutative iff*† *$S \subseteq S'$ iff S'' is commutative.*

(vi) *If S is commutative so are the subalgebras $A(S), S''$.*

(vii) *If S is a maximal commutative subset then $S = S''$.*

PROOF. (i), (ii), (iii) an clear. By (ii) we have $S' \subseteq (S')'' = S'''$. On the other hand, since $S \subseteq S''$ (by (ii)), we get; using (iii), $S''' \subseteq S'$. Hence $S' = S'''$ which is (iv). For (v), we observe that

† iff = if and only if.

"S is commutative" iff $S \subseteq S'$. If $S \subseteq S'$ then by (iii), (iv), $S'' \subseteq S' \subseteq (S'')'$, so that S'' is commutative. On the other hand "S'' commutative" trivially implies "S commutative" (since $S \subseteq S''$).

For (vi), we note that by (v), S'' is commutative and so also $A(S) \subseteq S''$ (see (ii)).

Finally, for (vii), we have by (vi), $S \subseteq S''$ and S'' is commutative by (vi). The maximality of S implies that $S = S''$.

1.1.9. PROPOSITION. *Every commutative subset S of an algebra A is contained in a maximal*[†] *commutative subalgebra $A_m(S)$ of A such that*

$$S \subseteq A(S) \subseteq S'' \subseteq A_m(S).$$

In particular, each element a of A is contained in a maximal commutative subalgebra $A_m(a)$. If A is unital with unity e then $e \in A_m(S)$.

PROOF. By 1.1.8 (vi), $A(S), S''$ are commutative subalgebras, and $S \subseteq A(S) \subseteq S''$. Since the union of any linearly ordered (with respect to inclusion) family of commutative subalgebras is a commutative subalgebra we can apply Zorn's lemma to obtain a maximal commutative subalgebra $A_m(S) \supseteq S''$.

If A has unity e then $S_1 = S \bigcup \{e\}$ is commutative and hence

$$A_m(S_1) \supseteq A_m(S) \supseteq S.$$

By maximality of $A_m(S)$ we have $A_m(S_1) = A_m(S)$, $e \in A_m(S_1) = A_m(S)$.

1.1.10. Let A, A^* be two algebras (over the same field F). A mapping $\varphi : A \to A^*$ is called a *homomorphism* if it is linear, and multiplicative, i.e. $\varphi(ab) = \varphi(a)\varphi(b)$ for all $a, b \in A$. An injective or $1-1$ homomorphism φ is called a *monomorphism* and a surjective or onto homomorphism is called an *epimorphism*. A homomorphism φ which is both $1-1$ and onto is called an

[†] i.e. if S_1 is a commutative subalgebra of A with $A_m(S) \subseteq S_1$ then $S_1 = A_m(S)$.

isomorphism (of A onto A^*). Sometimes we use the term "isomorphism into" for a monomorphism and "homomorphism onto" for an epimorphism.

1.1.11. Given two algebras A_1, A_2 (over F) we have the direct product algebra $A = A_1 \times A_2$:

$$A = \{(a_1, a_2) : a_1 \in A_1, a_2 \in A_2\}.$$

The algebra operations on A are given by:

$$\begin{aligned}(a_1, a_2) + (b_1, b_2) &= (a_1 + b_1, a_2 + b_2)\\ \lambda(a_1, a_2) &= (\lambda a_1, \lambda a_2)\\ (a_1, a_2)(b_1, b_2) &= (a_1 b_1, a_2 b_2)\end{aligned}$$

where $a_1, b_1 \in A_1; a_2, b_2 \in A_2;\ \lambda \in F$.

1.1.12. An algebra A (over F) can be extended canonically into a unital algebra A_1 called the *unitization* of A. We take for A_1 the carterian product linear space (over F) given by:

$$A_1 = F \times A = \{(\lambda, a) : \lambda \in F, a \in A\}.$$

Write $(1, 0) = e_1, (0, a) = \dot{a}$. Then:

$$(\lambda, a) = \lambda e_1 + \dot{a}, (\mu, b) = \mu e_1 + \dot{b}.$$

Define multiplication in A_1 by:

$$(\lambda e_1 + \dot{a})(\mu e_1 + \dot{b}) = \lambda\mu e + \lambda\dot{b} + \mu\dot{a} + \dot{a}\dot{b}.$$

It is easy to check that under this multiplication A_1 is an algebra and that the map $a \mapsto \dot{a}$ is a monomorphism. We identify $\dot{a}$ with a and write.

$(\lambda, a) = \lambda e_1 + a$, so that $A_1 = Fe_1 + A$. We observe that A as a linear subspace of A_1 has codim† 1. Further it is clear from the definition of multiplication in A_1 that A is both a left as well as a right ideal of A_1.

If B is a subalgebra of a unital A with its unity $e \notin B$ then $B_1 = Fe + B$ is unital subalgebra of A called the *unitization* of

† codim = codimension = $\dim(A_1/A)$.

B in A.

We note that the above construction for A_1 can be carried out even when A has a unity e; of course $e_1 \neq e$ (since $e_1 \notin A$). Finally, we remark that a ring R has a unitization R_1.

It is given by

$$R_1 = \mathbb{Z}^{\dagger} \times R = \mathbb{Z}e_1 + R$$

with

$$(me_1 + a)(ne_1 + b) = mne_1 + mb + na + ab\,(m, n \in \mathbb{Z}).$$

1.1.13. Let R be a ring and '$\circ$' denote the binary operation, called circle operation, on R given by

$$a \circ b = a + b + ab \qquad (a, b \in R).$$

We denote R with this binary (multiplication) operation by $S_\circ = S_\circ(R)$; $S_\circ$ is a multiplicative system.

1.1.14. LEMMA. *Let R be unital, with unity e, and $S_\circ$ as defined above. Denote by $S = S(R)$ the underlying multiplicative semi-group structure of R. Then the map.*

$$\tau = \tau(S) : x \in S \mapsto x - e \in S_\circ$$

is an isomorphism of S onto $S_\circ$ (as multiplicative structures). Hence $S_\circ$ is a semi-group with '0 ' as the identity element.

PROOF. Clearly τ is bijective. Further

$$\tau(a) \circ \tau(b) = (a - e) \circ (b - e) = ab - e = \tau(ab).$$

COROLLARY. *Let R be any ring. Then $S_\circ = S_\circ(R)$ is a semi-group with '0' as identity.*

PROOF. Consider the unitization R_1 of R. Then $e_1 + R = \{e_1 + x : x \in R\}$, under multiplication, is a subsemi-group S^* of $S(R_1)$. By considering the restriction of the isomorphism (see 1.1.14), $\tau : S(R_1) \to S_\circ(R_1)$ to S^* we get the desired conclusion for $S_\circ$.

1.1.16. If $a \circ b = 0$ then a is called a *l.q.i.* (=*left quasi-*

---

† $\mathbb{Z} = \{0, \pm 1, \pm 2, \ldots\}$ (the ring of integers).

inverse) of b and b a *r.q.i.* (=*right quasi-inverse*) of a; also then b (respy. a) is said to be *l.q.* (respy. *r.q.*) *invertible*. If a is both l.q. invertible and r.q. invertible it is called *q. invertible* or *q. regular* (the meanings of l.q. regular and r.q. regular are clear).

1.1.17. LEMMA. (a) *If $a \in R$ has a l.q.i. a'_l and a r.q.i. a'_r then $a'_l = a'_r = a'$ (say) and a' is q.i. of a. Moreover, $a \leftrightarrow a'$.*

(b) *The set G_q of q. invertible elements of R is a group under the multiplication '$\circ$'; G_q is a subsemi-group of $S_\circ$.*

PROOF. (i) The proof of the first statement is similar to that of 1.1.4 (a); for the second we note that

$$a \circ a' = 0 = a' \circ a \Rightarrow aa' = a'a.$$

(ii) The proof is similar to that of 1.1.4 (b).

1.1.18. LEMMA. *Suppose that a is an invertible (respy. q. invertible) element of a unital ring (respy. ring) R, $x \in R$ and $x \leftrightarrow a$. Then $x \leftrightarrow a^{-1}$ (respy. $x \leftrightarrow a'$). In particular a^{-1} (respy. a') $\in \{a\}''$ (the double commutant).*

PROOF. $xa^{-1} = (a^{-1}a)xa^{-1} = a^{-1}(ax)a^{-1} = a^{-1}(xa)a^{-1} = a^{-1}x$. Similarly, $x \circ a' = a' \circ x$.

1.1.19. COROLLARY. *Let A_m be a maximal, commutative subalgebra of an algebra A. If $b \in A_m$ is invertible (respy. q. invertible) in A then b^{-1} (respy. b') $\in A_m$.*

PROOF. By 1.1.18, $b^{-1} \leftrightarrow A_m$ and so $S = A_m \bigcup \{b^{-1}\}$ is commutative. By 1.1.8 (vi), the subalgebra $A(S)$ is commutative. Since $A(S) \supseteq S \supseteq A_m$, by maximality of A_m, $A_m = S = A(S)$, so that $b^{-1} \in S \subseteq A_m$. Similarly $b' \in A_m$.

1.1.20. LEMMA. *Let R be a unital ring with unity e. Then an element a of R has a l.q.i. (respy. r.q.i.) b iff $e+a$ has $e+b$ as a l.i. (respy. r.i.). In particular a is q. invertible iff $e+a$ is invertible and then we have $(e+a)^{-1} = e + a'$, a' being the q.i. of a. Moreover, $\tau^{-1} : a \mapsto e + a$ is an isomorphism of G_q onto G_i.*

PROOF. The statements follow from the identity

$(e+a)(e+b) = e + (a \circ b)$.

1.1.21. LEMMA. *If A_1 is the unitization of an algebra (or ring) A and $a \in A$ has a l.q.i. (or a r.q.i.) b_1 in A_1 then $b_1 \in A$. Hence a is q. invertible in A_1 iff its q. invertible in A.*

PROOF. Write $b_1 = \alpha e_1 + b$, where $b \in A$ and $\alpha \in F$ or $\mathbb{Z}$ according as A is an algebra over F or a ring. Then

$$a \circ b_1 = 0 \Rightarrow a + \alpha e_1 + b + \alpha a + ab = 0 \Rightarrow \alpha = 0 \Rightarrow b_1 = b.$$

1.1.22. Remark. Related to $\circ$-operation in R is the operation denoted by $\times$ and given by $a \times b = a + b - ab$ $(a, b \in R)$. This operation is again associative and has other similar properties of $\circ$. If $S_\times$ denotes the multiplicative system in R corresponding to $\times$, then the map $a \mapsto -a$ is easily seen to be an isomorphism of $S_\circ$ onto $S_\times$. The operations $\circ, \times$ are mutually connected by:

$$a \circ b = -(-a \times -b); \; a \times b = -(-a \circ -b).$$

The operation $\times$ has been introduced by Hille (following a suggestion of Jacobson) and called by him as cross-product. Some authors like Rickart, Bonsall-Duncan adopt the definition of the cross-product for the circle operation.

1.1.23. LEMMA. *Every np. (=nilpotent) element a of a ring R is q. invertible with*

$$a' = -a + a^2 + \cdots \pm a^{k-1} \; (a^k = 0, \; a^{k-1} \neq 0). \qquad (*)$$

Further, if R is unital with unity e, then $e + a$ is invertible with

$$(e+a)^{-1} = a - a^2 + \cdots \pm a^{k-1}. \qquad (**)$$

PROOF. It is straightforward to check that a' as defined in $(*)$ is the q.i. of a, and $(e+a)^{-1}$ as defined in $(**)$ is the inverse of $e + a$.

1.1.24. LEMMA. *If φ is a homomorphism of a ring R then φ preserves $\circ$-operation and hence also l. (or r.) q. invertibility.*

If R is unital with unity e then $\varphi(e)$ is the unity of $\varphi(R)$ and φ preserves l. or r. invertibility.

PROOF.

$$\begin{aligned}\varphi(a\circ b)=\varphi(a+b+ab) &= \varphi(a)+\varphi(b)+\varphi(a)\varphi(b)\\ &= \varphi(a)\circ\varphi(b)\ (a\cdot b\in R).\end{aligned}$$

Further, since $\varphi(0)=0$, $a\circ b=0\Rightarrow\varphi(a)\circ\varphi(b)=0$, completing the proof of the first statement. The second statement is an immediate consequence of φ being a homomorphism.

1.1.25. LEMMA. *In a ring, if $a\circ b$ (respy. $b\circ a$) $=a$ and a is l.q. (respy. r.q.) invertible then $b=0$.*

PROOF. Suppose that $a\circ b=a$. Then

$$b=0\circ b=(a'_l\circ a)\circ b=a'_l\circ(a\circ b)=a'_l\circ a=0.$$

Similarly, $b\circ a=a\Rightarrow b=0$.

1.1.26. LEMMA. *Let A be an algebra and $u\in A$ an idempotent. Then:*

(i) *For any $\lambda\neq-1,\lambda u$ is q. invertible with $(\lambda u)'=-\lambda(1+\lambda)^{-1}u$. In particular $u'=-\frac{1}{2}u$. Also, when e exists, $e'=-\frac{1}{2}e$.*

(ii) *If $-u$ is q. invertible then $u=0$. Hence, in a unital $A,-e$ is not q. invertible.*

(iii) *If A has unity e then $e+u$ is invertible with $(e+u)^{-1}=e-\frac{1}{2}u$ and $(\lambda e)^{-1}=\lambda^{-1}e\,(\lambda\neq0)$.*

PROOF. (i) $\lambda u-\lambda(1+\lambda)^{-1}u-\lambda^2(1+\lambda)^{-1}u=0$, whence $(\lambda u)'=-\lambda(1+\lambda)^{-1}u$.

(ii) $(-u)\circ u=-u+u-u^2=-u$. Hence, by 1.1.25, $u=0$.

(iii) $(e+u)^{-1}=e+u'=e-\frac{1}{2}u$. Since $(\lambda e)(\lambda^{-1}e)=e$ and $\lambda e\leftrightarrow\lambda^{-1}e,(\lambda e)^{-1}=\lambda^{-1}e$.

1.1.27. PROPOSITION. (a) *In any ring R we have the identities:*

(i) $ba\circ(-ba-bxa)=-b(ab\circ x)a$

(ii) $(-ba - bxa) \circ ba = -b(x \circ ab)a$

(iii) $-a^2 \circ b = a \circ (-a \circ b)$

(iv) $b \circ -a^2 = (b \circ -a) \circ a$
where $a, b, x \in R$.

(b) *In a unital ring* R *with unity* e *we have:*

(v) $(bxa + e)(e - ba) = bx(e - ab)a + e - ba$

(vi) $(e - ba)(bxa + e) = b(e - ab)xa + e - ba$
where $a, b, x \in R$.

PROOF. (i)

$$\begin{aligned} \text{LHS}^{\dagger} &= ba + (-ba - bxa) + ba(-ba - bxa) \\ &= -b(x + ab + abx)a = -b(ab \circ x)a = \text{RHS}^{\dagger} \end{aligned}$$

(ii) Similar to (i).

(iii)

$$\begin{aligned} \text{RHS} &= a + (-a \circ b) + a(-a \circ b) \\ &= a - a + b - ab + a(-a + b - ab) \\ &= -a^2 + b - a^2 b = -a^2 \circ b = \text{LHS}. \end{aligned}$$

(iv) Similar to (iii).

(v) LHS $= bx(a - aba) + e - ba = bx(e - ab)a + e - ba =$ RHS.

(vi) Similar to (v).

1.1.28. COROLLARY. (i) ab *is l.q. (respy. r.q.) invertible iff* ba *is l.q. (respy. r.q.) invertible. In particular,* ab *is q. invertible iff* ba *is q. invertible.*

(ii) $e - ab$ *is l. (respy. r.) invertible iff* $e - ba$ *is l. (respy. r.) invertible. In particular,* $e - ab$ *is invertible iff* $e - ba$ *is invertible.*

PROOF. (i) By taking $x = (ab)'_l$ (respy. $(ab)'_r$) in (ii) (respy. (i)) we conclude that "ab is l.q. invertible (respy. r.q. invertible)" $\Rightarrow$ "ba is l.q. invertible (respy. r.q. invertible)". This plus symmetry consideration proves (i).

(ii) Taking $x = (e - ab)_l^{-1}$ (respy. $(e - ab)_r^{-1}$ in (v) (respy (vi)) of 1.1.27, and using symmetry we get the desired conclusions.

† LHS = Left Hand Side; RHS= Right Hand Side.

1.1.29. LEMMA. *In a unital ring R with unity e, if ab has a r.i. (respy. l.i.) c then bc (respy. ca) is a r.i. (respy. l.i.) of a (respy. b). Similarly, in any ring R, if $a \circ b$ has a r.q.i. (respy. l.q.i.) c then $b \circ c$ (respy. $c \circ a$) is a r.q.i. (respy. l.q.i.) of a (respy.) b.*

PROOF. Clear.

1.1.30. COROLLARY. *The elements ab, ba are invertible iff a, b are invertible. In particular, if $a \leftrightarrow b$ and ab invertible then a, b are invertible.*

Similarly, $a \circ b$, $b \circ a$ are q. invertible iff a, b are q. invertible, and when $a \leftrightarrow b$, $a \circ b$ is q. invertible iff a, b are q. invertible.

PROOF. If a, b an invertible then as already seen, ab is invertible with $b^{-1}a^{-1}$ as its inverse. On the other hand if ab, ba are invertible than by applying 1.1.29, 1.1.4 (a) we conclude that a, b are invertible. Hence the statements concerning invertible elements. The proof of the statements concerning q. invertible elements is similar.

1.1.31. PROPOSITION. *Let R be a ring and $x, a \in R$. If x', $(a + x'a)'_l$ and $(a + ax')'_r$ exist then $(x + a)'$ exists and*

$$(x + a)' = (a + x'a)'_l \circ x' = x' \circ (a + ax)'_r.$$

PROOF. We have

$$\begin{aligned} (a + x'a)'_l \circ x' \circ (x + a) &= (a + x'a)'_l \circ (x' + x + a + x'x + x'a) \\ &= (a + x'a)'_l \circ (x' \circ x + a + x'a) \\ &= (a + x'a)'_l \circ (a + x'a) = 0 \\ &\qquad\qquad (\text{since } x' \circ x = 0) \end{aligned}$$

Similarly,

$$\begin{aligned} (x + a) \circ x' \circ (a + ax')'_r &= (x + a + x' + xx' + ax') \circ (a + ax)'_r \\ &= (a + ax') \circ (a + ax)'_r = 0. \end{aligned}$$

It follows that $x + a$ is both l.q. invertible and r.q. invertible and so q. invertible. The required conclusions are now clear.

1.1.32. COROLLARY. *With the hypothesis in* 1.1.31 *we have* $(x + a)' - x' = (a + x'a)'_l + (a + x'a)'_l x' = (a + x'a)'_r + x'(a + x'a)'_r$.

1.1.33. COROLLARY. *If* $x \leftrightarrow a$ *and* $x', (a + ax')'$ *exist, then* $(x + a)'$ *exists with*

$$(x + a)' = x' \circ (a + ax')' = (a + ax')' \circ x'.$$

PROOF. Since $x \leftrightarrow a$, by 1.1.18, $x' \leftrightarrow a$, so that

$$(a + x'a)'_l = (a + ax')'_l = (a + ax')'_r = (a + x'a)'_r.$$

The required result now follows by 1.1.31.

§ 2. Ideals and Radical

1.2.1. Let R be a ring. If I is both a l.(=left) ideal and a r.(=right) ideal of R it is called a *bi-ideal.* We will use the term ideal for denoting any one of these: a l. ideal, a r. ideal or a bi-ideal. When R is commutative or anti-commutative, there is, of course, no distinction between a l. or a r. ideal and the term ideal has an unambiguous meaning.

If a ring R is an algebra (over a field F) then we have also the algebra ideals i.e. ring ideals which are also linear subspaces. While every algebraic ideal is a ring ideal the reverse is not always true. Thus, if the 1-dim real linear space $\mathbb{R}$ is given the trivial algebra structure ($\mathbb{R}^2$ † $= \{0\}$) the subset $\mathbb{Z}$ of $\mathbb{R}$ is a ring ideal which is not an algebra ideal. This distinction vanishes for a unital algebra A since we have the relation $\lambda a = (\lambda e)a$ ($\lambda \in F; a \in A$; $e =$ unity of A).

Let R be a ring (or algebra). Recall that an element $a \neq 0$ of R is called a *l.z.d.*(=left zero divisor) if there is an element $b \neq 0$ with $ab = 0$. Similarly, $a \neq 0$ is a *r.z.d.*(=right zero divisor) if there is a $c \neq 0$ with $ca = 0$.

Given a subset S of R the *left annihilator* $\mathcal{A}_l(S)$ and *right annihilator* $\mathcal{A}_r(S)$ are defined by:

$\mathcal{A}_l(S) = \{x \in R : xa = 0, \forall\ a \in S\}$

$\mathcal{A}_r(S) = \{y \in R : ay = 0, \forall\ a \in S\}$.

---

† In a ring R, for two subsets S_1, S_2 we write $S_1S_2 = \{xy : x \in S_1, y \in S_2\}$; also for a subset $S, S^2 = SS$.

If $S = \{a\}$ then we write $\mathcal{A}_l(a), \mathcal{A}_r(a)$ for the corresponding annihilator ideals. If $S = R$ we denote the two annihilators by $\mathcal{A}_l$ and $\mathcal{A}_r$.

A ring R (respy. algebra A) is called a ring (respy. algebra) *with trivial multiplication* if R^2 (respy. A^2) $= \{0\}$.

1.2.4. LEMMA. (i) *$\mathcal{A}_l(S)$ (respy. $\mathcal{A}_r(S)$ is a l. (respy. r.) ideal of R.*

(ii) *If I is a l. (respy. r.) ideal of R then $\mathcal{A}_l(I)$ (respy. $\mathcal{A}_r(I)$) is a bi-ideal.*

(iii) *$\mathcal{A}_l, \mathcal{A}_r$ are nilpotent bi-ideals: $\mathcal{A}_l^2 = \mathcal{A}_r^2 = \{0\}$.*

PROOF. (i) Clear.

(ii) Let I be a l. ideal, $x \in \mathcal{A}_l(I), y \in A$ and $a \in I$.

Then $xy.a = x.ya = 0, xy \in \mathcal{A}_l(I)$, proving $\mathcal{A}_l(I)$ is a bi-ideal. Similarly, $\mathcal{A}_r(I)$ is a bi-ideal.

(iii) If $a, b \in \mathcal{A}_l$ then $ab \in aA = \{0\}$, so that $\mathcal{A}_l^2 = \{0\}$. Similarly, $\mathcal{A}_r^2 = \{0\}$.

1.2.5. PROPOSITION. *If a ring R (respy. algebra A) $\neq \{0\}$ has no l.-ideals $\neq R, \{0\}$ (respy. no l.(algebra) ideals $\neq A, \{0\}$ then R (respy. A) is*[†] *either a division ring (respy. division algebra) or a ring (respy. algebra) with trivial multiplication and with underlying additive group of R (respy. linear space of A) of prime order (respy. of dim 1). If R (respy. A) is unital then R (respy. A) is a division ring (respy. algebra).*

PROOF. Assume that R (respy. A) has a l.z.d. a, so that $ab = 0$ with $a \neq 0, b \neq 0$. Then $I_l = \mathcal{A}_l(b) \neq \{0\}$ so that the hypothesis implies that $I_l = R$ (respy. A). Thus

$$Rb = \{0\} \quad (\text{respy. } Ab = \{0\}).$$

Since R (respy. A) is a r. ideal its right annihilator I is a bi-ideal with $b \in I$, and since $I \neq \{0\}$ we conclude that $I = R$ (respy. A). Therefore R^2 (respy. A^2) $= \{0\}$. It follows that the additive subgroup of R (respy. subspace of A) is cyclic of prime order (respy. of dim.1).

† The same conclusions hold if a similar condition is put on r. ideals instead of l. ideals.

Next suppose that R (respy. A) has no l.z.d., so that $R^* = R\backslash\{0\}$ (respy. $A^* = A\backslash\{0\}$) † is closed under multiplication. If $a \in R^*$ (respy. $a \in A^*$) then Ra (respy. Aa) is a l. ideal $\neq \{0\}$. Hence, by our hypothesis,

$$Ra = R(\text{ respy. } Aa = A) \tag{$*$}$$

It follows that there is an element $e \neq 0$ with $ea = a$. Then $(e^2 - e)a = a - a = 0$, so that $e^2 = e$ since R (respy. A) has no l.z.d.. For arbitrary $x \in R^*$ (respy. A^*),

$e(ex - x) = ex - ex = 0$ and hence $ex = x$, so that e is a l. unity of R^* (respy. A^*). Again, by $(*)$, there is an element $b \in R^*$ (respy. $b \in A^*$) such that $ba = e$, so that a has a l. inverse (b). It follows that R^* (respy. A^*) is a group †† and R (respy. A) is a division ring (respy. algebra). Finally, if R (respy. A) is unital then clearly R^2 (respy. A^2) $\neq \{0\}$ and R (respy. A) is a division ring (respy. algebra).

1.2.6. A l. ideal (respy. r. ideal) I of an algebra (or a ring) A is called *regular* or modular if there is an element u in A such that

$$xu - x \text{ (respy. } ux - x) \in I \text{ for all } x \in A.$$

The condition can be expressed briefly by writing $A(u-1)$ (respy. $(u-1)A) \subseteq I$.

The element u is called a *relative* r. unity (respy. *relative* l. unity) for I; u is not unique (see 1.2.8). If u is both a relative l. unity and a relative r. unity it is called a *relative bi-unity* or just *relative unity*. We also use sometimes the word relative unity for a relative r or l. unity. The precise sense of the usage will be clear from the context. When A is commutative, relative r. unity and l. unity concepts coincide and relative unity has an unambiguous meaning.

1.2.7. Remark. In a unital A every ideal I is regular, with unity e as a relative unity (since $xe - x = ex - x = 0 \in I$). On

† If S, T are two sets, we denote the difference set consisting of all elements in S which are not on T by $S\backslash T$.

†† A semi-group with a l. unity in which every element has a l. inverse is a group.

the other hand if A has trivial multiplication then no ideal $\neq A$ is regular (since $xu - x, ux - x = -x$ for all x.)

In the ring of even integer $2\mathbb{Z}$, the (principal) ideal $6\mathbb{Z}$ is regular (with relative unity $u = 4$) while the ideal $4\mathbb{Z}$ is not regular.

1.2.8. PROPOSITION. (a) *If u is a relative r. (respy. l.) unity for a regular l. (respy. r.) ideal I and $a \in I$ then $u+a$ is a relative r. (respy. l.) unity for I.*

(b) *Let u, v be relative r. (respy. l.) unities for I. Then uv, and hence u^n $(n = 1, 2, \cdots)$, are relative r. (respy. l.) unities for I. Further,*

$$x(u - v)(respy.\ (u - v)x) \in I \quad (x \in A).$$

(c) *For a regular bi-ideal I, every relative r. or l. unity for I is a relative (bi-) unity. Moreover, if u, v are relative unities then $u - v \in I$. Finally, u is a relative unity for I iff $u^{\#} = u + I$ is a unity of the quotient $A^{\#} = A/I$.*

PROOF. (a) Let I be a l. ideal. Then

$$x(u + a) - x = (xu - x) + xa \in I + I = I \quad (x \in A).$$

The proof when I is a r. ideal is similar.

(b) Assume that I is a l. ideal. Since

$$xuv - x = xu.v - xu + xu - x \in I + I = I$$

uv is a relative r. unity for I. Taking $v = u$ and using induction we get u^n is a relative r. unit for I.

Further,

$$x(u - v) = xu - x - (xv - x) \in I + I = I.$$

The corresponding results when I is a r. ideal are proved similarly.

(c) Suppose that u (respy. v) is a relative r. (respy. l.) unity for I. Then

$$u - v = (vu - v) - (vu - u) \in I + I = I.$$

Therefore, by (a), $u = v + (u - v)$ is a relative l. unity.

Similarly, v is a relative r. unity. Finally, observe that for $x \in A$,

$$xu - x, ux - x \in A \text{ iff } x^{\#}u^{\#} = x^{\#}, u^{\#}x^{\#} = x^{\#},$$

where $x \mapsto x^{\#}$ is the canonical quotient homomorphism.

1.2.9. LEMMA. *Let A be an algebra (or a ring) and I a regular l. (respy. r.) ideal with relative unity u. Then:*

(i) *If $u \in I$ then $I = A$*

(ii) *If $I \neq A$ then $-u$ is not l.q. (respy. r.q.) invertible*

(iii) *Any l. (respy. r.) ideal J containing I is regular with u as a relative unity for J.*

PROOF. It is enough to prove the results when I is a l. ideal.

(i) Since $u \in I$, for any $x \in A, xu \in I$ and hence $x = x - xu + xu \in I + I = I$, whence $I = A$.

(ii) Suppose that $-u$ is l.q. invertible with a as its l.q.i.. Then $a - u - au = 0$, so that $u = a - au \in I$, and so by (i), $I = A$ – a contradiction.

(iii) Obvious.

1.2.10. LEMMA (Krull). *Let $I \neq A$ be a regular ideal. Then there is a maximal regular ideal M of the same type (l., r. or bi-) as I with $I \subseteq M$.*

PROOF. If u is a relative unity for I, by 1.2.9(i), $u \notin I$. Apply Zorn's lemma, to the poset† of ideals (all of the same type as I) $\neq A$, to obtain M (note that if $\{I_\alpha\}$ is any chain of ideals in the poset then $u \notin \bigcup I_\alpha$, since $u \notin$ any I_α).

1.2.11. COROLLARY. *In a unital A every ideal $\neq A$ is contained in a maximal ideal.*

1.2.12. LEMMA. *Let A be unital with unity e. Then:*

† For definition see 2.1.1.

(i) *If* $I \neq A$ *is an ideal then* $e \notin I$.

(ii) *If* I_l *(respy.* I_r*) is a l. (respy. r.) ideal* $\neq A$ *then* I_l *(respy.* I_r *cannot contain any l. (respy. r.) invertible element of* A.

(iii) $Aa = A$ *(respy.* $aA = A$*) iff* a *is a l. (respy. r.) invertible element.*

PROOF. (i) Since e is a relative unity for I this follows from 1.2.9 (i).

(ii) If $a \in I_l$ is l. invertible then $e = a_l^{-1}a \in I_l$, so that $I_l = A$. Similarly $I_r = A$ if I_r contains a r. invertible element.

(iii) If $Aa = A$ then there is a $b \in A$ with $ba = e$, whence a is l. invertible. Conversely, if a is l. invertible then since $a \in Aa$ we must have $Aa = A$ by (ii). Similarly we can prove $aA = A$ iff a is r. invertible.

1.2.13. LEMMA. *In a unital* A *an element* a *is l. (respy. r.) invertible iff* a *is not contained in any l. ideal (respy. r. ideal)* $\neq A$.

PROOF. This is obtained by combining (ii), (iii) of 1.2.12 (noting that Aa (respy. aA) is a l. (respy. r.) ideal).

1.2.14. COROLLARY. *An element* $a \in A$ *is l. (respy. r.) invertible iff* a *does not belong to any maximal l. (respy. r.) ideal of* A. *In particular, when* A *is commutative,* a *is invertible iff it does not belong to any maximal ideal.*

PROOF. This follows from 1.2.13, 1.2.10.

1.2.15. LEMMA. (a) *If* I *is a regular l. (respy. r.) ideal of* A *and* J *a regular bi-ideal of* A *then* $I \bigcap J$ *is a regular l. (respy. r.) ideal of* A

(b) *If* I *is a bi-ideal of* A *and* J *a regular bi-ideal of* I *then* J *is a bi-ideal of* A. *Further,* J *is regular if* I *is regular.*

PROOF. (a) Let I be a regular l. ideal with relative r. unity u and J a regular bi-ideal with relative unity v. Write $w = v + u - vu$. Then, for $x \in A$, where

$$xw - x = xu - x - x(vu - v) \in I + I = I. \tag{1}$$

Also,

$$xw - x = xv - x - (xv - x)u \in J + Ju = J. \tag{2}$$

It follows from (1), (2) that $I \bigcap J$ is l. regular with w as as a relative r. unity.

(b) Let u be a relative unity for J in I. If $x \in A$ then $ux \in I$ and hence, if $a \in J$ then $uxa \in J$. Also $xa \in I$, so that $u.xa - xa \in J$. Therefore $xa = uxa - (uxa - xa) \in J + J = J$, proving J is a l. ideal. Similarly, J is a r. ideal. Assume now that I is also regular with relative unity v. Set $w = u + v - vu$. Then $xw - x = xv - x - (xv - x)u \in J$ (since $xv - x \in I$). Thus J is regular with relative unity w.

1.2.16. LEMMA. *Let $\varphi : A \to A^*$ be an epimorphism Then:*

(i) *For any ideal I of $A, I^* = \varphi(I)$ is an ideal of A^* of the same type (l., r. or bi.) as I.*

(ii) *If I is regular with relative unity u then I^* is regular with relative unity $u^* = \varphi(u)$.*

PROOF. (i) Clear.

(ii) The regularity of I^* follows from the identity $\varphi(x)\varphi(u) - \varphi(x) = \varphi(xu - x)$ when I is a l. ideal and from the analogous identity when I is a r. ideal.

1.2.17. LEMMA. *Let $\varphi : A \to A^*$ be a homomorphism (not necessarily epimorphism). Then:*

(i) *For an ideal I^* of $A^*, I = \varphi^{-1}(I^*)$ is an ideal of A of the same type.*

(ii) *If I^* is regular with a relative unity u^* belonging to $\varphi(A)$, then I is regular with any $u \in \varphi^{-1}(u^*)$ as a relative unity.*

(iii) *If φ is an epimorphism and M^* is a maximal regular ideal of A^* then $M = \varphi^{-1}(M^*)$ is a maximal regular ideal of A.*

PROOF. (i), (ii): Clear.

(iii) It is enough to observe that the map

$$I^* \mapsto \varphi^{-1}(I^*) \quad (I^* \text{ an ideal of } A^*)$$

is a bijection between the ideals of A^* and the ideals of A containing $\ker \varphi$.

1.2.18. Remark. The condition "$u^* \in \varphi(A)$" in (ii) of 1.2.17 cannot be omitted. This is shown by the following counterexample. A is an algebra (over F) with trivial multiplication and A_1 its unitization and φ the natural injection $A \to A_1$. If $a \in A$ and $I = Fa$ then I is an ideal of both A_1 and A. As an ideal of A_1, I is regular (since A_1 is unital) but as an ideal of A it is not regular (see 1.2.7).

1.2.19. LEMMA. *Let A be an algebra (or ring) and $a \in A$. Write*

$$I_l(a) = \{xa + x : x \in A\}, I_r(a) = \{ax + x : x \in A\}.$$

Then $I_l(a)$ (respy. $I_r(a)$) is a regular l. (respy. r.) ideal with $-a$ as a relative r. (respy. l.) unity.

PROOF. It is easily seen that $I_l(a)$ is a l. ideal and $I_r(a)$ is a r. ideal. Further, since $x(-a) - x = (-x)a - x$, the regularity for $I_l(a)$ follows. The regularity for $I_r(a)$ also follows similarly.

1.2.20. COROLLARY. *$a \in A$ is l.q. (respy. r.q.) invertible iff $I_l(a)$ (respy. $I_r(a)$) $= A$.*

PROOF. If $I_l(a) = A$, then since $-a \in I_l(a)$ there is a $b \in A$ with
$-a = ba + b$, whence $b \circ a = 0$, so that a is l.q. invertible. Conversely, if a has a l.q.i. b. then we have
$-a = ba + b \in I_l(a)$, so that, by 1.2.19,

$$I_l(a) = A.$$

The proof of the assertion for $I_r(a)$ is similar.

1.2.21. LEMMA. (cf. [22, p.173]) (a) *$a \in A$ is l.q. (respy. r.q.) invertible iff to each maximal regular l. (respy. r.) ideal M_l (respy. M_r) there is an element b possibly depending on M_l (respy. M_r) such that $b \circ a \in M_l$ (respy. $a \circ b \in M_r$)*

(b) *$a \in A$ is l.q. (respy. r.q.) invertible iff $-a$ is not a relative r. (respy. l.) unity for any maximal regular l. (respy. r.) ideal of A.*

PROOF. (a) If a has a l.q.i. b then $b \circ a = 0 \in M_l$. Next assume that the stated condition is satisfied for all M_l. If a is not l.q. invertible then by 1.2.20, $I_l(a) \neq A$. By 1.2.10, there is an M_l with $I_l(a) \subseteq M_l$. By our assumption we can find a b with $boa \in M_l$. It follows that

$$-a = b + ba - (boa) \in I_l(a) + M_l \subseteq M_l + M_l = M_l.$$

By 1.2.19, $-a$ is a relative unity for $I_l(a)$ and so also for M_l. Hence, by 1.2.9(i), $M_l = A$ – which is impossible. So a is l.q. invertible, as required

The proof of the statement concerning r.q. invertible is similar

(b) The statement to be proved for l.q. invertibility is clearly equivalent to: a is not l.q. invertible iff $-a$ is a relative r. unity of some maximal regular left ideal M_l.

If $-a$ is not l.q. invertible then by 1.2.20, $I_l(a) \neq A$ and so $I_l(a) \subseteq$ (some) M_l. By 1.2.19, $-a$ is a relative unity for $I_l(a)$ and so also for M_l. Conversely, if $-a$ is a relative r. unity for some M_l then by 1.2.9(ii), $a = -(-a)$ is not l.q.i.

The proof of the statement for r.q. invertibility is similar.

1.2.22. A l. or r. *ideal* I is called *q. invertible* if every element of I is q. invertible. Denote the intersection of all maximal regular l. ideals (respy. the intersection of all maximal regular r. ideals) of A by $\sqrt[l]{A}$ (respy. $\sqrt[r]{A}$). If there are no maximal regular l. ideal (respy. r. ideal) in A we define $\sqrt[l]{A}$ ($\sqrt[r]{A}$) $= A$.

1.2.23. LEMMA. (a) *If every element of a l. (respy. r.) ideal I is l.q. (respy. r.q.) then I is q. invertible.*

(b) *The image, under an epimorphism, of a q.i. ideal is a q.i. ideal (of the same type).*

PROOF. (a) It suffices to prove the result when I is a l. ideal. For $a \in I$, let a_l' be a *l.q.i.* Since $a_l' oa = 0$ we have $a_l' = -a - a_l' a \in I$. By our hypothesis, a_l' has a *l.q.i.* $b : boa_l' = 0$. By 1.1.17 (a) – applied to a_l' - we get $a = b$, which implies that a is q. invertible with $a' = a_l'$.

(b) If $\varphi : A \to A^*$ is an epimorphism and I an ideal then by 1.2.16, $I^* = \varphi(I)$ is an ideal of A^* of the same type as I. By 1.1.24, I^* is q. invertible.

1.2.24. PROPOSITION. (a) $\sqrt[l]{A}$ *(respy.* $\sqrt[r]{A}$*) is a q. invertible* *l. (respy. r.) ideal containing every q. invertible l. (respy. r.) ideal of A. In particular, every element of* $\sqrt[l]{A}$ *or* $\sqrt[r]{A}$ *is q. invertible.*

(b) $\sqrt[l]{A} = \sqrt[r]{A} = \sqrt{A}$ *(say);* $\sqrt{A}$ *is a bi-ideal containing every q. invertible l. or r. ideal of A.*

(c) $\sqrt{A}$ *contains every nil – in particular nilpotent – l. or r. ideal of A.*

(d) *If* $u \neq 0$ *is an idempotent of A then* $u \notin \sqrt{A}$*. In particular, when A is unital with unity e,* $e \notin \sqrt{A}$*.*

PROOF. (a) Since $\sqrt[l]{A} = \bigcap M_l$, if $a \in \sqrt[l]{A} \subseteq M_l$ then $a \circ a = a + a + a^2 \in M_l$, whence by 1.1.21 (a), a is l.q. invertible. So by 1.2.23 (a) $\sqrt[l]{A}$ is q. invertible. Let I_l be any q. invertible l. ideal of A. If $I_l \not\subseteq \sqrt[l]{A}$ then there is an M_l with $I_l \not\subseteq M_l$, so that we have: $(*) M_l + I_l = A$. If u be the relative r. unity for M_l then by $(*)$ we have

$$-u = a + b \quad (a \in M_l, b \in I_l).$$

Since $b \in I_l$ its $q.i.$ b' exists and therefore

$$\begin{aligned} u = -a - b &= -a - (-b' - b'b) = -a + b' + b'(-u - a) \\ &= -a + (b' - b'u) - b'a \in M_l, \end{aligned}$$

which is impossible (since $M_l \neq A$). Hence $I_l \subseteq M_l$, as required. The corresponding result for $\sqrt[r]{A}$ is proved similarly.

(b) For $a \in \sqrt[l]{A}$ and $x \in A$ we claim that the principal l. ideal $\{ax\}_l = Aax + \mathbb{Z}ax$ is q. invertible. If y is any element of this ideal then

$$y = bax + nax = (ba + na)x \quad (b \in A, n \in \mathbb{Z}).$$

Since $a \in \sqrt[l]{A}, y_0 = x(ba + na) \in \sqrt[l]{A}$ and hence y_0 is q. invertible. By 1.1.28(i), y is q. invertible, so that the ideal $\{ax\}_l$ is q. invertible. By (a), $\{ax\}_l \subseteq \sqrt[l]{A}$. In particular $ax \in \sqrt[l]{A}$, whence $\sqrt[l]{A}$ is a bi-ideal. Similarly, $\sqrt[r]{A}$ is a bi-ideal. Since $\sqrt[r]{A}$ (respy. $\sqrt[l]{A}$) is a q.i.l. (respy. r.) ideal we must have by (a), $\sqrt[r]{A} \subseteq \sqrt[l]{A}$ (respy. $\sqrt[l]{A} \subseteq \sqrt[r]{A}$). Hence $\sqrt[l]{A} = \sqrt[r]{A}$.

(c) By 1.1.23, every nil ideal is q. invertible and so contained in $\sqrt{A}$ (by (b)).

(d) If $u \in \sqrt{A}$ then $-u \in \sqrt{A}$, and $-u$ is q. invertible. But then by 1.1.26(ii), $u = 0 -$ a contradiction.

1.2.25. The bi-ideal $\sqrt{A}$ is called the Jacobson radical or *radical* of A. Following Neumark [22,p.173] an element of $\sqrt{A}$ is called an *essentially nilpotent* element of A. In view of 1.1.24, an element a is *essentially* nilpotent iff the principal l. ideal $\{a\}_l = Aa + \mathbb{Z}a$ (or equivalently, the principal r. ideal $\{a\}_r$) is q. invertible. Further, if A is commutative then every nilpotent element a is essentially nilpotent (since $\{a\}_l = \{a\}_r$ is a nil ideal).

1.2.26. LEMMA. *If* $\varphi : A \to A^*$ *is an epimorphism then* $\varphi(\sqrt{A}) \subseteq \sqrt{A^*}$.

PROOF. By 1.1.24, $\varphi(\sqrt{A})$ is a q. invertible bi-ideal of A^*, whence by 1.2.24 (b), $\varphi(\sqrt{A}) \subseteq \sqrt{A^*}$.

1.2.27. An algebra A in which $\sqrt{A} = A$ is called a *radical* algebra. Note that an algebra is a radical algebra iff every element is q. invertible. Any algebra A_0, with trivial multiplication ($A_0^2 = \{0\}$) is a radical algebra (since every $a \in A_0$ is q. invertible with $a' = -a$).

An algebra $A \neq \{0\}$ is called *s.s.* (=semi-simple) if $\sqrt{A} = \{0\}$

1.2.28. LEMMA. *In a s.s algebra A we have*

$$\mathcal{A}_l = \mathcal{A}_r = \{0\}.$$

PROOF. By 1.2.4 (iii), 1.1.24 (c), $\mathcal{A}_l, \mathcal{A}_r \subseteq \sqrt{A} = \{0\}$.

§ 3. Characters and Hypermaximal ideals

1.3.1. Let A be an algebra over a field F. A homomorphism χ of A *onto* F (as a 1-dim algebra) is called a *character* of A; $\chi^{-1}(0) = \ker \chi$ is called the *kernel* of χ.

We denote by $\triangle = \triangle(A)$ the set of all characters of A; $\triangle$ may be empty.

Let X be a linear space over F. A linear map $\psi : X \to F$ is called a *linear functional.*

1.3.2. LEMMA. *Let A, A^* be two algebras over the same field F and $\varphi : A \to A^*$ an epimorphism. If $\chi^* \in \triangle(A^*)$ then $\chi = \chi^* \circ \varphi \in \triangle(A)$.*

PROOF. Clear.

1.3.3. LEMMA. *A homomorphism $\chi : A \to F$ is a character iff it is non-trivial* (*i.e.* $\chi \neq 0$ †).

PROOF. If $\chi \neq 0$ there is an element $a_0 \in A$ with $\chi(a_0) = \alpha \neq 0$. If $\beta \in F$ then $\chi(\beta\alpha^{-1}a_0) = \beta$, proving χ is surjective, whence χ is a character. The converse is trivial.

1.3.4. LEMMA. *A subspace B of a LS (= linear space) A is of codim 1 iff it is the kernel $\psi^{-1}(0)$ of a non-trivial linear functional ψ on A.*

PROOF. If $\psi \neq 0$ there is an $a_0 \in A$ with $\psi(a_0) = \lambda \neq 0$. Write $\lambda^{-1}a_0 = u$; then $\psi(u) = 1$. For $x \in A$ we can write $x = \psi(x)u + y$, where $y = x - \psi(x)u \in \ker\psi = B$ (say). It follows that $A = Fu + B$, so that codim $B = 1$. Conversely, if codim $B = 1$, there is an $a_0 \in A \backslash B$ such that $A = Fa_0 + B$. If $x \in A$ and $x = \lambda a_0 + b$, define $\psi(x) = \lambda$. It is clear that ψ is a linear functional on A with $\ker\psi = B$.

1.3.5. LEMMA. *A regular l. or r. ideal I, of codim 1, of an algebra A is a bi-ideal, which is moreever, maximal as a l. or a r. ideal, or also, as a bi-ideal.*

PROOF. Let I be a codim 1 regular l. ideal with u as a relative r. unity. Since codim $I = 1$ we have $Fu + I = A$. If $y \in A$ then $y = \alpha u + b$, with $\alpha \in F, b \in I$. Therefore, if $a \in I$,

$$ay = a(\alpha u + b) = \alpha(au - a) + \alpha a + ab \in I,$$

whence I is a r. ideal so a bi-ideal. The proof of this when I is a r. ideal is similar. The maximality assertions are immediate consequences of the codim 1 property of I.

† '0' denotes the zero-functional mapping everything to 0.

1.3.6. DEFINITION. A codim 1 regular ideal I of A is called a *hypermaximal* ideal; by 1.3.5 every hypermaximal ideal is a bi-ideal.

1.3.7. If an algebra A has hypermaximal ideals we denote their intersection by $\sqrt[h]{A}$. We set $\sqrt[h]{A} = A$ if A has no hypermaximal ideal. We call $\sqrt[h]{A}$ the *hyper-radical* of A, and we say that A is h.s.s. (=hyper semi-simple) if $\sqrt[h]{A} = \{0\}$.

1.3.8. LEMMA. $\sqrt{A} \subseteq \sqrt[h]{A}$.

PROOF. By 1.3.5 every hypermaximal ideal is maximal regular l. ideal. The inclusion relation now follows from 1.2.24 (b) and the definition of $\sqrt[l]{A}$.

1.3.9. PROPOSITION. *The hypermaximal ideals of an algebra A are precisely the kernels $M_\chi = \ker\chi$ of characters χ of A. Moreover, the correspondence $\chi \mapsto M_\chi$ is a bijection between the set $\triangle$ of characters and the set $\mathfrak{M}$ of hypermaximal ideals.*

PROOF. Since χ is a homomorphism, M_χ is a bi-ideal, and since χ is surjective there is a $u \in A$ with $\chi(u) = 1$. Further, for any $x \in A$, we have $(*)\, \chi(xu - x) = \chi(x) \cdot 1 - \chi(x) = 0$, whence M_χ is regular with u as relative unity. Also, by 1.3.4, codim $M_\chi = 1$, whence M_χ is hypermaximal. If M is any hypermaximal ideal the quotient $A^\# = A/M$ has dim 1. If u is relative unity for M, by 1.2.8 (c), $u^\# = u + M$ is unity of $A^\#$. If $x^\# \in A^\#$ then $x^\# = \alpha_x u^\#$ where $\alpha_x \in F$, since dim $A^\# = 1$. The map $\chi = \chi_M : x \mapsto \alpha_x$ is clearly a character of A with $\ker\chi = M = M_\chi$. Finally, it is clear from the construction of χ_M from M that $M_1 = M_2 \Rightarrow \chi_{M_1} = \chi_{M_2}$. Hence $\chi \mapsto M_\chi$ is a bijection.

1.3.10. COROLLARY. (i) *The relative unities for $M = M_\chi$ are precisely the elements u in A satisfying $\chi(u) = 1$. In particular, if A has unity e then $\chi(e) = 1$ and further $\chi(a) \neq 0$ for any invertible a.*

(ii) *The set of all relative unities (= bi-unities) for M_χ is the unity coset E of $A^\# = A/M_\chi$.*

PROOF. (i) It is clear from the relation $(*)$ in the proof of 1.3.9 that any u with $\chi(u) = 1$ is a relative unity for $M = M_\chi$.

Conversely, let v be any relative unity for M. Choose any element $a_0 \in A \backslash M$; then $\chi(a_0) \neq 0$. Since $a_0 v - a_0 \in M$ we get

$$\chi(a_0)\chi(v) = \chi(a_0),\ \chi(v) = 1.$$

If a is any invertible element then

$$\chi(a)\chi(a^{-1}) = \chi(e) = 1, \text{ whence } \chi(a) \neq 0.$$

(ii) Since $M = M_\chi$ is a bi-ideal, by 1.2.8, $E = u^{\#}$ is the set of relative unities for M, u being a relative unity for M. Also, by 1.2.8 (c), E is the unity coset of $A^{\#}$.

1.3.11. LEMMA. *Every character χ of A vanishes on $\sqrt{A}$.*

PROOF. By 1.3.8, $\sqrt{A} \subseteq \sqrt[h]{A} \subseteq \ker \chi$, hence the result.

1.3.12. Remark. If A is a radical algebra then $\triangle(A) = \emptyset$ (if $\triangle = \triangle(A) \neq \emptyset, \chi \in \triangle$ then $\sqrt{A} \subseteq \ker \chi \subset A$, so that A is not a radical algebra). Again, for the matrix algebra $A = M_n(F)(n > 1), \triangle(A) = \emptyset$, since A being simple has no hypermaximal ideals.

1.3.13. EXAMPLES OF HYPERMAXIMAL IDEALS:

(a) Let G be a finite group of order n and $A = F[G]$ the group algebra of G over a field F of characteristic 0. Write $a = \sum g$, where the sum in A is over all elements g of G; $a \in A$. Clearly, $ag = A$. for each $g \in G$, whence $a^2 = na$. If we write $u = \frac{1}{n}a$, then $u^2 = u$. Writing $v = e - u$, it is easy to see that $M = Av$ is hypermaximal (since it has a 1-dim direct summand $Au = Fu$).

(b) Let $L^1(G)$ be the group algebra of a locally compact Hausdorff group G. Consider the functional

$$\Lambda(f) = \int_G f(t)dt$$

where the integral is the (left invariant) Haar integral. Clearly Λ is linear. Further

$$\Lambda(f * g) = \int f * g(t)dt = \int \left(\int f(s)g(s^{-1}t)ds \right) dt$$

$$= \int f(s)ds \int g(s^{-1}t)dt$$
$$= \int f(s)ds \int g(s^{-1}t)d(s^{-1}t) = \Lambda(f)\Lambda(g),$$

where we have used the left invariance of the Haar measure. Finally, $\Lambda \not\equiv 0$. For, if U is a nucleus † of G with $\bar{U}$ compact, and $t_0 \in G \backslash \bar{U}$, we can, by local compactness of G, choose a continuous f with $f(t_0) = 0$, $f(t) = 1$ on $\bar{U}$. Then

$$\Lambda(f) = \int_G f(t)dt \geqslant \int_U f(t)dt = \mu_l(u) > 0,$$

where μ_l denotes the left Haar measure of G.

Thus Λ is a character and $M = \ker \Lambda$ is a hypermaximal ideal (M is called the augmentation ideal).

(c) Let F be a field and F^S denote the algebra, under pointwise operations, of all F-valued of functions $f = f(s)$ on a set S. If $s_0 \in S$ then the map $\chi_{s_0} : f \mapsto f(s_0)$ is easily seen to be a character and $M_{s_0} = \ker \chi_{s_0} = \{f \in F^S : f(s_0) = 0)\}$ is a hypermaximal ideal.

1.3.14. LEMMA. *Let J be a bi-ideal of A. If M is a hypermaximal ideal of J then M is a bi-ideal of A. If J is regular in A then M is also regular in A.*

PROOF. These follow from 1.2.15 (b) since a hypermaximal ideal is a regular bi-ideal.

The first assertion can also be proved directly in the following way.
Let χ be the character of J determined by $M : M = \ker \chi$.
Take $u \in J$ with $\chi(u) = 1$. If $x \in A, a \in M \subseteq J$ then $xa \in J$ and

$$\chi(xa) = \chi(u)\chi(xa) = \chi(uxa) = \chi(ux)\chi(a) = 0.$$

Therefore $xa \in M$, similarly $ax \in M$. Thus M is a bi-ideal of A.

---

† for definition see 2.1.3.

1.3.15. PROPOSITION. *Let A be an algebra and J a bi-ideal of A. Then:*

(i) *Every character χ of J satisfies the condition $\chi(xa) = \chi(ax)$ $(x \in A,\ a \in J)$.*

(ii) *χ can be uniquely extended to a character $\hat{\chi}$ of A.*

PROOF. If $M = \ker \chi$, and u an element of J with $\chi(u) = 1$, then M is a hypermaximal ideal of J with u as a relative unity. Take $x \in A$. Then $ux, xu \in J$ and by hypermaximality of M we have the relations.

$$ux = m + \alpha u;\ xu = m_1 + \beta u \quad (m, m_1 \in M). \tag{1}$$

Then

$$uxu = mu + \alpha u^2;\ uxu = um_1 + \beta u^2 \tag{2}$$

since $mu, um_1 \in M$ we obtain from (2)

$$\alpha = \alpha\chi(u^2) = \chi(\alpha u^2) = \chi(uxu) = \chi(\beta u^2) = \beta.$$

So (1) gives

$$\chi(ux) = \alpha = \beta = \chi(xu).$$

Now extend χ on J to $\hat{\chi}$ on A by setting

$$\hat{\chi}(x) = \chi(ux) = \chi(xu) \quad (x \in A).$$

Clearly $\hat{\chi}$ is linear. Further

$$\begin{aligned} \hat{\chi}(xy) &= \chi(uxy)\chi(u) = \chi(ux.yu) = \chi(ux)\chi(yu) \\ &= \chi(ux)\chi(uy) = \hat{\chi}(x)\hat{\chi}(y). \end{aligned}$$

This proves that $\hat{\chi}$ is a character of A which is an extension of χ. Further, if χ_1 is any character of A extending χ then

$$\chi_1(x) = \chi_1(x)\chi(u) = \chi_1(x)\chi_1(u) = \chi_1(xu) = \chi(xu) = \hat{\chi}(x),$$

which shows that the extension $\hat{\chi}$ is unique. Finally,

$$\chi(xa) = \hat{\chi}(xa) = \hat{\chi}(x)\hat{\chi}(a) = \hat{\chi}(a)\hat{\chi}(x) = \hat{\chi}(ax) = \chi(ax)$$

where $x \in A, a \in J$.

1.3.16. COROLLARY. *Every hypermaximal ideal* M *of* J *satisfies the symmetry condition for* $x \in A, a \in J$,

$$xa \in M \Leftrightarrow ax \in M.$$

PROOF. This follows from 1.3.15 (i) since $M = \ker \chi$ for some character χ of J.

1.3.17. COROLLARY. *Every character* χ *of an algebra* A *can be extended uniquely to a character* χ_1 *of its unitization* A_1.

PROOF. If $x_1 \in A_1, x_1 = \lambda e_1 + x\ (\lambda \in F, x \in a)$ then $\chi_1(x_1) = \lambda + \chi(x)$ is easily checked to yield, a character of χ_1 of A. The uniqueness of χ_1 follows from 1.3.15 (ii) (taking $J = A$) or can also be seen directly using the fact that for any character χ' of A_1 we have $\chi'(e_1) = 1$.

1.3.18. LEMMA. *Let* $A = A_1 \oplus A_2$ *be a direct sum of subalgebras* A_1, A_2. *If* $\triangle = \triangle(A), \triangle_j = \triangle(A_j)\ (j = 1, 2)$ *then* $\triangle$ *is the disjoint union*

$\triangle = \triangle_1^0 \dot{\bigcup} \triangle_2^0$ *of subsets* $\triangle_j^0$ *such that*

$\triangle_j^0 | A_j = \triangle_j$ (i.e. *if* $\chi \in \triangle_j^0$ *then* $\chi | A_j \in \triangle_j$.

PROOF. Let $\chi \in \triangle$ and $\chi(u) = 1\ (u \in A)$. Since $u = u_1 + u_2$ with $u_j \in A_j$, and $u_1 u_2 = u_2 u_1 = 0$ (since $A_1 A_2 = A_2 A_1 = A_1 \bigcap A_2 = \{0\}$) we obtain

$$1 = \chi(u) = \chi(u_1) + \chi(u_2),\ \chi(u_1)\chi(u_2) = 0.$$

From these equations it follows that precisely one of $\chi(u_1), \chi(u_2)$ is 1 and the other 0. If $\chi(u_1) = 1, \chi(u_2) = 0$ then for $y \in A_2, u_1 y = 0$, so that

$$0 = \chi(u_1 y) = \chi(u_1)\chi(y) = \chi(y).$$

This shows that $\chi \in \triangle_1^0$. Similarly, if $\chi(u_2) = 1$ then $\chi \in \triangle_2^0$.

1.3.19. LEMMA. (cf.[24,p.233]). *Let* A *be a unital algebra and* χ *a linear functional with* $\chi(e) = 1$.

The following two statements are equivalent:

(i) χ *is a character.*

(ii) $M = \ker \chi$ *is closed for squares,* i.e. $x \in M \Rightarrow x^2 \in M$.

PROOF. (i) $\Rightarrow$ (ii), since M is an ideal by 1.3.9.

To prove that (ii) $\Rightarrow$ (i), assume that (ii) holds. Since χ is a linear functional $\neq 0$, codim $M = 1$. For $x, y \in A$ we have

$$x = \chi(x)e + a, \quad y = \chi(y)e + b \quad (a, b \in M).$$

Computing from these relations the product xy and substituting it for xy in $\chi(xy)$ and simplifying we get

$$\chi(xy) = \chi(x)\chi(y) + \chi(ab). \tag{1}$$

Taking $y = x, b = a$ in (1) and using hypotheses (ii) we obtain

$$\chi(x^2) = \chi(x)^2 \text{ for all } x \in A. \tag{2}$$

Replacing in (2), x by $x+y$ we obtain $\chi((x+y)^2) = (\chi(x) + \chi(y))^2$ which reduces to

$$\chi(xy + yx) = 2\chi(x)\chi(y). \tag{3}$$

It follows that:

$$x \text{ or } y \in M \Rightarrow xy + yx \in M. \tag{4}$$

Now we have the identity

$$(xy - yx)^2 + (xy + yx)^2 = 2((xy)^2 + (yx)^2) = 2(xyx.y + y.xyx) \tag{5}$$

where $x, y \in A$. If $y \in M$, then by (4), $xy + yx$ as well as the end expression in (5) belong to M. Therefore, by applying χ to both sides of (5) and by using linearity of χ and relations (2),(4) we obtain

$$xy - yx \in M. \tag{6}$$

From (4), (6) we conclude that if $y \in M$ then $xy \in M$

This means that in (1) we must have $\chi(ab) = 0$, so that (1) reduces to

$$\chi(xy) = \chi(x)\chi(y)$$

proving χ is a character, as desired.

§ 4. Extension of Ideals

1.4.1. Let J be a bi-ideal of an algebra A. For a l. ideal J_l of J we set

$$\hat{J}_l = \{a \in A : Ja \subseteq J_l\}. \tag{*}$$

Similarly, for a r. ideal J_r of J we set

$$\hat{J}_r = \{a \in A : aJ \subseteq J_r\}. \tag{**}$$

1.4.2. PROPOSITION. (i) *$\hat{J}_l$ is a l. ideal of A with $J_l \subseteq \hat{J}_l$; $\hat{J}_r$ is a r. ideal of A with $J_r \subseteq \hat{J}_r$.*

(ii) *If I' is a regular l.(respy. r.) ideal of J with relative r. (respy. l.) unity u, then $\hat{I}'$ is also regular with u as a relative r. (respy. l.) unity. Further, if $I' \neq J$ then $J \not\subseteq \hat{I}'$, hence in particular $\hat{I}' \neq A$.*

(iii) *If I is a l. (respy. r.) ideal of A and $I' = J \bigcap I$ then $I \subseteq \hat{I}'$.*

(iv) *If M' is a maximal regular l. (respy. r.) ideal of J then $\hat{M}'$ is a maximal regular l. (respy. r.) ideal of A such that $M' = J \bigcap \hat{M}'$. Also, if M' is a maximal regular bi-ideal of J then $\hat{M}'$ is a maximal regular bi-ideal of A.*

(v) *If M is a maximal regular l. (respy. r.) ideal of A such that $J \not\subseteq M$ then M has a relative r. (respy. l.) unity j belonging to J; further $M' = J \bigcap M$ is a maximal regular l. (respy. r.) ideal of J with j as a relative r. (respy. l.) unity and $M = \hat{M}'$. Finally, if M is a maximal regular bi-ideal of A with $J \not\subseteq M$ then $M' = J \bigcap M$ is a maximal regular bi-ideal of J.*

PROOF. We shall prove only the results for l. ideals.

(i) Clearly $\hat{J}_l$ is a linear subspace of A. For $a \in \hat{J}_l, x \in A$ we have $J \cdot xa = Jx \cdot a \subseteq Ja \subseteq J_l, xa \in \hat{J}_l$, so that $\hat{J}_l$ is a l. ideal of A.

(ii) If $x \in A$ and $a \in J$ then $ax \in J$. By regularity for I', $ax - axu \in I'$. This implies that $J(x - xu) \subseteq I'$, whence $x - xu \in \hat{I}'$, $\hat{I}'$ is regular. Further, by 1.2.8 (b) u^2 is a relative r. unity of I', so that if $I' \neq J$ then $u^2 \notin I'$. It follows that $Ju \not\subseteq I', u \notin \hat{I}'$, so that $J \not\subseteq \hat{I}'$.

(iii) Let I be l. regular with relative r. unity u. If $x \in I$ then $Jx \subseteq Ax \subseteq I$ and $Jx \subseteq J$ (since J is a bi-ideal). Therefore $Jx \in J \bigcap I = I', x \in \hat{I}'$, so that $I \subseteq \hat{I}'$.

(iv) Assume that M' is a maximal regular l. ideal of J with r. unity u. By (ii), $\hat{M}'$ is regular with relative r unity u. If I is a l. ideal of $A, I \neq A$, with $I \supseteq \hat{M}' \supseteq M'$ then I is regular with u as relative r. unity and $u \notin I$ (since $I \neq A$). Further, $J \bigcap I \supseteq M'$, and since $u \notin J \bigcap I$, $J \bigcap I \neq J$. By maximality of $M', J \bigcap I = M'$. Since $I \supseteq \hat{M}' \supseteq M'$ we get $J \bigcap \hat{M}' = M'$. For $x \in I$, $xu = x - (x - xu) \in I$ (using regularity of I); also $xu \in J$ (since $u \in J$). Thus $xu \in J \bigcap I = M'$. It follows that $Jxu \subseteq JM' \subseteq M'$, whence $xu \in \hat{M}'$. Consequently $x = xu - (xu - x) \in \hat{M}'$. This proves $I = \hat{M}', \hat{M}'$ is maximal.

(v) Let M be a maximal regular l. ideal with relative r. unity u. Since $J \nsubseteq M$, by maximality of M we have $J + M = A$. Therefore u has a representation $(*)$ $u = j + m$ $(j \in J, m \in M)$, so that for any $x \in A$

$$x - xj = x - x(u - m) = x - xu + xm \in M.$$

Thus $j \in J$ is a relative r. unity for M. By restricting x to J we get $x - xj \in J \bigcap M = M'$ showing that M' is regular (in J) with relative r. unity j. Since $u \notin M$ it follows from $(*)$ that $j \notin M, j \notin M', M' \neq J$. Consider now a l. ideal I' of J with $I' \neq J, I' \supseteq M'$. Then $A \supseteq \hat{I}' \supseteq \hat{M}'$. Also, by (iii), $M \subseteq \hat{M}'$. By maximality of M we obtain $\hat{I}' = \hat{M}' = M$. It follows that $M' = J \bigcap M = J \bigcap \hat{I}' \supseteq I'$, so that $I' = M'$ and M' is maximal. Finally, by (iv), $M' = J \bigcap \hat{M}' = J \bigcap M$.

1.4.3. PROPOSITION. *Let J be a bi-ideal of an algebra A. Then there is a bijection between the maximal regular l. (respy. r.) ideals M of A with $J \nsubseteq M$, and the maximal regular l. (respy. r.) ideals M' of J given by*

$$M' = J \bigcap M, M = \hat{M}'.$$

In particular, M' is hypermaximal iff M is hypermaximal. Finally, the map $M \to M' = J \bigcap M$ is a bijection between the maximal regular bi-ideals of A and those of J.

PROOF. The first statement follows from 1.4.2 ((iv),(v)). To prove the particular case statement assume that M is hypermaximal and χ the character determining it. Let χ' be the character of J obtained by restriction of χ. Then it is clear that $M' = J \bigcap M = \ker \chi'$ and hence M' is a hypermaximal ideal of J. Conversely, if M' is hypermaximal ideal of J and χ' the character determining it then χ' can be extended, by 1.3.15, uniquely to a character $\chi = \hat{\chi}'$ of A. If $M = \ker \chi$ then M is hypermaximal and clearly we have $J \bigcap M = M'$. Finally, by 1.4.2 ((iv), (v)) M' is maximal regular bi-ideal iff M is a maximal regular bi-ideal.

1.4.4. COROLLARY. $\sqrt{J} = J \bigcap \sqrt{A}$.

PROOF.

$$\begin{aligned} \sqrt{J} &= \bigcap M' = \bigcap (J \bigcap \hat{M}') \\ &= J \bigcap (\bigcap \hat{M}') = J \bigcap \sqrt{A} \end{aligned}$$

(since in the computation of the right hand end term above the maximal regular l. ideals of A which contain J can be clearly dropped).

1.4.5. COROLLARY. *If A_1 is the unitization of an algebra A then*

$$\sqrt{A} = A \bigcap \sqrt{A_1}.$$

PROOF. By 1.4.4 (since A is a bi-ideal of A_1).

1.4.6. Let A be an algebra and A_1 its unitization. Let I_l (respy. I_r) be a regular l. (respy. r.) ideal of A with relative r. (respy. l.) unity u. By taking A_1 for A and $J = A$ in 1.4.1, we can form $\hat{I}_l$ (respy. $\hat{I}_r$). We also set

$$\check{I}_l = \{x_1 \in A_1 : x_1 u \in I_l\}; \check{I}_r = \{x_1 \in A : u x_1 \in I_r\}.$$

1.4.7. PROPOSITION. (i) *$\check{I}_l$ (respy. $\check{I}_r$) is a l. (respy. r.) ideal of A_1 with*

$$I_l \subseteq \check{I}_l \subseteq \hat{I}_l; \quad I_r \subseteq \check{I}_r \subseteq \hat{I}_r. \qquad (*)$$

Further, if I_l *(respy.* I_r*)* $\neq A$ *then* $\check{I}_l$ *(respy.* $\check{I}_r$*)* $\neq A_1$.

(ii) *Given a l. (respy. r.) ideal* I_1 *of* A_1 *with* $I_1 \not\subseteq A$, *the ideal* $I = A \bigcap I_1$ *is regular and we can choose a relative r. (respy. l.) unity* u *for* I *such that* $e_1 - u \in I_1$

(iii) *If* I *is a regular l. (respy. r.) ideal* $\neq A$ *with relative r. (respy. l.) unity* u *then*

$$\check{I} = I + A_1(e_1 - u) \ (respy.\ I + (e_1 - u)A_1), \qquad (**)$$

$I = A \bigcap \check{I}$. *Further, if* $I_1 \neq A$ *is a l. (respy. r.) ideal of* A_1 *with* $I = A \bigcap I_1$ *then* $I_1 \subseteq \check{I}$.

(iv) *For a regular bi-ideal* I *with a relative (bi-) unity* u *we have*

$$\check{I} = I + A_1(e_1 - u) = I + (e_1 - u)A = I + F(e_1 - u),$$

and $\check{I}$ *is bi-ideal.*

(v) *If* M *is a maximal regular l. (respy. r.) ideal of* A *then* $\check{M}$ *is a maximal regular l. (respy. r.) ideal of* A_1 *and* $\check{M} = \hat{M}, M = A \bigcap \check{M}$.

PROOF. As before we will prove all results only for l. ideals.

(i) It is clear from the definition of $\check{I}_l$ that it is a subspace of A_1. Further, if $x_1 \in \hat{I}_l$ (so that $x_1 u \in I_l$) and $y_1 = \alpha e_1 + y \, (y \in A)$, we have

$$y_1 x_1 u = \alpha x_1 u + y \cdot x_1 u \in I_l + I_l = I_l.$$

It follows that $y_1 x_1 \in \check{I}_l$, showing that $\check{I}_l$ is a l. ideal of A_1. If $x_1 \in \check{I}_l$ and $y \in A$ then $y_1 x u \in I_l$. Since $y x_1 \in A$ and I_l is regular we get

$$y_1 x_1 = y x_1 u - (y x_1 u - y x_1) \in I_l,$$

whence $A x_1 \subseteq I_l$, so that $x_1 \in \hat{I}_l$, and $\check{I}_l, \subseteq \hat{I}_l$. Also, if $I_l \neq A$ then $u \notin I_l, e_1 \notin \check{I}_l$, and $\check{I}_l \neq A_1$.

(ii) Let I_1 be a l. ideal and choose an element.

$$a_1 = \lambda e_1 + a \ (\lambda \in F, \lambda \neq 0, a \in A) \text{ in } I_1 \backslash A.$$

Then $u = -\lambda^{-1}a \in A$, and for any $x \in A$ we have

$$xu - x = -\lambda^{-1}xa_1 \in A \bigcap I_1 = I,$$

so that u is a relative r. unity for I_1. Also,

$$e_1 - u = \lambda^{-1}a_1 \in I_1.$$

(iii) If $x_1 = \lambda e_1 + x \in A_1 (x \in A)$, then

$$(x_1 - x_1u)u = x_1u - x_1u \cdot u \in I = I_l \text{ (since } x_1u \in A).$$

It follows that $x_1 - x_1u \in \check{I}_l$, so that $A_1(e_1 - u) \subseteq \check{I}_l$. This inclusion with (first half of) $(*)$ yields: $I_l + A_1(e - u) \subseteq \check{I}_l$. To prove the reverse inclusion relation consider an $x_1 \in \check{I}_l$ (so that $x_1u \in I_l$). Then

$$x_1 = x_1u + x_1 - x_1u \in I_l + A_1(e_1 - u),$$

completing the proof of the relation $(**)$. Further, it is clear that $I_l \subseteq A \bigcap \check{I}_l$. On the other hand, if $a \in A, a \in \check{I}_l$ then $au \in I_l$ and so $a \in I_l$ (since u is a relative r. unity of I_l). Thus $I_l = A \bigcap \check{I}_l$. Finally, consider the ideal I_1. By (ii) we can choose for $I = A \bigcap I_1$ a relative r. unity u such that $e_1 - u \in I_1$. If $y_1 = \lambda e_1 + y \in I_1$ then

$$y_1u = \lambda u + yu \in A.$$

Also,

$$y_1u = y_1 - y_1(e_1 - u) \in I_1 + I_1 = I_1.$$

So $y_1u \in A \bigcap I_1 = I$, whence $y_1 \in \check{I}$. Thus, $I_1 \subseteq \check{I}$, as required.

(iv) If $x_1 \in A_1, x_1 = \lambda e_1 + x \ (x \in A)$ then

$$\begin{aligned} x_1(e_1 - u) &= \lambda(e_1 - u) + x - xu \in F(e_1 - u) + I \\ (e_1 - u)x_1 &= \lambda(e_1 - u) + x - ux \in F(e_1 - u) + I \end{aligned}$$

Hence $I + A_1(e_1 - u) = I + F(e_1 - u) = I + (e_1 - u)A_1 = \check{I}$ (say). Clearly $\check{I}$ is a bi-ideal.

(v) By $(*)$ we have $\check{M} \subseteq \hat{M}$. Again, by 1.4.2 (iv), $M = A \bigcap \hat{M}$. It now follows, by (iii), that $\hat{M} \subseteq \check{M}$. Combining the two inclusions we get $\check{M} = \hat{M}$, and $M = A \bigcap \hat{M} = A \bigcap \check{M}$.

1.4.8. COROLLARY. *The correspondence $I_1 \mapsto I = A \bigcap J_1$ is injective over each of the following sets of ideals of A_1: the set of all bi-ideals $I_1 \not\subseteq A$; the set of maximal l. ideals $M_1 \not\subseteq A$; the set of maximal r. ideals $M_1 \not\subseteq A$. Also, every maximal regular l. (respy. r.) ideal M of A is of the form $M = A \bigcap \check{M}$, where*

$$\check{M} = \{x_1 \in A : x_1 u \ (respy.\ ux_1) \in M\}.$$

PROOF. Suppose $I_1, J_1 \not\subseteq A$ are bi-ideals of A_1 with $A \bigcap I_1 = A \bigcap J_1 = I$ (say). If $I = A$, clearly $I_1 = J_1 = A_1$ (since codim $A = 1$). We may therefore assume that $I \neq A$. By 1.4.7 (ii) we can choose relative r. unities u, v for I such that $e_1 - u \in I_1, e_1 - v \in J_1$. By 1.2.8 (c), $v - u \in I \subseteq J_1$. If $x_1 = \lambda e_1 + y$ $(\lambda \in F, y \in A)$ is an element of I_1 then we can write:

$$x_1 = \lambda(e_1 - v) + \lambda(v - u) + \lambda u + y, \tag{1}$$

$$\lambda u + y = x_1 u - (yu - y). \tag{2}$$

It follows from (2) that

$$\lambda u + y \in x_1 u + I \subseteq I_1 + I = I_1 (\text{ since } x_1 u \in I_1)$$

so that

$$\lambda u + y \in A \bigcap I_1 = I \subseteq J_1. \tag{3}$$

Since $e_1 - v, v - u \in J_1$, it follows from (1),(3) that $x_1 \in J_1$. Thus, $I_1 \subseteq J_1$, and by symmetry considerations we conclude that $I_1 = J_1$.

The injectiveness over the set of maximal regular l. (respy. r.) ideals M follows from 1.4.3, 1.4.7(v). The final assertion concerning the form of M is a consequence of 1.4.7(v) and 1.4.2 ((iii), (iv)).

1.4.9. COROLLARY. (a) *A is a hypermaximal ideal of A_1.*

(b) *The correspondence $M_1 \mapsto M = A \bigcap M_1$ is a bijection between the hypermaximal ideals $M_1 \neq A$ of A_1 and the hypermaximal ideals M of A; moreover $M_1 = \check{M}$.*

(c) $\sqrt[h]{A} = A \bigcap \sqrt[h]{A_1}$.

PROOF. (a) This follows from the construction of A_1 (see 1.1.12).

(b) By 1.4.7 (v), $\check{M} = \hat{M}$, $M = A \bigcap \check{M}$. The bijectivity of the map $M \mapsto \hat{M} = M_1$ is ensured by 1.4.3.

(c) This follows from (b) and the definitions of $\sqrt[h]{A}$, $\sqrt[h]{A_1}$.

1.4.10. Remark. The character of A_1 determined by the hypermaximal ideal A is denoted by χ_0 and called the *distinguished* or canonical character of A_1. It is given by

$$\chi_0(\lambda e + x) = \lambda \qquad (x \in A).$$

It follows from 1.4.9 and 1.3.17 that there is a bijection $\chi \to \chi_1$ between the sets of characters Δ and $\Delta_1 \backslash \{\chi_0\}$, where χ_1 is the unique extension of χ to A_1.

1.4.11. PROPOSITION. *Let A be a radical algebra and A_1 its unitization. Then:*

(i) *no ideal $I \neq A$ is regular;*

(ii) $\sqrt{A_1} = A$;

(iii) *A_1 is a local algebra having A as its unique maximal ideal.*

PROOF. (i) If A has a regular l. (or r.) ideal $\neq A$ then there is a maximal regular l. (or r.) ideal M, whence $\sqrt{A} \subseteq M \neq A$, contradicting A is a radical algebra.

(i), (iii): If A_1 has a maximal l. (or r.) ideal $M_1 \neq A$ then $M = A \bigcap M_1$ is a regular ideal of A, $\neq A$, contradicting (i). On the other hand, A is a maximal (l. or r. or bi-ideal) of A_1. Thus $\sqrt{A_1} = A$ and A is the unique maximal ideal of A_1.

§ 5. Regular Representation and Primitive Ideal

1.5.1. Let A be an algebra over a field F and X a linear space over F. A homomorphism $\varphi : A \to \mathcal{E}(X)$ (=the algebra of linear endomorphisms† of X) is called a *linear representation*

† i.e. linear maps of X into itself.

or just a *representation* of A in X. It is well-known that a representation φ of A gives rise to an A-module structure on $X : a.x = \varphi(a)x$. Conversely, if X is an A-module it yields a natural representation $a \mapsto l_a$, where l_a is the linear map on X given by

$$l_a : x \mapsto ax \quad (a \in A, x \in X)$$

In particular we have for an algebra A the representation.

$$\mathcal{L} : a \mapsto l_a (a \in A), \text{ where } l_a \text{ is now given by}$$

$$l_a : x \mapsto ax \quad (x \in A).$$

$\mathcal{L}$ is called the (left) ***regular representation*** of A

The kernel of the homomorphism $\mathcal{L}$ is given by $\ker \mathcal{L} = \{a \in A : aA = 0\}$ =the left annihilator $\mathcal{A}_l$. $\mathcal{L}$ is faithful (i.e. is a monomorphism) iff $\mathcal{A}_l = \{0\}$. In particular, by 1.2.28, $\mathcal{L}$ is faithful whenever A is s.s. . .

1.5.2. If I_l is a regular l. ideal of A then the regular representation $\mathcal{L}$ induces a representation $\mathcal{L}^\#$ in the quotient space $A^\# = A/I_l$. $\mathcal{L}^\#$ is given by : $\mathcal{L}^\# : a \mapsto l_a^\#$, with $l_a^\#(x + I_l) = ax + I_l(x + A)$. $\mathcal{L}^\#$ is called the *regular representation of A in $A^\#$*

The representation $\mathcal{L}^\#$ is called *irreducible* if $A^\#$ is a simple † ($\mathcal{L}^\#-$) module. For $\mathcal{L}^\#$ to be irreducible it is a necessary and sufficient that I be a maximal regular l. ideal of A.

If $\mathcal{L}^\#$ is the regular representation of A in $A^\# = A/I$, we write

$$(I_l : A) = \ker \mathcal{L}^\# = \{a \in A : l_a^\# = 0\} = \{a \in A : aA \subseteq I_l\}$$

Similarly, we write for a regular r. ideal I_r,

$$(I_r : A) = \{a \in A : Aa \subseteq I_r\}.$$

1.5.3. LEMMA. *Let I be a regular l. (or r.) ideal of A. Then*

† A module E over a ring R is called a simple R-module if $RE \neq \{0\}$ and has only E and $\{0\}$ as its submodules.

$(I : A)$ is a bi-ideal of A with $J = (I : A) \subseteq I$; also J is the largest bi-ideal (of A) contained in I. If I is a regular bi-ideal then $(I : A) = I$.

PROOF. That $J = (I : A)$ is a bi-ideal is clear. Assume now that I is a regular l. ideal. Then for $x \in J, xA \subseteq I$, so that in particular $xu \in I$, where u is a relative right unity for I.

$$x = x - xu + xu \in I, \qquad J \subseteq I.$$

Further, if J' is a bi-ideal of A with $J' \subseteq I$ then $J'A \subseteq J' \subseteq I$, whence $J' \subseteq (I : A) = J$. The proofs of these statements when I is a r. ideal are similar. Finally, when I itself is a bi-ideal we have clearly $(I : A) = J = I$.

1.5.4. Let A be an algebra (or even a ring). An ideal P is called *l.* (=left) *primitive* if there is a maximal regular l. ideal M_l of A with $P = (M_l : A)$ Similarly, if there is a maximal r. ideal M_r of A with $P = (M_r : A)$ then P is called r. primitive. By 1.5.3 a l. or r. primitive ideal is a bi-ideal In general a l. primitive ideal of a ring need not be r. primitive.[†] If an ideal P is both l. and r. primitive then it is called *bi-primitive.* If A is commutative then all these concepts evidently coincide.

In the sequel we shall use the term *primitive* for l. primitive.

We call an algebra (or ring) A primitive if the zero ideal $\{0\}$ is primitive; equivalently, if there is a maximal regular l. ideal M_l with $\{0\} = (M_l : A)$.

1.5.5. LEMMA. *Let A, A^* be algebras (or rings) and $\varphi : A \to A^*$ be an epimorphism. If P is a primitive ideal of A with* $\ker \varphi \subseteq P$, *then $\varphi(P)$ is primitive.*

PROOF. If $P = (M : A)$, where M is a maximal regular l. ideal of A, then $\varphi(M)$ is a maximal regular l. ideal of A^* and $\varphi(P) = (\varphi(M) : A)$, whence $\varphi(M)$ is primitive.

1.5.6. COROLLARY. *If P is a primitive ideal of A and $\theta : A \to A/P = A^*$ the canonical homomorphism, then $\theta(P) = \{0^*\}$*

† See [4', *pp.*473 – 75]

is primitive.

1.5.7. PROPOSITION. *If A is an algebra (or ring) and $A \neq \sqrt{A}$, then*

$$\sqrt{A} = \bigcap (M_l : A) = \bigcap (M_r : A)$$

where M_l (respy. M_r) runs through all maximal regular l. (respy. r.) ideals of A. In particular A is the intersection of all (l.) primitive ideals of A as well as the intersection of all r. primitive ideals.

PROOF. By 1.2.22, 1.2.24 (b) we have $\sqrt{A} \subseteq M_l$, whence by 1.5.3, $\sqrt{A} \subseteq (M_l : A)$. Also, since M_l is regular we have $(M_l : A) \subseteq M_l$. Therefore

$$\sqrt{A} \subseteq \bigcap (M_l : A) \subseteq \bigcap M_l = \sqrt{A}.$$

so that $\sqrt{A} = \bigcap (M_l : A)$, Similarly, $\sqrt{A} = \bigcap (M_r : A)$.

1.5.8. LEMMA (cf. [27, p.144]). *A maximal regular l. (respy. r.) ring ideal of an algebra A is an algebra ideal. Further, every primitive ring ideal of A is an algebra ideal. Hence the algebra radical of A coincides with the ring radical of A.*

PROOF. Let M be a maximal regular l. ring ideal of A and write $M_1 = \{x \in A; Ax \subseteq M\}$. Clearly, M_1 is a l. algebra ideal and $M \subseteq M_1$. Further, if u is a relative r. unity for M then $u \notin M_1$. (since $u \in M_1 \Rightarrow u^2 \in M \Rightarrow M = A$, by 1.2.8 (b), 1.2.9 (i)). By ,maximality of M we conclude that $M_1 = M$, whence M is an algebra ideal. Similarly, the proof when M is a r. ideal.

Now if P is a primitive ring ideal of A then $P = (A : M)$ where M is a maximal regular l. ideal. If $a \in P, \alpha \in F$ then

$$\alpha a A = a.\alpha A \subseteq aA \subseteq M,$$

so that $\alpha a \in P$, where P is an algebra ideal.

1.5.9. LEMMA. (Jacobson). *Every primitive ideal P is prime.*†

† A bi-ideal P in a ring R is called prime or a prime ideal if $P \neq R$ and for any bi-ideals I, J $IJ \subseteq P \Rightarrow I \subseteq P$ or $J \subseteq P$.

PROOF. (cf. 21, p.54). Suppose that $IJ \subseteq P = (M : R)$ and $J \not\subseteq P$. Then $JR \not\subseteq M$ and hence, by maximality of $M, JR + M = R$. It follows that

$$IR \subseteq I(JR + M) \subseteq IJ + M \subseteq P + M = M$$

so that $I \subseteq (M : R) = P$.

1.5.10. PROPOSITION. (a) *Every maximal regular bi-ideal M is bi-primitive, in particular prime. Furthermore, the quotient $A^{\#} = A/M$ is unital and simple. If A is commutative then $A^{\#}$ is a division algebra.*

(b) *If A is commutative then the primitive ideals of A are precisely the maximal regular ideals.*

PROOF. (a) Let M_l be a maximal regular l. ideal with $M \subseteq M_l$. Then

$$M \subseteq (M_l : A) \subseteq M_l.$$

Since $(M_l : A)$ is a bi-ideal, by maximality of $M, M = (M_l : A)$, whence M is (l.) primitive. Similarly it is r. primitive. Further, the bijection, between the bi-ideals of $A^{\#}$ and the bi-ideals of A containing M, shows that $A^{\#}$ is simple. $A^{\#}$ is further unital since M is regular. Finally, when A is commutative it follows from 1.2.5 that $A^{\#}$ is a division algebra.

(b) In view of (a) we have only to show that a primitive ideal I is maximal regular. Since I is primitive there is a maximal regular ideal M with $I = (M : A)$ But since now M is a bi-ideal, by 1.5.3, $(M : A) = M$. Thus, $I = M$ and I is maximal regular.

1.5.11. The intersection of all maximal regular bi-ideals of an algebra A is called the *strong radical* and is denoted by $\sqrt[s]{A}$; if A has no maximal regular bi-ideal we define A to be $\sqrt[s]{A} : \sqrt[s]{A} = A$.

1.5.12. PROPOSITION. *In any algebra A we have*

$$\sqrt{A} \subseteq \sqrt[s]{A} \subseteq \sqrt[h]{A}.$$

PROOF. The first inclusion relation follows since every maximal regular bi-ideal is (by 1.5.10 (a)) primitive; the second is a consequence of every hypermaximal ideal being (by 1.3.5) a maximal regular bi-ideal.

1.5.13. Let A be an algebra. An A-module X is said to be *cyclic*, with *generator* x_0, if $x_0 \in X$ and $Ax_0 = X$. Also, an A-module X is said to be *irreducible* if (i) $Ax \neq \{0\}$ and (ii) the only submodules of X are $\{0\}$ and X.

If X is an A-module, for $x_0 \in X$ we write $\ker x_0 = \{a \in A : ax_0 = 0\}$. Similarly, for $a_0 \in A$ we write $\ker a_0 = \{x \in x : a_0 x = 0\}$. Clearly $\ker x_0$ is a l. ideal of A and $\ker a_0$ a submodule of X.

1.5.14. LEMMA. *Let X be an A-module. Then:*

(i) *If X is cyclic with generator x_0 then* $\ker x_0$ *is a regular l. ideal of A.*

(ii) *If X is irreducible then it is cyclic with any non-zero element x_0 as generator. Further, $M = \ker x_0$ is a maximal regular l. ideal of A.*

PROOF. (i) Since $X = Ax_0$ there is an element $u \in A$ with $x_0 = ux_0$. Then $(a - au_0)x_0 = ax_0 - ax_0 = 0$, so that $a - au \in \ker x_0$ whence $\ker x_0$ is regular with u as a relative r. unity.

(ii) The set $X_0 = \{x \in X : Ax_0 = \{0\}\}$ is clearly a submodule of X. By irreducibility of X we have $X_0 = \{0\}$ or X. But condition (i) in the definition of irreducibility rules out $X_0 = X$ and so we must have $X_0 = \{0\}$. Thus, if $x_0 \in X, x_0 \neq 0$ then $Ax_0 \neq \{0\}$ and so $Ax_0 = X$, proving X is cyclic with generator x_0. Also, by (i), $M = \ker x_0$ is a regular l. ideal. It remains to prove that M is maximal. If L is a l. ideal of A with $L \supset M$, we can choose an element $b \in L \backslash M$, and then $bx_0 \neq 0$. It follows that $Abx_0 = X$, whence there is an $a_0 \in A$ with $a_0 bx_0 = x_0$. But then, for any $a \in A$,

$$a - aa_0 b \in \ker x_0 = M \subset L. \tag{$*$}$$

Since $b \in L$ and L is a l. ideal we conclude from $(*)$ that $a \in L$, which means $L = A$, proving M is maximal.

1.5.15. If A is an algebra and I_l a l. ideal of A, then the quotient $A^{\#} = A/I_l$ is canonically a left A-module:

$$a(x + I_l) = ax + I_l \quad (a, x \in A).$$

The corresponding representation $\mathcal{L}^{\#}$ is given by, $\mathcal{L}^{\#} : a \mapsto l_a^{\#}$, where $l_a^{\#}$ is the linear transformations on $A^{\#}$ such that $l_a^{\#}(x + I_l) = ax + I_l$ (cf.1.5.2). $\mathcal{L}^{\#}$ is faithful (i.e. 1-1) if $(I_l : A) = \{0\}$.

1.5.16. LEMMA. *Let $I = I_l$ be a regular l. ideal of an algebra. Then:*

(i) *The quotient module $A^{\#} = A/I$ is cyclic.*

(ii) *$A^{\#}$ is irreducible iff I is maximal.*

PROOF. (i) If u is a relative r. unity for I then $x + I = xu + I = x(u + I)$, so that $A^{\#}$ is cyclic with $u + I$ as generator.

(ii) If $\pi : A \to A^{\#}$ is the canonical module homomorphism given by $x \mapsto x + I$, then clearly $\pi^{-1}(X_0)$, where X_0 ranges through all submodules of $X = A^{\#}$, are precisely the l. ideals of A containing I. Hence the irreducibility of X is equivalent to the maximality of I.

1.5.17. LEMMA. *Let X be a cyclic A-module with generator x_0 and $I = \ker x_0$. Then*

$$\Phi : a + I \mapsto ax_0$$

is a module isomorphism of $A^{\#} = A/I$ onto X.

PROOF. Φ is well-defined since $a_1 + I = a_2 + I \Rightarrow a_1 x_0 = a_2 x_0$. Since every element of X has the form ax_0, Φ is surjective. Finally, Φ is a module homomorphism since

$$\Phi(b(a + I)) = \Phi(ba + I) = bax_0 = b\Phi(a + I)$$

1.5.18. Let A be an algebra and X an A-module. Write $\mathcal{D} = \mathcal{D}(X) =$ the set of all A-endomorphisms (i.e. endomorphism T of the additive group of X satisfying $Tax = aTx$ for all $a \in A$). Clearly $\mathcal{D}$ is a unital algebra, with the identity map of X as the unity element.

1.5.19. LEMMA (Schur). *For an irreducible A-module $\mathcal{D}$ is a division algebra.*

PROOF. If $T \in \mathcal{D}, T \neq 0$ then $TX, \ker X$ are submodules with $TX \neq \{0\}, \ker T \neq X$. By irreducibility of X, $TX = X, \ker T = \{0\}$, so that T is invertible and $T^{-1} \in \mathcal{D}$.

1.5.20. Let X be an irreducible A-module and $\mathcal{D} = \mathcal{D}(X)$. If $x_1, \cdots, x_n \in X$ these vectors are said to be $\mathcal{D}$-independent if for any $T_1, \cdots, T_n \in D$,

$$T_1 x_1 + \cdots + T_n x_n = 0 \Rightarrow T_1 = \cdots = T_n = 0.$$

Since $\mathcal{D}$ is a division algebra, X can be regarded as a linear space over $\mathcal{D}$. Then clearly $\mathcal{D}$-independence is the same as linear independence in this linear space.

1.5.21. THEOREM (Jacobson's Density Theorem). *Given $\mathcal{D}$-independent vectors $x_1, \cdots, x_n$ and arbitrary vectors $y_1, \cdots, y_n$ all in irreducible module X, there is an a in A with $ax_j = y_j (1 \leqslant j \leqslant n)$.*

PROOF. See [4, p.123].

§ 6. Real and Complex Algebras

1.6.1. Let $\mathbb{K}$ denote either the real field $\mathbb{R}$ or the complex field $\mathbb{C}$. An algebra over $\mathbb{K}$ is called *real* or *complex* according as $\mathbb{K}$ is $\mathbb{R}$ or $\mathbb{C}$. Every complex algebra A becomes a real algebra $A^{[\mathbb{R}]}$ if we restrict the scalars for A to $\mathbb{R}$. A real algebra A is said to have a *complex structure* if A can be made into a complex algebra $A^{[\mathbb{C}]}$ such that $(A^{[\mathbb{C}]})^{[\mathbb{R}]} = A$.

1.6.2. LEMMA. *A real algebra A has a complex structure iff A admits a linear endomorphism J such that: $(*)$ $J(xy) = (Jx)y + x(Jy)$; $J^2 = -I$, where I is the identity map of A.*

PROOF. If A is a complex algebra, J defined by $Jx = ix$ $(i = \sqrt{-1})$ satisfies the above condition $(*)$. Conversely, if there is a J satisfying $(*)$ a complex structure can be defined by setting

$$(\alpha + i\beta)x = \alpha x + \beta Jx \quad (\alpha, \beta \in \mathbb{R}, \alpha \in A).$$

1.6.3. COROLLARY. *A real unital algebra with unity e has* *a complex structure iff there is an element j in the centre of A such that $j^2 = -e$.*

PROOF. If A is complex we can take $j = ie$. Conversely, if an element j satisfying the above condition exists then $Jx = jx$ will define a linear endomorphism satisfying $(*)$ of 1.6.2.

1.6.4. Let A be a real linear space and $\tilde{A}$ the set of ordered pairs $(x + y)$ with $x, y \in A$. We denote (x, y) by $x + iy$, where $i = \sqrt{-1}$. For $z = (x, y) = x + iy$, we write $\operatorname{Re} z = x$, $\operatorname{Im} z = y$ The set $\tilde{A}$ becomes a complex linear space if we define addition and multiplication in $\tilde{A}$ by:

$$(x_1 + iy_1) + (x_2 + iy_2) = (x_1 + x_2) + i(y_1 + y_2) \qquad (1)$$

$$(\alpha + i\beta)(x + iy) = (\alpha x - \beta y) + i(\alpha y + \beta x) \qquad (2)$$

where $x_j, y_j \ \ (j = 1, 2), x, y \in A; \alpha, \beta \in \mathbb{R}$.

We identify $x \in A$ with $x + i0 = (x, 0)$.

In particular $0 = 0 + i0 = (0, 0)$.

For $z \in \tilde{A}, z = x + iy$, we set $\bar{z} = x - iy$. Then it is straight forward to check that the map $\omega : z \mapsto \omega(z) = \bar{z}$ (called conjugation) has the properties:

$$\overline{z_1 + z_2} = \bar{z}_1 + \bar{z}_2; \ \overline{\lambda z} = \bar{\lambda}\bar{z}; \ \bar{\bar{z}} = z$$

$(z_1, z_2, z \in \tilde{A}, \lambda \in \mathbb{C})$.

If now A is a real algebra then $\tilde{A}$ can be made into an algebra by defining multiplication in $\tilde{A}$ via the equation

$$z_1 z_2 = (x_1 x_2 - y_1 y_2) + i(x_1 y_2 + x_2 y_1) \qquad (3)$$

$(z_j = x_j + iy_j \ (j = 1, 2) \in \tilde{A})$.

The complex algebra $\tilde{A}$ is called the *complexification* of the real algebra A. Also it is easy to check that conjugation preserves multiplication: $\overline{z_1 z_2} = \bar{z}_1 \bar{z}_2$. For any subset $\tilde{S} \subseteq \tilde{A}$ we write $\bar{\tilde{S}}$ for $\omega(\tilde{S})$. We also write

$$\operatorname{Re} \tilde{S} = \{ \operatorname{Re} z : z \in \tilde{S} \}, \ \operatorname{Im} \tilde{S} = \{ \operatorname{Im} z : z \in \tilde{S} \}.$$

We call a subset $\tilde{S}$ *self-conjugate* if $\bar{\tilde{S}} = \tilde{S}$.

1.6.5. PROPOSITION. (a) *A is a subalgebra of* $(\tilde{A})^{[\mathbb{R}]}$.

(b) $\tilde{A}$ *is commutative iff* A *is commutative.*

(c) $\tilde{A}$ *is unital iff* A *is unital and then both* $\tilde{A}, A$ *have the same unity element.*

PROOF. (a) Clear.

(b) This follows from (3) of 1.6.4.

(c) If A has a unity e it is easily checked that e is also a unity of $\tilde{A}$. On the other hand if $\tilde{e}$ is a unity of $\tilde{A}$ then it is easily seen that the element $e =$ Re $\tilde{e}$ is a unity of A and consequently also of $\tilde{A}$. By uniqueness of the unity, $\tilde{e} = e$, completing the proof.

1.6.6. LEMMA. *Suppose that* $a, b, c, d, \in A$ *and*

$$z = (a + ib) \circ (c + id) \text{ in } \tilde{A}.$$

Then

$$\bar{z} = (a - ib) \circ (c - id)$$

PROOF. Write $z_1 = a + ib, z_2 = c + id$. Then

$$\begin{aligned} \bar{z} = \overline{z_1 \circ z_2} &= \overline{(z_1 + z_2 + z_1 z_2)} = \bar{z}_1 + \bar{z}_2 + \bar{z}_1 \bar{z}_2 \\ &= \bar{z}_1 \circ \bar{z}_2. \end{aligned}$$

1.6.7. LEMMA. *If* $c + id$ *(respy.* $a + ib$*) has* $a + ib$ *(respy.* $c + id$*) as l.q.i. respy. r.q.i) then* $c - id$ *(respy.* $a - ib$*) has* $a - ib$ *(respy.* $c - id$*) as l.q.i. (respy. r.q.i). In particular,* $a - ib$ *is q. invertible iff* $a + ib$ *is q. invertible and then we have* $(a - ib)' = c - id$ *if* $(a + ib)' = c + id$.

PROOF. This follows from 1.6.6, by taking $z = 0$ (noting that $\bar{0} = 0$)

1.6.8. LEMMA. $(a + id) \circ c = 0$ *(respy.* $c \circ (a + ib) = 0) \Rightarrow a \circ c$ *(respy.* $c \circ a) = 0$, *where* a, b, c *are elements of* A.

PROOF. This is obtained by equating real parts on both sides of the equation (after expanding).

1.6.9. LEMMA. *Let* A *be unital. Then* $a - ib$ *is invertible in* $\tilde{A}$ *iff* $a + ib$ *is invertible in* $\tilde{A}$, *and then we have*

$$(a - ib)^{-1} = c - id \text{ if } (a + ib)^{-1} = c + id.$$

PROOF. If $z_1z_2 = e$, then $\bar{z}_1\bar{z}_2 = \overline{z_1z_2} = \bar{e} = e$. The lemma now readily follows.

1.6.10. LEMMA. *Let A be a real algebra and $\tilde{A}$ its complexification. Then:*

(i) *An element $a \in A$ is q. invertible in $\tilde{A}$ iff it is q. invertible in A.*

(ii) *If A is unital, then $a \in A$ is invertible in $\tilde{A}$ iff it is invertible in A.*

PROOF. (i) since $A \subset \tilde{A}$ it is enough to prove that "a is q. invertible in $\tilde{A}$" $\Rightarrow$ "a is q. invertible in A". Suppose that

$$a \circ (c + id) = 0 = (c + id) \circ a.$$

By 1.6.8, $a \circ c = 0 = c \circ a$, † proving a is q. invertible in A.

(ii) The proof is similar to that of (i)

1.6.11. LEMMA. *If $\tilde{I}$ is a l. (respy. r.) ideal of $\tilde{A}$ then $\omega(\tilde{I})$ is a l. (respy. r.) ideal of $\tilde{A}$. Further, if $\tilde{I}$ is regular with relative r. (respy. l.) unity $\tilde{u}$ then $\omega(\tilde{I})$ is regular with r. (respy. l.) unity $\omega(\tilde{u})$.*

PROOF. Easy consequences of the properties of ω.

1.6.12. PROPOSITION. *Let A be a real algebra and $\tilde{A}$ its complexification. Then:*

(i) *If S is a subspace of A, $\tilde{S} = S + iS$ is a subspace of $\tilde{A}$.*

(ii) *If I is a l. (respy. r.) ideal of A, $\tilde{I} = I + iI$ is a l. (respy. r.) ideal of $\tilde{A}$. Moreover, $\tilde{I}$ is regular whenever I is regular and every relative r. (respy. l.) unity for I is also a relative r. (respy. l.) unity for $\tilde{I}$. If M is a maximal regular l. (respy. r.) ideal of A then $\tilde{M} = M + iM$ is a maximal regular l. (respy. r.) ideal of $\tilde{A}$. Finally, $\tilde{M}$ is hypermaximal whenever M is hypermaximal.*

---

† Since c is also a q. inverse of a *in* $\tilde{A}$, by uniqueness of q.i., $c = c + id$, or $d = 0$.

(iii) *If* $\tilde{S}$ *is a subspace of* $\tilde{A}$ *then* $Re\,\tilde{S} = \mathrm{Im}\,\tilde{S} = S$ *(say) is a subspace of* A *such that* $\tilde{S} \subseteq S + iS, A \bigcap \tilde{S} \subseteq S$. *Further,* $\tilde{S} = S + iS$ *iff* $\tilde{S}$ *is self-conjugate, and then* $S = A \bigcap \tilde{S}$.

(iv) *If* $\tilde{I}$ *is a l. (respy. r.) ideal of* $\tilde{A}, I = \mathrm{Re}\,\tilde{I}$ *is a l. (respy. r.) ideal of* A. *Also, if* $\tilde{I}$ *is regular with relative r. (respy. l.) unity* $\tilde{u}$ *then* I *is regular with relative r. (respy. l.) unity* $u = \mathrm{Re}\,\tilde{u}$.

(v) *Every self-conjugate regular l. (respy. r.) ideal* $\tilde{I}$ *of* $\tilde{A}$ *has a relative r. (respy. l.) unity* u *belonging to* A, *which is also a relative r. (respy. l.) unity for* $I = \mathrm{Re}\,\tilde{I} = A \bigcap \tilde{I}$. *Further, if* $\tilde{I} \neq \tilde{A}$ *then* $I \neq A$. *Finally, the correspondence*

$$I \mapsto I + iI = \tilde{I} \text{ (say)}$$

is a bijection between the set of regular l. (respy. r.) ideals of A *and the set of self-conjugate regular l. (respy. r.) ideals of* $\tilde{A}$; $I = A \bigcap \tilde{I}$.

PROOF. (i) This follows from (1),(2) of 1.6.4 .

(ii) The first statement is a consequence of (1), (3) of 1.6.4 . If I is a regular l. ideal with a relative r. unity u then

$$(x + iy)u - (x + iy) = xu - x + i(yu - y) \in I + iI = \tilde{I},$$

proving u is also a relative r. unity for $\tilde{I}$ (the proof of the corresponding result when I is a r. ideal is similar). If M is a regular l. ideal of A with a relative r. unity u, then as seen above u is a relative r. unity for $\tilde{M} = M + iM$. Since $u \notin M, u \notin \tilde{M}$ and so $\tilde{M} \neq \tilde{A}$. By 1.2.10 there is a maximal regular l. ideal $\tilde{M}_1$ of $\tilde{A}$ with $\tilde{M} \subseteq \tilde{M}_1$. It follows that $M \subseteq \mathrm{Re}\,\tilde{M}_1$, whence by the maximality of M we have $M = \mathrm{Re}\,\tilde{M}_1$. So

$$\tilde{M}_1 \subseteq \mathrm{Re}\,\tilde{M}_1 + i\,(\mathrm{Im}\,\tilde{M}_1) \subseteq M + \mathrm{i}M = \tilde{M},$$

whence $\tilde{M}_1 = \tilde{M}$ and $\tilde{M}$ is maximal (the proof of this when M is a *r.* ideal is similar).

Suppose now that M is hypermaximal. Then $\tilde{M}$ is (maximal) regular. If $z = x + iy \in \tilde{A}$ $(x, y \in A)$ and u is a relative unity for M, we have

$$x = \alpha u + m, y = \beta u + m_1 ,$$

where $\alpha, \beta \in \mathbb{R}$ and $m, m_1 \in M$. Therefore

$$z = (\alpha + i\beta)u + (m + im_1),$$

so that $\tilde{A} = \mathbb{C}u + \tilde{M}$, whence $\tilde{M}$ is hypermaximal.

(iii) It follows from (1),(2) of 1.6.4 and the definition of $\text{Re } \tilde{S}$ that it is a subspace of A. Also, since for $z \in \tilde{A}$, we have $\text{Re } \tilde{z} = \text{Im } iz$, $\text{Im } z = \text{Re } (-iz)$ it follows that $\text{Re } \tilde{S} = \text{Im } \tilde{S} = S$ (say). Further, since $z = \text{Re } \tilde{z} = i \text{ Im } z$ we have $\tilde{S} \subseteq S + iS$. Clearly, $A \bigcap \tilde{S} \subseteq S$. Finally, if $\tilde{S}$ is self-conjugate and $z \in \tilde{S}$ then $\bar{z} \in \tilde{S}$,

$$\text{Re } z = \frac{1}{2}(z + \bar{z}) \in \tilde{S}$$

so that S, and hence, $iS \subseteq \tilde{S}$, whence $\tilde{S} = S + iS, S = A \bigcap \tilde{S}$.

(iv) By (iii), I is a subspace of A. Also, it follows from (3) of 1.6.4 that it is a l. (respy. r.) ideal. Assume now that $\tilde{I}$ is a regular l. ideal with $\tilde{u} = u + iv$ $(u, v \in A)$ as a relative r. unity. Then, for $x \in A$, we have

$$x(u + iv) - x = xu - x + ixv \in \tilde{I},$$

so that $xu - x \in I = \text{Re } \tilde{I}$, proving I is regular with $u = \text{Re } \tilde{u}$ as a relative r. unity. The proof of the corresponding result when $\tilde{I}$ is a r. ideal is similar.

(v) Since $\tilde{I}$ is self-conjugate, by (iii),

$$\tilde{I} = I + iI, I = \text{Re } \tilde{I}, I = A \bigcap \tilde{I}.$$

By (iv), I is regular and by (ii), any relative r. unity for I is a relative r. unity for I is a relative r. unity for $\tilde{I}$. The bijectiveness of the map $I \mapsto I + iI$ is immediate.

1.6.13. LEMMA. *Let $\tilde{I}$ be a regular l. (respy. r.) ideal of $\tilde{A} = A + iA$. Then $I = \text{Re } \tilde{I} = A$ iff $\tilde{I}$ has a relative r. (respy. l.) unity of the form $iv (v \in A)$.*

PROOF. It suffices to prove the result when $\tilde{I}$ is a l. ideal. Assume that $\tilde{u} = iv$ is a relative r. unity for $\tilde{I}$. By 1.6.12 (iv), $\text{Re } \tilde{u} = 0$ is a relative r. unity for I. Since $0 \in I, I = A$.

Conversely, assume that $I = A$. Let $\tilde{u} = u + iv_0$ be a relative r. unity for $\tilde{I}$. Since $u \in A = I = \text{Re } \tilde{I}$ we must have $u + iy_0 \in \tilde{I}$

for some $y_0 \in A$. Then, for any $z = x + iy \in \tilde{A}$ and $v = v_0 - y_0$ we have

$$ziv - z = zi(v_0 - y_0) - z = z(u + iv_0) - z - z(u + iy_0) \in \tilde{I} + \tilde{I} = \tilde{I}$$

whence iv is a relative r. unity for $\tilde{I}$.

1.6.14. Let A be a real algebra and $\tilde{A}$ its complexification. If f is a functional (= real-valued function) on A, the extension $\tilde{f}$ of f given by:

$$\tilde{f}(x + y) = f(x) + if(y) \qquad (x, y \in A)$$

is called the *canonical* extension of f (to $\tilde{A}$). It is straightforward to verify that $\tilde{f}$ is linear functional (respy. character) of $\tilde{A}$ if f is a linear functional (respy. character) of A

A character $\tilde{\chi}$ of $\tilde{A}$ is called *real* if $\chi = \tilde{\chi}|A$ is real-valued (and then χ is a character of A). Note that the canonical extension $\tilde{\chi}$ of a character χ of A is real.

1.6.15. PROPOSITION. (a) *Every real character $\tilde{\chi}$ of $\tilde{A}$ satisfies: $\tilde{\chi}(\bar{z}) = \overline{\tilde{\chi}(z)}$ $(z \in \tilde{A})$*

(b) *A character $\tilde{\chi}$ of $\tilde{A}$ is real iff* $\ker \tilde{\chi}$ *is self-conjugate.*

(c) *There is a bijection between the set of all characters of A and the set of all real characters of $\tilde{A}$, given by $\chi \mapsto \tilde{\chi}$, where χ is a character of A and $\tilde{\chi}$ its canonical extension to $\tilde{A}$.*

PROOF. (a)

$$\begin{aligned} \tilde{\chi}(\bar{z}) &= \tilde{\chi}(x - iy) = \chi(x) - i\chi(y) \\ &= \overline{(\chi(x) + i\chi(y))} = \overline{\tilde{\chi}(z)} \end{aligned}$$

(b) If $\tilde{\chi}$ is real then it follows from (a) that $\ker \tilde{\chi}$ is self-conjugate. Conversely, if $\ker \tilde{\chi} = \tilde{M}$ is self-conjugate then by 1.6.12 (iii), $\tilde{M} = M + iM$, where $M = \operatorname{Re} \tilde{M}$. Therefore, by 1.6.12(v), $\tilde{M}$ has a relative unity for M. By 1.3.9, $\tilde{M}$ is hypermaximal. It follows that if $x \in A$ then

$$x = (\alpha + i\beta)u + m_1 + im_2, \text{ for some } \alpha, \beta \in \mathbb{R}; m_1, m_2 \in M.$$

By taking real parts we get from the above equation

$$x = \alpha u + m_1, \text{ whence } \tilde{\chi}(x) = \alpha \in \mathbb{R},$$

so that $\tilde{\chi}$ is real.

(c) If $\tilde{\chi}$ is a real character of $\tilde{A}$ and $\chi = \tilde{\chi}|A$, then χ is a character of A and further $\tilde{\chi}$ is the canonical extension of χ. The bijectiveness of the map $\chi \mapsto \tilde{\chi}$ is now clear.

1.6.16. PROPOSITION. *The complexification $\tilde{A}$ of a real s.s. algebra is s.s.*

PROOF. Let M be any maximal regular l. ideal. By 1.6.12 (ii), $\tilde{M} = M + iM$ is a maximal regular l. ideal of $\tilde{A}$. If $z = x + iy \in \sqrt{\tilde{A}}$ then $z \in \tilde{M}$, whence $x, y \in M$. Since M is an arbitrary maximal ideal we obtain

$$x, y \in \bigcap M = \sqrt{A} = \{0\}. \text{ Thus, } z = 0 \text{ and } \sqrt{\tilde{A}} = \{0\}.$$

1.6.17. DEFINITION. An algebra A over $\mathbb{K}$ is called ***formally real*** if it satisfies the condition †

$$x, y \in A \text{ and } x^2 + y^2 = 0 \Rightarrow x = y = 0. \qquad (*)$$

A formally real algebra $A(\neq \{0\})$ is necessarily real. For, if A is complex, $x \in A$ and $x \neq 0$ then $x^2 + (ix)^2 = 0$, so that it is not formally real.

1.6.18. REMARK. Any real algebra whose elements are real-valued function on a set S is formally real $(f(s)^2 + g(s)^2 = 0 \Rightarrow f(s) = g(s) = 0)$.

1.6.19. PROPOSITION. *Let A be a formally real algebra. Then A as well as its complexification contain no non-zero nilpotent element.*

PROOF. It follows from the definition of formal reality that $0 = x^2 = x^2 + 0^2 \Rightarrow x = 0 (x \in A)$. Now suppose that $y \in A, y^n = 0, y^{n-1} \neq 0 (n \geqslant 3)$. If $n = 2k$ then $y^{2k} = (y^k)^2 = 0 \Rightarrow y^k = 0$, contradicting $y^{n-1} \neq 0$. Similarly, if $n = 2k + 1$ then since $y^{n+1} = 0$ we get $y^{k+1} = 0$, again contradicting $y^{n-1} \neq 0$. Thus A has no non-zero nilpotent element.

---

† This condition which is weaker than the Artin-Schreier condition for formal reality in the case of fields has been used by Arens and appears to be due to him.

Next suppose that $z = x + iy\,(x, y \in A)$ and $z^2 = 0$. Then we get $x^2 - y^2 = 0, xy + yx = 0$ It follows that

$$x^4 = x^2.y^2 = x(xy)y = x(-yx)y = -(xy)^2,$$

$(x^2)^2 + (xy)^2 = 0$, whence by formal reality of A, $x^2 = 0$, $y^2 = x^2 = 0$, so that $x, y, z = 0$. We can show as above that the condition "$z^2 = 0 \Rightarrow z = 0$" is sufficient to conclude that $\tilde{A}$ contains no non-zero nilpotent element.

1.6.20. PROPOSITION. (a) *The complexification $\tilde{A}$ of a commutative real division algebra A is a division algebra iff A is formally real.*

(b) *Every commutative real division algebra which is not formally real has complex structure.*

PROOF. (a) Suppose that A is formally real. If $z \in \tilde{A}, z = x+iy \neq 0$ then by formal reality of $A, x^2+y^2 \neq 0$ and so invertible (since A is a division algebra). Since we have

$$z\bar{z} = \bar{z}z = x^2 + y^2$$

by 1.1.30, z is invertible and $\tilde{A}$ is a division algebra.

Conversely, suppose that $\tilde{A}$ is a division algebra and for some $x, y \in A$ we have $x^2 + y^2 = 0$. Then, writing $z = x + iy$ we get

$$z\bar{z} = x^2 + y^2 = 0,$$

so that $z = 0$, since $\tilde{A}$ is a division algebra. Thus, $x = y = 0$ and A is formally real.

(b) Since A is not formally real there exist in it elements $x, y \neq 0$ with $x^2 + y^2 = 0$. Writing $j = yx^{-1} = x^{-1}y$, we get $e + j^2 = 0$, or $j^2 = -e$. Therefore, by 1.6.3, A has complex structure.

1.6.21. LEMMA. *If A is a commutative formally real algebra and M a maximal regular ideal, than the quotient $A^{\#} = A/M$ is formally real.*

PROOF. If $x^{\#2} + y^{\#2} = 0^{\#}$ then $x^2 + y^2 \subseteq M \subseteq \tilde{M}$, where $\tilde{M}$ is the complexification of M; $\tilde{M}$ is also maximal regular. Since

$$(x + iy)(x - iy) = x^2 + y^2 \in \tilde{M},$$

$\tilde{M}$ is prime (by 1.5.9, 1.5.10 (a)) and $\tilde{M}$ is self-conjugate we conclude that $x \pm iy \in \tilde{M}$, whence

$$x, y \in M, x^{\#} = 0^{\#}, y^{\#} = 0^{\#},$$

proving $A^{\#}$ is formally real.

§ 7. Spectrum and Quasi-spectrum

1.7.1. DEFINITION. Let A be a unital algebra over a field F. For $x \in A$ set

$$\sigma_A(x) = \{\lambda \in F : x - \lambda e \text{ is not invertible }\}$$

where e is the unity of A. The set $\sigma(x) = \sigma_A(x)$ of scalars is called the *spectrum of* x and its set-complement $\rho(x) = F \backslash \sigma(x)$ the *resolvent set of* $x; \sigma(x)$ or $\rho(x)$ can be empty.

1.7.2. LEMMA. *Let A be a unital algebra. If $x \in A$ is invertible, then either $\sigma(x) = \sigma(x^{-1}) = \emptyset$, or $\sigma(x) \neq \emptyset$ and $\sigma(x^{-1}) = \{\lambda^{-1} : \lambda \in \sigma(x)\}$.*

PROOF. Since x and (hence) x^{-1} are invertible we have $0 \in \rho(x), 0 \in \rho(x^{-1})$. If $\lambda(\neq 0) \in \rho(x)$ then there is a y such that

$$y(x - \lambda e) = e = (x - \lambda e)y.$$

This can be rewritten as

$$y(-\lambda x)(x^{-1} - \lambda^{-1}e) = e = (x^{-1} - \lambda^{-1}e)(-\lambda x)y$$

which shows that $x^{-1} - \lambda^{-1}e$ has a l. inverse as well as a r. inverse and hence (by 1.1.4) it is invertible. Thus $\lambda^{-1} \in \rho(x^{-1})$. Further, since $(x^{-1})^{-1} = x$ we conclude that

$$\lambda(\neq 0) \in \rho(x) \text{ iff } \lambda^{-1} \in \rho(x^{-1}).$$

The lemma now follows by passing to the set-complements of $\rho(x), \rho(x^{-1})$ in F.

1.7.3. PROPOSITION. *Let A be a unital algebra over F with e as its unity and $x \in A$ be such that $\sigma(x) \neq \emptyset$. Then:*

(i) $\sigma(0) = \{0\}$;

(ii) $\sigma(e) = \{1\}$;

(iii) $\sigma(\mu x) = \mu\sigma(x)(\mu \in F)$;

(iv) $\sigma(\mu e + x) = \mu + \sigma(x), (\mu \in F)$;

(v) *For any polynomial* P *over* F, *we have*

$$\sigma(p(x)) \supseteq P(\sigma(x))^{\dagger}. \tag{$*$}$$

When F *is algebraically closed – in particular, when* $F = \mathbb{C}$ *– there is equality in* $(*)$, *i.e. we have*

$$\sigma(P(x)) = P(\sigma(x)). \tag{$**$}$$

PROOF. Since $0 - \lambda e = -\lambda e$ is invertible for $\lambda \neq 0$, and $e - \lambda e = (1-\lambda)e$ is invertible for $\lambda \neq 1$ we obtain (i),(ii). For(iii), the identity $\mu x - \lambda e = \mu(x - \mu^{-1}\lambda e), (\mu \neq 0, \lambda \in F)$ shows that $\lambda \in \sigma(\mu x)$ iff $\mu^{-1}x \in \sigma(x)$. Also, if $\mu = 0$, (iii) follows from (i). Thus we have (iii). Next, (iv) follows from the identity

$$(\mu e + x) - \lambda e = x - (\lambda - \mu)e.$$

It remains to prove(v). If $P = \alpha_0$ is a constant polynomial then $(*)$ holds since both sides reduce to $\{\alpha_0\}$. Next assume that P is nonconstant. If $\lambda \in \sigma(x)$ then $y = x - \lambda e$ is not invertible. So, by 1.2.12 (iii), one at least of $Ay,\ yA \neq A$, whence there is a l. or r. ideal $J(= Ay$ or $yA)$ with $y \in J,\ J \neq A$. If

$$P(x) = \alpha_n x^n + \cdots + \alpha_0 e \quad (\alpha_j \in F)$$

and

$$z_k = x^{k-1} + \lambda x^{k-2} + \cdots + \lambda^{k-1} e,$$

then

$$x^k - \lambda^k e = (x - \lambda e) z_k = y z_k = z_k y \in J, \text{ where } y = x - \lambda e,$$

so that

$$P(x) - P(\lambda)e = \sum_{k=1}^{n} \alpha_k (x^k - \lambda^k e) \in J.$$

† We write $P(\sigma(x))$ for $\{P(\lambda) : \lambda \in \sigma(x)\}$

Since $J \neq A, P(x) - P(\lambda)e$ is not invertible and $P(\lambda) \in \sigma(P(x))$, proving $(*)$.

Assume now that F is algebraically closed and that $P = P(X)$ is a polynomial of degree $n > 0$, over F. Then we have the factorization $P(X) - \lambda = \alpha(X - \mu_1) \cdots (X - \mu_n)$, where $\alpha, \mu_j \in F$ and μ_j are the zeros of the polynomial $P - \lambda$. It follows that $P(\mu_j) = \lambda \, (j = 1, \cdots, n)$. Further we have the factorization

$$y = P(x) - \lambda e = \alpha(x - \mu_1 e) \cdots (x - \mu_n e).$$

If $\lambda \in \sigma(P(x))$, y is not invertible, whence by virtue of 1.1.30, some $x - \mu_k e$ is not invertible. It follows that $\mu_k \in \sigma(x)$ and $\lambda = P(\mu_k) \in P(\sigma(x))$. Thus, $\sigma(P(x)) \subseteq P(\sigma(x))$. Combining this with $(*)$ we get $(**)$. It remains to deal with the case where $\deg P = 0$. In this case, if $P(X) = \alpha_0$ (say), $(**)$ holds since as noted in the proof of the first assertion in (v), each side of $(**)$ reduces to $\{\alpha_0\}$.

1.7.4. LEMMA. *Let A be unital, with unity e, and $x \in A$. Then $\lambda(\neq 0) \in \sigma(x)$ iff $-\lambda^{-1}x$ is not q. invertible.*

PROOF. The identity

$$x - \lambda e = -\lambda(e - \lambda^{-1}x) \quad (x \in A, \lambda \in F, \lambda \neq 0)$$

and 1.1.20 show that $\lambda \in \rho(x)$ iff $-\lambda^{-1}x$ is q. invertible. The required result now follows (by taking the negation of the statements on both sides of "iff").

1.7.5. DEFINITION. Let A be an algebra over F, not necessarily without a unity, and A_1 its unitization. For $x \in A$ we write

$$\sigma'(x) = \sigma'_A(x) = \sigma_{A_1}(x)$$

and call $\sigma'(x)$ the *quasi-spectrum* of x; its complement

$$\rho'(x) = F \backslash \sigma'(x)$$

is called the *quasi-resolvent* set of x.

1.7.6. LEMMA. $0 \in \sigma'(x)$ *for any $x \in A$, so that $\sigma'(x)$ is always nonempty*

PROOF. If possible let x have an inverse $y + \lambda e_1$ $(y \in A, \lambda \in$

F) in A_1. Then $e_1 = x(y + \lambda e_1) = xy + \lambda x \in A$ which is impossible. Hence $0 \in \sigma_{A_1}(x) = \sigma'(x)$.

1.7.7. PROPOSITION. (i) $\sigma'(0) = 0$; (ii) $\sigma'(\mu x) = \mu\sigma'(x) (\mu \in F)$; (iii) *If P is a constant-free polynomial over F then*

$$\sigma'(P(x)) \supseteq P(\sigma'(x))$$

with the inclusion relation becoming equality when F is algebraically closed, so that then

$$\sigma'(P(x)) = P(\sigma'(x)).$$

PROOF. These follow from 1.7.3 since for any element a in A we have $\sigma'(a) = \sigma_{A_1}(a)$.

1.7.8. LEMMA (Kaplansky). *Let A be an algebra $x \in A$. Then $\lambda \in \sigma'(x)\backslash\{0\}$ iff $-\lambda^{-1}x$ is not q. invertible.*

PROOF. Since

$$x - \lambda e_1 = -\lambda(e_1 - \lambda^{-1}x) \qquad (\lambda \neq 0)$$

where e_1 is the unity of the unitization A_1 of A, we obtain:

$$\begin{aligned} \lambda(\neq 0) \in \sigma_{A_1} \quad &\Leftrightarrow \quad x - \lambda e_1 \text{ is not invertible in } A_1 \\ &\Leftrightarrow \quad (e_1 - \lambda^{-1}x) \text{ is not invertible in } A_1 \\ &\Leftrightarrow \quad -\lambda^{-1}x \text{ is not q. invertible in } A_1 \text{ (by 1.1.20)} \\ &\Leftrightarrow \quad -\lambda^{-1}x \text{ is not q. invertible in } A \text{ (by 1.1.21).} \end{aligned}$$

1.7.9. COROLLARY. *If $u \neq 0$ is an idempotent then $\sigma'(u) = \{0,1\}$. Further, if A is unital, with unity e, and $u \neq 0, e$ then $\sigma(u) = \{0,1\}$.*

PROOF. The first statement follows from 1.7.8, 1.1.26 (i). For the second we note that by $(**)$ of 1.7.21, we have

$$\sigma(u) \bigcup \{0\} = \sigma'(u) = \{0,1\}.$$

If $0 \notin \sigma(u)$ then u is invertible and since $u^2 = u$ we get $u = e$, contradicting the assumption on u. Thus $0 \in \sigma(u)$,

$$\sigma(u) = \sigma(u) \bigcup \{0\} = \sigma'(u) = \{0,1\}.$$

1.7.10. LEMMA. *If u is a relative unity for a regular l. or r. ideal $I \neq A$ then $1 \in \sigma'(u)$.*

PROOF. By 1.2.9 (ii) $-u$ is not q. invertible, whence by 1.7.8, $1 \in \sigma'(u)$.

1.7.11. COROLLARY. *If χ is any character of A and $x \in A$ then $\chi(x) \in \sigma'(x)$.*

PROOF. Since always $0 \in \sigma'(x)$ we may assume that $\chi(x) = \lambda \neq 0$. Then $\chi(\lambda^{-1}x) = 1$, so that by 1.3.10, $u = \lambda^{-1}x$ is a relative (bi-) unity of M_χ, whence by 1.7.10, $1 \in \sigma'(u) = \sigma'(\lambda^{-1}x) = \lambda^{-1}\sigma'(x)$, so that $\lambda \in \sigma'(x)$.

1.7.12. LEMMA. *Assume that $x \in A$ is q. invertible with q.i. x'. Then*

$$\sigma'(x') = \left\{-\frac{\lambda}{1+\lambda} : \lambda \in \sigma'(x)\right\}.$$

PROOF. First note that $0 \in \sigma'(x)$ and $-\frac{0}{1+0} = 0 \in \sigma'(x')$. Since x is q. invertible, by 1.7.8, $-1 \notin \sigma(x)$. For $\lambda \neq 0$ we have the (easily verifiable) identity:

$$x' \circ (-\lambda^{-1}x) = (-\lambda^{-1}x) \circ x' = \frac{1+\lambda}{\lambda}x'.$$

It follows, by 1.7.8 and 1.1.30, that $-\frac{\lambda}{1+\lambda} \in \sigma'(x)$ iff $\lambda \in \sigma'(x)$, completing the proof.

1.7.13. DEFINITION. An element x in A is called *quasi-nilpotent* or *q. nilpotent* if $\sigma'(x) = \{0\}$. The set of q. nilpotent elements of A is denoted by $A^{qn}; 0 \in A^{qn}$ (by 1.7.7 (i)).

1.7.14. Remark. In the algebra $M_n(F)$ of all $n \times n$ matrices over a field F, a q. nilpotent element is nothing but a nilpotent matrix. This is a consequence of the fact that a matrix is nilpotent iff all its eiginvalues are 0.

1.7.15. PROPOSITION. *Let A be an algebra. Then:*

(i) *An element x in A is q. nilpotent iff λx is q. invertible for every $\lambda \in F$, in particular, a q. nilpotent element is q. invertible.*

(ii) *A nilpotent or essentially nilpotent element is q. nilpotent.*

(iii) $\sqrt{A} \subseteq A^{qn} \subseteq \sqrt[h]{A}$.

PROOF. (i) This follows from 1.7.8, 1.7.13.

(ii) If x is nilpotent so is λx, and hence by 1.1.23, λx is q. invertible, whence by (i), x is q. nilpotent. Again, if x is essentially nilpotent then by definition $x \in \sqrt{A}$, and hence also $\lambda x \in \sqrt{A}$, λx is q. invertible (by 1.2.24 (a)) so that by (i) x is q. nilpotent.

(iii) The first inclusion follows from (ii). For the second, we note that if $x \in A^{qn}$, $\sigma'(x) = \{0\}$. If $\chi \in \Delta$, $\chi(x) \in \sigma'(x)$ (by 1.7.11) so that $\chi(x) = 0$, whence $x \in \sqrt[h]{A}$.

1.7.16. Remark. Every element of a radical algebra A $(A = \sqrt{A})$ is q. nilpotent, so that $A = A^{qn}$. On the other hand, in the one-dimensional algebra F, 0 is the only q. nilpotent element (since, by 1.7.20, $\sigma'(\lambda) = \sigma(\lambda) \bigcup \{0\} = \{\lambda, 0\}$).

1.7.17. DEFINITION. An algebra A is called *quasi-semi-simple* or *q.s.s.* if 0 is the only q. nilpotent element of A, i.e. $A^{qn} = \{0\}$.

1.7.18. Remark. In view of 1.7.15 (iii), every q.s.s. algebra is s.s. Note that the matrix algebra $M_n(F)$ is s.s. (actually simple) but not q.s.s. when $n \geqslant 2$ (since there are non-zero nilpotent matrices in $M_n(F), n \geqslant 2$).

1.7.19. LEMMA. *Let A, B be algebras (over the same field), $\varphi : A \to B$ a homomorphism and $x \in A$. Then we have:*

$$\rho'_A(x) \subseteq \rho'_B(\varphi(x)); \; \sigma'_B(\varphi(x)) \subseteq \sigma'_A(x). \qquad (*)$$

$$\rho_A(x) \subseteq \rho_B(\varphi(x)); \; \sigma_B(\varphi(x)) \subseteq \sigma_A(x) \qquad (**)$$

whenever A, B are unital and B has as unity $\varphi(e)$, e being the unity of A.

PROOF. If $\lambda \in \rho'_A(x)$ then $-\lambda^{-1}x$ is q. invertible and so by 1.1.24,

$$-\lambda^{-1}\varphi(x) = \varphi(-\lambda^{-1}x)$$

is q. invertible, so that $\lambda \in \rho'_B(\varphi(x))$. This proves the first inclusion relation in $(*)$ and the second one follows by taking set-complements of both sides in F.

For proving $(**)$, we observe that by using the first relation in $(*)$ of 1.7.21, the first conclusion in $(*)$ above can be written as

$$\rho_A(x)\backslash\{0\} \subseteq \rho_B(\varphi(x))\backslash\{0\}. \qquad (*')$$

If $0 \in \rho_A(x)$, x is invertible and then $\varphi(x)$ is also invertible, and $0 \in \rho_B(\varphi(x))$. The inclusions in $(**)$ now readily follow from $(*')$.

1.7.20. COROLLARY. *Let A be a subalgebra of an algebra B and $x \in A$. Then we have:*

$$\rho'_A(x) \subseteq \rho'_B(x);\ \sigma'_B(x) \subseteq \sigma'_A(x). \qquad (*)$$

$$\rho_A(x) \subseteq \rho_B(x);\ \sigma_B(x) \subseteq \sigma_A(x) \qquad (**)$$

whenever B is unital and A is a subunital † *algebra of B.*

PROOF. This follows from 1.7.19 by taking φ to be the inclusion map $A \to B$.

1.7.21. PROPOSITION. *If A is a unital algebra over F and $x \in A$, then:*

$$\rho'(x) = \rho(x)\backslash\{0\}; \qquad (*)$$

$$\sigma'(x) = \sigma(x) \bigcup \{0\}. \qquad (**)$$

PROOF. Let A_1 be the unitization of A and e_1, e the unities of A_1, A respectively. If $\lambda \in \rho_{A_1}(x)$ then for some $y \in A$, $\mu \in F$ we have the equations

$$(x - \lambda e_1)(y + \mu e_1) = e_1 = (y + \mu e_1)(x - \lambda e_1). \qquad (1)$$

Multiplying the terms in (1) by e from the left and then by e from the right we obtain

$$(x - \lambda e)(y + \mu e) = e = (y + \mu e)(x - \lambda e). \qquad (2)$$

It follows from (2) that $\lambda \in \rho_A(x)$. Thus,

---

† i.e. if e is the unity of B then $e \in A$.

$$\rho'(x) = \rho_{A_1}(x) \subseteq \rho_A(x) = \rho(x).$$

On the other hand, if $\lambda \in \rho_A(x), \lambda \neq 0$ then since

$$x - \lambda e = -\lambda(e - \lambda^{-1}x),$$

$-\lambda^{-1}x$ is q. invertible in A and hence also in A_1, whence $\lambda \in \rho'(x)$. Further $0 \notin \rho_{A_1}(x)$, for otherwise either of the equations (1) with $\lambda = 0$ shows that $e_1 \in A$, which is impossible. This completes the proof of $(*)$, and $(**)$ follows from $(*)$ by taking set-complements in F.

1.7.22. PROPOSITION. *Let x, y be elements of an algebra A. Then we have*

$$\sigma'(xy) = \sigma'(yx), \tag{$*$}$$

$$\sigma(xy) \bigcup \{0\} = \sigma(yx) \bigcup \{0\} \tag{$**$}$$

whenever A is unital.

PROOF. First assume that A is unital with unity e. If $\lambda \in \sigma(xy)$, $\lambda \neq 0$ then $xy - \lambda e$, and hence also

$$(\lambda^{-1}x)y - e = -(e - (\lambda^{-1}x)y)$$

is not invertible. By 1.1.28 (ii), 1.1.14 (c), $-(e - y(\lambda^{-1}x)) = y\lambda^{-1}x - e$ is not invertible, so that $\lambda \in \sigma(yx)$. Thus

$$\sigma(xy) \bigcup \{0\} \subseteq \sigma(yx) \bigcup \{0\}.$$

Interchanging x, y we get the reverse inclusion. By combining the two inclusions we obtain $(**)$. To obtain $(*)$ it is enough to apply $(**)$ to the unitization A_1 of A (remember that for any $a \in A$, $0 \in \sigma'(a)$).

1.7.23. LEMMA. *Let A be a unital algebra (over F) with unity e, and χ a linear functional on A such that $\chi(e) = 1$. Then the following conditions are equivalent*

(i) *Every element $x \in \ker \chi$ is not invertible, or equivalently, $\chi(x) \neq 0$ for every invertible element x.*

(ii) $\chi(x) \in \sigma(x)$.

PROOF. (i) $\Rightarrow$ (ii). If $\chi(x) = \lambda$ then $x - \lambda e \in \ker\chi \neq A$, whence it is not invertible, so that $\lambda \in \sigma(x)$.
(ii) $\Rightarrow$ (i). If $x \in \ker\chi$ then $0 = \chi(x) \subseteq \sigma(x)$, whence x is not invertible.

1.7.24. LEMMA. *Let A be a unital algebra and χ a character of A. Then $\chi(x) \in \sigma(x)$ $(x \in A)$.*

PROOF. If $\chi(x) = \lambda$ then $x - \lambda e \in \ker\chi = M \neq A$ and therefore $x - \lambda e$ is not invertible, and so $\lambda \in \sigma(x)$.

1.7.25. LEMMA. *Suppose that A is an algebra over F, $a \in A$ and $\lambda \in F$. If there is a non-zero element $b \in A$ such that $ab = \lambda b$, then $\lambda \in \sigma'(a)$.*

PROOF. In the unitization A_1 of A the above condition can be rewritten as

$$(a - \lambda e_1)b = 0.$$

Since $b \neq 0$, $a - \lambda e_1$ is not invertible and consequently $\lambda \in \sigma'(a)$.

1.7.26. PROPOSITION. *Let A be an algebra and $a \in A$. Then we have:*

(i) *The double commutant $A_0 = \{a\}''$ is a commutative subalgebra of A containing a such that*

$$\sigma'_{A_0}(a) = \sigma'_A(a). \qquad (*)$$

(ii) *If $A_m = A_m(a)$ is a maximal commutative subalgebra of A, containing a then*

$$\sigma'_{A_m}(a) = \sigma'_A(a). \qquad (**)$$

(iii) *If A is unital with unity e then*

$$\sigma_{A_0}(a) = \sigma_{A_m}(a) = \sigma_A(a). \qquad (***)$$

PROOF. First let A have a unity e. By 1.1.8 (vi), A_0 is a commutative subalgebra and clearly $e \in A_0$. If $a - \lambda e$ has an inverse b in A then by 1.1.18

$$b \leftrightarrow \{a - \lambda e\}' = \{a\}', \text{ whence } b \in \{a\}'' = A_0.$$

Therefore $\rho_A(a) \subseteq \rho_{A_0}(a)$, so that by (**) of 1.7.20, $\rho_A(a) = \rho_{A_0}(a)$. Also, we have (see 1.1.9) $A_0 \subseteq A_m \subseteq A$. It follows (using 1.7.20) that $\rho_A(a) = \rho_{A_m}(a) = \rho_A(a)$. Therefore, by taking set-complements in F we obtain $(***)$.

Next let A be non-unital. By applying, $(***)$ to the subalgebras $(A_0)_1 = A_0 + Fe_1$, $(A_m)_1 = A_m + Fe_1$ of the unitization A_1 of A we obtain $(*)$ and $(**)$.

1.7.27. PROPOSITION. *Let A be an algebra and B a q. inverse closed*[†] *subalgebra of A.*
Then

$$\sigma'_B(x) = \sigma'_A(x) \qquad (x \in B).$$

Similarly, if A is unital and B a inverse closed subunital algebra of A then

$$\sigma_B(x) = \sigma_A(x) \qquad (x \in A).$$

PROOF. The first assertion follows from 1.7.8, B being q. inverse closed in A. The second is a consequence of B being inverse closed in A and the definition of the spectrum.

§ 8. Extended Spectrum and Extended Quasi-spectrum

1.8.1. DEFINITION. Let A be a unital real algebra, $\tilde{A}$ its complexification and $x \in A$. We write

$$\tilde{\sigma}(x) = \tilde{\sigma}_A(x) = \sigma_{\tilde{A}}(x)$$

and call $\tilde{\sigma}(x)$ the *extended spectrum* of x. Similarly, writing $\tilde{\sigma}'(x) = \sigma_{(\tilde{A})_1}(x)$, where $(\tilde{A})_1$ denotes the unitization of $\tilde{A}$, we call $\tilde{\sigma}'(x)$ the *extended quasi-spectrum* of x; its set-complement $\tilde{\rho}'(x) = \mathbb{K} \backslash \tilde{\sigma}'(x)$ is called the *extended quasi-resolvent* set of x.

1.8.2. LEMMA. *If A is a real algebra and $x \in A$ then*

$$\sigma'_A(x) = \tilde{\sigma}'_A(x) \bigcap \mathbb{R}. \qquad (*)$$

† A subalgebra (respy. subunital algebra) B of A is called q. inverse (respy. inverse) closed if for any $x \in B$, the q. inverse x' (respy. inverse x^{-1}) of x in A exists $\Rightarrow x'$ (respy. x^{-1}) $\in B$.

If A is unital we have also

$$\sigma_A(x) = \tilde{\sigma}_A \bigcap \mathbb{R}. \qquad (**)$$

In particular, $\sigma'_A(x) \subseteq \tilde{\sigma}'_A(x)$, $\sigma_A(x) \subseteq \tilde{\sigma}_A(x)$.

PROOF. We have $\lambda(\neq 0) \in \tilde{\sigma}'_A(x) \bigcap \mathbb{R} = \sigma'_{\tilde{A}}(x) \bigcap \mathbb{R}$ iff $-\lambda^{-1}x$ is not q. invertible in $\tilde{A}$ iff $-\lambda^{-1}x$ is not q. invertible in A iff $\lambda \in \sigma'_A(x)$. This prove $(*)$ (since '0' clearly belongs to both sides of $(*)$). The equality $(**)$ follows by applying $(*)$ to A_1 and using 1.7.21 .

1.8.3. PROPOSITION. *Let A be a real algebra and $x \in A$. Then $\tilde{\sigma}'(x)$ is a symmetric*† *subset of $\mathbb{C}$; also when A is unital, $\tilde{\sigma}(x)$ is symmetric.*

PROOF. We have

$$\begin{aligned} \lambda(\lambda \neq 0) \in \tilde{\sigma}'(x) \quad &\text{iff} \quad -\lambda^{-1}x \text{ is not q. invertible in } \tilde{A} \\ &\text{iff} \quad \overline{(-\lambda^{-1}x)} = -\bar{\lambda}^{-1}x \\ &\qquad \text{is not q. invertible in } \tilde{A} \text{ (see 1.6.7.)} \\ &\text{iff} \quad \bar{\lambda} \in \tilde{\sigma}'(x), \end{aligned}$$

and $\bar{0} = 0$, where bar denotes the conjugation in $\tilde{A}$ (see 1.6.4). Thus $\tilde{\sigma}'(x)$ is symmetric. The symmetry of $\tilde{\sigma}(x)$ is an immediate consequence of $(**)$ of 1.7.21 and the symmetry of $\tilde{\sigma}'(x)$.

1.8.4. PROPOSITION. *Let A be a real algebra, $\tilde{A}$ its complexification, and $x \in A$. Then:*

(i) *If $\lambda = \alpha + i\beta \neq 0$ $(\alpha, \beta \in \mathbb{R})$, $-\lambda^{-1}x$ is q. invertible in $\tilde{A}$ iff $\frac{x^2 - 2\alpha x}{|\lambda^2|}$ is q. invertible in A.*

(ii) *If A is unital with unity e, then $x - \lambda e$ is invertible in $\tilde{A}$ iff $(x - \alpha e)^2 + \beta^2 e$ is invertible in A.*

PROOF. Write $z = -\lambda^{-1}x$; then $\bar{z} = -\bar{\lambda}^{-1}x$ where bar denotes the conjugation in $\tilde{A}$. By 1.6.7, z is q. invertible iff $\bar{z}$ is q. invertible. Since $z \leftrightarrow \bar{z}$, by 1.1.30, $z \circ \bar{z}$ is q. invertible iff z (hence $\bar{z}$) is q. invertible. Since

† A subset S of $\mathbb{C}$ is said to be symmetric if $\lambda \in S \Rightarrow \bar{\lambda} \in S$, where $\bar{\lambda}$ denotes the complex conjugate of λ.

$$z \circ \bar{z} = \frac{x^2 - 2\alpha x}{|\lambda|^2},$$

(i) follows. The result (ii) can be proved similarly by using the identity

$$(x - \lambda e)(x - \bar{\lambda} e) = (x - \alpha e)^2 + \beta^2 e$$

and 1.1.30.

1.8.5. COROLLARY. *If* $\lambda = \alpha + i\beta (\alpha, \beta \in \mathbb{R}),\ \lambda \neq 0$ *then* $\lambda \in \tilde{\sigma}'(x)$ *iff* $y = (x^2 - 2\alpha x)(\alpha^2 + \beta^2)^{-1}$ *is not q. invertible in* A. *Similarly, when* A *is unital,* $\lambda \in \tilde{\sigma}(x)$ *iff* y *is not q. invertible in* A, *and* $0 \in \tilde{\sigma}(x)$ *iff* x *is not invertible in* A.

PROOF. The first assertion is an immediate consequence of 1.8.4 (i). Since $\tilde{\sigma}'(x) = \tilde{\sigma}(x) \bigcup \{0\}$ the second assertion follows from the first together with the observation that if x is not invertible in A it is also not invertible in $\tilde{A}$ $((x_1 + iy_1)x = e = x(x_1 + iy_1) \Rightarrow x_1 x = e = x x_1$, where $x, x_1, y_1 \in A)$.

1.8.6. PROPOSITION. (Rickart). *Let* A *be a complex algebra and* $A^{[\mathbb{R}]}$ *denote* A *as a real algebra. Then, for* $x \in A$, *we have*

$$\tilde{\sigma}_{A^{[\mathbb{R}]}}(x) = \sigma'_A(x) \bigcup \overline{\sigma'_A(x)},$$

where bar denotes complex conjugation.

PROOF. We denote the general element of $\tilde{A}^{[\mathbb{R}]}$ by $x + jy$, where $j^2 = -1$; $x_1, y \in A^{[\mathbb{R}]} = A$ (as a set). By 1.6.10, if $x \in A = A^{[\mathbb{R}]}$ has a q.i. in $\tilde{A}^{[\mathbb{R}]}$, it has also a q.i. in A. So, by 1.7.8, $\sigma'_A(x) \subseteq \tilde{\sigma}_{A^{[\mathbb{R}]}}(x)$ and hence by 1.8.3.

$$\overline{\sigma}'_A(x) \subseteq \tilde{\sigma}'_{A^{[\mathbb{R}]}}(x).$$

To complete the proof it is enogh to show that

$$\tilde{\sigma}'_{A^{[\mathbb{R}]}}(x) \subseteq \tilde{\sigma}'_A(x) \bigcup \overline{\sigma'_A(x)}.$$

Suppose that

$$\lambda \notin \sigma'_A(x) \bigcup \overline{\sigma'_A(x)}, \qquad \mu = -\lambda^{-1}.$$

Then $\mu x, \bar{\mu}x$ are q. invertible. Write

$$(\mu x)' = y, \ (\bar{\mu}x)' = y_1.$$

If $\mu = \gamma + i\delta$ $(\gamma, \delta \in \mathbb{R})$, then in $\tilde{A}^{[\mathbb{R}]}$ we have

$$\mu x = \gamma x + j\delta x, \qquad \bar{\mu}x = \gamma x - j\delta x.$$

It follows from 1.6.8 that we have

$$(\gamma x)' = (\mu x)' = y = (\bar{\mu}x)' = y_1.$$

Therefore $\mu x = y' = y_1' = \bar{\mu}x$, whence $\delta = 0$, $\mu = \gamma$. Thus

$$\mu x = \gamma x \in A^{[\mathbb{R}]} = A, \qquad (\mu x)' = y \in A$$

so that μx is q. invertible in A. It follows that $\lambda \in \tilde{\sigma}_{A^{[\mathbb{R}]}}$, completing the proof.

1.8.7. DEFINITION. An element x of a real algebra A is called *extended quasi-nilpotent* or *ext. q. nilpotent* if $\tilde{\sigma}_A(x) = \{0\}$, i.e. if x is q. nilpotent in $\tilde{A}$.

1.8.8. Remark. By (**) 1.8.2, every ext. q. nilpotent element is q. nilpotent. On the other hand, in the real algebra $A = \mathbb{C}^{[\mathbb{R}]}$, the element i is *q.* nilpotent but not *ext. q. nilpotent.* For, $\sigma_A'(i) = \sigma_A(i) \bigcup \{0\} = \{0\}$ since $\sigma_A(i) = \emptyset$ ($i - \alpha$ being invertible in A for every real α}. But, by 1.8.7,

$$\tilde{\sigma}_A'(i) = \sigma_{\mathbb{C}}'(i) \bigcup \overline{\sigma_{\mathbb{C}}'(i)} = \{0, \pm i\}.$$

1.8.9. LEMMA. *Let A be a real algebra. Then every element of $\sqrt{A}$ is ext. q. nilpotent.*

PROOF. Take $x \in \sqrt{A}$ and let $\lambda = \alpha + i\beta \neq 0$. Then, $\sqrt{A}$ being a bi-ideal, we have $y = (x^2 - 2\alpha x)(\alpha^2 + \beta^2)^{-1} \in \sqrt{A}$, so that y is q. invertible. It follows from 1.8.5 that $\lambda \notin \tilde{\sigma}'(x)$, whence $\tilde{\sigma}'(x) = \{0\}$, as desired.

1.8.10. DEFINITION. Let A be an algebra over $\mathbb{K}$ and $x \in A$. The spectral radius $r(x)$ is defined by

$$r(x) = r_A(x) = \sup\{|\lambda| : \lambda \in \sigma'(x)\}.$$

If A is unital and $\sigma(x) \neq \emptyset$ then we also have

$$r(x) = \sup\{|\lambda| : \lambda \in \sigma(x)\}.$$

Similarly, we have the *extended* spectral radius

$$\tilde{r}(x) = \sup\{|\lambda| : \lambda \in \tilde{\sigma}'(x)\}$$

for an element x of a real algebra A. When A is unital and $\tilde{\sigma}(x) \neq \emptyset$ we have

$$\tilde{r}(x) = \sup\{|\lambda| : \lambda \in \tilde{\sigma}(x)\}.$$

1.8.11. Remark. Clearly we have the inequalities:

$$0 \leqslant r(x) \leqslant \infty; \; r(x) \leqslant \tilde{r}(x).$$

1.8.12. LEMMA. *Let A be an algebra over $\mathbb{K}$, and $x \in A$. Then:*

(i) $r(0) = 0$.

(ii) $r(\mu x) = |\mu| r(x) (\mu \in \mathbb{K})$; *in particulars,* $r(-x) = r(x)$.

(iii) $r(x^n) = r(x)^n$, *provided A is a complex algebra.*

(iv) *If $u \neq 0$ is an idempotent of A then $r(u) = 1$; in particular,* $r(e) = 1$ (*whenever unity e exists*).

(v) $r(xy) = r(yx) \qquad (x, y \in A)$.

Proof. (i) By 1.7.7 (i), $\sigma'(0) = 0$ so that $r(0) = 0$.

(ii) By 1.7.7 (ii), $\sigma'(\mu x) = \mu\sigma'(x)$ and hence the result.

(iii) By 1.7.7 (iii), $\sigma'(x^n) = [\sigma'(x)]^n$, hence the result.

(iv) By 1.7.9, $\sigma'(u) = \{0, 1\}$, whence the result.

(v) An immediate consequence of $(*)$ of 1.7.22.

1.8.13. DEFINITION. Let A be a unital complex algebra

with unity e. A is called a *Liouville algebra*[†] if every element $x \in A$, $x \notin \mathbb{C}e$, $r(x) = \infty$. If A has no unity then A is called a Liouville algebra if for every $x \neq 0$, $r(x) = \infty$.

1.8.14. EXAMPLE. The algebra $A = \mathbb{C}[X]$ of all complex polynomials is a Liouville algebra. Here if $x \in A$, $x \notin \mathbb{C}e$ then $\sigma(x) = \mathbb{C}$, $r(x) = \infty$.

1.8.15. PROPOSITION. *Every Liouville algebra A is q.s.s., in particular s.s..*

PROOF. If A is unital and $x \notin \mathbb{C}e$ then $r(x) = \infty$ and x is not q. nilpotent. Also, if $x = \lambda e$, $\sigma(x) = \{\lambda\}$ and $r(x) = 0 \Rightarrow \lambda = 0$. Thus A is q.s.s. When A is not unital, by definition of the Liouville algebra, if $x \neq 0$, $r(x) = \infty$ and so x is not q. nilpotent. Thus A is q.s.s.

1.8.16. PROPOSITION. *Let A be a unital algebra over $\mathbb{K}$ such that the spectrum $\sigma(x)$ of every element x of A is non-empty and bounded. Then the Heisenberg commutator equation*

$$[x, y] = xy - yx = \mu e \qquad (e = \text{ unity of } A)$$

has no solution, for any $\mu \in \mathbb{K}\backslash\{0\}$.

PROOF. Suppose, to the contrary, there are elements $x, y \in A$ with

$$xy = yx + \mu e \text{ for some } \mu.$$

It follows that

$$\sigma(xy) = \sigma(yx + \mu e) = \mu + \sigma(yx). \tag{1}$$

By 1.7.22, we have

$$\sigma(xy) \bigcup \{0\} = \sigma(yx) \bigcup \{0\}. \tag{2}$$

From (1), (2) we obtain

$$\{\mu + \sigma(yx)\} \bigcup \{0\} = \sigma(yx) \bigcup \{0\}. \tag{3}$$

[†] This notion appears to have been first studied by Birtel [5′].

If $0 \in \sigma(yx)$ then (2) becomes

$$\{\mu + \sigma(yx)\} \bigcup \{0\} = \sigma(yx). \tag{4}$$

It follows that

$$\mu + 0 = \mu \in \sigma(yx).$$

By repeatedly using (4) we get $n\mu \in \sigma(yx)(n \geqslant 1)$. But this contradicts the boundedness of $\sigma(yx)$ since $|n\mu| = n|\mu| \to \infty$ (as $n \to \infty$). Therefore $0 \notin \sigma(yx)$. By a similar argument (with x, y interchanged and μ replaced by $-\mu$) we also get $0 \notin \sigma(xy)$. Thus, the equation (2) becomes just

$$\sigma(xy) = \sigma(yx). \tag{2'}$$

So (1) can be rewritten as

$$\sigma(xy) = \mu + \sigma(yx). \tag{1'}$$

Since $\sigma(xy) \neq \emptyset$ there is a $\lambda(\neq 0)$ in $\sigma(xy)$.

Then, by (1′), $\mu + \lambda \in \sigma(xy)$. Once again by (1′),

$$\mu + (\mu + \lambda) = 2\mu + \lambda \in \sigma(xy),$$

and so on. Thus, $n\mu + \lambda \in \sigma(xy)$ for all $n \geqslant 0$. Since

$$|n\mu + \lambda| \geqslant n|\mu| - |\lambda| \to \infty, \text{ as } n \to \infty,$$

once again we are led to contradict the boundedness of $\sigma(xy)$. Thus the Heisenberg equation has no solutions as asserted.

1.8.17. Remark. The result 1.8.16 is an algebraic generalization of the well-known fact that the Heisenberg commutation relation of quantum mechanics: $PQ - QP = \frac{h}{2\pi i}$ (h denoting Planck's constant) has no bounded operator solution.

§ 9. Strictly Real Algebras

1.9.1. DEFINITION. An element x of a real algebra A is called *strictly real* if its extended quasi-spectrum is real: $\tilde{\sigma}'(x) \subseteq \mathbb{R}$; by 1.8.2, this condition is equivalent to: $\tilde{\sigma}'(x) = \sigma'(x)$.

If A is unital the "strictly real" condition can also be stated as: $\tilde{\sigma}(x) \subseteq R$, or, $\tilde{\sigma}(x) = \sigma(x)$.

An *algebra* A is called strictly real if all its elements are strictly real.

1.9.2. Remark. In any real algebra A the element 0 or more generally, a nilpotent element x is strictly real (by 1.7.15 (ii), x is q. nilpotent in $\tilde{A}$, so $\sigma'_{\tilde{A}}(x) = \{0\}$). Again, if A is unital with unity e, the elements $\lambda e\,(\lambda \in \mathbb{R})$ are strictly real $(\sigma'_{\tilde{A}}(\lambda e) = \lambda\sigma'_{\tilde{A}}(e) = \lambda)$.

1.9.3. PROPOSITION. (a) *Every real radical algebra A – in particular real algebra A_0 with trivial multiplication – is strictly real.*

(b) A_0 *is not formally real.*

PROOF. (a) Recall that every element a of a radical algebra is q. invertible (cf. 1.2.27). It follows from 1.8.4 (i), 1.7.8 that $\tilde{\sigma}'_A(x) = \{0\}$, whence A is strictly real.

(b) This follows from 1.6.19 since every element of A_0 is nilpotent.

1.9.4. Remark. In A_0 we have an example of a strictly real algebra which is not formally real.

1.9.5. LEMMA. *The unitization A_1 of a real algebra A is strictly real iff A is strictly real.*

PROOF. If A is strictly real, and $x_1 \in A_1$,

$$x_1 = \lambda e + x \ (x \in A, \lambda \in \mathbb{R})$$

then $\tilde{\sigma}(x_1) = \tilde{\sigma}(\lambda e + x) = \lambda + \tilde{\sigma}(x) \subseteq \mathbb{R}$ (since $\tilde{\sigma}(x) \subseteq \mathbb{R}$, $\lambda \in \mathbb{R}$), proving A_1 is strictly real.

Conversely, if A_1 is strictly real and $x \in A$ we have

$$\tilde{\sigma}'_A(x) = \sigma'_{\tilde{A}}(x) = \sigma_{(\tilde{A})_1}(x) = \sigma_{\tilde{A}_1}(x) = \tilde{\sigma}_{A_1}(x) \subseteq \mathbb{R},$$

which proves that A is strictly real.

1.9.6. PROPOSITION. *A real algebra A is strictly real iff for any $x \in A$, x^2 is q. invertible.*

PROOF. Suppose that A is strictly real. By 1.9.5 A_1 is

strictly real, whence

$$\sigma_{(\tilde{A})_1}(x) = \sigma_{\tilde{A}_1}(x) \subseteq \mathbb{R}.$$

It follows that $\tilde{\sigma}'(x^2) = \sigma_{(\tilde{A})_1}(x^2) = \sigma_{(\tilde{A})_1}(x)^2 \geqslant^{\dagger} 0$, whence $-1 \notin \tilde{\sigma}(x^2)$, so that x^2 is q. invertible in $\tilde{A}$ and so also in A (by 1.6.10 (ii)).

Conversely, suppose that A satisfies that condition for all x in A. We have to show that $\tilde{\sigma}'(x) \subseteq \mathbb{R}$. If possible let $\tilde{\sigma}'(x)$ contain a number $\alpha + i\beta$ with $\beta \neq 0$ $(\alpha, \beta \in \mathbb{R})$. Write

$$y = (\alpha x^2 - (\alpha^2 - \beta^3)x)/\beta(\alpha^2 + \beta^2); \text{ then } y \in A.$$

By 1.7.7 (iii), $\tilde{\sigma}'(y) \ni \{\alpha(\alpha + i\beta)^2 - (\alpha^2 - \beta^2)(\alpha + i\beta)\}/\beta(\alpha^2 + \beta^2) = i$, so that $\tilde{\sigma}'(y^2) \ni i^2 = -1$. It follows that y^2 is not q. invertible in $\tilde{A}$ and so also not q. invertible in A, contradicting our supposition on A. Thus, $\tilde{\sigma}'(x) \subseteq \mathbb{R}$, as desired.

1.9.7. COROLLARY. *Let A be unital with unity e. Then A is strictly real iff for each $x \in A$, $e + x^2$ is invertible.* ††

PROOF. This is an immediate consequence of 1.1.20 and 1.9.6.

1.9.8. COROLLARY. *Every epimorphic image of a strictly real algebra is strictly real.*

PROOF. An immediate consequence of 1.1.24 and 1.9.6.

1.9.9. GELFAND'S EXAMPLE OF A NON-STRICTLY REAL ALGEBRA

The algebra consists of all real-valued functions $f = f(t)$ on $[-1, 1]$ which are holomorphically extendable to the closed unit disc. It is not strictly real since the function $1/(t^2 + 1)$ does not belong to the algebra, as its analytic extension $1/(z^2 + 1)$ has poles at $\pm i$.

† i.e. if $\lambda \in \tilde{\sigma}'(x^2)$ then $\lambda \geqslant 0$.

†† The condition "$e + x^2$ is invertible for all $x \in A$" was used by Gelfand [7', p.147] to define the notion of strict reality for unital commutative real Banach algebras; strictly real Banach algebras were called by him just real Banach algebras.

1.9.10. Remark. The algebra in 1.9.9 is however formally real (see 1.6.18). Thus we have here an example of a formally real algebra which is not strictly real (cf. 1.9.4).

1.9.11. LEMMA. *Let A_m be a maximal commutative subalgebra of a strictly real algebra. Then A_m is strictly real.*

PROOF. If $x \in A_m \subseteq A$. By 1.1.19, $(x^2)' \in A_m$, whence x^2 is q. invertible in A_m, A_m is strictly real.

1.9.12. Remark. A real algebra A which has a complex structure is not strictly real. To see this, we may assume, by 1.9.5, that A is unital. Since

$$a = e + (ie)^2 = 0, \tag{$*$}$$

a is not invertible and so by 1.9.7, A is not strictly real. Thus in particular $\mathbb{C}$ as a real algebra is not strictly real. The relation $(*)$ also shows that the real algebra $\mathbb{H}$ of Hamilton quaternions is not strictly real (note here however that $\mathbb{H}$ has no complex structure).

1.9.13. PROPOSITION. *Every strictly real q.s.s. algebra A is formally real.*

PROOF. For, suppose that $x^2 + y^2 = 0$ $(x, y \in A)$.
Then

$$\sigma'_{\tilde{A}}(x^2) = -\sigma'_{\tilde{A}}(y^2). \tag{$*$}$$

By strict reality, $\sigma'_{\tilde{A}}(x^2) = \sigma'_{\tilde{A}}(x)^2 \geqslant 0$, $\sigma'_{\tilde{A}}(y^2) \geqslant 0$. It follows from $(*)$ that

$$\sigma'_{\tilde{A}}(x)^2 = \sigma'_{\tilde{A}}(y)^2 = 0, \quad \text{whence}$$

$\sigma'_A(x) = \sigma'_{\tilde{A}}(x) = 0, \sigma'_A(y) = \sigma'_{\tilde{A}}(y) = 0$. Since A is q.s.s. we must have $x = y = 0$, proving that A is formally real.

1.9.14. PROPOSITION. *A formally real division algebra is strictly real. Conversely, a strictly real commutative division algebra is formally real.*

PROOF. Let A be a formally real division algebra and $x \in A$. Then by formal reality $e + x^2 = e^2 + x^2 \neq 0$, whence it is invertible. So, by 1.9.7, A is strictly real.

Suppose now that A is a strictly real commutative division algebra and $x, y \in A$, $x^2 + y^2 = 0$. If $x \neq 0$ then we have

$$e + x^{-2}y^2 = e + (x^{-1}y)^2 = 0,$$

contradicting strict reality. Therefore $x = 0$ and similarly $y = 0$. Thus A is formally real.

1.9.15. PROPOSITION. (Kaplansky [8′, p.405]). *A primitive strictly real algebra A is a division algebra.*

PROOF. Since A is primitive there is a maximal regular l. ideal M_l with $(M_l : A) = \{0\}$ (see 1.5.4). Then $X = A/M_l$ is a faithful A-module (see 1.5.15) which is further irreducible (by 1.5.16 (ii)). By 1.5.19 we have the division algebra $\mathcal{D} = \mathcal{D}(X)$. Denote by dim X the dimension of X as a linear space over $\mathcal{D}$. If dim $X > 1$ there are two $\mathcal{D}$-independent vectors $x, y \in X$. By density theorem (1.5.21) there is an $a \in A$ with $ax = y, ay = -x$. Then $a^2x = ay = -x$. Since A is strictly real, a^2 has a q.i. $b : a^2 + b + ba^2 = 0$. It follows that

$$a^2x + bx + ba^2x = 0, \text{ i.e. } -x + bx - bx = 0,$$

i.e. $x = 0$-impossible. So dim $X = 1$ and X is a division algebra. Since X is a faithful A-module, A is a division algebra.

1.9.16. PROPOSITION. *Let A be a strictly real algebra and $\tilde{A}$ its complexification. Then:*

(i) *If $\tilde{I} \neq \tilde{A}$ is a regular l. (respy. r.) ideal of $\tilde{A}$ then $I =$ Re $\tilde{I}$ is a regular l. (respy. r.) ideal of A with $I \neq A$.*

(ii) *If $\tilde{M}$ is a maximal regular l. (respy. r.) ideal of $\tilde{A}$ and $M = Re\,\tilde{M}$ then M is a maximal regular l. (respy. r.) ideal of A, and $\tilde{M} = M + iM$, $M = A \bigcap \tilde{M}$; in particular $\tilde{M}$ is self-conjugate.*

(iii) *The correspondence $M \mapsto \tilde{M} = M + iM$ is a bijection between the set of maximal regular l. (respy. r.) ideals of A and those of $\tilde{A}$.*

PROOF. (i) In view of 1.6.12 (iv) we have only to show that $I \neq A$. For this, by 1.6.13, it is enough to show that $\tilde{I}$ does not have any relative r. (respy. l.) unity of the form iv $(v \in A)$. Suppose that $\tilde{I}$ has a relative r. (respy. l.) unity of the form iv. Then, by 1.2.8. (b), $-v^2 = (iv)^2$ is a relative r. (respy. l.) unity for $\tilde{I}$. Since $\tilde{I} \neq \tilde{A}$, by 1.2.9, v^2 is not q. invertible in $\tilde{A}$. On the other hand, since A is strictly real v^2 is q. invertible in A and so also in $\tilde{A}$. This contradiction proves that $\tilde{I}$ has no relative unity of the form iv, and (so) $I \neq A$.

(ii) By 1.6.12 (iii), $\tilde{M} \subseteq M+iM$. Let $\tilde{u} = u+iv$ $(u, v \in A)$ be a relative r. (respy. l.) unity for $\tilde{M}$. By 1.6.12 (iv), u is a relative r. (respy. l.) unity for M. Since, by (i), $M \neq A$, $u \notin M$ whence $u \notin M+iM$, $M+iM \neq \tilde{A}$, so that by maximality of $\tilde{M}$ we have $\tilde{M} = M+iM$. By 1.6.12 (iii), $\tilde{M}$ is self-conjugate.

(iii) If M is a maximal regular l. (respy. r.) ideal then by 1.6.12 (ii), $M+iM$ is a maximal regular l. (respy. r.) ideal of $\tilde{A}$. Also, by (ii) above, every maximal regular l. (respy. r.) ideal of $\tilde{A}$ is of the form $M+iM$. The bijection assertion in now clear.

1.9.17. COROLLARY. *Every maximal regular (l. or r.) ideal of $\tilde{A}$ is self-conjugate. Hence $\sqrt{\tilde{A}}$ is self-conjugate. Also, $\sqrt{A} = A \bigcap \sqrt{\tilde{A}}$.*

PROOF. By 1.9.15 (ii), every maximal regular (l. or r.) ideal of $\tilde{A}$ is of the form $\tilde{M} = M+iM$ whence $\tilde{M}$ is self-conjugate. Since $\sqrt{\tilde{A}} = \bigcap \tilde{M}, \sqrt{\tilde{A}}$ is self-conjugate. Finally,

$$\sqrt{A} = \bigcap M = \bigcap (A \bigcap \tilde{M}) = A \bigcap (\bigcap \tilde{M}) = A \bigcap \sqrt{\tilde{A}}.$$

1.9.18. PROPOSITION. *Let A be a strictly real algebra $\tilde{A}$ its complexification. Denote by Δ, $\tilde{\Delta}$ the set of characters of A, $\tilde{A}$ respectively. Then the map*

$\chi \mapsto \tilde{\chi}$, where $\tilde{\chi}$ is the canonical extension of χ is a bijection between Δ and $\tilde{\Delta}$.

PROOF. In view of 1.6.14 (c), it is enough to prove that every character $\tilde{\chi}$ of $\tilde{A}$ is real. But this readily follows from 1.6.15 (b) and 1.9.17 .

CHAPTER II

TOPOLOGICAL PRELIMINARIES

§ 1. Topological Groups and Linear Spaces

2.1.1. Recall first the notion of a poset. A set $\mathcal{A}$ together with an ordering relation $\prec$, between certain pairs of its elements, which is reflexive, transitive, and anti-symmetric or proper $(a \prec b, b \prec a \Rightarrow a = b)$ is called a *poset* (or partially ordered set). $\mathcal{A}$ is called *directed* (above) if for any $\alpha, \beta \in \mathcal{A}$ there is a $\gamma \in \mathcal{A}$ with $\alpha, \beta \prec \gamma$.

Let S be a set and $(x_\alpha)(\alpha \in \mathcal{A})$ a *net* in S (i.e. $x_\alpha \in S$ for each $\alpha \in \mathcal{A}$, where $\mathcal{A}$ is a directed set) A net $(x_{\alpha'})$, $(\alpha' \in \mathcal{A}')$ in S is called a subnet † of a net (x_α) in S if $\mathcal{A}' \subseteq \mathcal{A}$ (so that $(x_{\alpha'})$ is a subset of (x_α)) and for each $\alpha \in \mathcal{A}$ there is an $\alpha' \in \mathcal{A}'$ with $\alpha \prec \alpha'$.

Let S be a topological space. A net (x_α) in S is said to converge to x in S, in symbols $x_\alpha \to x$, if for any neighbourhood U of x there is a $\beta = \beta(U)$ in $\mathcal{A}$ such that $x_\alpha \in U$ for all $\alpha \succ \beta$ ††. It is clear that if a net $x_\alpha \to x$ then every subnet $x_{\alpha'} \to x$. Further, in a Hausdorff space a convergent net has a unique limit, i.e., if $x_\alpha \to x, y$ then $x = y$. (If $x_\alpha \to x$ and $y \neq x$ we can choose neighbourhoods U, V of x, y respectively with $U \bigcap V = \emptyset$. Since $x_\alpha \to x$ there is an α_0 such that $\{x_\alpha : \alpha \succ \alpha_0\} \subseteq U$ and so disjoint with V, whence $x_\alpha \not\to y$ (i.e. x_α does not converge to y.))

Suppose now a net (x_α) in S does not converge to $x : x_\alpha \not\to x$. Then there is a neighbourhood U_0 of x such that for each α there is an α' with $\alpha' \succ \alpha, x_{\alpha'} \notin U_0$. The $x_{\alpha'}$ as α varies clearly form a subnet $(x_{\alpha'})$ which is disjoint with U_0. It follows that no subnet of $(x_{\alpha'})$ converges to x. Thus we have the following useful result: $(*)$ $x_\alpha \to x$ iff *every* subnet $(x_{\alpha'})$ of (x_α) has a subnet $x_{\alpha''} \to x$.

---

† We have adopted a restricted definition for the subnet since that would suffice for our purposes. For a more general definition see [16,p.70].

†† i.e. $\beta \prec \alpha$.

2.1.2. Remark. In terms of net convergence the continuity of a map $f : S \to S'$, where S, S' are topological spaces, can be expressed by the condition: $x_\alpha \to x$ in $S \Rightarrow f(x_\alpha) \to f(x)$ in S'.

2.1.3. A group G together with a topology on it is called a *topological group* or TG if the maps

$$m^\# : (x, y) \mapsto xy,\ i^\# : x \to x^{-1} \qquad (x, y \in G)$$

are continuous. Any neighbourhood of the identity element e of G is called a *nucleus.*

If S is a subset of G we write $S^{-1} = \{a^{-1} : a \in S\}$. We call S *symmetric* if $S^{-1} = S$. Also, for two subsets S_1, S_2 of G we write

$$S_1S_2 = \{ab : a \in S_1, b \in S_2\}.$$

2.1.4. PROPOSITION. *Let G be a TG. Then:*

(i) *The maps*

$$l_a : x \mapsto ax,\ r_a : x \mapsto xa,\ i^\# : x \mapsto x^{-1}$$

where $a, x \in G$, are homeomorphisms.

(ii) *If S is a subset of G and $a \in G$ then $\overline{aS} = a\overline{S}$, $\overline{Sa} = \overline{S}a$, $\overline{S^{-1}} = (\overline{S})^{-1}$, where bar denotes closure.*

(iii) *For any open subset O of G and subset S of G, OS SO are open.*

(iv) *If U is an open nucleus*† *and $a \in G$ then aU, Ua are open neighbourhoods of a, and evey open neighbourhood of a has these forms.*

(v) *If U is an open nucleus so is U^{-1} and $V = U \bigcap U^{-1}$ is an open symmetric nucleus such that $V \subseteq U$.*

(vi) *Any open nucleus U contains a symmetric open nucleus W with $W^2 \subseteq U$. Any such W has the property that its closure. $\overline{W} \subseteq U$. Hence every* TG *is regular.*††

PROOF. (i) It follows from the definition of a TG that $l_a, r_a, i^\#$ are continuous. Since $l_a^{-1} = l_{a^{-1}}, l_a^{-1}$ is continuous and

† i.e. a nucleus which is also an open set.

†† A topological space S is called *regular* if for every point s in S any neighbourhood of s contains the closure of some neighbourhood of s.

hence l_a is a homeomorphism. Similarly r_a is a homeomorphism. Finally, since $(i^{\#})^2 = 1, i^{\#-1} = i^{\#}$, so that $i^{\#}$ is a homeomorphism.

(ii) This is a consequence of (i).

(iii) Since l_a is a homeomorphism $l_a(O) = aO$ is open. Hence $SO = \bigcup_a aO(a \in S)$ is open. Similarly OS is open.

(iv) These follow from the homeomorphism property of the maps l_a, r_a.

(v) By homeomorphism property of $i^{\#}, U^{-1}$ is an open nucleus. Hence V is an open nucleus which moreover is clearly symmetric.

(vi) The first assertion follows from the continuity of the map $m^{\#}$ at (0,0) and (v). To prove the second assertion take an element $b \in \overline{W}$, where bar denotes closure. We must have $W \bigcap bW \neq \emptyset$, whence there are elements $w, w_1 \in W$ with $w = bw_1$ Therefore $b = ww_1^{-1} \in WW^{-1} = W^2 \subseteq V$. This proves that $\overline{W} \subseteq V$. It follows that for any element $a \in G, \overline{aW} = a\overline{W} \subseteq aU$, whence G is regular.

2.1.5. LEMMA. *Let G be a TG with identity element e Then:*

(i) *$\overline{e} = \overline{\{e\}}$ is the intersection of all nuclei of G and $\overline{e}$ is a normal subgroup of G.*

(ii) *For $x \in G$, $\overline{x} = x\overline{e}$.*

(iii) *G is T_1 iff $\overline{e} = e$.*

PROOF. If U is any nucleus, by 2.1.4(vi) there is a nucleus W with $\overline{W} \subseteq U$. Thus $\overline{e} \subseteq \overline{W} \subseteq U$. If $a \notin \overline{e}$ there is a symmetric nucleus V with $e \notin Va$. Then $a^{-1} \notin V$, and so by symmetry of $V, a \notin V$. Therefore $\overline{e}$ is the intersection of all nuclei. If $a, b \in \overline{e}$ then $ab \in a\overline{e} = \overline{ae} = \overline{a} \subseteq \overline{e}, a^{-1} \in \overline{e}^{-1} = \overline{e^{-1}} = \overline{e}$. Further, if $x \in G$ then $x^{-1}\overline{e}x = \overline{x^{-1}ex} = \overline{e}$. Thus $\overline{e}$ is a normal subgroup, completing the proof of (1).

We have $\overline{x} = \overline{xe} = x\overline{e}$ which is (ii), and (iii) follows from (ii).

2.1.6. LEMMA. *A T_1 group G is T_3, in particular it is T_2 (= Hausdorff).*

PROOF. By 2.1.4(vi), G is regular T_1 and so T_3 (by definition of T_3). Suppose that $a, b \in G$ and $a \neq b$. By T_1-property there is a neighbourhood V of a with $b \notin V$. By regularity choose a neighbourhood V_a of a with $\overline{V_a} \subseteq V$. Set $V_b = G \backslash \overline{V_a}$. Then V_b is a neighbourhood of b. Clearly, $V_a \bigcap V_b = \emptyset$, whence G is T_2.

2.1.7. THEOREM. (Birkhoff-Kakutani). *Every first countable TG G admits a one-sided invariant (for multiplication) semi-metric d which induces the topology of G. Further, the semi-metric d is a metric iff G is Hausdorff.*

PROOF. See [21, pp.34-36].

2.1.8. PROPOSITION. (Żelazko). *Let G be a group endowed with a complete metric topology such that l_a, r_a are continuous for each $a \in G$. Then the map $x \mapsto x^{-1}$ is continuous.*

PROOF. First we show that if a sequence $y_n \to e$ (in G) then $y_n^{-1} \to e$. For establishing this it is enough to show that (y_n) has a subsequence $(\overline{y}_k = y_{n_k})$ such that $\overline{y}_k^{-1} \to e$ (see 2.1.1). Define inductively the subsequence $(\overline{y_k})$ of (y_n) as follows.

Set $\overline{y_1} = y_1$. Suppose that $\overline{y_1}, \cdots \overline{y_k}$ has been chosen such that

$$d(p_r, p_{r+1}) < \frac{1}{2^{r+1}}, d(p_r \overline{y}_s^{-1}, p_{r+1} \overline{y}_s^{-1}) < \frac{1}{2^{r+1}} \tag{1}$$

for $r = 1, 2, \cdots, k-1$, $s = 1, 2, \cdots, k$, where $p_r = \overline{y_1} \cdots \overline{y_r}$ (product). Then we can choose $\overline{y}_{k+1}$ so that (1) is satisfied for $r = 1, \cdots, k$, $s = 1, \cdots, k+1$. This choice is possible by taking n_{k+1} sufficiently large, since there are only a finite number of inequalities to be satisfied, $y_n \to e$, and multiplication on the left or right is continuous. From(1), by using the triangle inequality for d we obtain

$$d(p_r, p_{r+s}) < \frac{1}{2^{r+1}} \left(1 + \frac{1}{2} + \frac{1}{2^2} + \cdots \right) < \frac{2}{2^{r+1}} = \frac{1}{2^r},$$

$$d(p_r \overline{y}_s^{-1}, p_{r+1} \overline{y}_s^{-1}) < \frac{1}{2^r}. \tag{2}$$

It follows by the completeness of d that

$$p_k \to p \text{ (say)}, \tag{3}$$

$$p_k \overline{y}_s^{-1} \to q_s \text{ (say)}. \tag{4}$$

Then

$$q_s = \lim_{k\to\infty} p_k \overline{y}_s^{-1} = (\lim_{k\to\infty} p_k)\overline{y}_s^{-1} = p\overline{y}_s^{-1}. \tag{5}$$

Further, by using (2),(3),(5), we get

$$d(p,p_k) \leqslant 2^{-k},\ d(q_s, p_k\overline{y}_s^{-1} \leqslant 2^{-k}). \tag{6}$$

It follows that

$$\begin{aligned} d(p,q_s) &\leqslant d(p,p_s\overline{y}_s^{-1}) + d(q_s,p_s\overline{y}_s^{-1}) = d(p,p_{s-1}) + d(q_s,p_s\overline{y}_s^{-1}) \\ &\leqslant \frac{1}{2^{s-1}} + \frac{1}{2^s} < \frac{1}{2^{s-2}}, \end{aligned}$$

whence $q_s \to p$. Therefore, by (5),

$$\overline{y}_k^{-1} = p^{-1}q_k \to p^{-1}p = e.$$

Next, if $x_n \to x_0$ in G then $x_n x_0^{-1} \to e$, so that $x_0 x_n^{-1} = (x_n x_0^{-1})^{-1} \to e$, whence $x_n^{-1} \to x_0^{-1}$. Thus the map $x \mapsto x^{-1}$ is continuous, proving the proposition.

2.1.9. The letter $\mathbb{K}$ will denote as before $\mathbb{C}$ or $\mathbb{R}$. A linear space or LS X over $\mathbb{K}$ is called a *topological linear space* or TLS if X is equipped with a topology such that:

(T1) The map $(x,y) \mapsto xy$ of $X \times X \to X$ is continuous.

(T2) The map $(\lambda, x) \mapsto \lambda x$ of $\mathbb{K} \times \mathbb{K} \to X$ is continuous.

By taking $\lambda = -1$ in (T2) we obtain

(T3) The map $x \mapsto -x$ is continuous.

In terms of net convergence the continuity conditions (T1)-(T3) can also be put in the forms:

(T1′) If $x_\alpha \to x, y_\beta \to y$ in X then $x_\alpha + y_\beta \to x + y$ in X.

(T2′) If $\lambda_\alpha \to \lambda$ in $\mathbb{K}$, and $x_\beta \to x$ in X then $\lambda_\alpha x_\beta \to \lambda x$ (in X)

(T3′) If $x_\alpha \to x$ in X then $-x_\alpha \to -x$.

In a TLS a neighbourhood of 0 is called a *nucleus*.

2.1.10. LEMMA. *For each $\lambda \in \mathbb{K}$, the map $x \mapsto \lambda x\,(x \in X)$ is continuous, and, for each $\lambda \neq 0$ it is a homeomorphism.*

PROOF. The first half of the statement follows from (T 2). For the second half we observe that if m_λ denote the map $x \mapsto \lambda x$, then for $\lambda \neq 0, m_\lambda^{-1} = m_{\lambda^{-1}}$ and so m_λ^{-1} is also continuous.

2.1.11. Remark. While every group G is a TG under either the discrete † or the indiscrete † topology, a LS X (over $\mathbb{K}$) is a TLS under the indiscrete topology but not under the discrete topology (if $x \neq 0$ in X then $\frac{1}{n}x \neq 0$ and so $\frac{1}{n}x \not\to 0 = 0 \cdot x$, though $\frac{1}{n} \to 0$).

2.1.12. THEOREM (Tychnoff). *Every n-dimensional Hausdorff TLS is linearly homeomorphic to $\mathbb{K}^n$.*

PROOF. See [14, p.13].

2.1.13. LEMMA. *The underlying additive group of a* TLS *X is a TG. Further, $\overline{0} = \overline{\{0\}}$ is a closed subspace of $X \cdot X$ is Hausdorff iff $\{0\}$ is closed.*

PROOF. The first statement follows from (T_1), (T_3) of 2.1.9. For the second statement we note that, by 2.1.5(i), $\overline{0}$ is a closed subgroup. That it is a subspace follows from the continuity of scalar multiplication $(\lambda\overline{0} \subseteq \overline{\lambda 0} = \overline{0})$. Finally, the last statement follows from 2.1.5(iii), 2.1.6.

2.1.14. LEMMA. *Every TLS X is path connected and hence connected.*

PROOF. For $x, y \in X, \{(1-\lambda)x + \lambda y : 0 \leqslant \lambda \leqslant 1\}$ is a path joining x and y.

2.1.15. A subset S of LS X is called *symmetric* if $x \in S \Rightarrow -x \in S\,(S = -S)$. It is called *balanced* if $x \in S, \lambda \in \mathbb{K}, |\lambda| \leqslant 1 \Rightarrow \lambda x \in S$. Note that a balanced set is symmetric and that if S is a balanced set and $|\lambda| \leqslant |\mu|$ then λS †† $\subseteq \mu S$ $(\lambda x = \mu(\lambda\mu^{-1})x, |\lambda\mu^{-1}| \leqslant 1)$. A subset S is called *absorbing* if to each $x \in X$ there is a real number $\epsilon = \epsilon_x > 0$ such that $\lambda x \in S$ for all λ with $0 < |\lambda| \leqslant \epsilon$. Trivially the set X is absorbing; on the

† The discrete topology on a set S is that in which every subset is open and the indiscrete topology that in which empty set and S are the only open subsets.

†† If S is a subset of a LS X and $\lambda \in \mathbb{K}$ then $\lambda S \doteq \{\lambda x : x \in S\}$.

other hand $\{0\}$ can never be absorbing (unlesss $X = \{0\}$). Also, if S is absorbing and $\alpha \neq 0$ then αS is absorbing ($\lambda x \in \alpha S$ iff $\lambda\alpha^{-1}x \in S$).

2.1.16. LEMMA. *Let X be a TLS. Then:*

(i) *The closure $\overline{S}$ of a balanced set S in X is balanced.*

(ii) *Every nucleus U is absorbing.*

(iii) *Every nucleus U contains a balanced open nucleus as well as a balanced closed nucleus.*

PROOF. (i) This follows from the inclusion $\lambda\overline{S} \subseteq \overline{\lambda S}$.

(ii) Since the map $(\lambda, x) \mapsto \lambda x$ is continuous at $(0, x)$, there is an $\epsilon > 0$ and a neighbourhood $x + V$ of x (V a nucleus; cf.2.1.4(v)) such that for $|\lambda| \leqslant \epsilon$ and $y \in x + V$ we have $\lambda y \in U$. In particular, $\lambda x \in U$ for $|\lambda| \leqslant \epsilon$ whence U is absorbing.

(iii) From the continuity of the map $(\lambda, x) \mapsto \lambda x$ at (0,0) we obtain an $\epsilon > 0$ and an open nucleus V such that if $|\lambda| \leqslant \epsilon$ and $x \in V$ then $\lambda x \in U$. Write $\Lambda_\epsilon = \{\lambda \in \mathbb{K} : |\lambda| \leqslant \epsilon\}$ and $W = \Lambda_\epsilon V$. Since $W = \bigcup \lambda V$, when λ ranges in $\Lambda_\epsilon \backslash \{0\}, W$ is open (using 2.1.10 each λV is open). Further, if $\lambda \in \Lambda_\epsilon, |\mu| \leqslant 1$ and $x \in V$ then $|\mu\lambda| = |\mu|\,|\lambda| \leqslant \epsilon$. $\mu\lambda \in \Lambda_\epsilon$ and $\mu\lambda x \in W$, proving W is balanced. Also it is clear that $0 \in W \subseteq U$. Finally, by 2.1.4 (vi) there is a nucleus W_1 with $\overline{W_1} \subseteq U$. By what has been just proved there is a balanced nucleus $W_2 \subseteq W_1$. Then $\overline{W}_2 \subseteq \overline{W}_1 \subseteq U$ and $\overline{W}_2$ is balanced (by (i)).

2.1.17. PROPOSITION. *Every TLS X has a basis $\mathcal{U}$ of nuclei with the following properties:*

(i) *Each $U \in \mathcal{U}$ is balanced and absorbing.*

(ii) *If $U_1, U_2 \in \mathcal{U}$, there is a $U_3 \in \mathcal{U}$ with $U_3 \subseteq U_1 \bigcap U_2$.*

(iii) *If $U \in \mathcal{U}$ there is a $V \in \mathcal{U}$ with $V + V \subseteq U$.*

(iv) *If $U \in \mathcal{U}$ and $\lambda \in \mathbb{K}$ there is a $V \in \mathcal{U}$ with $\lambda V \subseteq U$.*

Conversely, given a nonempty family $\mathcal{U}$ of subsets of a LS X such that $\mathcal{U}$ satisfies (i)-(iii), *it determines a unique topology in X making it a TLS having $\mathcal{U}$ as a basis of nuclei.*

PROOF. See [28, p.96].

2.1.18. Let X be a TLS and X_0 a subspace of X. Then the quotient LS $X^{\#} = X/X_0 = \{x + X_0; x \in X\}$ carries a natural topology, viz., the quotient topology: a subset $S^{\#} \subseteq X^{\#}$ is open iff $\pi^{-1}(S^{\#})$ is open in X, where π is the canonical homomorphism $x \mapsto x^{\#} = x + X_0$.

2.1.19. LEMMA. *The quotient space $X^{\#}$ is a TLS which is Hausdorff iff X_0 is closed in X. The canonical map π is open and continuous.*

PROOF. By 2.1.17 we can choose a base $\mathcal{U}$ of nuclei of X. Write $\mathcal{U}^{\#} = \{$ all subsets $U^{\#}$ of $X^{\#}$ such that $\pi^{-1}(U^{\#}) \in \mathcal{U}\}$. Then it is easy to see that $\mathcal{U}^{\#}$ has the properties (i)-(iv) of 2.1.17, whence by this proposition $X^{\#}$ is a TLS. Further, by 2.1.13, $X^{\#}$ is Hausdorff iff $0^{\#} = X_0$ is closed. By definition of the quotient topology, π is continuous. It is also open as can be easily seen by using 2.1.4 (iii).

2.1.20. A subset S of a TLS X is called topologically bounded or *t. bounded* or just bounded if for every nucleus U there is a $\lambda = \lambda(U) \neq 0$ in $\mathbb{K}$ such that $S \subseteq \lambda U$.

2.1.21. LEMMA. (i) *Every subset of a bounded set is bounded.*

(ii) *The closure $\overline{S}$ of a bounded set S is bounded.*

(iii) *If S is a bounded set then to each nucleus U there is a positive real number $\epsilon = \epsilon(U)$ such that $S \subseteq \mu U$ for every $\mu \in \mathbb{K}$ with $|\mu| \geqslant \epsilon$. In particular, there is a positive integer n such that $S \subseteq nU$.*

PROOF. (i) Clear (if $S \subseteq \lambda U, S' \subseteq S$ then $S' \subseteq \lambda U$).

(ii) Given a nucleus U there is a nucleus V with $\overline{V} \subseteq U$ (see 2.1.4(vi)). Since S is bounded there is a μ with $S \subseteq \mu V$. It follows that $\overline{S} \subseteq \overline{\mu V} = \mu\overline{V} \subseteq \mu U$, whence $\overline{S}$ is bounded.

(iii) Choose a balanced nucleus $V \subseteq U$. By boundedness of S, there is a $\lambda \neq 0$ with $S \subseteq \lambda V$, so that if $x \in S, x = \lambda v\,(v \in V)$. Set $\epsilon = |\lambda|$. If $|\mu| \geqslant |\lambda|$ then $|\mu^{-1}\lambda| \leqslant 1, \mu^{-1}\lambda v \in V$ since V is balanced), so that

$$x = \lambda v = \mu \cdot \mu^{-1}\lambda v \in \mu V \subseteq \mu U.$$

Also, $S \subseteq nU$ for any integer $n \geqslant \epsilon$.

2.1.22. THEOREM (Mazur-Kolmogorov). *A subset S of a TLS X is bounded iff for each sequence (x_n) in S and sequence*

(λ_n) in $\mathbb{K}$ with $\lambda_n \to 0$, we have $\lambda_n x_n \to 0$.

PROOF. Suppose that S is bounded. By 2.1.21 (iii) there is an $\epsilon > 0$ with $S \subseteq \mu U$ for $|\mu| \geq \epsilon$. Write $\eta = \epsilon^{-1}$ and $\alpha = \mu^{-1}$. Then $\alpha S \subseteq U$ for $|\alpha| \leqslant \eta$. Since there is an integer N such that $|\lambda_n| \leqslant \eta$ for $n \geqslant N$, it follows that $\lambda_n x_n \in \lambda_n S \subseteq U$ for $n \geqslant N$, whence $\lambda_n x_n \to 0$. Conversely, if S is not bounded there is a nucleus U such that $S \nsubseteq \lambda U$ for any $\lambda \neq 0$ in $\mathbb{K}$. Therefore, in particular, $S \nsubseteq nU(n = 1, 2, \cdots)$. It follows that we can choose x_n in S such that $\frac{x_n}{n} \notin U$. Then clearly $\frac{x_n}{n} \not\to 0$ (though $\frac{1}{n} \to 0$) violating the stated condition, which completes the proof.

2.1.23. COROLLARY. *For S to be bounded it is sufficient that the following condition is satisfied:*

$$\textit{For any } x_n \in S, \ \frac{x_n}{n} \to 0.$$

2.1.24. DEFINITION. A TLS is said to be *locally bounded* if it has a bounded nucleus.

Note that if U is a bounded nucleus so is every nucleus $V \subseteq U$ (see 2.1.21(i)).

2.1.25. PROPOSITION. *Let X be a locally bounded TLS and U a bounded nucleus of X. If $\mu_n \in \mathbb{K}\backslash\{0\}$ and $\mu_n \to 0$ then $\{\mu_n U\}$ is a basis of nuclei. Thus, X is first countable and hence semi-metrizable. In particular, every locally bounded Hausdorff TLS is metrizable.*

PROOF. Suppose that V is any nucleus of X. Since U is bounded, there is, by 2.1.21 (iii), an $\epsilon > 0$ such that $U \subseteq \mu V$ for all $\mu \in \mathbb{K}$ with $|\mu| \geqslant \epsilon$. Since $\mu_n \to 0$ we can choose a sufficiently large n such that $|\mu_n|^{-1} \geqslant \epsilon$. The $U \subseteq \mu_n^{-1} V$ or $\mu_n U \subseteq V$. The statements regarding semi-metrizability and metrizability follow by applying 2.1.7 to the underlying additive TG of X.

2.1.26. LEMMA. *Let X, X^* be TLS's and $T : X \to X^*$ a linear transformation. If T is continuous at some point x_0 then it is continuous everywhere.*

Proof. Assume that T is continuous at x_0, and that $x_\alpha \to x$. Then $x_\alpha - x + x_0 \to x_0$, so that $Tx_\alpha - Tx + Tx_0 \to Tx_0$, whence

$Tx_\alpha \to Tx$.

2.1.27. LEMMA. *If* $T : X \to X^*$ *is a continuous (or more generally, sequentially continuous) linear transformation and* $S(\subseteq X)$ *a bounded set so is* $T(S)$.

Proof. If $x_n \in S$ then $\frac{x_n}{n} \to 0$, so that

$$\frac{Tx_n}{n} = T\left(\frac{x_n}{n}\right) \to T(0) = 0,$$

proving (by 2.1.23) that $T(S)$ is bounded.

2.1.28. Let X, X^* TLS's and $T : X \to X^*$ a linear transformation. T is said to be *t. bounded* or (sometimes) just bounded if it carries bounded sets to bounded sets. By 2.1.27, a continuous linear transformation is t. bounded.

2.1.29. PROPOSITION. *Let* X *be a first countable (or equivalently semi-metrizable) TLS and* X^* *any TLS. Then a linear transformation* $T : X \to X^*$ *is t. bounded iff it is continuous.*

PROOF. In view of remark in 2.1.28 we have to prove only the "if" part. Let T be bounded. If T is not continuous it is not continuous at 0 (by 2.1.26) so that there is a nucleus U^* such that $T^{-1}(U^*)$ is not a nucleus. Since X is first countable it has a countable decreasing sequence of open nuclei as basis: $U_1 \supseteq U_2 \supseteq \cdots$. By 2.1.10, $\frac{U_n}{n}$ is an open nucleus. Since $T^{-1}(U^*)$ is not a nucleus we have: $\frac{U_n}{n} \not\subseteq T^{-1}(U^*)$ $(n = 1, 2, \cdots)$. This means that there are elements $x_n \in U_n$ such that $\frac{x_n}{n} \notin T^{-1}(U^*)$ $(n = 1, 2, \cdots)$. Since $\{U_n\}$ is a decreasing basis and $x_n \in U_n$, $x_n \to 0$ and so x_n is bounded (see 2.3.7. (a)). Since T is bounded (Tx_n) is bounded. This means by 2.1.21, that $T\left(\frac{x_n}{n}\right) = \frac{Tx_n}{n} \to 0$, so that $\frac{Tx_n}{n} \in U^*$ for $n \geqslant N$. But then $\frac{x_n}{n} \in T^{-1}(U^*)$ $(n \geqslant N)$, contradicting the choice of x_n. Hence the proposition.

2.1.30. LEMMA. *Let* X *be a TLS and* f *a linear functional*† *on* X. *Then* f *is continuous iff* $\ker f$ *is closed.*

PROOF. It suffices to prove the "if" part. We may assume

† i.e. a function with values in $\mathbb{K}$.

that $f \neq 0$, $\ker f$ is closed. Then, for any $\epsilon > 0$, the translate $X_0 = \{x \in X : f(x) = \epsilon\}$ of $\ker f$ is closed, so that $U = X \backslash X_0$ is an open nucleus. By 2.1.16 (iii), there is a balanced nucleus $V \subseteq U$. We claim that for any $x \in V$, $|f(x)| < \epsilon$. If not there is a $x_0 \in V$ with $\delta = |f(x_0)| \geqslant \epsilon$. Write $y_0 = (\epsilon/\delta)x_0$. Then $|f(y_0)| = \epsilon$, so that $f(y_0) = \epsilon\theta$ where $\theta \in \mathbb{K}$, $|\theta| = 1$. Hence $f(\overline{\theta} y_0) = \epsilon$, so that $z_0 = \overline{\theta} y_0 \in X_0$. On the other hand, since $x_0 \in V$ and V is balanced,

$$z_0 = \overline{\theta} y_0 = \frac{\overline{\theta}\epsilon}{\delta} x_0 \in V$$

since $\left|\frac{\overline{\theta}\epsilon}{\delta}\right| \leqslant 1$. Therefore $z_0 \in X_0 \bigcap V$, contradicting $X_0 \bigcap V = \emptyset$. Thus, $|f(x)| < \epsilon \, (x \in V)$ proving f is continuous.

2.1.31. LEMMA. *Let X be a TLS with $\{U_\alpha\}$ as a basis of nuclei and f a continuous functional on X with $f(0) = 0$. Then, given $C > 0$, there is a $U_\Theta = U_{\alpha_0}$ such that $|f(x)| < C$ for all $x \in U_0$.*

PROOF. Set $G = \{x \in X : |f(x)| < C\}$. Since $f(0) = 0$, $0 \in G$ and so by continuity of f, G is an open nucleus. $\{U_\alpha\}$ being a basis we must have $G \supseteq$ some U_{α_0} which implies the lemma.

2.1.32. Let X be a TLS. A series $\sum_{n=1}^{\infty} x_n$ in X is said to *converge to* x if the sequence (s_n) of partial sums $s_n = x_1 + \cdots + x_n$ converges to x in $X : s_n \to x$. More generally, a generalized series $\sum_{\alpha \in \mathcal{A}} x_\alpha$, where $\mathcal{A}$ is any indexing set, is said to converge to x if the net s_F, where franges through all finite subsets of $\mathcal{A}$ and $s_F = \sum_{\alpha \in F} x_\alpha$, converges to x; x is called the *generalized sum* of the x_α's.

2.1.33. LEMMA. *If $\sum_{n=1}^{\infty} x_n$ converges in X then $x_n \to 0$.*

PROOF. Given a nucleus U, choose a symmetric nucleus W with $W + W \subseteq U$. Suppose that $\sum_{n=1}^{\infty} x_n = x$. Then there is an N_0

such that $\sum_{n=1}^{N} x_n - x \in W$ for $N \geqslant N_0$. It follows that

$$\begin{aligned} x_{N+1} &= \left(\sum_{n=1}^{N+1} x_n - x\right) - \left(\sum_{n=1}^{N} x_n - x\right) \\ &\in \quad W + W \subseteq U \qquad \text{for } N \geqslant N_0. \end{aligned}$$

It follows that $x_n \to 0$.

§ 2. Topological Algebras

2.2.1. DEFINITION. An algebra A over $\mathbb{K}$ is called a *topological algebra* or a TA if it is equipped with a topology such that:

(TA1) The map $(x, y) \mapsto x + y$ of $A \times A \to$ is continuous.

(TA2) The map $(\lambda, x) \mapsto \lambda x$ of $\mathbb{K} \times A \to A$ is continuous.

(TA3) The map $(x, y) \mapsto xy$ of $A \times A \to A$. is continuous.

In view of (TA1), (TA2) every TA is a TLS. Also, the condition (TA3) can be expressed in terms of net convergence as:

$$x_\alpha \to x, y_\beta \to y \Rightarrow x_\alpha y_\beta \to xy.$$

2.2.2. DEFINITION. An algebra A is called a *weak topological algebra* or a WTA if the underlying linear space of A is a TLS and further we have:

(TA 3′) The maps $l_a : x \mapsto ax, r_a : x \mapsto xa \, (x, a \in A)$ are continuous for all a.

Clearly (TA3) $\mapsto$ (TA 3′), so that a TA is a WTA.

Using 2.1.26, it is easy to see that in terms of net convergence the condition (TA 3′) can be expressed as

(TA 3″) If $x_\alpha \to 0$ then $ax_\alpha \to 0, x_\alpha a \to 0$ for each $a \in A$.

2.2.3. LEMMA. *For a WTA A to be a TA it is necessary and sufficient that*

(∗) *the map $(x, y) \mapsto xy$ is continuous at* $(0, 0)$.

PROOF. We have only to prove the sufficiency of the condition (∗). For this it is enough to show that in an A satisfying (∗), (TA3) holds.

Assume now that $(*)$ is satisfied, and in A $x_\alpha \to x_0$, $y_\beta \to y_0$. By (TA 3″) we have

$$x_0(y_\beta - y_0) \to 0; \tag{1}$$

$$(x_\alpha - x_0)y_0 \to 0. \tag{2}$$

Again, by $(*)$ we have

$$(x_\alpha - x_0)(y_\beta - y_0) \to 0. \tag{3}$$

From (1),(2),(3) we obtain (by adding the terms)

$$x_\alpha y_\beta - x_0 y_0 \to 0, \text{ or } , x_\alpha y_\beta \to x_0 y_0.$$

Thus,(TA3) holds and A is TA.

2.2.4. Examples. (i) Every algebra over $\mathbb{K}$ is a TA under the indiscrete topology.

(ii) The algebra $A = \mathbb{K}^S$ of all $\mathbb{K}$- valued functions (algebra operations being point (or coordinate)-wise is a TA under pointwise net convergence : i.e., if $f_\alpha, f \in A$ then $f_\alpha \to f$ if $f_\alpha(s) \to f(s) \, \forall s \in S$; this topology is called *weak* topology or topology of simple convergence. A is a commutative TA.

(iii) Let $\mathfrak{H}$ be a Hilbert space and $\mathbf{B} = \mathbf{B}(\mathfrak{H})$ the algebra of all bounded l.o.[†]'s on $\mathfrak{H}$. $\mathbf{B}$ is an algebra which not commutative if $\dim \mathfrak{H} > 1$. Under the weak or strong operator topology $\mathbf{B}$ is a WTA but not a TA (since multiplication is not jointly continuous; see [22, p.448]). The weak or strong operator topology is defined vic net convergence as follows. $T_\alpha \to T$ in the weak operator topology if $\langle T_\alpha x_1 y\rangle \to \langle T x_1 y\rangle$ for every $x, y \in \mathfrak{H}$, where $\langle\cdot\rangle$ denotes the inner product of $\mathfrak{H}$. Similarly, $T_\alpha \to T$ in the strong operator topology if $\|T_\alpha x - Tx\| \to 0$ for every $x \in \mathfrak{H}$, where $\|\cdot\|$ is the norm induced by the inner product: $\|x\| = \langle x, x\rangle^{\frac{1}{2}}$.

In the norm topology, $\mathbf{B}$ is a TA; $T_n \to T$ in the norm topology if $\|T_n - T\| \to 0$ (for definition of $\|T\|(T \in \mathbf{B})$, see 3.5.1).

† l.o.=linear operator.

2.2.5. LEMMA. *A WTA or a TA is a TG under addition. Under multiplication a TA A is a TSG.*†

PROOF. The first statement follows from 2.1.12; the second is a consequence of condition (TA3) of 2.2.1.

2.2.6. LEMMA. *In a unital TA A, the groups G_q, G_i are semi-topological*†† *groups and the map*

$$\tau^{-1} : a \in G_q \mapsto e + a \in G_i$$

is a t. isomorphism.‡

PROOF. By (TA1), (TA3), G_q and G_i are semi-topological groups. Also, by 1.1.20, τ^{-1} is an isomorphism. That τ^{-1} is topological follows from the fact that translation is a homeomorphism (see 2.1.4(i)).

2.2.7. LEMMA. *In a* TA (*or even* WTA) $A, \overline{0} = \overline{\{0\}}$ *is a closed bi-ideal of A.*

PROOF. By 2.1.13, $\overline{0}$ is a closed subspace of A. Further, from the continuity of l. or r. multiplication in A we obtain

$$x\overline{0} \subseteq \overline{x0} = \overline{0}, \overline{0}y \subseteq \overline{0}$$

which show that $\overline{0}$ is a bi-ideal, completing the proof.

2.2.8. LEMMA (a) *Any subalgebra of a* TA (*respy.* WTA) *is a* TA (*respy.* WTA).

(b) *Any direct product or direct sum of TA's (respy. WTA's) is a TA (respy. WTA).*

PROOF. (a) Clear.

---

† TSG= topological semi-group, i.e., a semi-group with a topology under which multiplication is continuous.

†† A semi-topological group is a group with a topology under which multiplication continuous (inversion may not be continuous).

‡ t. isomorphism = topological isomorphism, i.e. an (algebraic) isomorphism which is also a homeomorphism.

(b) Since the operations of the direct product are coordinatewise and a direct sum is a subalgebra of the direct product, the statements follow.

2.2.9. PROPOSITION. *The unitization A_1 of a TA (respy. WTA) A is a TA (respy. WTA) under the product topology of $A_1 = \mathbb{K} \times A$. Further, A is a closed bi-ideal of A_1 and A_1 is Hausdorff whenever A is Hausdorff.*

PROOF. Clearly A_1 is a TLS in either case. Suppose now that A is a TA and that in A_1 we have

$$\begin{aligned} x_{1\alpha} &= \lambda_\alpha e_1 + x_\alpha \to x_1 = \lambda e_1 + x \\ y_{1\beta} &= \mu_\beta e_1 + y_\beta \to y_1 = \mu e_1 + y \end{aligned}$$

where $x, y \in A$ and $\lambda_\alpha, \mu_\beta, \lambda, \mu \in \mathbb{K}$. Then

$$\lambda_\alpha \to \lambda, \mu_\beta \to \mu, x_\alpha \to x, y_\beta \to y.$$

It follows that

$$\begin{aligned} x_{1\alpha} y_{1\beta} &= \lambda_\alpha \mu_\beta e_1 + \lambda_\alpha y_\beta + \mu_\beta x_\alpha + x_\alpha y_\beta \\ &\to \lambda \mu e_1 + \lambda y + \mu x + xy = x_1 y_1, \end{aligned}$$

showing A is a TA.

Next suppose that A is a WTA, and $y_{1\alpha} = \mu_\alpha e_1 + y_\alpha \to \mu e_1 + y = y_1$. Then $l_{x_1}(y_{1\alpha}) = \lambda \mu_\alpha e_1 + \lambda y_\alpha + \mu_\alpha x + x y_\alpha \to l_{x_1}(\mu e_1 + y) = l_{x_1}(y_1)$. This shows that l_{x_1} is continuous. Similarly r_{x_1} is continuous. Hence A_1 is a WTA.

We have already noted that A is a bi-ideal of A_1 (see 1.1.12). That it is closed in A_1 is clear from the definition of the topology of A_1. Finally, if A is Hausdorff, $\{0\}$ is closed in A and so also in A_1 (since A is closed in A_1) and consequently A_1 is also Hausdorff (by 2.1.5,2.1.6).

2.2.10. PROPOSITION. *The complexification $\tilde{A}$ of a real TA (respy. WTA) A is a TA (respy. WTA) under the topology of $A \times A$.*

PROOF. The proof is on the same lines as the proof of the first part of 2.2.9.

2.2.11. PROPOSITION. *Let A be a TA and I a bi-ideal of*

A. Then the quotient algebra $A^{\#} = A/I$ *is a TA relative to the quotient topology. Moreover,* $A^{\#}$ *is Hausdorff iff* I *is closed.*

PROOF. In view of 2.1.19, it is enough to prove that multiplication in $A^{\#}$ is continuous. Let $U^{\#}$ be an open neighbourhood of $x^{\#}y^{\#} = (xy)^{\#}$. Then $U = \pi^{-1}(U^{\#})$, where π is the canonical map $A \to A^{\#}$, is an open neighbourhood of xy. Choose open neighbourhoods V, W, of x, y respy. such that $VW \subseteq U$. Then, since π is open (by virtue of 2.1.19), $V^{\#} = \pi(V), W^{\#} = \pi(W)$ are open neighbourhoods of $x^{\#}, y^{\#}$ respy. with $V^{\#}W^{\#} \subseteq U^{\#}$, proving the continuity of multiplication in $A^{\#}$.

2.2.12. COROLLARY. *Let* A *be a TA. Then* $A/\overline{0}$ *is a Hausdorff TA.*

PROOF. This is an immediate consequence of 2.2.7 and 2.2.11.

2.2.13. THEOREM. *Every WTA has a basis* $\mathfrak{U}$ *of nuclei satisfying* (i)-(iv) *of* 2.1.17 *and*

(v) *Given* $U \in \mathfrak{U}$ *and* $a \in A$ *there are* $V, W \in \mathfrak{U}$ *such that*

$$aV \subseteq U, \quad Wa \subseteq U.$$

Conversely, if $\mathfrak{U}$ *is a nonempty family of subsets of an algebra* A *satisfying* (i)-(iii) (*of* 2.1.17.), *and* (v) (*above*) *then* $\mathfrak{U}$ *is a basis of nuclei of a unique topology on* A *making it a WTA.*

PROOF. By 2.1.17 we can choose a basis $\mathfrak{U}$ of nuclei satisfying (i)-(iv). Then $\mathfrak{U}$ also satisfies (v) since l_a, r_a are continuous for each $a \in A$.

For the converse we observe that by 2.1.17, A is a TLS. To prove that A is a WTA it remains to show that each l_a and each r_a are continuous. By 2.1.26 it is enough to prove that l_a, r_a are continuous at 0. But this is precisely what condition (v) expresses.

2.2.14. THEOREM. *Every* TA A *has a basis* $\mathfrak{U}$ *of nuclei satisfying* (i)-(iv) *of* 2.1.17 (v) *of* 2.2.13, *and also*

(vi) *Given a* $U \in \mathfrak{U}$ *there is a* $V \in \mathfrak{U}$ *such that* $V^2 \subseteq U$.

Conversely, every algebra A *together with a family* $\mathfrak{U}$ *of subsets satisfying* (i)-(iii) *of* 2.1.17, (v) *of* 2.2.13, *and* (vi) *above is a TA under a unique topology having* $\mathfrak{U}$ *as a basis of nuclei.*

PROOF. If A is a TA then it is a WTA and so by 2.2.13, has a basis $\mathcal{U}$ of nuclei satisfying (i)-(v). This also satisfies (vi) since, multiplication in A is continuous.

The converse statement follows from the converse part of 2.2.13, and 2.2.3.

2.2.15. Remark. In terms of convergence of nets the conditions (v),(vi) above take the forms:

If $$x_\alpha \to 0 \text{ then } ax_\alpha \to 0, x_\alpha a \to 0; \qquad (v')$$

If $$x_\alpha \to 0, y_\beta \to 0 \text{ then } x_\alpha y_\beta \to 0. \qquad (vi')$$

2.2.16. LEMMA. *Let A be a TA.* We have:

(i) *If in $A, x_n \to 0$ and (y_n) is bounded then $x_n y_n \to 0$, $y_n x_n \to 0$.*

(ii) *If $S, T \subseteq A$ are bounded subsets so is ST.*

PROOF. (i) Given a nucleus U choose a nucleus V such that $V^2 \subseteq U$. Since (y_n) is bounded then is a scalar $\lambda \neq 0$ such that $\{y_n\} \subseteq \lambda V$. Since $x_n \to 0$ we can find N such that $x_n \in \lambda^{-1} V$ for $n \geqslant N$. Therefore

$$x_n y_n \in \lambda^{-1} V \cdot \lambda V = V^2 \subseteq U \quad \text{for } n \geqslant N,$$

whence $x_n y_n \to 0$.

(ii) Any sequence in ST is of the form $(x_n y_n)$ where $(x_n), (y_n)$ are sequences in S, T respy.. Since $\{x_n\} \subseteq S$ is bounded, $\frac{x_n}{n} \to 0$ (by 2.1.22). Since $\{y_n\} \subseteq T$ is bounded, by result (i), $\frac{x_n y_n}{n} \to 0$, whence by 2.1.23, ST is bounded.

2.2.17. LEMMA. *In a Hausdorff, TA (respy. unital TA) A, if the series $\sum_{n=1}^{\infty}(-1)^n x^n$ (respy. $\sum_{n=0}^{\infty}(e-x)^n$) converges then x is q invertible (respy. invertible) and its q inverse x' (respy. inverse x^{-1}) is given by*

$$x' = \sum_{n=1}^{\infty}(-1)^n x^n \ \left(x^{-1} = \sum_{n=0}^{\infty}(e-x)^n\right).$$

PROOF. Write $y = \sum_{n=1}^{\infty}(-1)^n x^n$. Then $x \circ y = x + \sum_{n=1}^{\infty}(-1)^n(x^n + x^{n+1}) = \lim_{N\to\infty}(-1)^N x^{N+1} = 0$ (by 2.1.33).

Similarly, $y \circ x = 0$. Therefore $x' = y$, proving the assertion concerning q invertibility. Again, if we set $e - x = z$, by 2.1.33, $z^n \to 0$. Therefore we have

$(e-z)\sum_{n=0}^{\infty} z^n = \lim_n(e - z^{n+1}) = e$, whence

$$(e-z)^{-1} = \sum_{n=0}^{\infty} z^n,\ x^{-1} = (e-z)^{-1} = \sum_{n=0}^{\infty}(e-x)^n,$$

completing the proof.

2.2.18. In a TA A we denote by $\Delta_c = \Delta_c(A)$ the set of all continuous characters; $\Delta = \Delta(A)$ will denote the set of *all* characters. Of course Δ_c (or even Δ) can be empty.

We set

$$\sqrt[ch]{A} = \left\{ \begin{array}{ll} \{\bigcap \chi^{-1}(0) : \chi \in \Delta_c\} & \text{if } \Delta_c \neq \emptyset \\ A & \text{if } \Delta_c = \emptyset. \end{array} \right\}$$

Since $\sqrt[ch]{A}$ is the intersection of all closed hypermaximal ideals (of A) it is closed and we have clearly the inclusion relation

$$\sqrt{A} \subseteq \sqrt[h]{A} \subseteq \sqrt[ch]{A}. \qquad (*)$$

2.2.19. DEFINITION. Following Michael, [20, p.48] we call a TA A *functionally continuous* if every character of A is continuous, i.e. if $\Delta_c = \Delta$.

In a functionally continuous algebra we have

$$\sqrt[h]{A} = \sqrt[ch]{A}. \qquad (**)$$

§ 3. Completions of Topological Linear spaces and Topological Algebras

2.3.1. DEFINITION. Let X be a TLS. Two nets $(x_\alpha),(y_\beta)$ in X are said to be *equivalent,* in symbols, $(x_\alpha)\sim(y_\beta)$ if they satisfy the condition:

(*) Given a nucleus U of X there are indices α_0,β_0 such that $x_\alpha - y_\beta \in U$ for $\alpha \succ \alpha_0, \beta \succ \beta_0$.

A net (x_α) is called a *Cauchy net* or a *C-net* if we have $(x_\alpha)\sim(x_\alpha)$, i.e. if for each nucleus U there is an α_0 such that $x_\alpha - x_\beta \in U$ for $\alpha,\beta \succ \alpha_0$. A TLS X is called *complete* if every C-net in X converges to some element in it.

2.3.2. LEMMA. *The notion of equivalence of nets in X has the properties:*

(i) *If $(x_\alpha)\sim(y_\beta)$ then $(y_\beta)\sim(x_\alpha)$.*

(ii) *If $(x_\alpha)\sim(y_\beta),(y_\beta)\sim(z_\gamma)$ then $(x_\alpha)\sim(z_\gamma)$.*

(iii) *If $(x_\alpha)\sim(y_\beta)$ and $(x_{\alpha'})$ is a subnet of (x_α) then $(x_{\alpha'})\sim(y_\beta)$.*

(iv) *Let (x_α) be a net and (x) a principal net. Then $(x_\alpha)\sim(x)$ iff $x_\alpha \to x$.*

(v) *If $(x_\alpha)\sim(y_\beta)$ and $\lambda\in\mathbb{K}$ then $(\lambda x_\alpha)\sim(\lambda y_\beta)$, in particular $(-x_\alpha)\sim(-y_\beta)$.*

(vi) *If $(x_\alpha)\sim(y_\beta),(z_\gamma)\sim(u_\delta)$ then $(x_\alpha - z_\gamma)\sim(y_\beta - u_\delta)$.*

PROOF. (i) Given a nucleus U select a symmetric nucleus $V\subseteq U$. Since $(x_\alpha)\sim(y_\beta)$ there are α_0,β_0 such that $x_\alpha - y_\beta \in V(\alpha\succ\alpha_0,\beta\succ\beta_0)$ whence

$$y_\beta - x_\alpha = -(x_\alpha - y_\beta)\in -V = V\subseteq U\ (\alpha\succ\alpha_0,\beta\succ\beta_0).$$

(ii) Given a nucleus U select a nucleus V with $V+V\subseteq U$, and choose $\alpha_0,\beta_0,\beta_1,\gamma_1$ such that

$$x_\alpha - y_\beta \in V(\alpha\succ\alpha_0,\beta\succ\beta_0), y_\beta - z_\gamma \in V(\beta\succ\beta_1,\gamma\succ\gamma_1).$$

Then, for $\alpha \succ \alpha_0, \gamma \succ \gamma_1$, we have

$$x_\alpha - z_\gamma = x_\alpha - y_{\beta'} + y_{\beta'} - z_\gamma \in V + V \subseteq U(\text{ where } \beta' \succ \beta_0, \beta_1).$$

(iii) Given U, choose α_0, β_0 such that $x_\alpha - y_\beta \in U(\alpha \succ \alpha_0, \beta \succ \beta_0)$. Since $(x_{\alpha'})$ is a subnet there is an $\alpha'_0 \succ \alpha_0$. Then $x_{\alpha'} - y_\beta \in U(\alpha' \succ \alpha'_0, \beta \succ \beta_0)$, whence $(x_{\alpha'}) \sim (y_\beta)$.

(iv) Clear.

(v) Given U, by 2.1.17(iv), there is a V with $\lambda V \subseteq U$. Choose α_0, β_0 such that $x_\alpha - y_\beta \in V(\alpha \succ \alpha_0, \beta \succ \beta_0)$. Then

$$\lambda x_\alpha - \lambda y_\beta = \lambda(x_\alpha - y_\beta) \in \lambda V \subseteq U.$$

(vi) Select a symmetric V with $V + V \subseteq U$. By hypothesis there are indices $\alpha_0, \beta_0, \gamma_0, \delta_0$ such that

$$x_\alpha - y_\beta \in V(\alpha \succ \alpha_0, \beta \succ \beta_0), z_\gamma - u_\delta \in V(\gamma \succ \gamma_0, \delta \succ \delta_0).$$

Then $(x_\alpha - z_\gamma) - (y_\beta - u_\delta) = x_\alpha - y_\beta - (z_\gamma - u_\delta) \in V + V \subseteq U$ $(\alpha \succ \alpha_0, \beta \succ \beta_0, \gamma \succ \gamma_0, \delta \succ \delta_0)$.

2.3.3. COROLLARY. (i) *If (x_α) is a C-net in X and $(x_{\alpha'})$ a subnet then it is a C-net such that $(x_{\alpha'}) \sim (x_\alpha)$.*

(ii) *If (x_α) is a C-net in X so is (λx_α); in particular $(-x_\alpha)$ is a C-net.*

(iii) *If $(x_\alpha), (y_\beta)$ are C-nets so is $(x_\alpha - y_\beta)$.*

PROOF. (i) From $(x_\alpha) \sim (x_\alpha)$, by 2.3.2 (iii), $(x_{\alpha'}) \sim (x_\alpha)$ and also by 2.8.2.((i), (iii)) we get $(x_{\alpha'}) \sim (x_{\alpha'})$.

(ii) Follows from (v) of 2.3.2.

(iii) Follows from (vi) of 2.3.2.

2.3.4. LEMMA. *Let X, X^* be TLS's and $T : X \to X^*$ a continuous linear transformation. Let $(x_\alpha), (y_\beta)$ be nets in X with $(x_\alpha) \sim (y_\beta)$. Then $(Tx_\alpha) \sim (Tx_\beta)$. In particular, (Tx_α) is a C-net whenever (x_α) is a C-net.*

PROOF. Given a nucleus U^* of X^*, there is, by continuity of T a nucleus U of X with $T(U) \subseteq U^*$. Since $(x_\alpha) \sim (y_\beta)$ there are indices α_0, β_0 such that $x_\alpha - y_\beta \in U(\alpha \succ \alpha_0, \beta \succ \beta_0)$. Then

$$Tx_\alpha - Ty_\beta = T(x_\alpha - y_\beta) \in T(U) \subseteq U^*,$$

proving $(Tx_\alpha) \sim (Ty_\beta)$.

2.3.5. DEFINITION. A net (x_α) in a TLS X is called *bounded* if it is bounded as a set (see 2.1.20). A net (x_α) in X is called *essentially bounded* if given a nucleus U of X there is an index $\alpha_0 = \alpha_0(U)$ and a scalar $\lambda = \lambda(U) \neq 0)$ such that

$$\{x_\alpha : \alpha \succ \alpha_0\} \subseteq \lambda U.$$

Trivially, a bounded net is essentially bounded.

2.3.6. PROPOSITION. *Every C-net – in particular a convergent net –* (x_α) *is essentially bounded.*

PROOF. Given a nucleus U choose a balanced nucleus V such that $V + V \subseteq U$. Since (x_α) is a C-net there is an α_0 such that $x_\alpha - x_{\alpha_0} \in V$ for $\alpha \succ \alpha_0$, so that

$$x_\alpha \in x_{\alpha_0} + V \qquad (\alpha \succ \alpha_0)$$

Since V is absorbing (by 2.1.16(ii)) and balanced (by choice) we can choose $\lambda \geqslant 1$ such that $x_{\alpha_0} \in \lambda V$. Then

$$x_\alpha \in \lambda V + V \subseteq \lambda V + \lambda V = \lambda(V + V) \subseteq \lambda U \qquad (\alpha \succ \alpha_0),$$

proving that (x_α) is essentially bounded.

2.3.7. PROPOSITION. (a) *Every essentially bounded sequence – in particular, a C-sequence or a convergent sequence – is bounded.*

(b) *Not every convergent net is bounded.*

PROOF. (a) Given a nucleus U choose a balanced nucleus V with $V \subseteq U$. Since (x_n) is essentially bounded there is an integer $N \geqslant 1$ and a scalar λ_1 such that $\{x_n : n \geq N\} \subseteq \lambda_1 V$. Using the absorption and balanced porperties of V we can choose a λ_2 with

$$x_j \in \lambda_2 V \qquad (j = 1, \cdots, N).$$

Set $\lambda = \max(|\lambda_1|, |\lambda_2|)$. Then

$$\{x_n\} \subseteq \lambda V \subseteq \lambda U,$$

proving the boundedness of the sequence (x_n).

(b) It is enough to construct a convergent net in $\mathbb{R}$ which is not bounded. Write

$$\mathcal{A} = \{\alpha_1, \alpha_2, \cdots; \beta_1, \beta_2, \cdots\}$$

and define an partial ordering in $\mathcal{A}$ by

$$\beta_1 \prec \beta_2 \cdots ; \alpha_m \prec \beta_n (m = 1, 2, \cdots ; n = m, m+1, \cdots).$$

Then $\mathcal{A}$ is a directed set (as can be easily verified). Define a net x_γ $(\gamma \in \mathcal{A})$ in $\mathbb{R}$ by setting

$$x_{\alpha_n} = n, x_{\beta_n} = \frac{1}{n}.$$

Then $x_\gamma \to 0$, but (x_γ) is unbounded in $\mathbb{R}$ (since $\frac{1}{n} x_{\alpha_n} = 1 \not\to 0$).

2.3.8. PROPOSITION. *Let A be a TA. Then we have:*

(i) *Let $(x_\alpha), (y_\beta)(z_\gamma)(u_\delta)$ be essentially bounded nets in A such that*

$$(x_\alpha) \sim (z_\gamma), (y_\beta) \sim (u_\delta).$$

Then $(x_\alpha y_\beta) \sim (z_\gamma u_\delta)$.

(ii) *If $(x_\alpha), (y_\beta)$ are C-nets then so is $(x_\alpha y_\beta)$.*

(iii) *If $x_\alpha \to 0$ and (y_β) is essentially bounded then $x_\alpha y_\beta \to 0$, $y_\beta x_\alpha \to 0$.*

PROOF. (i) Given a nucleus U(of A) we can find, using 2.1.17(iii), 2.2.14(vi), a nucleus V with

$$V^2 + V^2 \subseteq U. \tag{1}$$

By essential boundedness of the nets we have scalars, $\lambda, \ \mu \neq 0$ such that

$$\{y_\beta : \beta \succ \beta_1\} \subseteq \lambda V, \{z_\gamma : \gamma \succ \gamma_1\} \subseteq \mu V. \tag{2}$$

Since $(x_\alpha) \sim (z_\gamma), (y_\beta) \sim (u_\delta)$ we have $x_\alpha - z_\gamma \in \frac{1}{\lambda}V\,(\alpha \succ \alpha_0, \gamma \succ \gamma_2), y_\beta - u_\delta \in \frac{1}{\lambda}V\,(\beta \succ \beta_2, \delta \succ \delta_2)$ Choose $\beta_0 \succ \beta_1, \beta_2$ and $\gamma_0 \succ \gamma_1, \gamma_2$. Then

$$\begin{aligned} x_\alpha y_\beta - z_\gamma u_\delta &= (x_\alpha - z_\gamma)y_\beta + z_\gamma(y_\beta - u_\delta) \\ &\in \frac{1}{\lambda}V \cdot \lambda V + \mu V \cdot \frac{1}{\mu}V = V^2 + V^2 \subseteq U \\ &\qquad (\alpha \succ \alpha_0, \beta \succ \beta_0, \gamma \succ \gamma_0, \delta \succ \delta_0) \end{aligned}$$

proving $(x_\alpha y_\beta) \sim (z_\gamma u_\delta)$.

(ii) Since by hypothesis $(x_\alpha) \sim (x_\alpha), (y_\beta) \sim (y_\beta)$ and by 2.3.6, C-nets are essentially bounded we conclude using(i) $(x_\alpha y_\beta) \sim (x_\alpha y_\beta)$ i.e. $(x_\alpha y_\beta)$ is a C-net.

(iii) Given a nucleus U, choose a nucleus V such that $V^2 \subseteq U$. Since (y_β) is essentially bounded there is a $\lambda \neq 0$ and a β_0 with $\{y_\beta : \beta \succ \beta_0\} \subseteq \lambda V$. Since $x_\alpha \to 0$ we have $x_\alpha \in \lambda^{-1}V$ for $\alpha \succ \alpha_0$. It follows that

$$x_\alpha y_\beta \in V^2 \subseteq U \ (\alpha \succ \alpha_0, \beta \succ \beta_0).$$

Hence $x_\alpha y_\beta \to 0$. Similarly, $y_\beta x_\alpha \to 0$.

2.3.9. Let X be a TLS. The relation $\sim$ between nets in X is not only symmetric and transitive but also reflexive when confined to C-nets. So $\sim$ is an equivalence relation in the usual sense on the class of all C-nets (x_α) in X. Denote by $\hat{X}$ the set of the resulting equivalence classes $[(x_\alpha)]$. To each subset S of X we associate a subset $\hat{S}$ of $\hat{X}$ given by

$\hat{S} = \{\ [(x_\alpha)] \in \hat{X}$: for some representative (x_α) of the class we have $x_\alpha \in S$ for all $\alpha\}$.

If (x) is a principal net the corresponding class (denoted by) $[x] = [(x)]$ is called a ***principal class.***

2.3.10. LEMMA. *$\hat{X}$ is a* LS *and the map $j : x \mapsto [x]$ is*

linear. It is injective iff X *is Hausdorff.*

PROOF. If $\hat{x} = [(x_\alpha)], \hat{y} = [(y_\beta)] \in X, \lambda \in \mathbb{K}$ we define linear operations in $\hat{X}$ by:

$$\begin{aligned} \hat{x} + \hat{y} &= [(x_\alpha)] + [(y_\beta)] = [(x_\alpha + y_\beta)] \\ \lambda\hat{x} &= \lambda[(x_\alpha)] = [(\lambda x_\alpha)] \end{aligned}$$

These operations are well-defined in view of 2.3.2. The linearity of j is an immediate consequence of the definition of linear operations:

$$[x + y] = [x] + [y], \ [\lambda x] = \lambda[x].$$

Finally, if X is Hausdorff and $x, y \in X, x \neq y$ there is a nucleus U such that $x - y \notin U$, so that $(x) \not\sim (y), [x] \neq [y]$. On the other hand, if X is not Hausdorff it is not T_1 (see 2.1.6). Therefore there is an $x \neq 0$ with $x \in \overline{0}$. If U is any nucleus then by 2.1.5(i), $x \in \overline{0} \subseteq U$, whence $(x) \sim (0), [x] = [0]$, so that j is not injective.

2.3.11. LEMMA. (i) *If* $S_1 \subseteq S_2 (\subseteq X)$ *then* $\hat{S}_1 \subseteq \hat{S}_2$.

(ii) $\hat{\lambda S} = \lambda\hat{S}$.

(iii) *If* $x \in S$ *then* $[x] \in \hat{S}$.

(iv) *If* $[x] \in \hat{S}$ *then* $x \in \overline{S}$, *where bar denotes closure in* X.

(v) *If* (x_α) *is a C-net such that* $x_\alpha \in S$ *for* $\alpha \succ \alpha_1$ *then* $[(x_\alpha)] \in \hat{S}$.

PROOF. (i) - (iii): Clear.

(iv) If $[x] \in \hat{S}$ there is a net $(x_\alpha) \sim (x)$ with $x_\alpha \in S$ (for all α). By 2.3.2 (iv), $x_\alpha \to x$, whence $x \in \overline{S}$.

(v) Since $(x_\alpha : \alpha \succ \alpha_1)$ is a subnet of (x_α) it is, by 2.3.2(iv), equivalent to (x_α). Hence the result.

2.3.12. LEMMA. *If* U *is a nucleus of* X *then* $\hat{U}$ *is an absorbing subset of* $\hat{X}$. *If* U *is balanced so is* $\hat{U}$.

PROOF. Choose a balanced nucleus V with $V + V \subseteq U$. Suppose that $[\{y_\beta\}] \in \hat{X}$. Then $y_\beta - y_{\beta'} \in V (\beta, \beta' \succ \beta_0)$. Since V, as a nucleus, is absorbing, $y_{\beta_0} \in \lambda V$ for some $\lambda \geqslant 1$. It follows that for $\beta \succ \beta_0$

$$y_\beta = (y_\beta - y_{\beta_0}) + y_{\beta_0} \in V + \lambda V \subseteq \lambda V + \lambda V = \lambda(V + V) \subseteq \lambda U.$$

Hence, by 2.3.11 (v), $[\{y_\beta\}] \in \lambda\hat{U}$, proving U is absorbing. Now assume that U is balanced. If $[(x_\alpha)] \in \hat{U}$ with $x_\alpha \in U$, then since U is balanced, $\lambda x_\alpha \in U(|\lambda| \leqslant 1)$, so that $\lambda[(x_\alpha)] = [(\lambda x_\alpha)] \in \hat{U}$, proving $\hat{U}$ is balanced.

2.3.13. THEOREM. *$\hat{X}$ can be made into a* TLS *such that:*

(i) $j : x \mapsto [x]$ *is continuous; if X is Hausdorff j is a homeomorphism.*

(ii) $j(X)$ *is dense in* $\hat{X}$.

(iii) $\hat{X}$ *is complete.*

(iv) $\hat{X}$ *is Hausdorff.*

(v) $\hat{X}$ *has the following universal mapping property: given any complete Hausdorff TLS X^* and continuous linear map $\varphi : X \to X^*$, there is a unique continuous linear map $\hat{\varphi} : \hat{X} \to X^*$ such that $\varphi = \hat{\varphi} \circ j$.*

PROOF. Choose a basis $\mathcal{U}$ of nuclei of X satisfying (i)-(iv) of 2.1.17. By 2.3.12, the family $\hat{\mathcal{U}} = \{\hat{U} : U \in \mathcal{U}\}$ of subsets of $\hat{X}$ satisfies (i) of 2.1.17. Further, by monotonicity of the (set) map $S \mapsto \hat{S}$ (see 2.3.11(1)) it satisfies (ii) of 2.1.17. Finally, given $U \in \mathcal{U}$ we can choose a $V \in \mathcal{U}$ with $V + V \subseteq U$. Select $[(x_\alpha)], [(y_\beta)] \in \hat{V}$. We may assume that $(x_\alpha), (y_\beta) \subseteq V$. Then $[(x_\alpha) + (y_\beta)] = [(x_\alpha + y_\beta)] \in \hat{U}$ since $x_\alpha + y_\beta \in U$. Thus $\hat{U}$ satisfies (iii) of 2.1.17. We can now apply the converse part of 2.1.17 to conclude that $\hat{X}$ is a TLS with $\hat{\mathcal{U}}$ as a basis of nuclei.

It remains to prove the statements (i)-(v) above. By definition of $\hat{U}$ we have $j(U) \subseteq \hat{U}$, whence j is continuous. Let now X be Hausdorff. Then, by 2.3.10, j is 1-1. Let V be any nucleus of X and choose a nucleus W of X with $\overline{W} \subseteq V$. If $[x] \in \hat{W}$ then $x \in \overline{W} \subseteq V$. It follows that $j^{-1}(\hat{W}) \subseteq V$, whence j^{-1} is continuous. This completes the proof of (i). For proof of (ii) it is enough to observe that $j(x_\alpha) = [x_\alpha] \to [(x_\alpha)]$.

To prove (iii), Let $\hat{x_\gamma} = [(x^\gamma_{\alpha(\gamma)})]$ be a C-net in $\hat{X}$. Therefore, given a nucleus $\hat{U}$ of $\hat{X}$, for each γ there is a γ_U such that $\hat{x}_\gamma - \hat{x}_{\gamma'} \in \hat{U}$ for $\gamma, \gamma' \succ \gamma_U$, i.e., we have $x^\gamma_{\alpha(\gamma)} - x^{\gamma'}_{\alpha(\gamma')} \in U$ for all $\alpha(\gamma), \alpha(\gamma')$ with $\gamma, \gamma' \succ \gamma_U$. This mean that the net (with index γ) of principal classes

$$([x^\gamma_{\alpha(U)}]) \sim \hat{x}_\gamma \ (\alpha(U) = \alpha(\gamma_U)).$$

By considering $(x^\gamma_{\alpha(U)})$ as a net indexed by pairs (γ, U) with the ordering

$$(\gamma, U) \prec (\gamma', U') \text{ if } \gamma \prec \gamma', U' \subseteq U$$

it is clear that it converges to the element $[(x^\gamma_{\alpha(U)})]$ of $\hat{X}$. So $\hat{x}_\gamma \to [(x^\gamma_{\alpha(U)})]$, completing the proof of (iii).

For proving (iv) take $\hat{x} = [(x_\alpha)] \neq [0]$. Then $x_\alpha \not\to 0$. So we may assume that there is a nucleus U of X with the representative (x_α) of $\hat{x}$ disjoint with it. Choose a symmetric nucleus W with $W + W \subseteq U$. We claim that $\hat{W} \not\ni [(x_\alpha)]$. For, if $[(x_\alpha)] \in \hat{W}$ then there is a C-net $(y_\beta) \in W$ with $(x_\alpha) \sim (y_\beta)$. It follows that there are $x_{\alpha'}, y_{\alpha'}$ with $x_{\alpha'} - y_{\beta'} \in W$. Then

$$x_{\alpha'} = x_{\alpha'} - y_{\beta'} + y_{\beta'} \in W + W \subseteq U$$

contradicting $U \bigcap (x_\alpha) = \emptyset$. Thus, $\hat{W}$ is a nucleus of $\hat{X}$ not containing $\hat{x}$, whence $\hat{X}$ is T_1 and so Hausdorff.

Finally, for proving (v), take an elements $[\{x_\alpha\}] \in \hat{X}$. By 2.3.4, $(\varphi(x_\alpha))$ is a C-net in X^* and so converges uniquely to $x^* \in X^*$. Set

$$\hat{\varphi}([\{x_\alpha\}]) = x^*.$$

This is well-defined, since if $(x_\alpha) \sim (y_\beta)$ then $(\varphi(x_\alpha)) \sim (\varphi(y_\beta))$, so that $\varphi(y_\beta) \to x^*$. Also,

$$\varphi(x) = \hat{\varphi}([x]) = \hat{\varphi} \circ j(x).$$

Finally, if ψ is a continuous linear map with $\psi \circ j = \hat{\varphi} \circ j$ then clearly $\psi = \hat{\varphi}$ on the dense subspace $j(X)$ and hence everywhere.

This completes the proof.

2.3.14. THEOREM. *If A is a TA then its completion $\hat{A}$ (as a TLS) is a complete TA.*

PROOF. By 2.3.13, $\hat{A}$ is a complete TLS. So we have only to show that $\hat{A}$ is a TA. If $\hat{x} = [(x_\alpha)], \hat{y} = [(y_\beta)]$ are two elements of $\hat{A}$ then, by 2.3.8(ii), $(x_\alpha y_\beta)$ is a C-net and so determines an element $[(x_\alpha y_\beta)]$ of $\hat{A}$. We define multiplication in $\hat{A}$ by setting $\hat{x}\hat{y} = [(x_\alpha y_\beta)]$ (that this product is well-defined is assured by 2.3.8). It is easy to see that under this multiplication $\hat{A}$ is an algebra.

To prove that $\hat{A}$ is a TA, consider a nucleus U and an element $a \in A$. Since l_a is continuous there is a nucleus V with $aV \subseteq U$. Suppose now that $\hat{x}_\gamma = [(x^\gamma_{\alpha(\gamma)})] \to 0$ in $\hat{A}$. Then there is a γ_0 such that for $\gamma \succ \gamma_0, x^\gamma_{\alpha(\gamma)} \in V$ if $\alpha(\gamma) \succ$ some $\alpha_\gamma(\gamma)$. It follows that

$$ax^\gamma_{\alpha(\gamma)} \in aV \subseteq U$$

whence

$$\hat{a}\hat{x}_\gamma \to 0. \tag{1}$$

Similarly, using continuity of r_a we get

$$\hat{x_\gamma}\hat{a} \to 0. \tag{2}$$

The conclusions (1),(2) imply that $\hat{A}$ is a WTA. Further, since A is a TA, given a nucleus U, there is a nucleus V such that

$$V^2 \subseteq U. \tag{3}$$

Assume now that

$$\hat{x_\gamma} = [(x^\gamma_{\alpha(\gamma)})] \to 0, \hat{y_\delta} = [(y^\delta_{\beta(\delta)})] \to 0.$$

Then we have for some γ_0, δ_0, $x^\gamma_{\alpha(\gamma)}, y^\delta_{\beta(\delta)} \in V^2 \subseteq U$, whence $\hat{x_\gamma}\hat{y_\delta} \to 0$, completing the proof that A is a TA.

CHAPTER III

SOME TYPES OF TOPOLOGICAL ALGEBRAS

§ 1. Quarter-norms

3.1.1. DEFINITION. A real-valued function $p = p(x)$ on an additive abelian group X is called a *subadditive* or *sad.* functional if it satisfies:

(Q1) $p(0) = 0$;

(Q2) $p(-x) = p(x)(x \in X)$;

(Q3) $p(x + y) \leqslant p(x) + p(y) \qquad (x, y \in X)$.

If p is a sad. functional then we have also

(Q4) $p(x) \geqslant 0 \qquad (x \in X)$

$$(0 = p(0) = p(x - x) \leqslant p(x) + p(-x) = p(x) + p(x) = 2p(x)).$$

If p satisfies

(Q5) $p(x) = 0 \Rightarrow x = 0$

then p is called *faithful*.

3.1.2. LEMMA. *Let X be an additive abelian group and p a sad. functional on X. Then:*

(i) $p(nx) \leqslant |n|p(x)$ *for any integer* n.

(ii) $|p(x) - p(y)| \leqslant p(x - y)$.

(iii) *If* $p(a) = 0$ *then* $p(x + a) = p(x)$ *for all* $x \in X$.

(iv) $d_p(x, y) = p(x - y)$ *is a semi-metric*† *on* X *which is invariant for translations:*

$$d_p(x + a, y + a) = d_p(x, y) \qquad (x, y, a \in X).$$

Also, d_p *is a metric iff* p *is faithful.*

---

† d is called a semi-metric or pseudo-metric if it satisfies all metric axioms except: $d(x, y) = 0 \Rightarrow x = y$.

If d is any semi-metric on X invariant for translations then $p(x) = d(x, 0)$ is a sad. functional.

PROOF. (i) If n is positive we have $p(nx) \leqslant p((n-1)x+x) \leqslant p((n-1)x) + p(x)$ from which the inequality follows for positive n by induction. If n is negative then $p(nx) = p(-n. -x) \leqslant -np(-x) = -np(x) = |n|p(x)$, completing the proof for all n.

(ii) $p(x) = p(x - y + y) \leqslant p(x - y) + p(y)$. Interchanging x, y we get $p(y) \leqslant p(y - x) + p(x) = p(x - y) + p(x)$. The required inequality now follows.

(iii) $p(x + a) \leqslant p(x) + p(a) = p(x) + 0 = p(x)$. Also $p(x) = p(x + a - a) \leqslant p(x + a) + p(-a) = p(x + a)$. Hence the required equality.

(iv) By (Q4), $d_p(x, y) \geqslant 0$ and by (Q1), $d_p(x, x) = 0$. The symmetry and triangle inequality properties for d_p follow from (Q2), (Q3). Further, we have clearly $d_p(x + a, y + a) = d_p(x, y)$. Finally, if p is faithful then $d_p(x, y) = p(x - y) = 0 \Rightarrow x = y$, so that p is a metric. Conversely, if d_p is a metric then p is faithful $(p(x) = d_p(x, 0) = 0 \Rightarrow x = 0)$.

Suppose now d is an invariant semi-metric and $p(x) = d(x, 0)$. Then $p(0) = d(0, 0) = 0$,

$$p(-x) = d(-x, 0) = d(x - x, x) = d(0, x) = d(x, 0) = p(x).$$

Finally,

$$\begin{aligned} p(x + y) &= d(x + y, 0) = d(x, -y) \leqslant d(x, 0) + d(0 - y) \\ &\leqslant p(x) + p(-y) = p(x) + p(y). \end{aligned}$$

3.1.3. LEMMA. *Let X be an additive abelian TG and p a sad. functional on X. If p is continuous at 0 then it is uniformly continuous on X.*

PROOF. Let $\{U_\alpha\}$ be a basis of nuclei of X. Then, for $x \in X$, $\{x + U_\alpha\}$ is a basis of neighbourhoods of x. Suppose now that p is continuous at 0. Then, given $\epsilon > 0$ there is a nucleus U_α such that $p(a) < \epsilon$ when $a \in U_\alpha$. Therefore, for any $x \in X$,

$$x + a \in x + U_\alpha, \qquad |p(x + a) - p(x)| < p(a) < \epsilon$$

which show that p is uniformly continuous on X.

3.1.4. The topology induced on X by the semi-metric d_p is called the *p-topology* and X with the p-topology is denoted by (X,p). A neighbourhood basis for a point x_0 in the p-topology is given by the family of sets $\{\mathcal{N}(x_0,\epsilon) : \epsilon > 0\}$, where

$$\mathcal{N}(x_0,\epsilon) = \{x \in X : p(x - x_0) < \epsilon\}.$$

Since $\mathcal{N}(x_0,\epsilon_1) \subseteq \mathcal{N}(x_0,\epsilon_2)$ for $\epsilon_1 < \epsilon_2$ it follows that the subfamily $\{\mathcal{N}(x_0,\frac{1}{n}) : n = 1,2,\ldots\}$ is also a neighbourhood basis at x_0, whence the p-topology is first countable. It is clear that a sequence $x_n \to x$ in the p-topology iff $p(x_n - x) \to 0$. We call the p-topology *complete* if it is complete with respect to d_p. As already seen in 3.1.2 (iv), d_p is a metric if p is faithful.

Two sad. functionals p_1, p_2 on X an said to be *equivalent,* in symbols, $p_1 \sim p_2$ if both of them induce the same topology on X.

3.1.5. LEMMA. *Let p be a sad. functional on an additive abelian group X. Then:*

(i) *X is a TG under the p-topology;*

(ii) *p is a continuous function on (X,p);*

(iii) $\ker p = \{x \in X : p(x) = 0\}$ *is a closed subgroup and* $\ker p = \overline{0}$.

(iv) *If $p_1 \sim p_2$ then* $\ker p_1 = \ker p_2$. *In particular p_1 is faithful iff p_2 is faithful.*

PROOF. (i) It follows from (Q3) that the map $(x,y) \mapsto x + y$ is a continuous and from (Q2) the map $x \mapsto -x$ is continuous. So X is a TG.

(ii) The continuity of p follows from 3.1.2 (ii).

(iii) If $x, y \in \ker p$ then $p(x - y) \leqslant p(x) + p(y) = 0 + 0 = 0$, whence $x - y \in \ker p$. So $\ker p$ is a subgroup. Further, it is closed since $\ker p = p^{-1}(0)$ and p is continuous. Since $0 \in \ker p$ and $\ker p$ is closed, $\overline{0} \subseteq \ker p$. On the other hand, if $p(x) = 0$, $p(0 - x) = p(-x) = p(x) = 0$, so that $0, 0, \cdots \to, x \quad x \in \overline{0}$. Thus, $\overline{0} = \ker p$.

(iv) By (iii), $\ker p_1 = \overline{0} = \ker p_2$.

3.1.6. The set $\mathfrak{S}$ of all sad. functionals on X can be partially ordered by: $p \prec q$ if $p(x) \leqslant q(x)$ for all $x \in X\,(p, q \in \mathfrak{S})$.

3.1.7. LEMMA. (a) *If p is a sad. functional and $\epsilon > 0$ then tp is a sad. functional.*

(b) *If $p_j (j = 1, \cdots, n)$ are sad. functionals on X then so are $p = p_1 + \cdots + p_n$, $q = p_1 \vee \cdots \vee p_n$* † $(q(x) = \max_j p_j(x))$

PROOF. (a) Clear.

(b) That p is sad. is clear. To prove that q is sad. it is clearly enough to show that it satisfies (Q3). Now, by definition of q, we have for given $x, y \in X$, an index j with $q(x+y) = p_j(x+y)$. Then

$$p_j(x+y) \leqslant p_j(x) + p_j(y) \leqslant q(x) + q(y),$$

so that $q(x+y) \leqslant q(x) + q(y)$.

3.1.8. DEFINITION. Let X be a LS and p a sad. functional on the underlying additive group of X. Then p is called a *quarter-norm* if it satisfies.

(Q6) If (x_n) is a sequence in X, $x \in X$, $p(x_n - x) \to 0$ and $\lambda_n \to \lambda$ in $\mathbb{K}$, then $p(\lambda_n x_n - \lambda_x) \to 0$ (i.e. if in the p-topology $x_n \to x$ and $\lambda_n \to \lambda$ in $\mathbb{K}$ then $\lambda_n x_n \to \lambda x$).

The condition (Q6) can be split up into three subconditions:

(Q6a) If $p(x_n) \to 0$, $\lambda_n \to 0$ then $p(\lambda_n x_n) \to 0$.

(Q6b) If $p(x_n - x) \to 0$ then $p(\lambda x_n - \lambda x) \to 0$, $\lambda \in \mathbb{K}$.

(Q6c) If $\lambda_n \to \lambda$ then $p(\lambda_n x - \lambda x) \to 0, x \in X$.

That (Q6) implies each of there subconditions is clear. Conversely, it follows from the identity

$$\lambda_n x_n - \lambda x = (\lambda_n - \lambda)(x_n - x) + \lambda x_n - \lambda x + \lambda_n x - \lambda x \qquad (*)$$

that the subconditions can be replaced by (Q6).

3.1.9. Remark. If p is a quarter-norm on X, define p^* on X by: $p^*(x) = p(x)$ or 1 according a $p(x) \leqslant 1$ or $p(x) > 1$. Then it is easy to see that p^* is also a quarter-norm; p^* is called a *reduced* quarter-norm, the *reduced form* of p. It is clear that

---

† q is the lattice sum (see 8.6.7) with respect to the partial order defined in 3.1.6.

$p^* \sim p$ and that p^* is bounded (as a functional): $p^*(x) \leqslant 1$ for all x in X, or briefly $p^* \leqslant 1$.

3.1.10. LEMMA. *Let p be a quarter-norm on X. Then $p^{**} = p/1{+}p$ is a quarter-norm which is also bounded: $p^{**} \leqslant 1$. Further, $p^{**} \sim p$.*

PROOF. For proving p^{**} is a quarter-norm only the subadditivity has to be checked (since the rest of the properties are clear). If $p(x+y) = 0$ then $p^{**}(x+y) = 0$ and then the subadditivity holds trivially. Next assume that $p(x+y) > 0$. Then $0 < p(x+y) \leqslant p(x) + p(y)$, whence

$$1 + \frac{1}{p(x+y)} \geqslant 1 + \frac{1}{p(x)+p(y)},$$

so that

$$\begin{aligned} p^{**}(x+y) &= \frac{1}{1+\frac{1}{p(x+y)}} \leqslant \frac{1}{1+\frac{1}{p(x)+p(y)}} = \frac{p(x)+p(y)}{1+p(x)+p(y)} \\ &\leqslant \frac{p(x)}{1+p(x)} + \frac{p(y)}{1+p(y)} = p^{**}(x) + p^{**}(y). \end{aligned}$$

The boundedness of p^{**} is evident from its definition. Also, the definition of p^{**} shows that $p^{**}(x_n - x) \to 0$ iff $p(x_n - x) \to 0$, whence $p^{**} \sim p$, completing the proof.

3.1.11. PROPOSITION. *Let X be a LS over $\mathbb{K}$. A sad. functional p on X is a quarter-norm iff X is a TLS under the p-topology.*

PROOF. This follows from 3.1.5 (i) and the observation that the condition (Q6) equivalent to the continuity of the map $(\lambda, x) \mapsto \lambda x$ in the p-topology.

3.1.12. COROLLARY. *A TLS X is quarter-normed iff it is first countable.*

PROOF. Since the p-topology is first countable (see 3.1.4) we have only to prove the "if" part. If X is first countable then by 2.1.7 it has an invariant (with respect to addition) metric d.

By 3.1.2, $p(x) = d(x, 0)$ is a sad. functional which is, by 3.1.11, a quarter-norm, so that X is quarter-normed.

3.1.13. Remark. It can be shown easily that if $X = (X, p)$ is a quarter-normed LS then it is complete as a TLS iff it is complete with respect to p (i.e. with respect to the semi-metric d_p). Such a quarter-norm p is called a *complete quarter-norm.*

3.1.14. Let X be a LS over $\mathbb{K}$ and p a faithful quarter-norm. Then X with the p-topology is called a pre-(F) space. The quarter-norm in this case is called a (F) *norm* and will be usually denoted in the sequel by $|\cdot|$. If the (F) norm $|\cdot|$ is complete then X is called a (F) space. Note that a (F) space is a complete metric linear space.

3.1.15. THEOREM (Open mapping theorem). *If T is a continuous linear transformation of a (F) space X onto a (F) space X^* then T is open (i.e. T maps every open set of X on an open set of X^*).*

PROOF. See [22, p.48].

3.1.16. Theorem (Closed graph theorem). *Let X, X^* be (F) spaces. If $T : X \to X^*$ is a linear map which is closed then it is continuous.*

(A map $T : X \to X^*$ is called closed if $x_n \to x$ in X and $Tx_n \to y$ in $X^* \Rightarrow y = Tx$.)

PROOF. See [22, p.50].

3.1.17. Let $\mathcal{P}$ be a family of quarter-norms p on a LS X. Then the $\mathcal{P}$-topology on X is that topology in which a net $x_\alpha \to x$ iff $p(x_\alpha - x) \to 0$ for each $p \in \mathcal{P}$. It is easy to see that X is a TLS under the $\mathcal{P}$-topology. If the family $\mathcal{P}$ consists of a single p then the $\mathcal{P}$-topology reduces to the p-topology defined in 3.1.4.

A neighbourhood basis for a point x_0 in X, in the $\mathcal{P}$-topology, is given by the family of all sets of the form.

$$\mathcal{N}(x_0 : p_1, \cdots, p_n, \epsilon) = \{x \in X : p_j(x - x_j) < \epsilon, j = 1, \cdots, n\}$$

where $p_1, \cdots, p_n \in \mathcal{P}$ and $\epsilon > 0$. We write $X = (X, \mathcal{P})$ to denote X with the $\mathcal{P}$-topology.

Two families $\mathcal{P}_1, \mathcal{P}_2$ of quarter-norms on X are said to be *equivalent,* in symbols, $\mathcal{P}_1 \sim \mathcal{P}_2$ if the $\mathcal{P}_1$-topology and the $\mathcal{P}_2$-topology are the same: $(X, \mathcal{P}_1) = (X, \mathcal{P}_2)$ (cf. equivalence of 2 sad.'s in 3.1.4).

3.1.18. LEMMA. *If $p_j (j = 1, \cdots n)$ are quarter-norms on X so are*

$$p = p_1 + \cdots + p_n, \qquad q = p_1 \vee \cdots \vee p_n$$

Further, $p \sim \{p_1, \cdots, p_n\} \sim q$.

PROOF. By 3.1.7 (b), p and q are sad. functionals. The verification that they are also quarter-norms is straightforward. Further, since

$$\begin{aligned} p(x_\alpha - x) \to 0 \quad &\Leftrightarrow \quad p_j(x_\alpha - x) \to 0 \ (j = 1, \cdots, n) \\ &\Leftrightarrow \quad q(x_\alpha - x) \to 0 \end{aligned}$$

the equivalence statement follows.

3.1.19. LEMMA. (a) *If p, p^* are quarter-norms on X such that there are constants $C_1, C_2 > 0$ satisfying*

$$C_1 p \leqslant p^* \leqslant C_2 p \qquad (*)$$

then $p \sim p^*$.

(b) *If p is a quarter-norm and $t > 0$ then tp is a quarter-norm with $tp \sim p$.*

PROOF (a) This is an immediate consequence of the inequalities

$$C_1 p(x_\alpha - x) \leqslant p^*(x_\alpha - x) \leqslant C_2 p(x_\alpha - x).$$

(b) That tp is a quarter-norm is clear. The equivalence assertion follows from (a) (taking $p^* = tp$, $C_1 = C_2 = t$).

3.1.20. If $\mathcal{P}$ is any family of quarter-norms (or more generally sad. functionals) on X we write

$$\ker \mathcal{P} = \cap \ker p \qquad (p \in \mathcal{P})$$

A net (x_α) is called a C-net with respect to a quarter-norm p if given $\epsilon > 0$ there is an $\alpha_0 = \alpha_0(\epsilon)$ such that $p(x_\alpha - x_\beta) < \epsilon$

for all $\alpha, \beta \succ \alpha_0$. It is called a C-net with respect to family $\mathcal{P}$ of quarter-norms if it is a C-net with respect to each p in $\mathcal{P}$. X is called $\mathcal{P}$-complete if every C-net with respect to $\mathcal{P}$ converges in X.

3.1.21. LEMMA. *Let $X = (X, \mathcal{P})$ be a TLS, where $\mathcal{P}$ is a family of quarter-norms on X. Then:*

(i) $B_{\epsilon,p} = \{x \in X : p(x) < \epsilon\}$ *is a symmetric nucleus of* X, *for each* $p \in \mathcal{P}$.

(ii) $\overline{0} = \overline{\{0\}} = \ker \mathcal{P}$, *where bar denotes closure in* X.

(iii) X *is Hausdorff iff* $\ker \mathcal{P} = \{0\}$.

(iv) X *is a complete TLS iff it is* $\mathcal{P}$*-complete.*

(v) *If f is a continuous functional on X with $f(0) = 0$ then* $\ker \mathcal{P} \subseteq \ker f$.

PROOF. (i) By definition of $\mathcal{P}$-topology, $B_{\epsilon,p}$ is a nucleus of X; further it is symmetric since $p(-x) = p(x) (x \in X)$.

(ii) Since $p(0) = 0$, for any p, $0 \in \ker \mathcal{P}$. Now $\ker \mathcal{P}$ is closed by virtue of 3.1.5 (iii). So, $\overline{0} \subseteq \ker \mathcal{P}$. If $x \in \ker \mathcal{P}$, $p(0 - x) = p(-x) = p(x) = 0$ $(\forall p)$, so that $0, 0 \cdots \to x$, whence $x \in \overline{0}$. Therefore $\overline{0} = \ker \mathcal{P}$

(iii) This follows from (ii) and 2.1.13.

(iv) This is clear from the definitions of the C-net and $\mathcal{P}$-topology.

(v) If $x \in \ker \mathcal{P} = \overline{0}$, we have $0, 0, \cdots \to x$. By continuity of f, the sequence $f(0), f(0), \cdots \to f(x)$. Since $f(0) = 0$, we get $0, 0, \cdots \to f(x)$. By uniqueness of limit property in $\mathbb{K}$, we conclude that $f(x) = 0$, completing the proof.

3.1.22. PROPOSITION. *Let $X = (X, p)$ be a quarter-normed LS and X_0 a subspace of X. Write $X^{\#} = X/X_0$ and define $p^{\#}$ on $X^{\#}$ by*

$$p^{\#}(x + X_0) = \inf\{p(x + a) : a \in X_0\}.$$

Then:

(i) *The canonical map* $\pi : x \mapsto x + X_0 = x^{\#} = \pi(x)$ *satisfies* $p^{\#}(\pi x) \leqslant p(x)$.

(ii) *The functional* $p^{\#}$ *is a quarter-norm.*

(iii) *The topology of the quotient space* $X^{\#}$ *is induced by* $p^{\#}$.

(iv) $p^{\#}$ *is faithful iff* X_0 *is closed in* X.

(v) *If* p *is complete so is* $p^{\#}$.

PROOF. (i) Evident.

(ii) We have

$$\begin{aligned} p^{\#}(x+y+X_0) &\leqslant p(x+y+a+b) \qquad (a,b\in X_0)\\ &\leqslant p(x+a)+p(y+b), \text{ so that}\\ p^{\#}(x+y+X_0) &\leqslant p^{\#}(x+X_0)+p^{\#}(y+X_0). \end{aligned}$$

Since $0\in X_0$ and $p(0)=0$, we get $p^{\#}(0^{\#})=0$.

Again $p(x)=p(-x)\Rightarrow p(x^{\#})=p(-x^{\#})$, where $x^{\#}=x+X_0$. Thus $p^{\#}$ is a sad. fnctional. Suppose now that $x_n^{\#}\to x^{\#}$, $\lambda_n\to\lambda$. Since

$$p^{\#}(x_n^{\#}-x^{\#})=p^{\#}((x_n-x)^{\#})\to 0,$$

using the definition of $p^{\#}$, we can first fix $x_n\in\pi^{-1}(x_n^{\#})$, $x\in\pi^{-1}(x^{\#})$ and then choose suitably $a_n\in X_0$ such that

$$p(x_n+a_n-x)<p^{\#}(x_n^{\#}-x^{\#})+\frac{1}{n}\ (n=1,2,\cdots).$$

So $p(x_n+a_n-x)\to 0$. Since p is a quarter-norm we obtain

$$p(\lambda_n x_n+\lambda_n a_n-\lambda x)\to 0,$$

whence by using the inequality in (i) we get $p^{\#}(\lambda_n x_n^{\#}-\lambda x^{\#})\to 0$, which shows $p^{\#}$ is a quarter-norm.

(iii) Consider the ball $B^{\#}=B^{\#}(x^{\#},\epsilon)=\{y^{\#}\in A^{\#}: p^{\#}(y^{\#}-x^{\#})<\epsilon\}$. We shall show that $B^{\#}$ is open in the quotient topology. Now

$$\pi^{-1}(B^{\#})=\pi^{-1}(B^{\#}(x^{\#},\epsilon))=\bigcup_{a\in X_0}B(x+a,\epsilon)$$

and hence open in A. Since, by 2.1.18, π is an open map, $B^{\#}=\pi(\pi^{-1}(B^{\#}))$ is open in the quotient topology. It follows that every open set of $(A^{\#}, p^{\#})$ is open in the quotient topology. Conversely, let $G^{\#}$ be an open set in the quotient topology and $x^{\#} \in G^{\#}$. By continuity of π, $\pi^{-1}(G^{\#})$ is open in A and so we have: $(*)$ $\pi^{-1}(-G^{\#}) \supseteq B(x, \epsilon)$ for any $x \in \pi^{-1}(x^*)$ and some $\epsilon = \epsilon_x > 0$. If $y^{\#} \in B^{\#}(x^{\#})$ then $p^{\#}((y-x)^{\#}) < \epsilon$, whence $p(y-x+a) < \epsilon$ for some $a \in X_0$. Then $y + a \in B(x, \epsilon) \subseteq \pi^{-1}(G^{\#})$ (using, $(*)$), and $y^{\#} = \pi(y+a) \in G^{\#}$, whence $G^{\#}$ is open in the $p^{\#}$-topology.

(iv) Now $p^{\#}(x+X_0) = 0 \Leftrightarrow x \in \overline{X}_0$ where bar denotes closure in the p-topology. It follows that $p^{\#}$ is faithful iff $x \in \overline{X}_0 \Rightarrow x \in X_0$, i.e. iff $X_0 = \overline{X}_0$.

(v) Suppose that p is complete and $(x_n^{\#})$ is a C-sequence of $X^{\#}$. We can find a subsequence $(x_{n_k}^{\#})$ with

$$p^{\#}(x_{n_k}^{\#} - x_{n_{k+1}}^{\#}) < \frac{1}{2^k}.$$

Choose inductively $x_k \in X$ with $\pi(x_k) = x_{n_k}^{\#}$ and $p(x_k, x_{k+1}) < \frac{1}{2^k}$. Them (x_k) is a C-sequence of X and so $x_k \to x$ (say) in X (since p is complete). By continuity of π,

$$x_{n_k}^{\#} = \pi(x_{n_k}) \to x^{\#},$$

and since $(x_{n_k}^{\#})$ is a subsequence of the C-sequence $(x_n^{\#})$ we have also $x_n^{\#} \to x^{\#}$, proving $p^{\#}$ is complete.

3.1.23. DEFINITION. Let X be a TLS of the form $X = (X, \mathcal{P})$, where $\mathcal{P}$ is a family of quarter-norms. A series $\sum_{n=1}^{\infty} x_n$ in X is said to *absolutely converge* if $\sum_{n=1}^{\infty} p(x_n)$ converges for each $p \in \mathcal{P}$.

3.1.24. LEMMA. *Let $X = (X, \mathcal{P})$ be sequentially complete (in particular, complete). Then every absolutely convergent series $\sum_{n=1}^{\infty} x_n$ converges in X.*

PROOF. This follows from the inequality.

$$p(x_{n+1} + \cdots + x_{n+k}) \leqslant p(x) + \cdots + p(x_{n+k})(p \in \mathcal{P}).$$

3.1.25 LEMMA. *Let $T : X \to Y$ be a continuous transformation (not necessarily linear) with $T0 = 0$, where $X = (X, \mathcal{P})$, $Y = (Y, \mathcal{Q})$ are TLS's and $\mathcal{P}, \mathcal{Q}$ families of quarter-norms. Then for $x \in \ker \mathcal{P}$, $q_\beta(Tx) = 0$ for all $q_\beta \in \mathcal{Q}$.*

By continuity of T,

$$Tx, Tx, \cdots \to T0 = 0.$$

So $q_\beta(Tx) \to q_\beta(0) = 0$, whence $q_\beta(Tx) = 0$.

§ 2. ρ-Seminorms

3.2.1. DEFINITION. Let X be a LS over $\mathbb{K}$ and ρ a real number such that $0 < \rho \leqslant 1$. A real-valued function $p = p(x)$ on X is called a ρ-[†] *seminorm* if it satisfies:

(Q3) $p(x + y) \leqslant p(x) + p(y)$ for all $x, y \in X$;

(Q7) $p(\lambda x) = |\lambda|^\rho p(x)$ for all $x \in X, \lambda \in \mathbb{K}$.

If in the above, $\rho = 1$ then p is called a *seminorm.* (Q7) is called ρ*-modulus homogeneity* (*modulus homogeneity* if $\rho = 1$) condition and ρ is called the *homogeneity index* of p.

A ρ-seminorm (respy. seminorm) p is called a ρ-*norm*(respy. *norm*) if p is faithful (i.e. satisfies (Q5) of 3.1.1).

3.2.2. Remark. The homogenity index of a $p \neq 0$ [††] is uniquely determined; for, if p is both ρ-seminormed and ρ'-seminormed then for an x_0 with $p(x_0) \neq 0$ we have $p(2x_0) = 2^\rho p(x_0) = 2^{\rho'} p(x_0)$, whence $\rho = \rho'$.

3.2.3 LEMMA. *If $p_j (j = 1, 2, \cdots, n)$ are ρ-seminorms on X then so are*

$$p = p_1 + \cdots + p_n, \quad q = p_1 \vee \cdots \vee p_n.$$

(q defined as in 3.1.7).

PROOF. Straightforward and so omitted.

[†] In literature the letter p has been used in the place of our ρ. We have adopted ρ to avoid conflict with p used by us with a different connotation.

[††] i.e. p is not identically 0.

3.2.4. PROPOSITION. *Every ρ-seminorm p, on X is a quarter-norm. In particular it satisfies* (Q1), (Q2), (Q4) *of* 3.1.1 *and also* (i), (ii) *of* 3.1.2.

PROOF. By taking in (Q7), $\lambda = 0, -1$ successively we get (Q1),(Q2), so that p is a sad. fnctional To show that it is a quarter-norm we have to prove (Q6). Using the identity $(*)$ of 3.1.8, (Q3) and (Q7) we get

$$p(\lambda_n x_n - \lambda x) \leqslant |\lambda_n - \lambda|^\rho p(x_n - x) + |\lambda|^\rho p(x_n - x) + |\lambda_n - \lambda|^\rho p(x),$$

from which (Q6) readily follows.

3.2.5. LEMMA. *Let X be a TLS p a ρ-seminorm on X and $V = \{x \in X : p(x) < 1\}$. Then p is continuous iff $0 \in V^o (=$ int $V)$*

PROOF. If p is continuous then since $p(0) = 0$ there is an open nucleus U such that if $x \in U$ then $p(x) < 1$. This implies that $0 \in U \subseteq V^o$. Conversely, if $0 \in V^o$ then, using 2.1.10, we get $\epsilon^{\frac{1}{\rho}} V^o$ is an open nucleus. If $x \in V^o, y = \epsilon^{\frac{1}{\rho}} x$ then $p(y) = \epsilon p(x) < \epsilon$. Hence p is continuous at 0 and so everywhere (by 3.1.3).

3.2.6. If p is a ρ-seminorm on a LS X then the resulting TLS, $X = (X, p)$ (cf. 3.1.4) is called a *ρ-seminormed* LS. If p is a ρ-norm then X is called a *ρ-normed* LS. We generally denote a ρ-norm by the symbol $\|\cdot\|$ instead of p. Further, if $p = \|\cdot\|$ is complete then X is called a *ρ-Banach* space (a *Banach* space if $\rho = 1$). A ρ-seminorm p and a ρ'-seminorm p' are said to be equivalent, $p \sim p'$, if they are equivalent as sad functionals (i.e. if they induce the same topology).

3.2.7. LEMMA. *Let $X = (X, p)$ be a ρ-seminormed LS. Then we have:*

(i) *If $B_r = \{x \in X : p(x) < r\}, \overline{B}_r = \{x \in X : p(x) \leqslant r\}$ then $\overline{B}_r$ is the (topological) closure of B_r in X, where $r \in \mathbb{R}, r > 0$. Further more, B_r (respy. $\overline{B}_r$) is a balanced open (respy. closed) nucleus of X.*

(ii) *Let (r_n) be a sequence of positive numbers with $r_n \to 0$. If*

$V_n = B_{r_n}, U_n = \overline{B}_{r_n}$ then $\{V_n\}, \{U_n\}$ are bases of nuclei.

PROOF. (i) Trivially, $B_r \subseteq \overline{B}_r$. By continuity of p (see 3.1.5(iii)) $\overline{B}_r$ is closed. For $x \in \overline{B}_r$, set $x_n = (1 - \frac{1}{n+1})^{\frac{1}{\rho}} x$. Since

$$p(x_n) \leqslant (1 - \frac{1}{n+1})r < r \qquad (n = 1, 2, \cdots)$$

$x_n \in B_r$ and clearly $x_n \to x$. It follows that $\overline{B}_r$ is the closure of B_r

(ii) If n is sufficiently large then we have

$$V_n, U_n \subseteq B_\epsilon = \mathcal{N}(0; p, \epsilon)$$

Hence the result.

3.2.8. LEMMA. *A ρ-seminormed LS $X = (X, p)$ is locally connected.*

PROOF. Since $\{B_r : r > 0\}$ is a basis of nuclei, it is enough to show that (each) B_r is connected. But $B_r = r^{\frac{1}{\rho}} B_1$ is homeomorphic to B_1 (see 2.1.10), so that suffices to prove that B_1 is connected. Now B_1 is path-connected (and hence connected), since if $x \in B_1$ then $p(x) < 1$ and $\{tx : 0 \leqslant t < 1\}$ is a path in B_1 joining 0 and x.

3.2.9. LEMMA. *If p is a ρ-seminorm $(0 < \rho \leqslant 1)$ on a LS X (over $\mathbb{K}$) and $0 < \rho' \leqslant \rho$. Then $q = p^{\frac{\rho'}{\rho}}$ is a ρ'-seminorm with $q \sim p$. In particular, if p is a seminorm, p^ρ is a ρ-seminorm with $p^\rho \sim p$.*

PROOF. Write $\frac{\rho'}{\rho} = t$, so that $0 < t \leqslant 1$, and $q = p^t$. If $x, y \in X$ then

$$\begin{aligned} q(x+y) &= p(x+y)^t \leqslant (p(x) + p(y))^t \leqslant p(x)^t + p(y)^t \\ &\qquad\qquad\qquad\qquad (\text{since } 0 < t \leqslant 1) \\ &\leqslant q(x) + q(y). \end{aligned}$$

Also, if $\lambda \in \mathbb{K}$,

$$q(\lambda x) = p(\lambda x)^t = |\lambda|^{\rho t} p(x)^t = |\lambda|^{\rho'} q(x).$$

Finally, it is clear that $q(x_n - x) = p(x_n - x)^t \to 0$ iff $p(x_n - x) \to 0$, proving $q \sim p$.

3.2.10. PROPOSITION. *Let $(X, p), (X^*, p^*)$ be respy. a ρ-seminormed LS and a ρ'-seminormed LS. Let $T : X \to X^*$ be a linear transformation. Then T is continuous iff there is a constant $C > 0$ such that*

$$p^*(Tx) \leqslant Cp(x)^{\rho'/\rho} \qquad (x \in X). \qquad (*)$$

PROOF. Suppose $(*)$ holds. Then

$$p^*(Tx_\alpha - Tx) = p^*(T(x_\alpha - x)) \leqslant Cp(x_\alpha - x)^{\rho'/\rho},$$

whence T is continuous. Conversely, suppose that T is continuous. By continuity of T at 0, given $\epsilon > 0$, we can find $\delta > 0$ such that

$$p^*(Tx) \leqslant \epsilon \qquad \text{whenever } p(x) \leqslant \delta. \qquad (**)$$

For any x with $p(x) \neq 0$, write $y = \delta^{1/\rho} x / p(x)^{\frac{1}{\rho}}$. Then $p(y) = \delta$, so that by $(**)$ we have $p^*(Ty) \leqslant \epsilon$, which reduces to

$$p^*(Tx) \leqslant Cp(x)^{\rho'/\rho}, \qquad \text{with } C = \epsilon/\delta^{\rho'/\rho}. \qquad (*)$$

Suppose next $p(x) = 0$, then for any integer n, $p(nx) = 0$, so that by $(**)$,

$$p^*(nTx) \leqslant \epsilon, \ \text{ i.e. } n^{\rho^*} p^*(Tx) \leqslant \epsilon.$$

Since n is arbitrary we conclude that $p^*(Tx) = 0$, so that the inequality $(*)$ holds in this case as well. The proof is complete.

3.2.11. COROLLARY. *Let p, p^* be respectively a ρ-seminorm and a ρ^*-seminorm on a LS X. Then*

(i) *p^* is continuous in the p-topology iff there is a constant $C > 0$ such that*

$$p^* \leqslant Cp^{\frac{\rho^*}{\rho}}. \qquad (*)$$

(ii) *$p^* \sim p$ iff there are constants $C, C^* > 0$ such that*

$$p^* \leqslant Cp^{\frac{\rho^*}{\rho}}, \ p \leqslant C^* p^{*\frac{\rho}{\rho^*}}. \qquad (**)$$

(iii) *Suppose that* $\rho^* = \rho$. *Then* $p^* \sim p$ *iff there are constants* $C_1, C_2 > 0$ *such that*

$$C_1 p \leqslant p^* \leqslant C_2 p. \qquad (***)$$

(iv) *If* p *is a* ρ*-seminorm on* X *and* $C > 0$, *then so is* $p^* = Cp$; *further,* $p^* \sim p$.

PROOF. (i) This follows from 3.2.10 by taking $X^* = X$ and $T = 1_X$ (the identity map)

(ii) This follows from (i) (by considering also the continuity condition for p in the p^*-topology).

(iii) This follows from $(**)$ (taking $C_1 = 1/C^*, C_2 = C$).

(iv) Clear.

3.2.12. DEFINITION. Let p be a ρ-seminorm on a LS X. A subset S of X is said to be *bounded* with respect to p or p*-bounded* if there is a constant $M > 0$ such that

$$p(x) \leqslant M \text{ for all } x \in S, \text{ i.e. } p|S \text{ is bounded.}$$

If $p = \|\cdot\|$ is a ρ-norm then instead of p-bounded we also say *norm* bounded or *n.*bounded.

3.2.13. PROPOSITION. *Let* $X = (X, p)$ *be a* ρ*-seminormed LS. A subset* S *of* X *is bounded (i.e. t. bounded) iff it is* p*-bounded.*

PROOF. Assume that S is bounded. Since $B = \{x \in X : p(x) < 1\}$ is a nucleus of X, there is a $\lambda_0 \neq 0$ such that $S \subseteq \lambda_0 B$. If $x \in S, x = \lambda_0 a\ (a \in B)$ then

$$p(x) = |\lambda_0|^\rho p(a) < |\lambda_0|^\rho \text{ so that } p|S \text{ is bounded}$$

Conversely, assume that $p(x) < M$ for all $x \in S$. Let V be any nucleus. Then we can choose $\epsilon > 0$ such that

$$B_\epsilon = \{x : p(x) < \epsilon\} \subseteq V.$$

Write $\eta = \epsilon/M$. If $x \in S$ then

$$p(\eta^{\frac{1}{\rho}} x) = \eta p(x) < \frac{\epsilon}{M} M = \epsilon,$$

so that $\eta^{\frac{1}{\rho}} S \subseteq B_\epsilon$, or $S \subseteq \eta^{-\frac{1}{\rho}} B_\epsilon \subseteq \lambda V$, where $\lambda = \eta^{-\frac{1}{\rho}}$. This proves S is bounded.

3.2.14. COROLLARY. *Every ρ-seminormed LS X is locally bounded.*

PROOF. B_1 is a bounded nucleus for X.

3.2.15. COROLLARY. *If $\sum_{n=1}^{\infty} x_n$ is a convergent series in $X = (X,p)$, then the sets $\{x_1, x_2, \cdots\}$ and $\{s_1, s_2, \cdots\}$ where $s_n = \sum_{j=1}^{n} x_j$, are p-bounded.*

PROOF. Since the sequences (s_n) of partial sums $s_n = \sum_{j=1}^{n} x_j$ converges, by 2.3.7(a), (s_n) is bounded and so, by 3.2.13 $\{s_1, s_2, \cdots\}$ is p-bounded. Suppose that $p(s_n) \leqslant M$ for all n. Then

$$p(x_n) = p(s_n - s_{n-1}) \leqslant p(s_n) + p(s_{n-1}) \leqslant 2M,$$

whence $\{x_1, x_2, \cdots\}$ is bounded.

3.2.16. THEOREM (Banach-Steinhauss). *Let X be a ρ-Banach space and $\{Y_\alpha : \alpha \in a\}$ be a family of ρ-normed linear spaces and $T_\alpha : X \to Y_\alpha$ a family of continuous linear transformations. If for each $x \in X$ there is a $C_x > 0$ with $\|T_\alpha x\| \leqslant C_x$ for all α then there is a $C > 0$ such that: $\|T_\alpha\|$* † *$\leqslant C$ for all α.*

PROOF.†† For each positive integer n, write $X_n = \{x \in X : \|T_\alpha(x)\| \leqslant n \text{ for all } \alpha\}$. The continuity of the T_α and the fact that $\|\cdot\|$ is a ρ-norm (see 3.5.7) imply that X_n is closed in X. Furthermore

$$\bigcup_{n=1}^{\infty} X_n = X(\text{ since } x \in X_n \text{ if } n \geqslant C_x).$$

† $\|T_\alpha\|$ is the same as the bound $|T_\alpha|$ of T_α defined in 3.5.1.

†† Follows closely the exposition of the proof of the theorem for Banach spaces given by Simmons [26, p.239].

Since X is a complete metric space, by Baire's category theorem there is an $X_{n_0} \neq \emptyset$. It follows that X_{n_0} contain a ball $S_0 = \overline{B}_0 = \{x \in X : \|x - x_0\| \leqslant r_0\}$

Therefore $\|T_\alpha(x)\| \leqslant n_0$ for all $x \in S_0$, and all α or briefly

$$\|T_\alpha(S_0)\| \leqslant n_0 \text{ for all } \alpha. \qquad (*)$$

Clearly $S = \dfrac{S_0 - x_0}{r_0^{\frac{1}{\rho}}}$ is the closed unit ball.

If $x \in S$, $x = \dfrac{y - x_0}{r_0^{\frac{1}{\rho}}}$ $(y \in S_0)$, then

$$\|T_\alpha(x)\| = \frac{\|T_\alpha(y) - T_\alpha(x_0)\|}{r_0} \leqslant \frac{n_0 + n_0}{r_0} = \frac{2n_0}{r_0} = C(\text{ say })$$

(using $(*)$ and noting $x_0 \in S_0$). By 3.5.2, $\|T_\alpha\| \leqslant C$, completing the proof.

§ 3. Quarter-normed Algebras; (F) Algebras

3.3.1. DEFINITION. Let A be an algebra (over $\mathbb{K}$) and p a quarter-norm on A as a LS. Then (A, p) is called a *quarter-normed algebra* if p satisfies the condition

(QA) If $p(x_n - x)$, $p(y_n - y) \to 0$ then.

$p(x_n y_n - xy) \to 0$ (here $(x_n), (y_n)$ are sequences of elements and x, y elements, of A).

It is clear that (QA) is precisely the condition needed to make A a TA under the p-topology. More generally, if $\mathcal{P}$ is any family of quarter-norms on A with each $p \in \mathcal{P}$ satisfying (QA) then A is a TA under the $\mathcal{P}$ topology; A with the topology will be referred to as a $\mathcal{P}$-algebra.

3.3.2. PROPOSITION (Arens). *Let A be a WTA whose topology is induced by a complete quarter-norm p. Then A is a (complete) quarter-normed algebra.*

PROOF. It is clearly sufficient to prove that A is a TA (under the p-topology). Set

$$U_n = \{x \in A : p(x) \leqslant \frac{1}{n}\} \qquad (n = 1, 2, \cdots).$$

Then U_n is closed (since p is continuous) and symmetric (since $p(-x) = p(x)$). Thus, by virture of 3.1.21 (i), $\{U_n\}$ is a basis of closed symmetric nuclei. We shall now show that for any closed symmetric nucleus U and element $x \in A$ there is an integer n_0 such that

$$xU_{n_0} \subseteq U + U.$$

Write

$$A_n = \{x \in A : xU_n \subseteq U\}.$$

Then A_n is closed. For, if $x_n \in A_n$ and $x_n \to x$ in A then for any $y \in U_n$, $x_n y \to xy$ (since r_y is continuous), so that $xy \in U$ (U being closed), whence $x \in A_n$ and A_n is closed. Since $l_x : y \mapsto xy$ $(x \in A)$ is continuous at $y = 0$ there is an n such that $xU_n \subseteq U$, so that $x \in A_n$. This means that $\bigcup_{n=1}^{\infty} A_n = A$. Since A is complete semi-metric it is Baire (i.e. of the second category).[†] Consequently there is an n_0 such that

$$A_{n_0} \supseteq \overline{B} = \{x \in X : p(x - x_0) \leqslant r_0\}$$

for some $x_0 \in X$ and $r_0 > 0$. If $x \in A$ and $p(x) \leqslant r_0$ then

$$y = x + x_0 \in \overline{B} \subseteq A_{n_0};$$

in particular $x_0 = 0 + x_0 \in A_{n_0}$. It follows that

$$xU_{n_0} = yU_{n_0} - x_0 U_{n_0} \subseteq U + U.$$

Given any nucleus V, we can find, using 2.1.14 (vi), 2.1.16 (iii), a closed balanced (hence symmetric) nucleus U with $U + U \subseteq V$.

Therefore

$$xU_{n_0} \subseteq U + U \subseteq V.$$

This means that if

$$p(x) \leqslant r_0, p(y) \leqslant \frac{1}{n_0} \qquad \text{(i.e. } y \in U_{n_0})$$

[†] See [16, pp.200-1].

then $xy \in xU_{n_0} \subseteq V$, proving that the map $(x, y) \mapsto xy$ is continuous at $(0,0)$, whence by 2.2.3, A is a TA.

3.3.3. DEFINITION. Let S be a (multiplicative) semi-group and p a *non-negative* real-valued function on S. The function p is called *submultiplicative* or *sm.* if

$$p(x, y) \leqslant p(x)p(y) \qquad \text{for all } x, y \in S.$$

The function p is called *almost submultiplicative* or *a.sm.* if there is a constant $C > 0$ such that

$$p(x, y) \leqslant Cp(x)p(y) \qquad (x, y \in S).$$

The smallest constant C satisfying the above inequality is denoted by $|p|$, and we have:

$$p(xy) \leqslant |p|p(x)p(y).$$

It is clear that p is sm. iff $|p| \leqslant 1$. Also, for any a.sm. p, if $p^*(x) = |p|p(x)$ then clearly p^* is sm.

3.3.4. LEMMA. (Żelazko). *Let p on S be sm. Then, for any $x \in S$ either $p(x^n) \geqslant 1$ for all $n \geqslant 1$, or $p(x^n) \to 0$.*

PROOF. Suppose that $p(x^{n_0}) < 1$ for some integer $n_0 > 0$. Then

$$p(x^{kn_0}) \leqslant p(x^{n_0})^k \to 0 \qquad \text{as } k \to \infty.$$

Write $M = \max\{p(x^n) : 1 \leqslant n \leqslant n_0\}$. Given $\epsilon > 0$, choose N_0 such that for $k \geqslant N_0$, $p(x^{kn_0}) < \frac{\epsilon}{M}$. Write $N = N_0 n_0$. Then for $n \geqslant N$, $n = qn_0 + r$, with $q \geqslant N$ and $r < n_0$. Hence

$$p(x^n) \leqslant p(x^{qn_0})p(x^r) < \frac{\epsilon}{M}.M = \epsilon,$$

and $p(x^n) \to 0$.

3.3.5. DEFINITION. Let p be a non-negative real-valued function on the semi-group S. Set

$$\nu_p(x) = \limsup_{n \to \infty} p(x^n)^{\frac{1}{n}}; \nu_p^*(x) = \sup_n p(x^n)^{\frac{1}{n}}.$$

Then clearly

$$0 \leqslant \nu_p(x) \leqslant \nu_p^*(x) \leqslant \infty.$$

Moreover, it is evident that $\nu_p(x) < \infty$ iff $\nu_p^*(x) < \infty$.

3.3.6. LEMMA. (Gelfand †). *For an a. sm. p we have*

$$\nu_p(x) = \lim_{n\to\infty} p(x^n)^{\frac{1}{n}} < \infty \qquad (x \in S).$$

If p is sm. then we have also $\nu_p(x) = \inf_n p(x^n)^{\frac{1}{n}}$.

PROOF. Assume first that p is sm. and set

$$c = \inf_n p(x^n)^{\frac{1}{n}} \ (\leqslant p(x) < \infty).$$

Then, given $\epsilon > 0$, there is an integer $k > 0$ such that

$$p(x^k)^{\frac{1}{k}} < c + \epsilon. \tag{1}$$

For any integer $n > 0$, we write

$$n = q(n)k + r(n) \tag{2}$$

where $q(n), r(n)$ are non-negative integers with $0 \leqslant r(n) < k$. Then

$$\frac{1}{k} = \frac{q(n)}{n} + \frac{r(n)}{nk},$$

so that

$$\frac{q(n)}{n} \to \frac{1}{k} \text{ as } n \to \infty \tag{3}$$

(since $r(n)/nk < \frac{1}{n}$). Using (2) and sm. property of p we get

$$p(x^n)^{\frac{1}{n}} = (p(x^{kq(n)}).x^{r(n)}))^{\frac{1}{n}} \leqslant p(x^k)^{\frac{q(n)}{n}} p(x)^{\frac{r(n)}{n}} \tag{4}$$

Using (1), we have, for sufficiently large n,

$$c \leqslant p(x^n)^{\frac{1}{n}} \leqslant p(x^k)^{\frac{1}{k}} + \epsilon < c + 2\epsilon \text{ (using (1))}. \tag{5}$$

† He obtained the result for a sm. norm.

Therefore $$\lim_{n\to\infty} p(x^n)^{\frac{1}{n}} = c.$$

Now suppose that p is a.sm. and write $p^* = |p|p$. Then p^* is sm. and so $\nu_{p^*}(x) = \lim_{n\to\infty} p^*(x^n)^{\frac{1}{n}}$ exists. But

$$\lim_{n\to\infty} p^*(x^n)^{\frac{1}{n}} = \lim_{n\to\infty} |p|^{\frac{1}{n}} p(x^n)^{\frac{1}{n}} = \lim_{n\to\infty} p(x^n)^{\frac{1}{n}},$$

since $|p|^{\frac{1}{n}} \to 1$, and so $\nu_p(x) = \lim_{n\to\infty} p(x^n)^{\frac{1}{n}} < \infty$.

3.3.7. PROPOSITION. *Let p* † *be an a.sm. function on S. Then $\nu(x) = \nu_p(x)$ $(x \in S)$ has the following properties:*

(i) $0 \leqslant \nu(x) \leqslant |p|p(x)$; *in particular* $p(x) = 0 \Rightarrow \nu(x) = 0$.

(ii) (a) $\nu(xy) = \nu(yx)$ (b) $\nu(x^k) = \nu(x)^k$.

(iii) *If $x \leftrightarrow y$ then $\nu(xy) \leqslant \nu(x)\nu(y)$.*

(iv) *If u is an idempotent with $p(u) > 0$ then $\nu(u) = 1$.*

(v) *If S has unity* †† *e and $p \neq 0$ then $\nu(e) = 1$, and for any invertible $x \in S$, $\nu(x) > 0$.*

(vi) *If p is sm. then:*

$0 \leqslant \nu(x) \leqslant p(x)$, *and* $\nu(x) = p(x)$ *iff* $p(x^2) = p(x)^2$ *iff* $p(x^n) = p(x)^n$ *(for integers $n \geqslant 1$).*

PROOF. (i) We have $p(x^n) \leqslant |p|^{n-1}p(x)^n$, whence

$$\nu(x) = \lim_{n\to\infty} p(x^n)^{\frac{1}{n}} \leqslant \lim_{n\to\infty} |p|^{1-\frac{1}{n}} p(x) = |p|p(x).$$

(ii) (a) By observing that

$$(xy)^n = x(yx)^{n-1}y (n \geqslant 2)$$

we obtain
$$p((xy)^n) \leqslant |p|^2 p(x)p((yx)^{n-1})p(y) \tag{1}$$

† As always we assume $p \neq 0$.

†† i.e. S is a monoid.

so that

$$p((xy)^n)^{\frac{1}{n}} \leqslant |p|^{\frac{2}{n}} p(x)^{\frac{1}{n}} [p(yx)^{n-1}]^{\frac{1}{n}} p(y)^{\frac{1}{n}}. \tag{2}$$

If $p(x)$ or $p(y) = 0$ then it follows from (1) that

$$p((xy)^n) = 0, \nu(xy) = 0, \nu(yx) = 0$$

(interchanging x, y)) so that in this case we have $\nu(xy) = \nu(yx)$.

Next consider the case $p(x), p(y) > 0$. By allowing $n \to \infty$ in (2) we obtain $\nu(xy) \leqslant \nu(yx)$, so that by symmetry consideration we conclude that $\nu(xy) = \nu(yx)$.

(b) $\nu(x^k) = \lim_n p(x^{kn})^{\frac{1}{n}} = \lim_n (p(x^{kn})^{\frac{1}{kn}})^k = \nu(x)^k$.

(iii) Since $x \leftrightarrow y$, $(xy)^n = x^n y^n$. Therefore

$$\begin{aligned} \nu(xy) &= \lim_n p(x^n y^n)^{\frac{1}{n}} \leqslant \lim_n |p|^{\frac{1}{n}} p(x^n) p(y^n)^{\frac{1}{n}} \\ &\leqslant \nu(x)\nu(y). \end{aligned}$$

(iv) $\nu(u) = \lim_n p(u^n)^{\frac{1}{n}} = \lim_n p(u)^{\frac{1}{n}} = 1$ (since $p(u) > 0$).

(v) Since $p \neq 0$ there is an x_0 in S with $p(x_0) > 0$. Then $0 < p(x_0) = p(ex_0) \leqslant |p| p(e) p(x_0)$, whence $p(e) > 0$. Taking $u = e$ in (iv) we get $\nu(e) = 1$. Again since $1 = \nu(e) \leqslant \nu(x)\nu(x^{-1}), \nu(x) > 0$.

(vi) The inequality herein follows from (1), since p being sm., $|p| \leqslant 1$. We now assume that $p(x^2) = p(x)^2$. By iteration we get

$$p(x^{2^k}) = p(x)^{2^k} \qquad (k = 1, 2, \cdots).$$

For arbitrary integer $n > 1$, choose integer k such that $n < 2^k$ and write $2^k = n + m (m > 0)$. Then

$$\begin{aligned} p(x)^n p(x)^m &= p(x)^{2^k} = p(x^{2^k}) \leqslant p(x^n) p(x^m) \\ &\leqslant p(x^n) p(x)^m \end{aligned}$$

whence $p(x)^n \leqslant p(x^n)$, so that $p(x)^n = p(x^n)$. Hence it follows from its definition that $\nu(x) = p(x)$. Conversely, if $\nu(x) = p(x)$ then

$$p(x)^n = \nu(x)^n = \nu(x^n) \leqslant p(x^n),$$

so that $p(x^n) = p(x)^n$. This completes the proof of the proposition.

3.3.8. If A is an algebra, by considering the underlying multiplicative semi-group of A we can speak of an a.sm. or a sm. quarter-norm or ρ-seminorm on A.

3.3.9. LEMMA. *Let p be an a.sm. quarter-norm on an algebra A. Then (A, P) is a quarter-normed algebra. More generally if $\mathcal{P}$ is a family of a.sm. quarter-norms on A then $(A, \mathcal{P})$ is a $\mathcal{P}$-algebra.*

PROOF. Using the identity

$$x_n y_n - xy = (x_n - x)(y_n - y) + (x_n - x)y + x(y_n - y) \qquad (*)$$

and a.sm. property of p we get

$$p(x_n y_n - xy) \leqslant |p|\{p(x_n - x)p(y_n - y) + p(x_n - x)p(y) + p(x)(p(y_n) - p(y))\}. \quad (**)$$

It follows from $(**)$ that p satisfies (QA) of 3.3.1, whence (A, p) is a TA. Similarly, since each $p \in \mathcal{P}$ satisfies $(**)$, $(A, \mathcal{P})$ is a TA.

3.3.10. LEMMA. *Let A be an algebra. Then we have:*

(i) *If p is an a.sm. quarter-norm (respy. ρ-seminorm) then $tp\,(t > 0)$ is also an a.sm. quarter-norm (respy. ρ-seminorm) with $t|tp| = |p|$, $tp \sim p$.*

(ii) *If p is a sm. quarter-norm (respy. ρ-seminorm) and $t \geqslant 1$ then tp is also a sm. quarter-norm (respy. ρ-seminorm).*

(iii) *If p is a.sm. then $p^* = |p|p$ is sm. and $p^* \sim p$.*

PROOF. (i) Clearly $p' = tp$ is a quarter-norm (respy. ρ-seminorm). Also,

$$p'(xy) = tp(xy) \leqslant t|p|p(x)p(y) \leqslant t^{-1}|p|p'(x)p'(y).$$

So we have $|p'| \leqslant t^{-1}|p|$. Similarly, by considering $p = t^{-1}p'$ we get $|p| \leqslant t|p'|$. Combining the two inequalities we obtain $|p| = t|tp|$. Also, by 3.1.19 (b), $tp \sim p$.

(ii) Since $t \geqslant 1$ we have $t^2 \geqslant t$ so that tp is sm.

(iii) By (i) p^* is a quarter-norm (respy. ρ-seminorm). Further,

$$p^*(xy) = |p|p(xy) \leqslant |p|^2 p(x)p(y) = p^*(x)p^*(y).$$

Finally, by 3.1.19 (b), $p^* \sim p$.

3.3.11. LEMMA. *If $p_j (j = 1, \cdots, n)$ are a.sm. quarter-norms (respy. ρ-seminorms) then $q = p_1 \vee \cdots \vee p_n$ is also an a.sm. quarter-norm (respy. ρ-seminorm) with $|q| \leqslant \max_j |p_j|$.*

PROOF. By 3.1.18 (respy. 3.2.3) q is a quarter-norm (respy. ρ-seminorm). Further, if $C = \max |p_j|$ then

$$p_j(xy) \leqslant |p_j| p_j(x) p_j(y) \leqslant C p_j(x) p_j(y) \qquad (j = 1, \cdots, n)$$

whence $q(xy) \leqslant Cq(x)q(y)$, so that $|q| \leqslant C$.

3.3.12. PROPOSITION. (Żelazko). *Let $A = (A, p)$ be a unital quarter-normed algebra with p a.sm. Then A is locally bounded.*

PROOF. Write $U = B_1 = \{x \in A : p(x) < 1\}$.
If $x_n \in U$, $\lambda_n \in \mathbb{K}$ with $\lambda_n \to 0$, then

$$p(\lambda_n x_n) = p_n(\lambda_n e.x_n) \leqslant |p|p(\lambda_n e)p(x_n) \leqslant |p|p(\lambda_n e) \to 0.$$

Hence, by 2.1.22, U is bounded and so A is locally bounded.

3.3.13. DEFINITION. A quarter-normed algebra $A = (A, p)$ is called a pre-(F) *algebra* if the underlying LS of A is a pre-(F) space (i.e. if p is faithful). A pre-(F) algebra is called a (F) *algebra* if p is complete.

If A is a pre-(F) or (F) algebra its quarter-norm p will be usually denoted by $|\cdot|$ and called a (F) *norm* (cf. remark in 3.1.14). If the (F) norm is sm. we call A a *sm. pre-*(F) algebra or *sm.* (F) algebra as the case may be.

3.3.14. Examples of (F) algebras

(i) Let $\mathbb{K}^\infty$ denote the algebra of all infinite sequences $x = (\alpha_n)$ of elements α_n from $\mathbb{K}$, the operations being coordinatewise, i.e. if $x = (\alpha_n)$, $y = (\beta_n)$ and $\lambda \in \mathbb{K}$ then

$$x + y = (\alpha_n + \beta_n),\ xy = (\alpha_n \beta_n),\ \lambda x = (\lambda \alpha_n).$$

Set

$$|x|_\infty = \sum_{n=1}^{\infty} \frac{1}{2^n} \frac{|\alpha_n|^{\frac{1}{n}}}{1+|\alpha_n|^{\frac{1}{n}}},$$

where $|\alpha|$ denotes the norm or absolute value of α in $\mathbb{K}(=\mathbb{R}$ or $\mathbb{C})$. It can be shown that $(\mathbb{K}^\infty, |\cdot|_\infty)$ is a commutative unital (F) algebra over $\mathbb{K}$.

(ii) Denote by $\mathbb{H}^\infty$ the real algebra of all sequences $x = (q_n)$ of elements from $\mathbb{H}$ (with coordinate-wise operations), where $\mathbb{H}$ stands for the algebra of Hamilton quaternions; $\mathbb{H} = \mathbb{R} + \mathbb{R}i + \mathbb{R}j + \mathbb{R}k$, $i^2, j^2, k^2 = -1$, $ij = k = -ji$, $jk = -kj$, $ki = -ik$.
Define $|x|_\infty$ as in example (i) with q_n replacing α_n, and $|q_n| = (\alpha_0^2 + \alpha_1^2 + \alpha_2^2 + \alpha_3^2)^{\frac{1}{2}}$ if $q_n = \alpha_0 + \alpha_1 i + \alpha_2 j + \alpha_3 k$. Then $(\mathbb{H}^\infty, |\cdot|_\infty)$ is a real unital (F) algebra which is not commutative.

(iii) Denote by $\mathcal{E}$ the algebra (under pointwise operations) of all entire functions $f = f(z)$ of a complex variable z; $\mathcal{E}$ is a unital complex algebra. The following metric $|\ |_G$ introduced by V.G.Iyer is an (F) metric on $\mathcal{E}$: if $f = f(z) = \sum_{n=0}^{\infty} \alpha_n z^n$, $|f|_G = \sup\{|\alpha_0|, |\alpha_n|^{\frac{1}{n}}, n \geqslant 1\}$ (see [6′, Theorem 1, Remark 2]). It is known that the convergence under this metric is the same as uniform convergence over compacta in $\mathbb{C}$ (ibid, Theorem 3). The algebra $\mathcal{E} = (\mathcal{E}, |\cdot|_G)$ is a unital (F) algebra. $\mathcal{E}$ is not locally bounded (ibid, Theorem 2) and hence, by 3.3.2, the (F) norm $|\cdot|_G$ is not a.sm.. Finally, $\mathcal{E}$ is separable since the set of polynomials with complex† rational† coefficient is dense in $\mathcal{E}$.

(iv) The entire functions form an algebra also with respect to pointwise linear operations and Hadamard multiplication: if $f = \sum_{n=0}^{\infty} \alpha_n z^n$, $g = \sum_{n=0}^{\infty} \beta_n z^n$ the Hadamard product $f \times g$ is defined by

† i.e. having rational real and imaginary parts.

$$f \times g = \sum_{n=0}^{\infty} \alpha_n \beta_n z^n.$$

(Note that pointwise multiplication of f and g corresponds to Cauchy multiplication of the associated power series.) We denote the above algebra by $\mathcal{E}^{(\times)}$ · $\mathcal{E}^{(\times)}$ has no unity element (since $1 + z + z^2 + \cdots$ is not entire). Also, we have clearly

$$|f \times g|_G \leqslant |f|_G |g|_G.$$

So $\mathcal{E}^{(\times)} = (\mathcal{E}^{(\times)}, |\cdot|_G)$ is a sm. (F) algebra which again is not locally bounded (note that both $\mathcal{E}, \mathcal{E}^{(\times)}$ have the same topology).

(v) Let $M = M([0,1])$ be the algebra of (equivalence classes) of almost everywhere defined Lebesgue measurable complex functions on $[0,1]$. For $f \in M$, if we set $|f| = \int_0^1 \frac{|f(t)|}{1+|f(t)|} dt$ then it is an (F) norm and M is an (F) algebra. The norm or metric convergence in M is the same as convergence in measure (see [28, pp.116-17]).

3.3.15. LEMMA. *The unitization A_1 of a quarter-normed algebra (A, p) is canonically a quarter-normed algebra (A_1, p_1) with*

$$p_1(\lambda e_1 + x) = |\lambda| + p(x) \ (x \in A,\ \lambda \in \mathbb{K}), \text{so that } p_1(e_1) = 1. \quad (*)$$

Further, A_1 is a pre-(F) algebra or (F) algebra according as A is a pre-(F) algebra or (F) algebra.

PROOF. We already know that A_1 is a TA under the product topology of $\mathbb{K} \times A$ (see 2.2.9). It remains to see that p_1 (defined as in $(*)$) induces the topology of A_1. But this readily follows since we have $p_1(\lambda e_1 + x_n - (\lambda e_1 + x)) = |\lambda_n - \lambda| + p(x_n - x)$. If p is faithful, and $p_1(\lambda e_1 + x) = 0$ then $|\lambda| = 0$, $p(x) = 0$, so that $\lambda = 0, x = 0$, whence p_1 is faithful. Again when p is complete, it is easy to see, using $(*)$, that p_1 is complete.

3.3.16. Remark. p_1 may fail to be sm. when p is sm. A counter example is provided by the unitization $(\mathcal{E}^{(\times)})_1$ of the algebra $\mathcal{E}^{(\times)}$ of 3.3.14 (iv). If $p_1 = |\cdot|_1$ then p_1 cannot be sm., for in that case by 3.3.12 $(\mathcal{E}^{(\times)})_1$ is locally bounded whence also $\mathcal{E}^{(\times)}$ which is impossible (see 3.3.14 (iv)). So $|\cdot|$ is not sm. (nor even a.sm.).

3.3.17. PROPOSITION. *The metric completion $\hat{A}$ of a sm. pre-(F) algebra $A = (A, p)$ is canonically a sm. (F) algebra $\hat{A} = (\hat{A}, \hat{P})$.*

PROOF. If $\hat{x}, \hat{y} \in \hat{A}$ and $\hat{x} = \lim x_n$, $\hat{y} = \lim y_n$ $(x_n, y_n \in A)$ we define

$$\hat{x} + \hat{y} = \lim(x_n + y_n),\ \hat{x}\hat{y} = \lim x_n y_n,\ \lambda\hat{x} = \lim \lambda x_n.$$

The above limits exist since the sequences defining them are C-sequences, as can be verified by using the sa. and sm. properties of p as well as the property (Q6b) of p (given in 3.1.8). We extend p to $\hat{p}$ on $\hat{A}$ by defining $\hat{p}(\hat{x}) = \lim p(x_n)$ where $x_n \in A$, $x_n \to \hat{x}$; the limit exists since $p(x_n)$ is a C-sequence (as can be seen using 3.1.2 (ii)). It is straightforward to verify that $\hat{p}$ is a (F) norm. Further, $\hat{p}$ is sm. since

$$p(\hat{x}\hat{y}) = \lim_{n\to\infty} p(x_n y_n) \leqslant \lim_{n\to\infty} p(x_n)p(y_n) \leqslant p(\hat{x})p(\hat{y}).$$

This complete the proof.

3.3.18. LEMMA. $(A, |\cdot|)$ *be a sm. pre-(F) algebra Then:*

(i) *An $x \in A$ with $|x| < 1$ is q. invertible iff the series $\sum_{n=1}^{\infty}(-1)^n x^n$ converges and then the q. inverse x' is given by*

$$x' = \sum_{n=1}^{\infty}(-1)^n x^n. \qquad (*)$$

In particular, if A is complete (i.e. A is an (F) algebra) then every x with $|x| < 1$ is q. invertible with its q. inverse given by $(*)$.

(ii) *Let A be unital with unity e. Then an element $x \in A$ with $|e - x| < 1$ is invertible iff the series $\sum_{n=0}^{\infty}(e - x)^n$ converges,* *and then the inverse* x^{-1} *is given by*

$$x^{-1} = \sum_{n=0}^{\infty} (e - x)^n. \qquad (**)$$

In particular, if A *is an* (F) *algebra every* x *with* $|e-x| < 1$ *is invertible with* x^{-1} *given by* $(**)$.

PROOF. (i) In view of 2.2.17 it is enough to prove the "only if" part, i.e. if x' exists then the series $\sum(-1)^n x^n$ converges to x'. Since $|x| < 1$, by 3.3.4, $|x^n| \to 0$, so that $x^n \to 0$ in A. Therefore

$$x \circ \sum_{n=1}^{N} (-1)^n x^n = (-1)^N x^N, \text{ whence, } \sum_{n=1}^{N} (-1)^n x^n = x' \circ (-1)^N x^N$$

Making $N \to \infty$ we get $\sum_{n=1}^{\infty} (-1)^n x^n = x' \circ 0 = x'$.

Now assume that $|\cdot|$ is complete. Since $|x| < 1$ the numerical series $\sum_{n=1}^{\infty} |x|^n$ converges, whence the series $\sum_{n=1}^{\infty} (-1)^n x^n$ converges absolutely and consequently, by 3.1.24, it converges and so has x' for its limit.

(ii) The proofs are similar to that of (i).

3.3.19. PROPOSITION. *Let* $A = (A, |\cdot|)$ *be a sm.* (F) *algebra–in particular a* ρ*-Banach algebra-and* $\nu = \nu_{|\cdot|}$. *If* $\nu(x) < 1$ *(in particular if* $|x| < 1$*) then* x *is q. invertible with*

$$x' = \sum_{n=1}^{\infty} (-1)^n x^n, \qquad (*)$$

where the series converges absolutely. Moreover, if x *is q. invertible and* $|y - x| < (1 + |x'|)^{-1}$ *then* y *is q. invertible. Hence, every q. invertible* x *has an open (ball) neighbourhood* $B(x, (1 + |x|)^{-1})$ *consisting of q. invertible elements.*

PROOF. Since $\nu(x) < 1$ we can choose η such that $\nu(x) < \eta < 1$. Then $|x^n|^{\frac{1}{n}} < \eta < 1$ for $n \geqslant N$. This implies that

$$\sum_{n=N}^{\infty} |x^n| \leqslant \sum_{n=N}^{\infty} \eta^n < \infty \qquad (\text{since } \eta < 1).$$

If follows that the series $\sum_{n=1}^{\infty}(-1)^n x^n$ converges absolutely and hence also in A (by 3.1.24). By 3.3.18 (i), $x' = \sum_{n=1}^{\infty}(-1)^n x^n$ is q. inverse of x.

Assume that $x \in A$ is q. invertible and $y \in A$ with $|y-x| < (1+|x'|)^{-1}$. Using $x' \circ x = 0$ we have
$x' \circ y = x' + y + x'y = (-x - x'x) + y + x'y = y - x + x'(y-x)$.

Therefore

$$|x' \circ y| < |y-x| + |x'||y-x| = (1+|x'|)|y-x| < 1.$$

Similarly, $|y \circ x'| < 1$. It follows that $x' \circ y$, $y \circ x'$ are q. invertible, whence by 1.1.30 that y is q. invertible, as required.

3.3.20. COROLLARY. *Let $A = (A, |\cdot|)$ be a unital sm. (F) algebra with unity e. If $\nu(x) < 1$ (in particular $|x| < 1$) then:*

(i) *$e \pm x$ are invertible with*

$$(e+x)^{-1} = \sum_{n=0}^{\infty}(-1)^n x^n \tag{$*$}$$

$$(e-x)^{-1} = \sum_{n=0}^{\infty} x^n \tag{$**$}$$

the series in $(*), (**)$ converging absolutely.

(ii) *If $\nu(e-y) < 1$ (in particular $|e-y| < 1$) then y is invertible with*

$$y^{-1} = \sum_{n=0}^{\infty}(e-y)^n \tag{$***$}$$

PROOF. The representation $(*)$ follows from $(*)$ of 3.3.19 by using the fact that $e + x$ is invertible iff x is q. invertible. The representation $(**)$ is, of course, got from $(*)$ by replacing x by $-x$. Finally the representation $(***)$ is got from $(**)$ by putting $x = e - y$.

3.3.21. COROLLARY. *If $x \in A$ is invertible and $|y-x| < |x^{-1}|^{-1}$, then y is invertible with*

$$y^{-1} = \sum_{n=0}^{\infty}[x^{-1}(x-y)]^n x^{-1}.$$

Hence, every invertible x has an open neighbourhood $B(x, |x^{-1}|^{-1})$ comprising invertible elements.

PROOF. Since

$$|e - x^{-1}y| = |x^{-1}(x-y)| \leqslant |x^{-1}|\,|x-y| < 1$$

$x^{-1}y$ is invertible (by 3.3.20 (ii)), and we have

$$y^{-1}x = (x^{-1}y)^{-1} = \sum_{n=0}^{\infty}(e - x^{-1}y)^n = \sum_{n=0}^{\infty}[x^{-1}(x-y)]^n.$$

Hence

$$y^{-1} = \sum_{n=0}^{\infty}[x^{-1}(x-y)]^n x^{-1}.$$

3.3.22. PROPOSITION. *Let p be an a.sm. quarter-norm on an algebra A and x, y q. invertible elements of A with q. inverses x', y' respy. Then:*

$$p(y'-x')[1-(1+|p|p(x'))|p|p(y-x)] \leqslant (1+|p|p(x'))^2 p(y-x). \quad (*)$$

$$p(y'-x')[1-(1+p(x'))p(y-x)] \leqslant (1+p(x'))^2 p(y-x) \text{ if } |p| \leqslant 1. \quad (**)$$

PROOF.[†] Write $y = x + a$, $y' = x' + b$. Then $0 = y \circ y' = (x+a)\circ(x'+b)$ which, on using $x \circ x' = 0$, reduces to $a + b + xb + ax' + ab = 0$.

Therefore

$$x \circ b = x - a - ax' - ab.$$

[†] The principle underlying the proof is essentially due to Arens who obtained (**) of 3.3.23 for normed algebras.

By pre-multiplying (with respect to $\circ$ operation) by x' we get

$$b = x' \circ (x - a - ax' - ab) = -a - ax' - ab - x'a - x'ax' - x'ab,$$

using $x' \circ x = 0$. It follows that

$$\begin{aligned} p(b) \quad &\leqslant \quad p(a) + |p|p(a)p(x') + |p|p(a)p(b) + |p|p(x')p(a) \\ &\quad +|p|^2|p(x')^2p(a) + |p|^2p(x')p(a)p(b), \end{aligned}$$

so that

$$\begin{aligned} p(b)[1 - |p|p(a) - |p|^2p(x')p(a)] \quad &\leqslant \quad p(a) + 2|p|p(a)p(x') \\ &\quad +|p|^2p(x')^2p(a) \\ &= (1 + |p|p(x'))^2p(a) \end{aligned}$$

which is same as $(*)$. When $|p| \leqslant 1$ it is clear that the inequality remains unchanged if we drop the terms $|p|, |p|^2$, whence we obtain $(**)$.

3.3.23. PROPOSITION. *If* $A = (A, p)$, *where* p *is an a.sm. quater-norm, is unital and* $x, y \in A$ *are invertible then we have:*

$$p(y^{-1} - x^{-1})[1 - |p|^2p(y - x)p(x^{-1})] \leqslant |p|^2p(x^{-1})^2p(y - x) \quad (*)$$

$$p(y^{-1} - x^1)(1 - p(y - x)p(x^{-1})) \leqslant p(x^{-1})^2p(y - x) \text{ if } |p| \leqslant 1. \quad (**)$$

PROOF. The proof of $(*)$ is similar to that of 3.3.22 and even simpler. Writing $a = y - x, b = y^{-1} - x^{-1}$ we get $e = yy^{-1} = (x + a)(x^{-1} + b)$ which reduces to $xb = -ax^{-1} - ab$, i.e. $b = -x^{-1}ax^{-1} - x^{-1}ab$. Applying p to both sides, using a.sm. property of p, and rearranging terms we get $(*)$. Once again $(**)$ follows from $(*)$ by here omitting the term $|p|^2$ in $(*)$.

3.3.24. COROLLARY (Bonsall-Duncan). *For a sm.* p, *if* $p(y - x) \leqslant \frac{1}{2}p(x^{-1})^{-1}$ *then*

$$p(y^{-1} - x^{-1}) \leqslant 2p(x^{-1})^2p(y - x).$$

Proof. We have $1 - p(y - x)p(x^{-1}) \geqslant 1 - \frac{1}{2} = \frac{1}{2}$, whence the required inequality follows from $(**)$ of 3.3.23.

§ 4. ρ-Seminormed Algebras; ρ-Banach Algebras

3.4.1. LEMMA. *If p is an a.sm. ρ-seminorm on an algebra A and q any ρ-seminorm on A with $q \sim p$. Then q is also a.sm.*

PROOF. By virtue of 3.2.11 (iii), there are constants $C_1, C_2 > 0$ with $C_1 p \leqslant q \leqslant C_2 p$.
Therefore

$$q(xy) \leqslant C_2 p(xy) \leqslant C_2 |p| p(x) p(y) \leqslant C_2 |p| C_1^{-2} q(x) q(y),$$

whence q is a.sm.

3.4.2. LEMMA. *Let p be a ρ-seminorm on an algebra A such that the map*
$m^{\#} : (x,y) \mapsto xy$ is continuous at $(0,0)$ in the p-topology, Then:

(i) ker p *is a closed bi-ideal of* A;

(ii) p *is a.sm.*

PROOF. (i) By 2.1.13, 3.1.21 (ii), $\ker p$ is a closed subspace of A. It remains to show that it is a bi-ideal. Since $m^{\#}$ is continuous at (0,0), given $\epsilon > 0$, there is a $\delta > 0$ such that

$$p(x), p(y) \leqslant \delta \Rightarrow p(xy) \leqslant \epsilon. \tag{1}$$

For a given $y \in A$ we can choose n such that

$$p\left(\frac{y}{n}\right) = \frac{1}{n^{\rho}} p(y) \leqslant \delta \tag{2}$$

(the choice being possible since $\rho > 0$).
If $p(x) = 0$ then

$$p(nx) = n^{\rho} p(x) = 0. \tag{3}$$

From (1), (2), (3) we obtain

$$p(xy) = p\left(nx \cdot \frac{y}{n}\right) \leqslant \epsilon.$$

Since ϵ is arbitrary, $p(xy) = 0$; similarly $p(yx) = 0$. This completes the proof.

(ii) Suppose that $x, y \in A$, $p(x), p(y) \neq 0$. Writing

$$x_1 = \delta^{\frac{1}{\rho}}/p(x)^{\frac{1}{\rho}}, \quad y_1 = \delta^{\frac{1}{\rho}}/p(y)^{\frac{1}{\rho}}$$

we find that $p(x_1) = p(y_1) = \delta$, whence by (1), $p(xy) \leqslant \epsilon$ which reduces to

$$p(xy) \leqslant Cp(x)p(y) \tag{4}$$

where $C = \epsilon/\delta^2$. If $p(x)$ or $p(y) = 0$, then by (i), $p(xy) = 0$, so that (4) holds trivially. Thus (4) holds for all x, y and p is a.sm.

3.4.3. COROLLARY. *Let p be a ρ-seminorm on an algebra A. Then A is a TA under the p-topology iff p is a.sm.*

PROOF. The "only if" part clearly follows from 3.4.2. The "if" part is a consequence of 3.3.9.

3.4.4. A TA A whose topology is induced by a ρ-seminorm p is called a *ρ-seminormed algebra* (*semi-normed algebra* if $\rho = 1$) and we write $A = (A, p)$. A complete ρ-seminormed algebra is called a *ρ-semi-Banach algebra* (*semi-Banach algebra* if $\rho = 1$). In view of 3.4.3 the p in (A, p) must be necessarily a.sm. . Such a conclusion is not possible if p were only a quarter-norm (cf. 3.3.16).

Since, by 3.3.10, $p^* = |p|p$ is a sm. ρ-seminorm with $p^* \sim p$, *we can assume when necessary,* without any loss of generality, that the ρ-seminorm p of (A, p) is sm.

The meaning of a ρ-normed or a ρ-Banach algebra is clear. Of course, when $\rho = 1$, a ρ-normed (respy. ρ-Banach) algebra is just a normed (respy. Banach) algebra.

Note that a ρ-normed algebra is a *pre-*(F) algebra and a ρ-Banach algebra an (F) algebra.

3.4.5. LEMMA. *Let (A, p) be a ρ-seminormed algebra. Then $\nu = \nu_p$ has the property*

$$\nu(\lambda x) = |\lambda|^{\rho}\nu(x).$$

PROOF. $\nu(\lambda x) = \lim_n p(\lambda^n x^n)^{\frac{1}{n}} = \lim_n |\lambda|^{\rho} p(x^n)^{\frac{1}{n}} =$

$|\lambda|^\rho \nu(x)$.

3.4.6. Examples of ρ-seminormed algebras and ρ-Banach algebras.

(i) Let S be any set. The set $\mathbb{K}^S$ of all $\mathbb{K}$-valued functions f on S, under pointwise operations, is a unital commutative algebra over $\mathbb{K}$. Fix a point s_0 in S and define $p_0(f) = |f(s_0)|$. Then p_0 is a sm. semi-norm on $\mathbb{K}^s$ and $(\mathbb{K}^s, p_0)$ is a sm. semi-Banach algebra. For $0 < \rho \leqslant 1$, if we set, $p_0^\rho(f) = |f(s_0)|^\rho$ then $(\mathbb{K}^s, p_0^\rho)$ is a ρ-semi-Banach algebra.

(ii) $\mathbb{K}^n$ is a Banach algebra under the norm

$$\|x\| = |\lambda_1| + \cdots |\lambda_n|, \text{ where } x = (\lambda_1, \cdots, \lambda_n) \in \mathbb{K}^n.$$

(iii) The set $\mathbb{H}^S$, where $\mathbb{H}$ is the 4-dim real algebra of Hamilton quaternions, is under pointwise-operations an algebra over $\mathbb{R}$ which is not commutative. Defining, as above, $p_0(f) = |f(s_0)|$, where $|\cdot|$ is now the quaternion norm (see 3.314(ii)) we obtain a real ρ-semi-Banach algebra $(\mathbb{H}^S, p_0^\rho)$ which is not commutative.

(iv) If $B = B(S, \mathbb{K})$ denotes the set of all bounded $\mathbb{K}$-valued functions on S the B is a Banach algebra over $\mathbb{K}$ under the sup (=supremum) norm: $\|f\|_\infty = \sup_{s \in S} |f(0)| (f \in B)$. It is commutative. Also $(B, \|\cdot\|_\infty^\rho)$ is a ρ-Banach algebra (the ρ-norm property of $\|\cdot\|_\infty^\rho$ follows from 3.2.9). Similarly, we have the real ρ-Banach algebra $B(S, \mathbb{H})$ of all bounded $\mathbb{H}$-valued functions with ρ-norm $\|\cdot\|_\infty^\rho : \|f\|_\infty^\rho = \sup_s |f(s)|^\rho$, where $|\cdot|$ denotes the absolute value in $\mathbb{H}$ (see 3.3.14(ii)) This algebra is not commutative.

(v) All sequence (x_n) of elements in $\mathbb{K}$ such that $\sum_{n=1}^{\infty} |x_n|^\rho$ $(0 < \rho \leqslant 1)$ form under coordinate-wise operations a commutative ρ-Banach algebra $l^\rho = l^\rho(\mathbb{K})$ under the ρ-norm $\|x\|_\rho = \sum_n |x_n|^\rho$, which is sm. $(|xy|_\rho = \sum_n |x_n y_n|^\rho = \sum_n |x_n|^\rho |y_n|^\rho \leqslant \|x\|_\rho \|y\|_\rho)$. More generally, if S is any set, denote by $l^\rho(S)$ the set of all $\mathbb{K}$-valued function f on S

such that the generalized sum $\sum_s |f(s)|^\rho < \infty$. Then $l^\rho(S)$ is a ρ-Banach algebra under the norm $\|f\|_\rho = \sum_s |f(s)|^\rho$, which again is sm.: $\|fg\|_\rho \leqslant \|f\|_\rho \|g\|_\rho$.

(vi) Let $L^1 = L^1[0,1]$ denote the space of (equivalent classes) of absolutely Lebesgue integrable functions on [0,1]. L^1 is a Banach space under the norm $\|f\| = \int_0^1 |f|(t)dt = \int_0^1 |f(t)|dt$. L^1 is closed for a multiplication '$*$' called convolution defined as follows:

$$f * g(s) = \int_0^s f(s-t)g(t)dt$$

($f * g$ exists for almost all s by Fubini's theorem and so $f * g \in L^1$).

It is straightforward to check that the operation $*$ is associative and commutative. Thus L^1 is a commutative Banach algebra; we have $\|f * g\| \leqslant \|f\|\|g\|$. (See [10,p 17]).

(vii) Let S be a compact Hausdorff space $C(S) = C(S, \mathbb{K})$ the algebra of $\mathbb{K}$-valued continuous functions on S. Then $C(S)$ is a Banach algebra under the 'sup' norm. If S is a locally compact Hausdorff space, denote by $C_0(S)$ the algebra of $\mathbb{K}$-valued continuous function f which vanish at ∞ (i.e. f has the property that for any $\epsilon > 0$ there is a compact set $K = K_\epsilon$ such that $f = 0$ on $S \backslash K$). $C_0(S)$ is also a Banach algebra under the sup norm; of course, if S is compact, $C_0(S) = C(S)$.

If S is locally compact Hausdorff and S_∞ is its one-point compactification then it is easy to see that $C(S_\infty)$ is the unitization of $C_0(S)$.

3.4.7. PROPOSITION (Żelazko). *Let G be a discrete TG. Then the set $L^\rho = L^\rho(G) = L^\rho(G, \mathbb{K}), 0 < \rho \leqslant 1$, of all $\mathbb{K}$-valued functions $x = x(s) (s \in G)$ such that $\sum_s |x(s)|^\rho < \infty$, is an algebra (over $\mathbb{K}$) under pointwise linear operations and convolution as multiplication: if $x, y \in L^\rho$ then its convolution product is defined by $x * y(s) = \sum_t x(t)y(t^{-1}s)$. For $s \in G$, set $x_s(t) = 1$ or 0 according as $s = t$ or $s \neq t$ Then $x_s \in L^\rho$, and the map $s \mapsto x_s$ is 1-1 and $x_s * x_t = x_{st}$ (so that G is multiplicatively embedded in*

$L^\rho(G)$). *Moreover,* L^ρ *is a* ρ*-Banach algebra under the* ρ*-norm* $\|x\|_\rho = \sum_s |x(s)|^\rho$.

PROOF. That L^ρ is a LS and $\|\cdot\|_\rho$ is a ρ-norm are clear (the subadditivity of $\|\cdot\|^\rho$ depends on the inequality $(s+t)^\rho \leqslant s^\rho + t^\rho$, if $s,\ t \geqslant 0, 0 < \rho \leqslant 1$). If $x, y \in L^\rho$ then

$$\begin{aligned}\|x * y\|_\rho &= \sum_s |\sum_t x(t)y(t^{-1}s)|^\rho \leqslant \sum_{s,t} |x(t)|^\rho |y(t^{-1}s)|^\rho \\ &\leqslant \sum_t |x(t)|^\rho (\sum_s |y(t^{-1}s)|^\rho) = \sum_t |x(t)|^\rho \|y\|^\rho \\ &\leqslant \|x\|_\rho \|y\|_\rho.\end{aligned}$$

It follows that L^ρ is closed for convolution and that $\|\cdot\|_\rho$ is sm. The proof of completeness of L^ρ is straightforward (and will be omitted). Thus L^ρ is a ρ-Banach algebra. Further it is clear that $x_s \in L^\rho$ and $s \mapsto x_s$ is $1-1$. Finally, by a simple computation, $x_s * x_t = x_{st}$.

3.4.8. COROLLARY. $L^\rho(G)$ *is commutative iff* G *is commutative.*

PROOF. If G is commutation then

$$\begin{aligned}x * y(s) &= \sum_t x(t)y(t^{-1}s) = \sum_t y(t^{-1}s)x(t) \\ &= \sum_u y(u)x(su^{-1}) = \sum_u y(u)x(u^{-1}s) = y * x(s).\end{aligned}$$

On the other hand, if L^ρ is commutative we have $x_{st} = x_s * x_t = x_t * x_s = x_{ts}$, $st = ts$ (since $s \mapsto x_s$ is 1-1) and G is commutative.

3.4.9. Remark. Zelazko [31, p.12, Theorem 29] has shown that if G is a locally compact group then for $0 < \rho < 1, L^\rho = L^\rho(G)$ the space of (left) Haar measurable function $x(s)$ on G such that $\int |x(s)|^\rho d_s < \infty$ is closed for convolution multiplication iff G is discrete. Thus, under convolution, L^ρ is a ρ-Banach algebra only when G is discrete. L^1, on the other hand, is always a Banach algebra (under convolution).

3.4.10. PROPOSITION. *Let* W^ρ *denote the set of all continuous complex-valued* 2π*-periodic functions* $f = f(t)$ *on* $\mathbb{R}$ *having* *a Fourier expansion* $f = \sum_{-\infty}^{\infty} \hat{f}(n)e^{int}$ *with* $\sum_{-\infty}^{\infty} |\hat{f}(n)|^{\rho} < \infty$. *Then* W^{ρ} *is under pointwise linear and multiplication operations a commutative* ρ*-Banach algebra (the Wiener-Żelazko algebra) with* ρ*-norm* $\|x\|_{\rho} = \sum_{\infty}^{\infty} |\hat{f}(n)|\|^{\rho}$.

PROOF. Clearly W^{ρ} can be identified with the algebra $L^{\rho}(\mathbb{Z}, \mathbb{C}) = L^{\rho}(\mathbb{Z}) = l^{\rho}(\mathbb{Z})$. By 3.4.7, $L^{\rho}(\mathbb{Z})$ is ρ-Banach which moreover, by 3.4.8 is commutative (since $\mathbb{Z}$ is commutative).

3.4.11. LEMMA. *Let* A *be an algebra (over* $\mathbb{K}$*) and* A_1 *its unitization. Then a* ρ*-seminorm* p *on* A *has a canonical extension* p_1 *on* A_1 *with* $p_1(e_1) = 1$. *If* p *is a.sm. so is* p_1 *and* $|p_1| = \max\{1, |p|\}$. *In particular,* p_1 *is sm. when* p *is sm. Further,* p_1 *is faithful or complete according as* p *is faithful or complete.*

PROOF. Define p_1 by setting for $x_1 \in A_1, x_1 = \lambda e_1 + x$ $(x \in A)$ $p_1(x_1) = |\lambda|^{\rho} + p(x)$. Clearly $p_1(e_1) = 1$ and p_1 is a ρ-seminorm on A_1. If p is a.sm. and $y_1 = \mu e_1 + y \in A_1$ then

$$\begin{aligned} p_1(x_1y_1) &= |\lambda\mu|^{\rho} + p(\lambda y + \mu x + xy) \\ &\leqslant |\lambda|^{\rho}|\mu|^{\rho} + |\lambda|^{\rho}p(y) + |\mu|^{\rho}p(x) + |p|p(x)p(y). \qquad (*) \end{aligned}$$

If $C = \max(1, |p|)$ then $(*)$ implies

$$p_1(x_1y_1) \leqslant C(|\lambda|^{\rho} + p(x))(|\mu|^{\rho} + p(y)) = Cp_1(x_1)p_1(y_1) \qquad (**)$$

whence $|p_1| \leqslant C$. On the other hand, $|p| \leqslant |p_1|$ and $1 = p_1(e_1) = p_1(e_1^2) \leqslant |p_1|p(e_1)^2 = |p_1|$, so that $|p_1| \geqslant C$. Hence $|p_1| = C = \max(1, |p|))$. If p is sm. then $|p| \leqslant 1$, so that $|p_1| = \max(1, |p|) = 1$, whence p_1 is sm. Finally, that p_1 is faithful (respy.complete) when p is faithful (respy. complete) follows as in the case of (F) norms (see proof of 3.3.15).

3.4.12. Remark. Note that the canonical extension of a ρ-seminorm p differs from the canonical extension of p as a quarter-norm (except when $\rho = 1$).

3.4.13. DEFINITION. A ρ-seminorm p on an algebra A is called *normalized* if p is sm., and $p(e) = 1$ whenever A has a unity e.

3.4.14. LEMMA. *Let* $(A_j, p_j)(j = 1, \cdots, n)$ *be* ρ-*seminormed algebras. Then the product TA* $A = A_1 \times \cdots \times A_n$ *(under coordinate-wise operations) is a* ρ*-seminormed algebra* $A = (A, q)$ *where* q *is given by:*

$$\text{for } x = (x_1, \cdots, x_n) \in A, q(x) = \max_j p_j(x_j).$$

Moreover, $|q| \leqslant \max |p_j|$. *In particular* q *is sm. if each* p_j *is sm. If the* A_j *are unital and* p_j *normalized then* A *is unital with* q *normalized.*

PROOF. It is clear from the definition of q that a sequence in A converges with respect to q iff their associated coordinate sequences converge in the respective factor spaces. But this precisely means that the q-topology is the same as the product topology. Further, if $x = (x_1, \cdots, x_n), y = (y_1, \cdots, y_n)$ are in A then

$$q(xy) = \max_j p_j(x_j y_j) \leqslant \max_j |p_j| p_j(x_j) p_j(y_j) \leqslant C q(x) q(y)$$

where $C = \max |p_j|$, so that $|q| \leqslant C$, as required. If p_j are sm. then $|p_j| \leqslant 1$, whence $|q| \leqslant C = \max |p_j| \leqslant 1$, so q is sm. . Finally, if e_j is the unity of $A_j (j = 1, \cdots, n)$ then $e = (e_1, \cdots, e_n)$ is unity of A and $q(e) = \max p_j(e_j) = 1$ (since each $p_j(e_j) = 1$).

3.4.15. PROPOSITION. *Let* p *be a quarter-norm on an algebra* A *and* I *be a bi-ideal of* A. *Write* $A^{\#} = A/I$. *Define* $p^{\#}$ *on* $A^{\#}$ *by*

$$p^{\#}(x + I) = \inf\{p(x + a) : a \in I\}.$$

Then:

(i) $p^{\#}$ *is a quarter-norm on* $A^{\#}$

(ii) *If* p *is a* ρ*-seminorm so is* $p^{\#}$

(iii) *If* p *is a.sm. then so is* $p^{\#}$ *with* $|p^{\#}| \leqslant |p|$; *hence* $p^{\#}$ *is sm. whenever* p *is sm.*

(iv) *If A is unital so is $A^{\#}$; if $\bar{I} \neq A$ (bar denoting closure) and p is normalized then so is $p^{\#}$.*

PROOF. (i) This follows from 3.1.22.

(ii) It is enough to check modulus homogenity condition for $p^{\#}$. Now

$$p^{\#}(\lambda x + I) = \inf\{p(\lambda x + a) : a \in I\} \leqslant |\lambda|^{\rho} p^{\#}(x + I). \quad (1)$$

If $\lambda \neq 0$ then

$$p^{\#}(x + I) = p^{\#}(\lambda^{-1}\lambda x + I) \leqslant |\lambda^{-1}|^{\rho} p^{\#}(\lambda x + I). \quad (2)$$

From (1),(2) we get the homogenity condition (when $\lambda \neq 0$). When $\lambda = 0$ the homogenity condition holds trivially since both sides are 0.

(iii) This follows from the inequality:

$$p^{\#}((x+I)(y+I)) \leqslant p((x+a)(y+b)) \leqslant |p|p(x+a)p(y+b)(a, b \in I).$$

(iv) Let A have unity e. Then $e^{\#} = e + I$ is the unity of $A^{\#}$. Assume now that p is normalized, so that $p(e) = 1$ and p is sm. (consequently also $p^{\#}$ is sm.) Since $\bar{I} \neq A, e \notin \bar{I}$ and so $p^{\#}(e^{\#}) \neq 0$. Since $p^{\#}(e^{\#}) = p^{\#}(e^{\#} \cdot e^{\#}) \leqslant p^{\#}(e^{\#})^2$ we conclude that $p^{\#}(e^{\#}) \geqslant 1$. On the other hand, $p^{\#}(e^{\#}) \leqslant p(e) = 1$. Thus, $p^{\#}(e^{\#}) = 1$, as required.

3.4.16. COROLLARY. *If (A, P) is a ρ-seminormed algebra then the quotient $A^{\#}$ is ρ-normed iff I is closed. In particular, $A/\ker p$ is a ρ-normed algebra with*

$$p^{\#}(x^{\#}) = p(x) \ (x \in A, x^{\#} = x + \ker p).$$

PROOF. The first statement follows from 3.1.22(iv). For the second it suffices to note that if $a \in \ker p$ then by 3.1.2(iii), $p(x + a) = p(x)$.

3.4.17. PROPOSITION. *The completion of a ρ-normed algebra (A, p) is a ρ-Banach algebra $(\hat{A}, \hat{p})$.*

PROOF. We may assume that p is sm. The construction of $\hat{A}$ and $\hat{p}$ is exactly as in 3.3.17. That $\hat{p}$ is now a ρ-seminorm is an easy consequence of the definition of $\hat{p}$: if $x_n \in A, \hat{x} \in \hat{A}$ and $x_n \to \hat{x}$ then $\hat{p}(\lambda x) = \lim_{n\to\infty} p(\lambda x_n) = \lim_{n\to\infty} |\lambda|^\rho p(x_n) = |\lambda|^\rho \hat{p}(x)$.

3.4.18. PROPOSITION. *Let $A = (A, p)$ be a real ρ-normed algebra which has a complex structure and let the resulting complex algebra $A^{[\mathbb{C}]}$ be denoted by $\check{A}$. Set for $x \in \check{A}(= A)$,*

$$\check{p}(x) = \sup_{0\leqslant\theta\leqslant 2\pi} p(\mathrm{e}^{i\theta}x).$$

Then $(A, \check{p})$ is a complex ρ-seminormed algebra with $p \leqslant \check{p}, |\check{p}| \leqslant |p|$. In particular, if p is sm. so is $\check{p}$.

PROOF. First note that $p(x) \leqslant \check{p}(x)$, since $\mathrm{e}^{i\theta}x = x$ when $\theta = 0$. Further we have

$$\begin{aligned}\check{p}(x+y) &= \sup_\theta p(\mathrm{e}^{i\theta}(x+y)) \leqslant \sup_\theta (p(\mathrm{e}^{i\theta}x) + p(\mathrm{e}^{i\theta}y)) \\ &\leqslant \sup_\theta p(\mathrm{e}^{i\theta}x) + \sup_\theta p(\mathrm{e}^{i\theta}y) = \check{p}(x) + \check{p}(y).\end{aligned}$$

Also, for $\alpha \in \mathbb{R}$,

$$\check{p}(\mathrm{e}^{i\alpha}x) = \sup_\theta p(\mathrm{e}^{i(\alpha+\theta)}x) = \check{p}(x).$$

For $\lambda \in \mathbb{C}$ we write $\lambda = |\lambda|\mathrm{e}^{i\alpha}$ and we get

$$\begin{aligned}\check{p}(|\lambda|\mathrm{e}^{i\alpha}x) &= \sup_\theta p(|\lambda|\mathrm{e}^{i\alpha}\mathrm{e}^{i\theta}x) = |\lambda|^\rho \sup_\theta p(\mathrm{e}^{i\theta}\mathrm{e}^{i\alpha}x) \\ &= |\lambda|^\rho \check{p}(\mathrm{e}^{i\alpha}x) = |\lambda|^\rho \check{p}(x).\end{aligned}$$

Finally,

$$\begin{aligned}\check{p}(xy) &= \sup_\theta p(\mathrm{e}^{i\theta}xy) \leqslant \sup_\theta |p| p(\mathrm{e}^{i\theta}x)p(y) \\ &\leqslant |p|\check{p}(x)p(y) \leqslant |p|\check{p}(x)\check{p}(y),\end{aligned}$$

whence $|\check{p}| \leqslant |p|$.

§ 5. Bounded Linear Transformations on ρ-seminormed LS's

3.5.1. DEFINITION. Let $X = (X, p)$, $X^* = (X^*, p^*)$ be respy. a ρ-seminormed and a ρ^*-seminormed LS's (with $p \neq 0$) and $T : X \to X^*$ a linear transformation. T is said to be ***n. bounded*** (= norm bounded) or sometimes just bounded if there is a constant $C > 0$ such that

$$p^*(Tx) \leqslant Cp(x)^{\frac{\rho^*}{\rho}} \text{ for all } x \in X. \tag{$*$}$$

The smallest C satisfying $(*)$ is denoted by $|T|$ and is called the *bound* of T; we have

$$p^*(Tx) \leqslant |T|p(x)^{\frac{\rho^*}{\rho}} \quad (x \in X). \tag{$**$}$$

We use $\|T\|$ for $|T|$ when the LS's X, X^* are ρ-normed and ρ^*-normed.

3.5.2. PROPOSITION. *For a linear transformation T, write*

$$\begin{aligned} C_1 &= \sup\{p^*(Tx)/p(x)^{\frac{\rho^*}{\rho}} : p(x) > 0\} \\ C_2 &= \sup\{p^*(Tx) : p(x) \leqslant 1\}, \\ C_3 &= \sup\{p^*(Tx) : p(x) = 1\}. \end{aligned}$$

If T is bounded, then $|T| = C_1 = C_2 = C_3$.

PROOF. We see from $(**)$ of 3.5.1 that $p(x) = 0 \Rightarrow p^*(Tx) = 0$. It follows that $p^*(Tx) \leqslant C_1 p(x)^{\frac{\rho^*}{\rho}}$ for all x, whence $|T| \leqslant C_1$. On the other hand, $(**)$ shows that if $p(x) > 0$, $p^*(Tx)/p(x)^{\frac{\rho^*}{\rho}} \leqslant |T|$, so that $C_1 \leqslant |T|$. Thus $|T| = C_1$.

It is clear from the definitions that $C_3 \leqslant C_1$. To prove the reverse inequality assume that $p(x) > 0$, $y = x/p(x)^{\frac{1}{\rho}}$. Then $p(y) = 1$, $p^*(Tx)/p(x)^{\frac{\rho^*}{\rho}} = p^*(Ty)$, proving $C_1 \leqslant C_3$, so that $C_1 = C_3$.

Finally, assume that $0 < p(x) \leqslant 1$. Then

$$p^*(Tx) \leqslant p^*(Tx)/p(x)^{\frac{\rho^*}{\rho}} = p^*(Ty) \leqslant C_3,$$

where $y = x/p(x)^{\frac{1}{\rho}}$, $p(y) = 1$. The inequality $p^*(Tx) \leqslant C_3$ is also satisfied when $p(x) = 0$ (since then $p^*(Tx) = 0$), whence $C_2 \leqslant C_3$. Since trivially $C_3 \leqslant C_2$ we conclude that $C_2 = C_3$, completing the proof of the proposition.

3.5.3. COROLLARY. *A linear transformation* T *is bounded iff* $C_3 < \infty$.

PROOF. If T is bounded then by 3.5.2, $C_3 = |T| < \infty$. Conversely, assume now that $C_3 < \infty$. We shall show that this implies $C_2 < \infty$.

Suppose to the contrary $C_2 = \infty$. Then there is a sequence $(x_n) \in X$ with $p(x_n) \leqslant 1$, $p^*(Tx_n) \to \infty$. If there are an infinity of these x_n, say $x_{n'}$ $(n' = 1, 2, \cdots)$ with $0 < p(x_{n'}) \leqslant 1$, then setting $y_{n'} = x_{n'}/p(x_{n'})^{\frac{1}{\rho}}$ we obtain $p(y_{n'}) = 1$ and

$$p^*(Ty_{n'}) = p^*(Tx_{n'})/p(x_{n'})^{\frac{\rho^*}{\rho}} \geqslant p^*(Tx_{n'}) \to \infty,$$

which contradicts that $C_3 < \infty$. On the other hand,if there are only a finite number of x_n with $0 < p(x_n) \leqslant 1$ then we may assume after omitting these that $p(x_n) = 0$ for all n, $p^*(Tx_n) \to \infty$. Since $p \neq 0$ there is a y with $p(y) = 1$. Set $z_n = x_n - y$; then by 3.1.2 (iii), $p(z_n) = 1$. Since

$$|p^*(Tx_n) - p^*(Ty)| \leqslant p^*(Tx_n - Ty) = p^*(Tz_n)$$

and $p^*(Tx_n) \to \infty$, we get $p^*(Tz_n) \to \infty$, again contradicting that $C_3 < \infty$. Therefore $C_2 < \infty$. It follows from this that for all x with $0 < p(x)$ we have

$$p^*(Tx) \leqslant C_2 p(x)^{\frac{\rho^*}{\rho}} \qquad (***).$$

If $p(x) = 0$ then $p(n^{\frac{1}{\rho^*}} x) = 0$, and since $C_2 < \infty$ we get

$$np^*(Tx) \leqslant C_2 \qquad (n = 1, 2, \cdots).$$

This clearly implies that $p^*(Tx) = 0$, so that $(***)$ holds for all x and T is bounded.

3.5.4. COROLLARY. *For* T *to be bounded it is sufficient that there exist constants* $C, D > 0$ *such that*

$$p(x) \leqslant C \Rightarrow p^*(Tx) \leqslant D. \qquad (*)$$

PROOF. Suppose that $(*)$ holds. If $p(x) = 1$, then $p(C^{\frac{1}{\rho}}x) = C$, so that by $(*)$, we have

$$p^*(TC^{\frac{1}{\rho}}x) \leqslant D, \text{ whence } p^*(Tx) \leqslant C^{-\frac{\rho^*}{\rho}} D.$$

Hence C_3 (in the notation of 3.5.2) $< \infty$, whence by 3.5.3, T is bounded.

3.5.5. PROPOSITION. *Let $T : X \to X^*$ be a linear transformation, where $X = (X,p)$ is a ρ-seminormed* LS *and $X^* = (X^*, p^*)$ a ρ^*-seminormed LS. Then the following statements are equivalent.*

(i) *T is n. bounded.*

(ii) *T is continuous.*

(iii) *T is t. bounded.*

PROOF. If T is (n.) bounded then we have

$$p^*(Tx_\alpha - Tx) = p^*(T(x_\alpha - x)) \leqslant |T| p^{\frac{\rho^*}{\rho}}(x_\alpha - x)$$

whence T is continuous. Conversely, if T is continuous; given $\epsilon > 0$ there is a $\delta > 0$ such that $p^*(Tx) \leqslant \epsilon$ whenever $p(x) \leqslant \delta$. Hence, by 3.5.4, T is (n.) bounded. Thus (i) $\leftrightarrow$ (ii). Since (X,p) is first countable the equivalence of (ii) and (iii) follows from 2.1.29. (We can also easily deduce directly the equivalence of (i) and (iii), using 3.2.13).

3.5.6. COROLLARY. *Let p, p^* be respectively a ρ-seminorm and a ρ^*-seminorm on the same LS X. Then $p \sim p^*$ iff there are contants $C, C^* > 0$ such that*

$$p^* \leqslant C p^{\frac{\rho^*}{\rho}}, \qquad p \leqslant C^* {p^*}^{\frac{\rho}{\rho^*}}.$$

PROOF. The above conditions clearly express the boundedness of the identitly maps: $(X,p) \to (X,p^*)$ and $(X,p^*) \to (X,p)$, whence by 3.5.5 they are continuous and so $p \sim p^*$.

3.5.7. PROPOSITION. *Let $X = (X,p)$* † *be a ρ-seminormed LS. Then the bounded linear operators T on X form a unital ρ-seminormed algebra $\mathbf{B} = \mathbf{B}(X)$ with the bound $|T|$ of T as the ρ-seminorm. Moreover, $|\cdot|$ is sm. and satisfies $|I| = 1$ (I being the identity operator). Finally, if p is a ρ-norm or a complete ρ-norm then so is $|\cdot|$.*

PROOF. That $|\cdot|$ is a ρ-seminorm follows from its definition and the ρ-seminorm properties of p. To prove that $|\cdot|$ is sm. We observe that

$$p(T_1T_2x) \leqslant |T_1|p(T_2x) \leqslant |T_1||T_2|p(x),$$

so that $|T_1T_2 \leqslant |T_1||T_2|$. Trivially, $|I| = 1$. Further, if p is a ρ-norm then $|T| = 0 \Rightarrow p(Tx) = 0 \Rightarrow Tx = 0$ (for all $x \in X$) $\Rightarrow T = 0$, so that $|\cdot|$ is also a ρ-norm.

Assume now that p is also complete. If (T_n) in $\mathbf{B}$ is a C-sequence then we have

$$\|T_n - T_m\| \leqslant \epsilon \qquad (n, m \geqslant n). \tag{1}$$

This clearly implies

$$\|T_nx - T_mx\| \leqslant \epsilon\|x\| \quad (n, m \geqslant N). \tag{2}$$

If follows from (2) that (T_nx) is a C-sequence in A and let us write $Tx = \lim T_nx$ (the limit existing since p is complete). Then T is linear. For, if $Ty = \lim T_ny$ then $Tx+Ty = \lim(T_nx+T_ny) = \lim T_n(x+y) = T(x+y)$. By uniqueness of limit property (X being Hausdorff) we get $T(x+y) = Tx + Ty$. Similarly, $T\lambda x = \lambda Tx$. Finally, by allowing, $m \to \infty$ in (1) we get $\|T_n - T\| \leqslant \epsilon$ $(n \geqslant N)$, so that $\lim T_n = T$, and $\mathbf{B}$ is complete.

3.5.8. PROPOSTION. *Let $A = (A,p)$ be a ρ-seminormed*

† We always assume $p \neq 0$.

algebra and $\mathbf{B} = \mathbf{B}(A)$ *(see 3.5.7) the* ρ*-seminormed algebra of all bounded linear operators on the underlying seminormed LS of* A. *Then the left regular representation* $l : x \to l_x$ *is a continuous homomorphism of* A *into* $\mathbf{B}$. *Also,* $|l_x| \leqslant |p|p(x)$; *in particular, if* p *is sm., then* $|l_x| \leqslant p(x)$.

The map l *is injective iff the annihilator ideal* $\mathcal{A}_l = \{0\}$. *In the unital case we have* $p(x) \leqslant |l_x|p(e)$, *whence* l *is a t. isomorphism. Besides, if* $p(e) = 1$ *then* $p(x) = |l_x|$.

PROOF. That l is a homomorphism is easily verifiable and in fact a standard result in algebra. Since

$$p(l_x y) = p(xy) \leqslant |p|p(x)p(y)$$

we obtain

$$\begin{aligned} |l_x| &\leqslant \sup\{p(xy) : p(y) \leqslant 1\} \leqslant |p|p(x), \text{ i.e.} \\ |l_x| &\leqslant |p|p(x) \end{aligned} \tag{1}$$

Hence $l_x \in \mathbf{B}$ and l is bounded and so continuous (by 3.5.5). If p is sm. then $|p| \leqslant 1$, so that (1) gives

$$|l_x| \leqslant p(x). \tag{2}$$

The injectiveness (or faithfulness) conlusions are clear. Further, in the unital case we have

$$p(x) = p(l_x e) \leqslant |l_x|p(e), \tag{3}$$

so that l^{-1} is bounded and so continuous. Thus in this case l is a t. isomorphism. Finally, if $p(e) = 1$, (3) gives $p(x) \leqslant |l_x|$ which together with (2) yields $|l_x| = p(x)$.

3.5.9. THEOREM (Bonsall-Duncan †). *Let* $A = (A, p)$ *be a* ρ*-seminormed algebra and* S *a bounded (multiplicative) subsemigroup of* A. *Then we can find a sm.* ρ*-seminorm* q *such that*

(i) $q \sim p$,

† They confine themselves [4, p.18] to the case where p is a sm. norm.

(ii) $q(s) \leqslant 1$ *for all* s *in* S. *Further, if* p *is a* ρ *-norm so is* q *and if* A *has unity* e *then* $q(e) = 1$ (*so that* q *is normalized*).

PROOF. By repalcing A by its unitization A_1 (if necessary) we may assume that A has a unity e. Write $S^1 = S \bigcup \{e\}$. Clearly, S^1 is a subsemigroup of A which is also bounded, say, $p(s) \leqslant M$ for all $s \in S^1$. Define

$$p_1(x) = \sup\{p(sx) : s \in S^1\}.$$

Since

$$p(sx) \leqslant |p|p(s)p(x) \leqslant |p|Mp(x),$$

it follows that $p_1(x) < \infty$ and further

$$p_1 \leqslant Cp \qquad (C = |p|M). \tag{1}$$

It is easy to see that p_1 is a ρ-seminorm. Since $e \in S^1$ we obtain:

$$p(x) = p(ex) \leqslant p_1(x). \tag{2}$$

From (1), (2) we obtain:

$$p_1 \sim p. \tag{3}$$

Further

$$\begin{aligned} p_1(xy) &= \sup\{p(sxy) : s \in S^1\} \leqslant |p| \sup\{p(sx) : s \in S^1\}p(y) \\ &\leqslant |p|p_1(x)p(y) \leqslant |p|p_1(x)p_1(y) \qquad \text{(using (2))}. \end{aligned}$$

Setting

$$q(x) = |l_x|_1 = \sup\{p_1(xy) : p_1(y) \leqslant 1\}$$

we find that

$$q(x) \leqslant |p_1|p_1(x). \tag{4}$$

Since

$$q(xy) = |l_{xy}|_1 = |l_x l_y|_1 \leqslant |l_x|_1 |l_y|_1 = q(x)q(y),$$

q is sm. If we write $a = e/p_1^{1/\rho}(e)$, then $p_1(a) = 1$ and $q(x) \geqslant p_1(xa) = p_1(x)/p_1(e)$, so that

$$p_1(x) \leqslant p_1(e)q(x). \tag{5}$$

From (3), (4), (5) we conclude that $q \sim p_1 \sim p$, which proves (i).

To prove (ii), we observe that if $s \in S^1$,

$$p_1(sx) = \sup\{p(tsx) : t \in S^1\} \leqslant p_1(x) \tag{6}$$

(since $ts \in S^1$). It follows that

$$\begin{aligned} q(s) \quad &= \quad \sup\{p_1(sy) : p_1(y) \leqslant 1) \leqslant \sup\{p_1(y) : p_1(y) \leqslant 1\} \\ & \qquad\qquad\qquad\qquad\qquad\qquad\qquad\qquad \text{(using (6))} \\ &\leqslant \quad 1. \end{aligned} \tag{7}$$

In particular, $q(e) \leqslant 1$. But $q(e) = q(e^2) \leqslant q(e)^2$, whence $q(e) \geqslant 1$. Therefore $q(e) = 1$, as desired.

Finally, by (3), p_1 is faithful when p is faithful and then, by virtue of 3.5.7, $q = |l_x|_1$ is a ρ-norm.

3.5.10. COROLLARY

(i) $\nu(x) = \nu_p(x) = \inf\{q(x) : q$ *is a sm.* ρ*-seminorm with* $q \sim p\}$.

(ii) *If* A *has unity* e *then* $\nu(x) = \inf\{q(x) : q$ *is a sm.* ρ*-seminorm with* $q \sim p,\ q(e) = 1\}$.

PROOF. (i) Since $\nu(x) \leqslant q(x)$ for all q we get

$$\nu(x) \leqslant \inf_q q(x) = t \text{ (say)}.$$

Suppose that $\nu(x) < t$, so that $\nu(x)/t < 1$. Since $\frac{n}{n+1} \to 1$ (as $n \to \infty$ we can choose a sufficiently large n that we have

$$\frac{\nu(x)}{t} < \frac{n}{n+1} < 1. \tag{1}$$

Set $y = ((n+1)/nt)^{\frac{1}{\rho}} x$.
Then $q(y) = ((n+1)/nt)q(x)$, so that we have

$$\inf_q q(y) = \frac{n+1}{nt} \inf_q q(x) = \frac{n+1}{nt} t = \frac{n+1}{n}. \tag{2}$$

Now, by 3.4.5,

$$\nu(y) = \frac{n+1}{nt}\nu(x) = \frac{n+1}{n}.\frac{\nu(x)}{t} < \frac{n+1}{n} \cdot 1 = \frac{n+1}{n} = \inf_q q(y) \tag{3}$$

where in the last step we have used (2).
Again,

$$\nu(y) = \frac{n+1}{nt}\nu(x) = \frac{n+1}{n}\cdot\frac{\nu(x)}{t} < \frac{n+1}{n}\cdot\frac{n}{n+1} = 1 \quad \text{(using (1))}.$$

Since $\nu(y) = \lim_n p(y^n)^{\frac{1}{n}} < 1$ we get $p(y^n) < 1$ for all $n \geqslant N_0$, whence it folows that the semigroup $S = \{y^n : n = 1, 2, \cdots\}$ is p-bounded. By 3.5.9, there is a sm. ρ-seminorm $q_0 \sim p$ and $q_0(y) \leqslant 1 < \frac{n+1}{n}$, contradicting (2). Therefore we must have $\nu(x) = t$, proving (i).

(ii) This can be proved in exactly the same way as (i).

3.5.11. PROPOSITION. (a) *Let* X *be a* ρ*-seminormed LS and* X^* *a* ρ^**-Banach space. Let* X_0 *be a dense subspace and* $T : X_0 \to X^*$ *a continuous linear tranformation. Then* T *can be uniquely extended to a continuous linear transformation*

$$\hat{T} : X \to X^* \text{ with } |\hat{T}| = |T|.$$

(b) *Suppose that* A *is a* ρ*-seminormed algebra,* A_0 *a dense subalgebra of* A *and* A^* *a* ρ*-Banach algebra. Then every continuous homomorphism* $\varphi : A_0 \to B$ *can be uniquely extended to a continuous homomorphism* $\hat{\varphi} : A \to B$ *with* $|\hat{\varphi}| = |\varphi|$.

PROOF. (a) Let $X = (X, p)$, $X^* = (X^*, p^*)$. Suppose that $x \in X$, $x_n \to x$ $(x_n \in X_0)$. Then

$$p^*(Tx_n - Tx_m) \leqslant |T| p(x_n - x_m)^{\frac{\rho^*}{\rho}} \to 0,$$

so by completeness of Y, $Tx_n \to x^*$. Define $\hat{T}x = x^*$. By using the uniqueness of limit property of convergent sequences in X^* (which is Hausdorff) it is easy to verify that $\hat{T}$ is well-defined (i.e. independent of the particular choice of the sequence $x_n \to x$) and linear. Further we have

$$p^*(\hat{T}x) = \lim p^*(Tx_n) \leqslant |T| p(x_n)^{\frac{\rho^*}{\rho}} \leqslant |T| p(x)^{\frac{\rho^*}{\rho}},$$

so that $|\hat{T}| \leqslant |T|$. On the other hand, $\hat{T}$ being an extension of T, $|T| \leqslant |\hat{T}|$. Thus $|\hat{T}| = |T|$. The continuity if $\hat{T}$ is an immediate consequence of its boundedness. Finally, the uniqueness of $\hat{T}$ is also clear.

(b) Extend φ (as a linear transformation) to $\hat{\varphi}$ on A as in (a); then $\hat{\varphi}$ is continuous linear. We have to show that $\hat{\varphi}$ is a homomorphism. If $x, y \in A$, (x_n), $(y_n) \in A_0$, $x_n \to x, y_n \to y$ then $\hat{\varphi}(xy) = \lim \varphi(x_n y_n) = \lim \varphi(x_n)\varphi(y_n) = \hat{\varphi}(x)\hat{\varphi}(y)$, proving $\hat{\varphi}$ is a homomorphism and completing the proof.

§ 6. Topological Algebras with Inverses

3.6.1. DEFINITION. Let A be an algebra (over $\mathbb{K}$) and $a \in A$. Denote by l_a° the map $x \mapsto a \circ x$ $(x \in A)$. Similarly r_a° is the map $x \mapsto x \circ a$.

3.6.2. LEMMA. *Let A be a WTA. Then l_a°, r_a° are continuous. If a is q. invertible then l_a°, r_a° are homeomorphisms of A.*

PROOF. Since $l_a^\circ(x) = a + x + l_a x$ $(l_a : x \mapsto ax)$ the continuity of l_a° follows from the continuity of addition and the continuity of l_a. The continuity of r_a° follows similarly.

If a is q. invertible with a' as $q.i$, so that $a' \circ a = \circ = a \circ a'$, then

$$l_{a'}^\circ l_a^\circ = l_{a' \circ a}^\circ = l_0^\circ = 1_A = l_a^\circ l_{a'}^\circ.$$

Therefore, l_a° is invertible with its inverse $(l_a^\circ)^{-1} = l_{a'}^\circ$ continuous. It follows that l_a° is a homeomorphism. Similarly, since $(r_a^\circ)^{-1} = r_{a'}^\circ, r_a^\circ$ is a homeomorphism.

3.6.3. LEMMA. *If A is a unital WTA and $a \in A$ is invertible then l_a, r_a are linear homeomorphisms.*

PROOF. Here we have $l_a^{-1} = l_{a^{-1}}, r_a^{-1} = r_{a^{-1}}$ and hence the result.

3.6.4. DEFINITION. A TA A is called a *C algebra* or a continuous algebra if the map $x \mapsto x'$ of $G_q \to G_q$ is continuous, where G_q denote the group of q. invertible elements of A.

3.6.5. PROPOSITION. (a) *A TA A is a C algebra iff G_q is a TG.*

(b) *A unital TA A is a C algebra iff its group G_i of invertible elements is a TG.*

(c) *In a unital C algebra the groups G_q, G_i are TG's and the map*

$$\tau^{-1} : x \in G_q \to e + x \in G_i$$

is a t. isomorphism.

PROOF. (a) Since A is a TA the map $(x, y) \mapsto xy$ is always continuous. It follows that G_q is a TG iff the map $x \mapsto x'$ is continuous, i.e. iff A is a C algebra.

(b) By 1.1.20, $y = e+x \in G_i$ iff $x \in G_q$ and $y^{-1} = (e+x)^{-1} = e + x'$. It follows that G_i is a TG iff $y \mapsto y^{-1}$ is continuous iff $x \mapsto x'$ is continuous iff G_q is a TG.

(c) The first half of the statement follows from (a),(b). Also, $\tau^{-1} : x \mapsto e + x$ is an isomorphism of G_q onto G_i. Again, τ, τ^{-1}, being translations in A, are continuous. Hence the second half of the statement.

3.6.6. LEMMA. *Every quarter-normed algebra (A, p) with an a.sm. p is a C algebra. In particular, any ρ-seminormed algebra is a C algebra.*

PROOF. The first statement is an immediate consequence of the inequality $(*)$ of 3.3.22. The second statement folllows from the first since, by 3.4.3, the ρ-seminorm p of a ρ-seminormed algebra is always a.sm..

3.6.7. DEFINITION. A TA A is called a Q *algebra* or algebra with q. inverses if there is an open neighbourhood $U(0)$ of 0 consisting of q. invertible elements. A unital TA A is called an I *algebra* or algebra with inverses if there is an open neighbourhood $U(e)$ of e consisting of invertible elements.

3.6.8. LEMMA (Michael). *Let A be a TA. Then:*

(i) *If S is a balanced subset of A comprising q. invertible elements then $r(s) \leqslant 1$ for every s in S, where $r(s)$ denotes the spectral radius of s.*

(ii) *Write $\mathfrak{S} = \{x \in A : r(x) \leqslant 1\}$. Then A is a Q algebra iff* $\operatorname{int} \mathfrak{S} = \mathfrak{S}^\circ \neq \emptyset$.

PROOF. (i) Suppose that $r(s) > 1$ for some $s \in S$. Then there is a $\lambda \in \sigma(s)$ with $|\lambda| > 1$. By 1.7.8, $s_0 = -\lambda^{-1}s$ is not q. invertible. But since $|\lambda^{-1}| < 1, s \in S$ and S is balanced we must have $s_0 \in S$, so that s_0 is q. invertible. This contradiction proves (i).

(ii) First assume that A is a Q algebra with $U(0)$ a nucleus comprising q. invertible elements. Choose a balanced open nucleus $V \subseteq U(0)$. By(i), $V \subseteq \mathfrak{S}$ so that $\mathfrak{S}^\circ \neq \emptyset$. Conversely, assume that in A we have $\mathfrak{S}^\circ \neq \emptyset$. Then $U = \frac{1}{2}\mathfrak{S}^\circ \neq \emptyset$ and open. If $x \in U$ then $2x \in \mathfrak{S}$, so that by 1.8.11(ii), $r(x) \leqslant \frac{1}{2} < 1$. It follows that $-1 \notin \sigma'(x)$ whence, by 1.7.8, x is q. invertible. Thus every $x \in U$ is q. invertible and A is a Q algebra.

3.6.9. LEMMA. *If A is a Q (respy I) algebra then each q. invertible (respy. invertible) element x has an open neighbourhood. $U^\circ(x) = l_x^\circ(U(0))$ (respy. $U(x) = l_x(U(e))$ comprising q. invertible (respy. invertible) elements. Hence the set G_q (respy. G_i) of q. invertible (respy. invertible) elements of a Q (respy I) algebra A is open.*

PROOF. Since G_q (respy. G_i) is a group we have $U^\circ(x) = x \circ U(0) \subseteq G_i$ (respy. $U(x) = xU(e) \subseteq G_i$) Further, $U^\circ(x)$ (respy. $U(x)$) is open since l_x° (respy. l_x) is, by 3.6.2 (respy.3.6.3) a homeomorphism.

3.6.10. PROPOSITION. *For a TA (respy. unital TA) A to be a Q (respy. I) algebra it is necessary and sufficient that G_q (respy. G_i) is open. In particular, a TA which is a division algebra is an I algebra iff it is Hausdorff.*

PROOF. The "necessity" follows from 3.6.9 and the "sufficiency" is immediate from the definition of a Q (respy. I) algebra (since $0 \in G_q, e \in G_i$). Since in a division algebra $G_i = A\backslash\{0\}$ the assertion concerning TA's which are division algebras is clear.

3.6.11. COROLLARY. *Let A be a TA under two topologies τ_1, τ_2. If τ_2 is finer than τ_1 and (A, τ_1) is a Q algebra (respy. I*

algebra) *then* (A, τ_2) *is also a* Q *algebra* (*respy.* I *algebra*)

Proof. Clear (G_q τ_1-open $\Rightarrow G_q$ τ_2-open).

3.6.12. LEMMA. *Let* A *be a unital* TA *with unity* e. *If there is an open neighbourhood* $U(0)$ *of 0 consisting of q. invertible elements then* $U(e)$ † $= U(0) + e$ *is an open neighbourhood of* e *consisting of invertible elements. Conversely, if* $V(e)$ *is an open neighbourhood of* e *consisting of invertible elements then* $V(0) = V(e) - e$ *is an open neighbourhood of 0 consisting of q. invertible elements.*

PROOF. By 1.1.20, $x \in A$ is q. invertible iff $x + e$ is invertible. Moreover, the translations $x \mapsto x + e, y \mapsto y - e$ of A are homeomorphisms. The statements of the lemma are now clear.

3.6.13. COROLLARY. *A unital algebra* A *is an* I *algebra iff it is a* Q *algebra.*

3.6.14. LEMMA. *A* Q *algebra* A *is commutative iff* G_q *is commutative. Also, an* I *algebra* A *is commutative iff* G_i *is commutative.*

PROOF. Suffices to prove the "if" parts. Suppose that G_q is commutative and $x, y \in A$. Since $\frac{x}{n}, \frac{y}{n} \to 0$ and G_q is open there are integers n_1, n_2 such that $\frac{x}{n_1}, \frac{y}{n_2} \in G_q$. Then $\frac{x}{n_1} \leftrightarrow \frac{y}{n_2}$, whence $x \leftrightarrow y, A$ is commutative. Again, if G_i is commutative so is G_q since G_q isomorphic to G_i (by 1.1.20). This completes the proof.

3.6.15. Remark. The above lemma can fail if the algebra is not a Q algebra. For example, consider the algebra $\mathcal{P}^* = \mathcal{P}^*(X, Y)$ of all polynomials $P = P(X, Y)$ over $\mathbb{K}$ in two non-commuting variables X, Y. This algebra can be normed by: $\|P\| =$ the maximum of the absolute values of the coefficients of the monomials of P. It is easy to check this gives a norm that is sm.: $\|PQ\| \leqslant \|P\|\|Q\|$. It is clear that the group of invertible elements $G_i = \mathbb{K}\backslash\{0\}$ which is commutative, but $\mathcal{P}^*$ is not commutative since $XY \neq YX$. Note here that G_i is not open in $\mathcal{P}^*$

† i.e., $U(e) = \{x + e : x \in U(0)\}$.

so that $\mathcal{P}^*$ is not a Q algebra.

3.6.16. THEOREM (Arens[†] -Banach). *A* (F) *algebra* $A = (A, |\cdot|)$ *is a* C *algebra iff its group* G_q *is a* G_δ [†] *set of* A.

PROOF (cf.[31,p.4]). First assume that A is a C algebra, so that the map

$$x \mapsto x' \text{ is continuous on } G_q. \qquad (*)$$

For $n = 1, 2, \cdots$ define

$W_n = \{x \in \overline{G_q} : \exists \text{ an } \eta = \eta(x, n) > 0$ such that if $y, z \in G_q, |y - x| < \eta, |z - x| < \eta$ then $|y' - z'| < \frac{1}{n}\}$, where the bar in $\overline{G_q}$ denotes closure. It is easy to see, using continuity of the map $(*)$, that W_n is open in $\overline{G_q}$. Also it is clear that $W_n \supseteq G_q$, so that $W = \bigcap_n W_n \supseteq G_q$. Thus, W is a G_δ in $\overline{G}$ and $\overline{G_q}$ as a close subset of the metric space A is a G_δ of A. It follows that W is a G_δ of A. We shall complete the proof of the "only if" part by showing that $G_q = W$.

Suppose that $x \in W \subseteq \overline{G_q}, x_n \in G_q$ and $x_n \to x$. Then $|x'_n - x'_m| \to 0$ as $n, m \to \infty$, so that $x'_n \to y \in A$ (since A is complete). It follows that $x \circ y = (\lim x_n) \circ (\lim x'_n) = \lim (x_n \circ x'_n) = 0$. Similarly $y \circ x = 0$. So $y = x', x \in G_q$, proves $W = G_q$.

It remains to prove the 'if' part and assume therefore that G_q is a G_δ of A. Since A is a complete metric space, G_q is topologically complete and consequently there is in G_q an equivalent metric under which it is complete. By applying proposition 2.1.8 to G_q (with this complete metric) we conclude that inversion is continuous in G_q, completing the proof.

3.6.17. COROLLARY. *A unital* (F) *algebra* A *is a* C *algebra iff the group* G_i *of its invertible elements is a* G_δ *of* A.

PROOF. It suffices to observe that the map $x \mapsto e + x$ is a homeomorphism of A which carries G_q onto G_i.

3.6.18. COROLLARY. *An* (F) *algebra which is a* Q *algebra is also a* C *algebra.*

---

† Arens proved (essentially) only the "if" part of the theorem and that under the additional assumption that G_q is seperable.

†† i.e. G_q is the intersection of a countable family of open sets of A.

PROOF. Since an open set is a G_δ this follows form 3.6.16.

3.6.19. Remark. There exist Q algebras which are not C algebras. For instance, the algebra $\mathfrak{M}$ of germs of meromorphic functions of one complex variable in the neighbourhoods of 0 in $\mathbb{C}$ is a commutative division algebra. This can be topologized so as to make it into a complete Hausdorff locally convex TA. It is known that $\mathfrak{M}$ is not a C algebra (see [29, p.136]). On the other hand since it is a Hausdorff division algebra, by 3.6.10 it is a Q algebra.

There are also C algebras which are not Q algebra. For example, if $\overline{\mathbb{B}}_1$ is the closed unit ball in $\mathbb{K}$ denote by $\mathbb{P}$ the algebra of the restriction of polynomials in one variable to $\overline{\mathbb{B}}_1$ with coefficients from $\mathbb{K}$. $\mathbb{P}$ is a normed algebra under the sup norm. By 3.6.6, $\mathbb{P}$ is a C algebra. On the other hand since the invertible elements in $\mathbb{P}$ are just the non-zero constants, $\mathbb{P}$ is clearly not an I (and hence also not a Q) algebra.

3.6.20. DEFINITION. A Q (respy. I) algebra is called a CQ (respy. CI) *algebra* or algebra with continuous q. inverses (respy. continuous inverses) if the map
$x \mapsto x'$ (respy. $x \mapsto x^{-1}$) from $U(0)$ (respy. $U(0)$) into A is continuous, where $U(0)$ (respy. $U(e)$) as in 3.6.7.

3.6.21. LEMMA. *An algebra A is CQ (respy CI) iff it is a C algebra and a Q (respy. I) algebra.*

PROOF. The "if" part follows from 3.6.5((a),(b)) For the "only if" part, in view of 3.6.13, it is enough to prove that a CQ algebra is a C algebra. Let A be a CQ algebra and $U(0)$ as in 3.6.7. If $y \in U(0)$ then $x \circ y \in U^\circ(x)$ (as defined in 3.6.9) and $(x \circ y)' = y' \circ x'$. The map $x \circ y \mapsto (x \circ y)'$ is continuous since it is the composite of the continuous maps $x \circ y \mapsto y = l^\circ_{x'}(x \circ y), y \mapsto y', y' \mapsto y' \circ x'$. This proves that A is a C algebra.

3.6.22. LEMMA. *Every CI algebra is a CQ algebra. Conversely, every unital CQ algebra is a CI algebra.*

PROOF. In view of 3.6.21, 3.6.13 it is enough to prove that the map $x \mapsto x' \, (x \in U(0))$ is continuous iff the map $(e + x) \mapsto$

$(e+x)^{-1}$ $(x \in U(0))$ is continuous. But this is immediate since $(e+x)^{-1} = e + x'$.

3.6.23. PROPOSITION. (a) *A sm. (F) algebra is a CQ algebra.*

(b) *Every ρ-Banach algebra is a CQ algebra and a unital ρ-Banach algebra a CI algebra.*

PROOF. (a) By virtue of 3.3.18, a sm. (F) algebra is a Q algebra and hence by 3.6.18, 3.6.21 it is a CQ algebra.

(b) By the remark in 3.4.4 we may assume that the norm of the ρ-Banach is sm. and the first half of the required result follows from (a). The second half follows from the first in view of 3.6.13, 3.6.21.

3.6.24. LEMMA. *Let B be a q. inverse-closed*[†] *subalgebra (respy. inverse-closed subunital*[††] *algebra) A. If A is a Q (respy. I) algebra then so is B; if A is CQ (respy. CI) then so is B.*

PROOF. If $G_q(B), G_q(A)$ denote the groups of q. invertible elements of B, A respy then we have clearly the relation $G_q(B) = B \bigcap G_q(A)$ Hence the statements regarding Q and CQ properties of B. The I and CI properties of B follows similarly from the corresponding relation $G_i(B) = B \bigcap G_i(A)$.

3.6.25. PROPOSITION. *Let $A = (A, p)$ be a sm. (F) algebra which is a Q algebra. If B is a closed subalgebra of A then B is also a Q algebra.*

PROOF. Since A is a Q algebra we can choose an open neighbourhood $U(0)$ of 0 in A such that $x \in U(0) \Rightarrow x$ is q. invertible and $p(x) < 1$. Suppose that $x \in U_0 = B \bigcap U(0)$. By 3.3.18, x exists and is given by $x' = \sum_{n=1}^{\infty} (-1)^n x^n$. Since $x^n \in B$ and B is a closed subspace, $x' \in B$, so that $x' \in U_0$. Thus, U_0 is an open

---

[†] i.e. if $x \in B$ has q.i. x' (respy. inverse x^{-1}) in A then x' (respy. x^{-1}) $\in B$.

[††] For definition see 1.1.6.

neighbourhood of 0 in B, comprising q. invertible elements and B is a Q algebra.

3.6.26. LEMMA. *Let A, A^* be TA's and $\varphi : A \to A^*$ a continuous open epimorphism. Then A^* is a Q (respy. CQ) algebra provided A is a Q (respy. CQ) algebra.*

PROOF. If A is a Q algebra then G_q is open in A and consequently $G^* = \varphi(G_q)$ is an open neighbourhood of 0 in A^*. By 1.1.24, every element $y \in G^*$ is q. invertible with $y' \in G^*$. Hence A^* is a Q algebra.

Next suppose that A is a CQ algebra. Then as above $G^* = \varphi(G_q)$ is an neighbourhood of 0 in A^*. For $y_0 \in G^*$, choose an open neighbourhood V' of y_0' with $V' \subseteq G^*$. Set $U' = \varphi^{-1}(V^{-1}) \bigcap G_q$. Then U' is an open set, so that $U = \{x \in A : x' \in U'\}$ is an open neighbourhood of x_0, where $\varphi(x_0) = y_0$. It follows that $V = \varphi(U)$ is an open neighbourhood of y such that $q^{\#}(V) \subseteq V'$, where $q^{\#}$ is the map $y \mapsto y'$ $(y \in G_q(A^*))$. This proves the continuity of $q^{\#}$ at y_0 an arbitary point of G^*, whence A^* is a CQ algebra.

3.6.27. COROLLARY. *In a Q (respy. CQ) algebra A if I is bi-ideal the quotient algebra $A^{\#} = A/I$ is a Q (respy CQ) algebra.*

PROOF. This follows from 3.6.26 since the canonical map $\pi : A \to A^{\#}$ is an open continuous epimorphism.

3.6.28. PROPOSITION. *Let A be a TA and A_1 its unitization. Then A_1 is a Q (respy. CQ) algebra iff A is a Q (respy. CQ) algebra.*

PROOF. First assume that A is a Q algebra and U an open neighbourhood of 0 comprising q. invertible elements. Choose a balanced open neighbourhood V of 0 with $V + V \subseteq U$. Then, for $0 \leqslant |\mu| \leqslant 1$, we have

$$(1+\mu)V \subseteq V + \mu V \subseteq V + V \subseteq U.$$

Choose μ such that $0 \leqslant |\mu| < \frac{1}{2}$. Then $|1+\mu| > 1 - |\mu| > \frac{1}{2}$. Writing $\lambda = \frac{1}{2}(1+\mu)^{-1}$ we get $|\lambda| < 1$, so that for $x \in V, \lambda x \in V$.

Since

$$y = x(1+\mu)^{-1} = 2\lambda x \in 2\lambda V \subseteq U$$

it follows that y is q. invertible. Therefore, $e_1 + y$ is invertible and hence also

$$(1+\mu)(e_1 + y) = (1+\mu)e_1 + x.$$

is invertible. It follows that

$$V_1 = \{(1+\mu)e_1 + x : 0 \leqslant |\mu| < \frac{1}{2}, x \in V\}$$

is an open neighbourhood of e_1 comprising invertible elements, whence A_1 is an I algebra and so also a Q algebra.

Conversely, assume that A_1 is an I algebra; then it is a Q algebra and so, by 3.6.10, the group G_q^1 of q. invertible elements of A_1 is open. If G_q denotes the group of q. invertible elements it is clear (see 1.1.21) that $G_q = A \bigcap G_q^1$, so that G_q is open in A. By 3.6.10, A is a Q algebra.

It remains to prove that A_1 is a CQ algebra iff A is a CQ algebra. But for this, by 3.6.21 and the result already proved it is enough to show that A_1 is a C algebra iff A is a C algebra. Now by virtue of 1.1.20, $y_1 = \lambda e + x \, (\lambda \in \mathbb{K}, x \in A)$ in A_1 is invertible (i.e. $y_1 \in G_i'$) iff $\lambda \neq 0$ and $\lambda^{-1}x \in G_q$, and then we have the relation.

$y_1^{-1} = \lambda^{-1}(e_1 + (\lambda^{-1}x)')$, where dash denotes q.i.. Again, if $x \in G_q$ then $e_1 + x \in G_i^1$ and $(e_1 + x)^{-1} = e_1 + x'$. It follows from these relations that the map $y_1 \mapsto y_1^{-1}$ in G_i^1 is continuous, which completes the proof of the proposition.

3.6.29. PROPOSITION. *The complexification $\tilde{A}$ of a commutative Q algebra A is a Q algebra.*

PROOF. Let U be a nucleus ($=$ a neighbourhood of 0) comprising q. invertible elements. Then by 2.1.17(iii) (applied twice) and 2.2.14(vi) we can choose nuclei V, W such that

$$V + V + V + V \subseteq U, \; W^2 \subseteq V, \; W \subseteq V.$$

If x, y in W then we have

$$u = 2x + x^2 + y^2 \in W + W + W^2 + W^2 \subseteq V + V + V + V \subseteq U$$

so that u is q. invertible in A and hence also in $\tilde{A}$. But then since

$(x+iy)\circ(x-iy) = u$, $x+iy$ is also q. invertible in $\tilde{A}$. It follows that $\tilde{W} = W + iW$ is a nucleus of $\tilde{A}$, comprising q. invertible elements, so that $\tilde{A}$ is a Q algebra.

3.6.30. COROLLARY. *The complexification $\tilde{A}$ of a commutative real I algebra A is an I algebra.*

PROOF. Since, by 3.6.13, a unital algebra is an I algebra iff it is a Q algebra, the corollary follows from the proposition.

3.6.31. PROPOSITION. *The complexification $\tilde{A}$ of a commutative real CI algebra A is a CI algebra.*

PROOF. In view of 3.6.21, 3.6.30 it is enough to prove that the map $(*)$ $z \mapsto z^{-1}$ of $\tilde{G}_i$ onto itself is continuous where $\tilde{G}_i$ is the group of invertible elements of $\tilde{A}$. Suppose that $z_\alpha, z \in \tilde{G}_i, z_\alpha \to z, z_\alpha = x_\alpha + iy_\alpha, z = x + iy$, where $x_\alpha, y_\alpha, x, y \in A$. By 1.6.7, $\overline{z_\alpha} = x_\alpha - iy_\alpha$ and $\overline{z} = x - iy$ lie in $\tilde{G}_i$ and hence the products

$a_\alpha = z_\alpha\overline{z_\alpha} = x_\alpha^2 + y_\alpha^2, a = z\overline{z} = x^2 + y^2$ also lie in $\tilde{G}_i$. But $a_\alpha, a \in A$, whence by 1.6.10, $a_\alpha, a \in G_i$. It follows that

$$z_\alpha^{-1} = \overline{z}_\alpha(x_\alpha^2 + y_\alpha^2)^{-1} = \overline{z}_\alpha a_\alpha^{-1} \to \overline{z}a^{-1} = \overline{z}(x^2+y^2)^{-1} = z^{-1},$$

proving the continuity of the map $(*)$

3.6.32. COROLLARY. *The complexification $\tilde{A}$ of a real commutative CQ algebra A is a CQ algebra.*

PROOF. By 3.6.28, the unitization A_1 of A is a CQ, and hence by 3.6.22, a CI algebra. By 3.6.31, $\tilde{A}_1 = (\tilde{A})_1$ is a CI algebra and hence CQ. So, by 3.6.28, $\tilde{A}$ is CQ.

3.6.33. WILLIAMSON'S ALGEBRA. The field $\mathbb{C}(X)$ of rational functions over $\mathbb{C}$ can be embedded in the algebra M of example (V) of 3.3.14, by identifying a rational funcetion $P(X)/Q(X)$ (P, Q polynomials over $\mathbb{C}$) with the almost everywhere defined function $P(t)/Q(t)$ in M. The topology thereby inherited from M makes $\mathbb{C}(X)$ into a Hausdorff topological field (more precisely, a commuttive Hausdorff division algebra in which inversion is continuous). In particular, $\mathbb{C}(X)$ is a CI algebra as also a pre- (F)

algebra. This method of topologizing $\mathbb{C}(X)$ is due to Williamson [14′, p.731]; we call $\mathbb{C}(X)$ with this topology as Williamson's algebra and denote it by $\mathcal{W}$.

§ 7. Topological Zero Divisors

3.7.1. DEFINITION. Let A be an algebra. Write G_q^l (respy. G_q^r) $=$ the set of l.q. invertible (respy. r.q. invertible elements) of A. Then $G_q = G_q^l \bigcap G_q^r$. Write

$$S_q^l = A \backslash G_q^l,\ S_q^r = A \backslash G_q^r,\ S_q = S_q^l \bigcup S_q^r.$$

The elements of S_q^l, S_q^r, S_q are called respectively *l.q. singular*, *r.q.singular*, and *q. singular*. When A is unital, we have also G_i^l (respy. G_i^r) $=$ the set of l. invertible (respy. r. invertible) elements of A. Then $G_i = G_i^l \bigcap G_i^r$.

Now write

$$S^l = A \backslash G_i^l,\ S^r = A \backslash G_i^r, S^{bi} = S^l \bigcap S^r,\ S = S^l \bigcup S^r.$$

The elements of S^l, S^r S^{bi}, S are called respectively *l. singular*, *r. singular*, *bi-singular*, and *singular*.

3.7.2. LEMMA. *Let A be any algebra. Then $\sqrt{A} \subseteq G_q$. If A is unital then $\sqrt{A} \subseteq S^{bi}$.*

PROOF. The first inclusion follows from 1.2.24 ((a), (b)). Let now A be unital. Then by 1.2.2.4, $\sqrt{A}$ is a bi-ideal with $e \notin \sqrt{A}$. So, no element of $\sqrt{A}$ is l. or r. invertible, whence $\sqrt{A} \subseteq S^{bi}$.

3.7.3. DEFINITION. (Arens). Let A be a TA. An element x in A is called a left (respy. right) topological zero divisor †, or a l.t.z.d. (respy. r.t.z.d.) if there is a closed set F such that

† The notion t.z.d. under the name generalized zero divisor was first introduced by Shilov for commutative Banach algebras [7′, p.208]; his definition is essentially based on the condition (**) of 3.7.5.

$$0 \notin F \text{ and } 0 \in \overline{xF} \text{ respy. } \overline{Fx}$$

where bar denotes closure.

An element which is both a l.t.z.d. and a r.t.z.d. is called a *bi-topological zero divisor* or a *bi-t.z.d.*. Obviously, when A is commutative every l.t.z.d. or r.t.z.d. is a bi-t.z.d. The same is true (as can be easily seen) when A is anticommutative.

We denote by $\mathfrak{Z}^{lt}$, $\mathfrak{Z}^{rt}$, $\mathfrak{Z}^{bit}$ respectively the sets of l.t.z.d's, r.t.z.d.'s and bi-t.z.d.'s of A.

3.7.4. Remark. In any TA whose topology is not indiscrete, 0 is a bi-t.z.d. For, if $x \in A\backslash\bar{0}$, † then $0 \notin \bar{x} = F$ (say). But $0 \in \overline{0F} = \overline{F0} = \bar{0}$.

3.7.5. LEMMA. *Let A be a TA. For an element $x \in A$ to be a l.t.z.d. (respy. r.t.z.d.) it is necessary and sufficient that*

$$\exists \text{ a net } (x_\alpha) \textit{ in } A \textit{ with } x_\alpha \not\to 0 \textit{ but} \qquad (*)$$

xx_α *(respy. $x_\alpha x$)* $\to 0$.

If A is first countable then $(*)$ can be replaced by:

$$\exists \text{ a sequence } (x_n) \textit{ in } A \textit{ with } x_n \not\to 0, xx_n \textit{ (respy. } x_n x) \to 0. \qquad (**)$$

PROOF. Suppose there is a closed set F such that $0 \notin F$ and $0 \in \overline{xF}$ (respy. $\overline{Fx}$). Choose $x_\alpha \in F$ such that xx_α (respy. $x_\alpha x) \to 0$, then the net (x_α) clearly satisfies the condition $(*)$. Conversely, suppose $(*)$ holds. Since $x_\alpha \not\to 0$ there is a subnet $(x_{\alpha'})$ such that none of its subnets converge to 0. This means that if $F = \overline{\{x_{\alpha'}\}}$ then F is a closed such that $0 \notin F$, $0 \in \overline{xF}$ (respy. $0 \in \overline{Fx}$), whence x is a l.t.z.d. (respy. r.t.z.d.). This completes the proof of the first statement in the lemma. The second statement is clear.

3.7.6. PROPOSITION. (a) *If $x \in A$ is a l.t.z.d. (respy. r.t.z.d.) and $y \in A$ then yx (respy. xy) is a l.t.z.d. (respy. r.t.z.d.). In particular, if x is a bi-t.z.d. and $x \leftrightarrow y$ then xy is a bi-t.z.d.*

† For an element $a \in A$ we write $\bar{a}$ for $\overline{\{a\}}$

(b) *If xy is a l.t.z.d. (respy. r.t.z.d.) then x or y is a l.t.z.d. (respy. r.t.z.d.). In particular, x^2 is a l.t.z.d. (respy, r.t.z.d.) $\Rightarrow x$ is a l.t.z.d. (respy. r.t.z.d.).*

PROOF. (a) It suffices to observe that $xx_\alpha \to 0$ (respy. $x_\alpha x \to 0$) implies that $yxx_\alpha \to 0$ (respy. $x_\alpha xy \to 0$).

(b) Suppose that $xyy_\alpha \to 0$, $y_\alpha \not\to 0$. If $yy_\alpha \to 0$ then y is a l.t.z.d. On the other hand, if $yy_\alpha \not\to 0$, writing $x_\alpha = yy_\alpha$ we get $x_\alpha \not\to 0$, $xx_\alpha \to 0$, whence x is a l.t.z.. Similarly the proof when xy is a r.t.z.d.

3.7.7. In a unital algebra A a formal sum $\lambda+x$ ($\lambda \in \mathbb{K}, x \in A$) can be identified with the element $\lambda e + x$. But in the case of an algebra without unity $\lambda + x$ can be treated only as a formal sum. The notion of t.z.d. can be extended to formal sums. Thus we call $\lambda + x$ a l.t.z.d. (respy. r.t.z.d.) if there is a net $x_\alpha \not\to x$ with $\lambda x_\alpha + xx_\alpha = (\lambda + x)x_\alpha \to 0$ (respy. $x_\alpha(\lambda + x) \to 0$).

3.7.8. LEMMA. *Let A be a TA and $x \in A$. If $1 + x$ is a l.t.z.d. (respy. r.t.z.d.) then x is l. (respy. r.) q. singular. If A is unital and x is a l.t.z.d. (respy. r.t.z.d.) then x is l. (respy. r.) singular.*

PROOF. Suppose that $x_\alpha \not\to 0$ and $x_\alpha + xx_\alpha \to 0$. Then $x \circ x_\alpha \to x$. If x'_l exists then by premultiplying by x'_l on both sides we get $x_\alpha \to 0$ – a contradiction. So x is l.q. singular. Similarly, when $1 + x$ is r.t.z.d., x is r.q. singular.

Next let A be unital with unity e. Then if x in A is a *l.t.zd.*, writing $x = e + y$ we see that $1 + y = e + y = x$ is a l.t.z.d. So by above y is l.q. singular, whence $x = e + y$ is l. singular.

3.7.9. LEMMA. *Let A be a Haudorff TA. Then:*

(i) *Every l.z.d. (respy. r.z.d*[†]*) is a l.t.z.d. (respy. r.t.z.d.).*

(ii) *When A is finite-dimensional, a l.t.z.d. (respy. r.t.z.d.) is nothing but a l.z.d. (respy. r.z.d.).*

(iii) *An element x in A is a l.t.z.d. (respy. r.l.z.d.) iff l_α (respy. r_x) is not a topological monomorphism.*

[†] l.z.d. (respy. r.z.d) = left (respy. right) zero divisor.

PROOF. (i) If $xy = 0$ and $x, y \neq 0$ then $F = \{y\}$ is closed (since A is Hausdorff), $0 \notin F$ and $0 \in \overline{\{0\}} = \overline{0} = \overline{xy} = \overline{xF}$, so that x is a l.t.z.d. . Similarly, y is a r.t.z.d. .

(ii) Since A is finite-dimensional we may assume that it is a normed algebra (see 3.4.6. (ii), 2.1.12). If x is a l.t.z.d., then by 3.7.13 there is a sequence (x_n) with $xx_n \to 0$, $\|x_n\| = 1$. Since closed unit ball of A is compact (A being homeomorphic to $\mathbb{K}^n$), we can choose a subsequence (y_n) of (x_n) with $y_n \to y$. Then $xy_n \to 0$, $xy_n \to xy$. So $xy = 0$ and x is a l.z.d. Similarly, if x is r.t.z.d. it is r.z.d. .

(iii) If l_x is not a monomorphism then x is a l.z.d. and so by (i) a l.t.z.d. Next suppose that l_x is a monomorphism which is not topological. Then l_x^{-1} is not continuous (since l_x is always continuous) so that there is a net $y_\alpha \to 0$, but $x_\alpha = l_x^{-1} y_\alpha \not\to 0$. Therefore $xx_\alpha = l_x x_\alpha = y_\alpha \to 0$. By 3.7.5, x is a l.t.z.d. This proves the "if" part. For the "only if" part assume that x is a l.t.z.d., so that there is a net (x_α) such that

$$x_\alpha \not\to 0, \quad xx_\alpha \to 0. \tag{$*$}$$

If l_x is not $1-1$ it is not a monomorphism, and if it is $1-1$, the condition $(*)$ implies that l_x^{-1} is not continuous. Thus, l_x is not a topological monomorphism in either case, completing the proof.

(The proof of the corresponding statement coverning r.t.z.d. is similar.)

3.7.10. Examples of t.z. divisiors

(i) In the Banach algebra $\mathsf{C} = \mathsf{C}$ $([0,1]\ \mathbb{K})$ the function $f_0(t) = t$ is not a zero-divisor. But it is a t.z.d. . To see this, define $f_n \in C$ by: $f_n(t) = 0$ if $t \geqslant \frac{1}{n}$ and $= 1 - nt$ if $t < \frac{1}{n}$. Since $0 \leqslant f_n \leqslant 1$ and $f_n(0) = 1$ we have $\|f_n\| = 1$. Also $f_0 f_n \to 0$ since $\|f_0 f_n\| \leqslant \frac{1}{n}$. Thus f_0 is a t.z.d. .

(ii) Write $l_1 = l_1(\mathbb{K})$, the space of all infinite sequences $x = (x_n)$ $(x_n \in \mathbb{K})$ with $\sum_{n=1}^{\infty} |x_n| < \infty$; l_1 is a Banach space under the norm $\|x\| = \sum_{n=1}^{\infty} |x_n|$. Denote by $\mathbf{B} = \mathbf{B}(l_1)$ the algebra of a all bounded linear operators on l_1; $\mathbf{B}$ is a Banach algebra (under operator norm). Then l.o's S, T, U defined by

$$Sx = (x_2, x_3, \cdots), Tx = (0, x_1, x_2, \cdots), Ux = (x_1, 0, 0, \cdots)$$

belong to B : $\|S\| \leqslant 1$, $\|T\| = 1$, $\|U\| \leqslant 1$. Further $ST = I$ ($I =$ identity operator), $UT = 0$. T as a r.z.d. is a r.t.z.d. . On the other hand T being l. invertible is not a l.t.z.d. .

3.7.11. DEFINITION. An element x in A is called a *symmetric topological zero divisor* or *s.t.z.d.* if there is a net (x_α) in A with $x_\alpha \to 0$, xx_α and $x_\alpha x \to 0$. Evidently a s.t.z.d. is a bi-t.z.d. P.G. Dixon has given an example of an element in a Banach algebra which is a bi-t.z.d. (actually a bi.z.d.) but not s.t.z.d. (see [4, p.13, Example 13]).

3.7.12. Remark. It is clear from the definitions that a l.t.z.d. or a r.t.z.d. x of A continues to be so with respect to any TA B which is an extension[†] of A. If both A, B are unital then by 3.7.8, x is singular in A as well as B. Following Lorch we call a singularity which continues to be a singularly in any unital extension a *permanent singularity.* Thus any t.z.d. is a permanent singularity. F. Quigley has shown that every unital Banach algebra A has a unital Banach algebra extension B such that every singular element of A is a zero-divisor of B (see [23, pp.25-27]). On the other hand Arens has proved that in commutative unital Banach algebra the permanently singular elements are precisely the t.z.d.'s. For a proof of this theorem (extended to ρ-Banach algebra) and other interesting connection between permanently singular elements and t.z.d's in TA's see [31, p.32 and pp.112-123].

3.7.13. LEMMA. *In a ρ-normed algebra $A = (A, \|\cdot\|)$ an element x is a l.t.z.d. (respy. r.t.z.d.) iff there is sequence (x_n) in A with $\|x_n\| = 1$, xx_n (respy. $x_n x) \to 0$. In particular, x is a s.t.z.d. if there is a sequence (x_n) with $\|x_n\| = 1$ and $xx_n \to 0$, $x_n x \to 0$.*

PROOF. The "if" part of the first statement follows form 3.7.5 (since $\|x_n\| = 1 \Rightarrow x_n \not\to 0$). For the, "only if" part assume that x is a l.t.z.d. Since A is first countable, by $(**)$ of 3.7.5 there is a sequence (x_n) with $x_n \not\to 0$, $xx_n \to 0$. We may assume (after passing to a subsequence if necessary) that $\|x_n\| \geqslant \eta > 0$ for all

---

[†] i.e. A is a subalgebra of B and the topology of A is the relative topology from B.

n and some $\eta > 0$. Set $y_n = x_n/\|x_n\|^{\frac{1}{\rho}}$. Then $\|y_n\| = \|1\|$, and $xy_n \to 0$ since $\|xy_n\| = \|xx_n\|/\|x_n\| \leqslant \eta^{-1}\|xx_n\|$, $\|xx_n\| \to 0$. The proof of the "only if" part when x is a r.t.z.d. is similar. The final assertion of the lemma is clear.

3.7.14. LEMMA. *Let A be a ρ-normed algebra and $\hat{A}$ its completion. If $x \in A$ is a l.t.z.d., r.t.z.d. or s.t.z.d. of $\hat{A}$ then it is accordingly the same of A.*

PROOF. We shall prove the result only when x is a l.t.z.d. (the proofs for the other two being similar). By 3.7.13 there is a sequence $(\hat{x}_n)$ in $\hat{A}$ with $\|\hat{x}_n\| = 1$, $x\hat{x}_n \to 0$. Choose $x_n \in A$ such that $\|x_n - \hat{x}_n\| < \frac{1}{2n}$ so that $x_n - \hat{x}_n \to 0$ (as $n \to \infty$). Then

$$\|x_n\| \geqslant \|\hat{x}_n\| - \|\hat{x}_n - x_n\| \geqslant 1 - \frac{1}{2n} \geqslant \frac{1}{2},$$

so that $x_n \not\to 0$. Also, $xx_n = x(x_n - \hat{x}_n) + x\hat{x}_n \to 0$. Therefore x is a l.t.z.d. of A.

3.7.15. PROPOSITION (Rickart). *Let A be a ρ-Banach algebra and $a \in A$. Then we have:*

(i) *a is a l.t.z.d. (respy. r.t.z.d.) iff*
   $\omega_l(a) = 0$ (respy. $\omega_r(a) = 0$), where

$$\omega_l(a) = \inf_{x \neq 0} \|ax\|/\|x\|, \omega_r(a) = \inf_{x \neq 0} \|xa\|/\|x\|.$$

(ii) *Suppose that l_a (respy. r_a) is $1-1$. Then a is not a l.t.z.d. (respy. r.t.z.d) iff aA (respy. Aa) is closed.*

(iii) *If a is not a l.t.z.d. (respy. r.t.z.d.) then aA (respy. Aa) is closed.*

PROOF. As usual we shall prove only all results pertaining to l.t.z.d. .

(i) If $\omega_l(a) = 0$, there is a sequence (a_n) in A which $\|aa_n\|/\|a_n\| \to 0$. Writing $x_n = \|a_n\|^{-\frac{1}{\rho}} a_n$, we get $\|x_n\| = 1$, $\|ax_n\| \to 0$, whence by 3.7.13, a is a l.t.z.d. Conversely, if $\|x_n\| = 1$, $\|ax_n\| \to 0$ then clearly $w_l(a) = 0$.

(ii) Since l_a is $1-1$, l_a^{-1} exists and $l_a^{-1}(aA) = A$. If aA is closed then aA, as a closed subspace, is a ρ-Banach space. But the open mapping theorem, l_a^{-1} is bounded. But this means that $\sup\limits_x \|x\|/\|ax\| < \infty$, so that $\omega_l(a) \neq 0$, and so by (i), a is not a l.t.z.d. . Conversely, suppose that a is not a l.t.z.d. and $ax_n \to b$. Then $\omega_l(a) > 0$, whence $\|l_a^{-1}\| = 1/\omega_l(a) < \infty$. Since

$$\|x_n - x_m\| \leqslant \|l_a^{-1}\| \, \|ax_n - ax_m\| \to 0,$$

as $n, m \to \infty$, by completeness of A, $x_n \to x$ (say). Then $ax_n \to ax$. By Hausdorff property of A, $ax = b$ and aA is closed.

(iii) This follow from (ii) since if a is not a l.t.z.d. it is also not a l.z.d. so that l_a is $1-1$.

3.7.16. LEMMA. *Let A be a ρ-Banach algebra containing an element a satisfying the conditions:*

(i) *There is a sequence (λ_n) in $\mathbb{K}$ with $\lambda_n \to \infty$ such that each $\lambda_n a$ is r. (respy. l.) q. invertible.*

(ii) *The element a is not a l.t.z.d. (respy. not a r.t.z.d.).*

Then A has a l. (respy. r.) unity u.

PROOF. Write $b_n = (\lambda_n a)'_r$. Then $\lambda_n a + b_n + \lambda_n a b_n = 0$, so that

$$ab_n = -a - \frac{b_n}{\lambda_n} \tag{1}$$

and hence also

$$a\frac{b_n}{\|b_n\|^{\frac{1}{\rho}}} = -\frac{a}{\|b_n\|^{\frac{1}{\rho}}} - \frac{b_n}{\lambda_n \|b_n\|^{\frac{1}{\rho}}}. \tag{2}$$

If $\|b_n\| \to \infty$ then in (2) the R.H.S. $\to 0$ and a would be a l.t.z.d., contradicting the hypothesis (ii). So we must have $\|b_n\| \leqslant C$ for all n, whence it follows from (1) that $ab_n \to -a$, $-a \in \overline{aA} = aA$ (by (ii) above and 3.7.15 (iii)). It follows that $a = au$ for some $u \in A$. Now for any $x \in A$ we have $a(ux - x) = 0$. Since a is not a l.z.d., we conclude that $ux = x$, proving u is a l. unity of A.

3.7.17. PROPOSITION. *Every element of a radical* ρ*-Banach algebra* A *is a bi-t.z.d.* .

PROOF. Suppose that A contains an element a which is not a bi-t.z.d. . Since every element of $\sqrt{A} = A$ is q. invertible, we can apply 3.7.16 to conclude that A contains a l. or a r. unity u, and hence a non-zero idempotent u. But this contradicts 1.2.24. Hence the proposition.

3.7.18. COROLLARY. *If* A *is a* ρ*-Banach algebra then every element of* $\sqrt{A}$ *is a bi-t.z.d.* .

PROOF. If $a \in \sqrt{A}$ then by 3.7.17, a is a bi-t.z.d. of $\sqrt{A}$ and so also of A.

3.7.19. Let A be a real TA and $\tilde{A}$ its complexification. Then the property of an element x in A being a l.t.z.d., r.t.z.d., or s.t.z.d. clearly carries over when x is regarded as an element of $\tilde{A}$. Further we have

3.7.20. LEMMA. (i) $x + ix$ *is a l.t.z.d., r.t.z.d., or s.t.z.d. of* $\tilde{A}$ *according as* x *is a l.t.z.d., r.t.z.d., or s.t.z.d. of* A.

(ii) *If* $x \in A$ *is a l.t.z.d.* (*respy. r.t.z.d.*) *of* $\tilde{A}$ *then it is already a l.t.z.d.* (*respy. r.t.z.d.*) *of* A.

PROOF. (i) Clear.

(ii) Assume that $x(x_\alpha + iy_\alpha) \to 0$, $x_\alpha + iy_\alpha \not\to 0 (x_\alpha, y_\alpha \in A)$. Then either $x_\alpha \not\to 0$ or $y_\alpha \not\to 0$, while $xx_\alpha \to 0$, $yy_\alpha \to 0$. Therefore x is a l.t.z.d. of A if it is of $\tilde{A}$. Similarly x is a r.t.z.d. of A if it is of $\tilde{A}$.

3.7.21. LEMMA. *Let* $A = (A, \|\cdot\|)$ *be a* ρ*-normed algebra* (*respy. a unital* ρ*-normed algebra*) *and* (x_n) *a sequence of q. invertible* (*respy. invertible*) *elements of* A *such that* (i) $x_n \to x$ (ii) $\sup_n \|x'_n\|$ (*respy.* $\sup_n \|x_n^{-1}\| = \infty$, *where* x'_n (*respy.* x_n^{-1}) *is q. inverse* (*respy. inverse*) *of* x_n. *Then* $1 + x$ (*respy.* x) *is a s.t.z.d.* .

PROOF. First consider the case where x_n are q.invertible. We assume (as we may) that $\|\cdot\|$ is sm. and by passing to a subsequence (if necessary) we can also assume that $\|x'_n\| \to \infty$. Define $y_n = x'_n/\|x'_n\|^{\frac{1}{\rho}}$, so that $\|y_n\| = 1$. Then, by using the relation $x_n + x'_n + x_n x'_n = 0$, we have

$$(1+x)y_n = y_n + xy_n = y_n + (x - x_n)y_n + x_n y_n = -\frac{x_n}{\|x'_n\|^{\frac{1}{\rho}}} + (x - x_n)y_n,$$

so that

$$\|(1+x)y_n\| \leqslant \frac{\|x_n\|}{\|x'_n\|} + \|x - x_n\|.$$

Since $x_n \to x$, $\|x'_n\| \to \infty$ we get $(1+x)y_n \to 0$. Similarly, $y_n(1+x) \to \infty$. So $1+x$ is a s.t.z.d. .

Next, consider the case where x_n are invertible. Writing $z_n = x_n - e$, $z = x - e$ we see that z_n are q.invertible, $z_n \to z$,

$$\sup \|z'_n\| = \sup \|x_n^{-1} - e\| = \infty.$$

By the first part (proved above) we have $xy_n = (e+z)y_n = (1+z)y_n \to 0$, and similarly $y_n x \to 0$. Also $\|y_n\| = 1$. So, by 3.7.13, x is a s.t.z.d.

3.7.22. LEMMA. *Let A be a unital TA, G_i its group of invertible elements and $x_\alpha, x \in G_i$, $x_\alpha \to x$. Then $x_\alpha^{-1} \to x^{-1}$ iff the net (x_α^{-1}) is essentially bounded.*

PROOF. If $x_\alpha^{-1} \to x^{-1}$ then (x_α^{-1}) being a C-net is essentially bounded (by 2.3.6). Conversely, if (x_α^{-1}) is essentially bounded then, by 2.3.8 (iii),

$$x_\alpha^{-1}(x_\alpha - x) \to 0, \text{ i.e. } x_\alpha^{-1}x \to e.$$

It follows that

$$x_\alpha^{-1} = (x_\alpha^{-1}x)x^{-1} \to ex^{-1} = x^{-1}.$$

3.7.23. PROPOSITION. *Let A be a unital Hausdorff complete TA which is a C algebra. Suppose that $x_\alpha \in G_i$ and $x_\alpha \to x$ in A. Then $x \in G_i$ iff (x_α^{-1}) is a C-net.*

PROOF. If $x \in G_i$ then, A being a C-algebra, $x_\alpha^{-1} \to x^{-1}$, so that (x_α^{-1}) is a C-net. On the other hand, if we suppose that

(x_α^{-1}) is a C-net then by completeness of A, $x_\alpha^{-1} \to y$ (say) $\in A$. Since (x_α) is convergent it is essentially bounded, and so by 2.3.8 (iii),

$$x_\alpha(x_\alpha^{-1} - y) \to 0, \text{ or, } x_\alpha y \to e.$$

Also, $x_\alpha y \to xy$, whence by uniqueness of limits (A being Hausdorff) we get $xy = e$. Similarly we can show that $yx = e$. Thus x is invertible, $x \in G_i$, completing the proof.

3.7.24. PROPOSITION. *Let A be a CI (respy. CQ) algebra, G_i (respy. G_q) its group of invertible (respy. q. invertible) elements. If*

$$(x_n) \in G_i \text{ (respy. } G_q) \text{ and } x_n \to x \in A$$

then $x \in G_i$ (respy. G_q) iff x_n^{-1} (respy. (x_n')) is bounded.

PROOF. First suppose that $x_n \in G_i$, $x_n \to x$, $x \in G_i$. Then $x_n^{-1} \to x^{-1}$ (since A is a C algebra) and hence (x_n^{-1}) is bounded. Next suppose that $x_n \in G_i$, $x_n \to x \in A$ and (x_n^{-1}) is bounded. Then by 2.3.8 (iii),

$$(x_n - x)x_n^{-1} \to 0, \; x_n^{-1}(x_n - x) \to 0 \text{ i.e. } xx_n^{-1} \to e, \; x_n^{-1}x \to e.$$

Therefore, since A is a I algebra, xx_n^{-1} and $x_n^{-1}x$ are invertible for sufficiently large n, whence by 1.1.30, x is invertible, $x \in G_i$. This completes the proof of the part relating to inverses.

To prove the part relating to q. inverses, consider the unitization A_1 of A and observe that

$$(e_1 + x_n)^{-1} = e_1 + x_n'.$$

It follows that (x_n') is bounded iff $((e_1 + x_n)^{-1})$ is bounded. Therefore, by the part proved, $e_1 + x \in G_i$ iff $((e_1 + x_n)^{-1}) = (e_1 + x_n')$ is bounded iff (x_n') is bounded, completing the proof.

3.7.25. COROLLARY (Gelfand). *Let A be a unital ρ*† *-Banach algebra, G_i its group of invertible elements, $x_n \in G_i$ and $x_n \to x \in A$. Then $x \in G_i$ iff there is a constant $C > 0$*

---

† Gelfand considered only the case $\rho = 1$.

such that $\|x_n^{-1}\| \leqslant C$ *for all* n, *where* $\|\cdot\|$ *denoting the* ρ*-norm of* A.

PROOF. Since a unital ρ-Banach algebra is a CI algebra (by 3.6.23(b)) and the boundedness in A is the same as the boundedness with respect to $\|\cdot\|$ (by 3.2.13), the corollary follows from 3.7.24.

3.7.26. COROLLARY. *In a unital* ρ*-Banach algebra* A *the limit of a convergent sequence of invertible elements is either an invertible element or is a s.t.z.d. . In particular, every element of* ∂G_i † *is a s.t.z.d. .*

PROOF. The first assertion follows by combining 3.7.25 and 3.7.21. The second follows from the first since G_i being open ∂G_i is disjoint with G_i.

3.7.27. COROLLARY. *Every element of the radical* $\sqrt{A}$ *of a unital* ρ*-Banach algebra is a s.t.z.d.* (*cf.* 3.7.17).

PROOF. If $a \in \sqrt{A}$ then since $na \in \sqrt{A}$, na is q. invertible and consequently $(e+na)^{-1}$ exists. It follows that $(\frac{e}{n}+a)^{-1} = n(e+na)^{-1}$ exists, whence by 3.7.26, $a = \lim_n(\frac{e}{n}+a)$ is either invertible or a s.t.z.d. . But a cannot be invertible since $a \in \sqrt{A}$, so a is a s.t.z.d. as required.

3.7.28. PROPOSITION. *Let* A *be a unital* ρ*-Banach algebra,* $x \in A$ *and* $\sigma(x)$ *the spectrum of* x. *If* $\sigma(x) \neq \emptyset$ †† *and* $\lambda \in \partial\sigma(x)$ *then* $x - \lambda e$ *is a s.t.z.d. .*

PROOF. Since $\lambda \in \partial\sigma(x)$ and $\sigma(x)$ is closed, there is a sequence $\lambda_n \in \rho(x) = \mathbb{K}\backslash\sigma(x)$ such that $\lambda_n \to \lambda$. Then $x - \lambda_n e \in G_i$, $x - \lambda e \notin G_i$. Since $x - \lambda_n e \to x - \lambda e$ it follows that $x - \lambda e \in \partial G_i$ and consequently, by 3.7.26, $x - \lambda e$ is a s.t.z.d. .

3.7.29. PROPOSITION (Rickart). *Let* $A = (A, \|\cdot\|)$ *be a*

---

† For a subset S of a topological space X we denote by ∂S the frontier of S, i.e., $\partial S(=\partial S') = \overline{S} \bigcap \overline{S}'$, where $S' = X\backslash S$ and bar denotes closure.

†† This condition is satisfied for every element x if A is complex or strictly real. Moreover, whenever $\sigma(x) \neq \emptyset$ we have also $\partial\sigma(x) \neq \emptyset$, since by 6.1.2, $\sigma(x) \neq \mathbb{K}$, is closed but not open (by connectedness of $\mathbb{K}$).

ρ-Banach[†] *algebra with* $\|\cdot\|$ *sm. and* $x_n \to x$ *in* A *with* $x_n \leftrightarrow x$ *(for all* n*). Let* $\nu = \nu_{\|\cdot\|}$*. Then:*

(i) *If* A *is unital,* x_n *invertible and* $\nu(x_n^{-1})$ *is bounded then* x *is invertible.*

(ii) *If* x_n *are q. invertible and* $\nu(x_n')$ *bounded then* x *is q. invertible.*

PROOF. (i) Suppose that $\nu(x_n^{-1}) \leqslant C$. Then we have

$$\begin{aligned}\nu(e - x_n^{-1}x) &= \nu(x_n^{-1}x_n - x_n^{-1}x) \leqslant \nu(x_n^{-1})\nu(x_n - x) \\ &\qquad \text{(using 3.3.7. (iii))} \\ &\leqslant C\|x_n - x\| \to 0.\end{aligned}$$

It follows that for sufficiently large n, $\nu(e - x_n^{-1}x) < 1$, whence by 3.3.20 (ii), $x_n^{-1}x$ is invertible, so that x is also invertible.

(ii) This can be deduced from (i) by passing to the unitization A_1 of A.

3.7.30. PROPOSITION. *Let* $A = (A, \|\cdot\|)$ *be a unital* ρ*-normed algebra with* G_i *its group of invertible elements. If* $x_n \in G_i, x_n \to x$ *in* A *and* x *is a t.z.d. then* $\sup\|x_n^{-1}\| = \infty$ *and* x *is a s.t.z.d.* .

PROOF. Let $\hat{A}$ be the completion of A. Then x is a t.z.d. of $\hat{A}$ and so not invertible in $\hat{A}$, whence by 3.7.26, $\sup\|x_n^{-1}\| = \infty$ and so by 3.7.21, x is a s.t.z.d. .

3.7.31. COROLLARY. *An element* $x \in \partial G_i \backslash G_i$ *is a s.t.z.d. iff* $\sup\|x_n^{-1}\| = \infty$*, where* $x_n \in G_i$ *and* $x_n \to x$.

PROOF. This follows from 3.7.21, 3.7.30.

3.7.32. PROPOSITION. *In a strictly, real unital* ρ*-Banach algebra* A *every singular element is a bi-t.z.d. and the square of a singular element a s.t.z.d.* .

PROOF. Since x is singular and A is strictly real we have

† Rickart considered only Banach algebras (i.e. $\rho = 1$ case).

$0 \in \sigma(x) = \tilde{\sigma}(x) \subseteq \mathbb{R}$, and consequently we have $0 \in \tilde{\sigma}(x)^2 = \tilde{\sigma}(x^2) = \sigma(x^2)$. Since $\tilde{\sigma}(x) \subseteq \mathbb{R}$, $\sigma(x^2) = \tilde{\sigma}(x)^2 \geqslant 0$. It follows that $0 \in \partial\sigma(x^2)$, so that by 3.7.28, x^2 is a s.t.z.d.. Hence, by 3.7.6 (b), x is a bi-t.z.d.

3.7.33. COROLLARY. *In (a strictly real) A we have*

$$S(= S^l \bigcup S^r) = S^{bi} = S^l = S^r = \mathfrak{Z}^{lt} = \mathfrak{Z}^{rt} = \mathfrak{Z}^{bit}.$$

PROOF. It suffices to observe that a bi-t.z.d. is, by 3.7.8, both l. singular and r. singular.

3.7.34. Remarks. In the Banach algebra of complex valued continuous functions on the unit interval, with sup norm, every singular element is a t.z.d. . On the other hand, in the Banach algebra $\mathcal{A}$ of all complex-valued continuous functions $f = f(z)$ on the closed unit $|z| \leqslant 1$ which are holomorphic on $|z| < 1$, the function $f_0(z) = z$ $(|z| \leqslant 1)$ is singular but not a t.z.d. (see [10, p.70] or [7's p.29]). Further the real algebra $\mathcal{A}^{[\mathbb{R}]}$ gives an example of a real Banach algebra containing a singular element (f_0) which is not a t.z.d.; note that $\mathcal{A}^{\mathbb{R}}$ is not strictly real (since it has complex structure), so that this example does not contradict 3.7.32.

3.7.35. PROPOSITION. *Let B be a complex or strictly real unital ρ-Banach algebra and A a closed subunital algebra of B. Then, for every $x \in A$,*

$$\sigma_B(x) \subseteq \sigma_A(x), \tag{$*$}$$

$$\partial\sigma_A(x) \subseteq \partial\sigma_B(x). \tag{$**$}$$

Further, if either $(\sigma_A(x))^\circ = \emptyset$ or $\rho_B(x)$ is connected we have

$$\sigma_A(x) = \sigma_B(x). \tag{$***$}$$

PROOF. The relation $(*)$ has already been obtained (see 1.7.20). For proving $(**)$ we note that if $\lambda \in \sigma_A(x)$ then by 3.7.28, $x - \lambda e$ is a s.t.z.d. of A and hence also of B. In particular $x - \lambda e$ is singular in B and so $\lambda \in \sigma_B(x)$. Thus $\partial\sigma_A(x) \subseteq \sigma_B(x)$. Again, by $(**)$ of 1.7.20, $\rho_A(x) \subseteq \rho_B(x)$. Therefore

$$\begin{aligned}\partial\sigma_A(x) &= \partial\sigma_A(x)\bigcap\sigma_B(x) = (\overline{\rho_A(x)}\bigcap\sigma_A(x))\bigcap\sigma_B(x)\\ &= \overline{\rho_A(x)}\bigcap\sigma_B(x) \subseteq \overline{\rho}_B(x)\bigcap\sigma_B(x) = \partial\sigma_B(x)\end{aligned}$$

which is $(**)$.

Assume now that $(\sigma_A(x))^o = \emptyset$. Then

$$\sigma_A(x) = \partial\sigma_A(x) \subseteq \partial\sigma_B(x) \subseteq \sigma_B(x)$$

(since $\sigma_B(x)$ is closed). Combining this inclusion with the inclusion $(*)$ we get $(***)$. Next let $\rho_B(x)$ be connected. Since $\rho_A(x) \subseteq \rho_B(x)$ we can write

$$\rho_B(x) = \rho_A(x)\dot{\bigcup}(\rho_B(x)\backslash\rho_A(x)) = \rho_A(x)\dot{\bigcup}(\sigma_A(x)\backslash\sigma_B(x)) \quad (1)$$

where $\dot{\bigcup}$ denotes disjoint union. Since $\partial\sigma_A(x) \subseteq \partial_B(x)$ every point of $\sigma_A(x)\backslash\sigma_B(x)$ is an interior point, i.e. $\sigma_A(x)\backslash\sigma_B(x)$ is open. The connectedness of $\rho_B(x)$ and (1) imply that $\sigma_A(x) = \sigma_B(x)$, proving $(***)$ in this case as well.

3.7.36. COROLLARY. *Let A be a closed subunital algebra of a unital complex ρ-Banach algebra B and $x \in A$. Then*

$$\sigma_A(x) \subseteq \mathbb{R} \text{ iff } \sigma_B(x) \subseteq \mathbb{R}$$

and when these inclusions hold we have actually $\sigma_A(x) = \sigma_B(x)$.

PROOF. Since always $\sigma_B(x) \subseteq \sigma_A(x)$ it is enough to prove the "if" part of the "iff" assertion. Assume therefore that $\sigma_B(x) \subseteq \mathbb{R}$. By $(**)$ of 3.7.35, $\partial\sigma_A(x) \subseteq \partial\sigma_B(x) \subseteq \mathbb{R}$, and by 6.1.2, $\sigma_A(x)$ is compact. It follows that $(\sigma_A(x))^\circ = \emptyset$, whence by 3.7.35, $\sigma_A(x) = \sigma_B(x)$, completing the proof.

3.7.37. PROPOSITION. *A closed subalgebra A of a strictly real ρ-Banach algebra B is strictly real.*

PROOF. First assume that B is unital with unity e and set $A_1 = A + \mathbb{R}e$ ($A_1 = A$ if $e \in A$). Consider the complexification $\tilde{B}$ of B and write $\tilde{A}_1 = A_1 + iA_1$; $\tilde{A}_1$ is a closed subunital algebra of $\tilde{B}$. Now $\sigma\tilde{_{A_1}}(x) = \sigma_{\tilde{A_1}}(x), \sigma\tilde{_B}(x) = \sigma_{\tilde{B}}(x)$. Since B is strictly $\sigma_{\tilde{B}}(x) \subseteq \mathbb{R}$. By 3.7.36, $\sigma\tilde{_{A_1}}(x) = \sigma_{\tilde{A_1}}(x) = \sigma_{\tilde{B}}(x) \subseteq \mathbb{R}$. It follows that A_1, and hence A (by 1.9.5) is strictly real.

If B has no unity, consider its unitization B_1 which again is strictly real. Also A is closed in B_1 (since B is closed B_1). Applying the result just obtained above to the unital B_1 we conclude that A is strictly real.

3.7.38. DEFINITION. (Żelazko). Let A be a TA. A pair (S,T) of subsets of A is called a *generalized topological divisor of zero* [31,p.73] or g.t.z.d. if $0 \notin \overline{S}, \overline{T}$ but $0 \in \overline{ST}$, where bar denotes the closure.

3.7.39. LEMMA. *A g.t.z.d.* (S,T) *with* $S = \{x\}$ *(respy.* $T = \{y\}$*) is a l.t.z.d.* x *(respy. r.t.z.d.* y*).*

PROOF. Write $\overline{T} = F$. Then we have $0 \neq F$ while $0 \in \overline{ST} = \overline{xT} \subseteq \overline{x\overline{T}} = \overline{xF}$, proving x is a l.t.z.d. . Similarly, y is a r.t.z.d. .

3.7.40. LEMMA. *A pair of nets* $(x_\alpha), (y_\beta)$ *such that* $x_\alpha \not\to 0, y_\beta \not\to 0$ *and* $x_\alpha y_\beta \to 0$ *determine a g.t.z.d.*

PROOF. Since $x_\alpha \not\to 0$ there exists a nucleus U and a subnet $(x_{\alpha'})$ of (x_α) which lies outside U. It follows that if $S = \{x_{\alpha'}\}$ then $0 \notin \overline{S}$. Similarly there is a subnet $(y_{\beta'})$ of (y_β) such that $0 \notin \overline{T}$, where $T = \{y_{\beta'}\}$. Since $(x_{\alpha'} y_{\beta'})$ is a subnet of $(x_\alpha y_\beta)$ and $x_\alpha y_\beta \to 0$ we get $0 \in \overline{ST}$, proving (S,T) is a g.t.z.d.

3.7.41. PROPOSITION (Żelazko). *A commutative unital Hausdorff TA which admits no g.t.z.d. must be a* C *algebra.*

PROOF. Suppose that A is not a C algebra. Then we can find a net (x_α) in G_i with

$$x_\alpha \to x_0 \in G_i \ (1); x_\alpha^{-1} \not\to x_0^{-1}. \qquad (2)$$

By 3.7.22, (x_α^{-1}) is not essentially bounded and so by 2.3.6, it is not a C-net. It follows that there is a nucleus U and subnets $(x_\beta), (x_{\beta'})$ of the net (x_α) such that $x_{\beta'}^{-1} - x_\beta^{-1} \notin U$, so that

$$x_{\beta'}^{-1} - x_\beta^{-1} \not\to 0. \qquad (3)$$

As subnets of (x_α) it follows from (1) that

$$x_\beta \to x_0, \; x_{\beta'} \to x_0$$

whence

$$x_{\beta'} x_\beta \to x_0^2 \neq 0 \qquad \text{(since } x_0^2 \in G_i\text{)}.$$

Since A is Hausdorff we have

$$x_{\beta'} x_\beta \not\to 0. \tag{4}$$

On the other hand, by using the commutativity of A we get

$$x_{\beta'} x_\beta (x_{\beta'}^{-1} - x_\beta^{-1}) = x_\beta - x_{\beta'} \to 0. \tag{5}$$

From (4), (3), (5) we conclude, by 3.7.40, that A has a g.t.z.d., contradicting the hypothesis. Hence A is a C algebra.

3.7.42. Remark. For more information on g.t.z.d. see [31, pp.73-77].

3.7.43. Remark. For the class of topological algebras called locally sm. convex (defined in 4.4.11) there is a weaker definition of t.z.d. due to Michael. For results concerning such t.z.d.'s see [20, pp.43-48].

CHAPTER IV

LOCALLY PSEUDO-CONVEX SPACES AND ALGEBRAS

§ 1. ρ-convexity

4.1.1. DEFINITION. Following Landsberg [9', p.104] a subset S of a LS X (over $\mathbb{K}$) is called *ρ-convex*, where $0 < \rho \leqslant 1$, if

$$x, y, \in S;\ \alpha, \beta \in \mathbb{R}, \alpha, \beta \geqslant 0 \text{ and } \alpha^\rho + \beta^\rho = 1 \qquad (*)$$

$\Rightarrow \alpha x + \beta y \in S$.

It is called ***absolutely ρ-convex*** if

$$x, y \in S; \alpha, \beta \in \mathbb{K} \text{ and } |\alpha^\rho|^\rho + |\beta|^\rho \leqslant 1 \qquad (**)$$

$\Rightarrow \alpha x + \beta y \in S$.

Evidently when $\rho = 1$, ρ-convex (respy. absolutely ρ-convex) is just convex (respy. absolutely convex) in the usual sense.

Let S be a subset of X. An element of the form $\sum_j \alpha_j x_j = \sum_{j=1}^{n} \alpha_j x_j$, with $x_j \in S$, $\alpha_j \geqslant 0$ and $\sum_j \alpha_j^\rho = 1$, is called a *ρ-convex linear combination* of elements of S. The meaning of an ***absolutely ρ-convex*** linear combination of elements of S is clear. For any subset S of X we denote by $C_\rho(S)$ (respy. $|C_\rho|(S)$) the set of all ρ-convex (respy. ***absolutely*** ρ-convex) linear combination elements of S; $C_\rho(S)$ (respy. $|C_\rho|(S)$) is called the ***ρ-convex*** (respy. ***absolutely ρ-convex***) ***hull*** of S. Obviously, $C_\rho(S) \subseteq |C_\rho(S)|$.

4.1.2. LEMMA. (a) *If S is ρ-convex (respy. absolutely ρ-convex) then $S = C_\rho(S)$ (respy. $|C_\rho|(S)$).*

(b) *For any subset S of X, $C_s(S)$ (respy. $|C_\rho|(S)$) is the smallest ρ-convex (respy. absolutely ρ-convex) subset of X containing S.*

(c) *If* S *is* ρ*-convex (respy. absolutely* ρ*-convex) so is* λS $(\lambda \in \mathbb{K})$.

PROOF. (a) Let S be ρ-convex and $x = \sum_{j=1}^{n} \alpha_j x_j$ with $\alpha_j \geqslant 0$, $x_j \in S$ and $(*)$ $\sum_{j=1}^{n} \alpha_j^\rho = 1$. We shall show that $x \in S$ by induction on n the length of the combination). If $n = 2$, $x \in S$ by the definition of ρ-convexity. Assume now that $x \in S$ whenever the length is $n-1$. For the α_j^ρ occuring in $(*)$ write $\sum_{j=1}^{n-1} \alpha_j^\rho = \beta^\rho$ (say), so that we $\sum_{j=1}^{n-1} \frac{\alpha_\rho}{\beta} x_j \in S$. Therefore, since $\beta^\rho + \alpha_n^\rho = 1$, it follows that $\sum_{j=1}^{n} \alpha_j x_j = \beta \left(\sum_{j=1}^{n-1} \frac{\alpha_j}{\beta} x_j \right) + \alpha_n x_n \in S$. Thus, $C_\rho(S) \subseteq S$ whence $C_\rho(S) = S$ (since always $S \subseteq C_\rho(S)$). The assertion concerning $|C_\rho|(S)$ is proved similarly.

(b) It follows from (a) that if S_1 is a ρ-convex set with $S_1 \supseteq S$ then $S_1 = C_\rho(S_1) \supseteq C_\rho(S)$. We will now show that $C_\rho(S)$ is ρ-convex. Suppose that

$$x = \sum_{j=1}^{m} \alpha_j x_j, \; y = \sum_{k=1}^{n} \beta_k y_k \; (x_j, y_k \in S)$$

with $0 \leqslant \alpha_j, \; \beta_k; \; \sum_{j=1}^{m} |\alpha_j|^\rho = 1, \; \sum_{k=1}^{n} |\beta_k|^\rho = 1$.
If $0 \leqslant \lambda, \mu; \lambda^\rho + \mu^\rho = 1$ then

$$\lambda x + \mu y = \sum_{j=1}^{m} \lambda \alpha_j x_j + \sum_{k=1}^{n} \mu \beta_k y_k.$$

Since

$$\sum_{j=1}^{m} (\lambda \alpha_j)^\rho + \sum_{k=1}^{n} (\mu \beta_k) = \lambda^\rho \left(\sum_j \alpha_j^\rho \right) + \mu^\rho \left(\sum_k \beta_k^\rho \right) = \lambda^\rho + \mu^\rho = 1$$

we conclude that $\lambda x + \mu y \in C_\rho(S)$ which proves that $C_\rho(S)$ is ρ-convex. The proof of the assertion concerning $|C_\rho|(S)$ is similar.

(c) Clear.

4.1.3. Remark. If $\sum_{j=1}^{n} \alpha_j x_j$, $\sum_{j=1}^{n} \beta_j x_j$ are absolutely ρ-convex linear combinations of elements and $|\lambda|^\rho + |\mu|^\rho \leqslant 1$, then $\sum_{j=1}^{n} (\lambda\alpha_j + \mu\beta_j) x_j$ is also an absolutely ρ-convex linear combination of elements.

In fact we have

$$\begin{aligned}
\sum_j |\lambda\alpha_j + \mu\beta_j|^\rho &\leqslant \sum_j (|\lambda|\ |\alpha_j| + |\mu||\beta_j|)^\rho \\
&\leqslant \sum_j (|\lambda|^\rho\ |\alpha_j|^\rho + |\mu|^\rho |\beta_j|^\rho) \\
&\qquad\qquad (\text{since } 0 < \rho \leqslant 1) \\
&= |\lambda|^\rho \sum_j |\alpha_j|^\rho + |\mu|^\rho \sum_j |\beta|^\rho \leqslant |\lambda|^\rho + |\mu|^\rho \leqslant 1.
\end{aligned}$$

On the other hand, the analogous assertion for ρ-convex linear combination can fail if $\rho < 1$. For example, take $n = 2$; $\alpha_1 = \alpha_2 = \frac{1}{4}$; $\beta_1 = \frac{1}{9}$, $\beta_2 = \frac{4}{9}$; $\lambda = \mu = \frac{1}{4}$; $\rho = \frac{1}{2}$.

Then

$$\gamma_1 = (\lambda\alpha_1 + \mu\beta_1)^{\frac{1}{2}} = \frac{\sqrt{13}}{12},\ \gamma_2 = (\lambda\alpha_2 + \mu\beta_2)^{\frac{1}{2}} = \frac{5}{12}$$

so that $\gamma_1 + \gamma_2 = \frac{\sqrt{13}+5}{12} < 1$, showing that $\gamma_1^2 x + \gamma_2^2 x_2$ is not a ρ-convex linear combination.

4.1.4 LEMMA. *A subset S (of a LS X) is absolutely ρ-convex iff it is ρ-convex and balanced. Also, if S is any balanced subset then $C_\rho(S)$ is absolutely ρ-convex, so that $|C_\rho|(S) = C_\rho(S)$.*

PROOF. If S is absolutely ρ-convex it is trivially ρ-convex. Moreover, it is also balanced since $|\lambda| \leqslant 1 \Rightarrow |\lambda|^\rho \leqslant 1$. Conversely, suppose that S is ρ-convex and balanced. If $x_j \in S$, $\sum_j |\alpha_j|^\rho \leqslant 1$

then $\sum_j |\frac{\alpha_j}{\beta}|^\rho = 1$, where $\beta \geqslant 0$, $\beta^\rho = \sum_j |\alpha_j|^\rho \leqslant 1$. We assume that all $|\alpha_j| > 0$ and set $\beta_j = \alpha_j/|\alpha_j|$. Then $\beta_j x_j \in S$ (since S is balanced) and so by ρ-convexity

$$\sum_j \frac{|\alpha_j|}{\beta}\beta_j x_j \in S, \text{ i.e. } \frac{1}{\beta}\sum_j \alpha_j x_j \in S.$$

By balanced property of S we get $\sum_j \alpha_j x_j \in S$, completing the proof of the first assertion. For the second we assume that S is balanced. If $\alpha_j \geqslant 0$, $\sum_j \alpha_j^\rho = 1$, $|\lambda| \leqslant 1$ and $x_j \in S$ then

$$\lambda \sum_j \alpha_j x_j = \sum_j \alpha_j(\lambda x_j) \in C_\rho(S) \text{ (since } \lambda x_j \in S, S \text{ being balanced).}$$

Hence $C_\rho(S)$ is balanced and $C_\rho(S) = |C|_\rho(S)$.

4.1.5. LEMMA. *If S is ρ-convex and $\alpha, \beta > 0$ then*

$$\alpha^{\frac{1}{\rho}} S + \beta^{\frac{1}{\rho}} S \subseteq (\alpha+\beta)^{\frac{1}{\rho}} S.$$

If S is convex (i.e. $\rho = 1$) then $\alpha S + \beta S = (\alpha+\beta)S$.

PROOF. For $x, y \in S$

$$\begin{aligned} \alpha^{\frac{1}{\rho}} x + \beta^{\frac{1}{\rho}} y &= (\alpha+\beta)^{\frac{1}{\rho}} \left[\frac{\alpha^{\frac{1}{\rho}}}{(\alpha+\beta)^{\frac{1}{\rho}}} x + \frac{\beta^{\frac{1}{\rho}}}{(\alpha+\beta)^{\frac{1}{\rho}}} y \right] \\ &\in (\alpha+\beta)^{\frac{1}{\rho}} S \text{ (by } \rho \text{ -convexity).} \end{aligned}$$

If $\rho = 1$ the above gives $\alpha S + \beta S \subseteq (\alpha+\beta)S$. On the other hand, since $(\alpha+\beta)x = \alpha x + \beta y$ the reverse inclusion relation also holds.

4.1.6. LEMMA. *If U is an open subset of a TLS X then $|C_\rho|(U)$ is open.*

PROOF. Consider an element $x = \sum_{j=1}^{n} \alpha_j a_j$ $(a_j \in U)$,

$\alpha_j \neq 0, \sum_j |\alpha_j|^\rho \leqslant 1$. Since U is open we can choose a balanced open nucleus V such that $a_j + V \subseteq U\,(j = 1, \cdots, n)$. Write $\sum_j |\alpha_j| = \alpha (> 0)$. For $a \in V$ we have

$$x + \alpha a = \sum_j \alpha_j a_j + \sum_j |\alpha_j| a = \sum_j \alpha_j (a_j + \alpha_j^{-1} |\alpha_j| a).$$

Since V is balanced, $\alpha_j^{-1}|\alpha_j| a \in V, a_j + \alpha_j^{-1}|\alpha_j| a \in U$, so that $x + \alpha a \epsilon |C_\rho|(U), x + \alpha V \subseteq |C_\rho|(U)$. Since x is an arbitrary element of $|C_\rho|(U)$ we conclude that $|C_\rho|(U)$ is open.

4.1.7. LEMMA. *If S is absolutely ρ-convex $(0 < \rho \leqslant 1)$ and $0 < \rho' < \rho$ then it is also absolutely ρ'-convex. In particular, if S is absolutely convex it is absolutely ρ-convex for every ρ with $0 < \rho \leqslant 1$.*

PROOF. Suppose that $\sum |\alpha_j|^{\rho'} \leqslant 1$. Then

$$\sum |\alpha_j|^\rho = \sum |\alpha_j|^{\rho' \cdot \frac{\rho}{\rho'}} \leqslant \left(\sum |\alpha_j|^{\rho'}\right)^{\frac{\rho}{\rho'}} \qquad (\text{since } \frac{\rho}{\rho'} \geqslant 1)$$

$$\leqslant 1^{\frac{\rho}{\rho'}} = 1.$$

Therefore, every absolutely ρ'-convex linear combination is an absolutely ρ-convex linear combination, whence the lemma.

4.1.8. DEFINITION. Let S be an absorbing subset of a LS X. For $x \in X$, $0 < \rho \leqslant 1$, set

$$\begin{aligned} p(x) &= p_\rho(x) = p_{\rho,S}(x) = p_S(x) \\ &= \inf_\alpha \{\alpha > 0 : x \in \alpha^{\frac{1}{\rho}} S\} \\ &= \inf_\beta \{\beta^\rho : \beta > 0, x \in \beta S\}. \end{aligned}$$

The non-negative realvalued function p is called the *ρ-gauge* (*gauge* if $\rho = 1$). The gauge of S is also known as the Minkowski functional of S.

It is clear from the definition of the ρ-gauge that we have:

$$p_{S_2} \leqslant p_{S_2} \text{ if } S_1 \subseteq S_2, \qquad (*)$$

$$p_X = 0, \tag{**}$$

$$p_{\alpha S} = \frac{1}{\alpha} p_S \qquad (\alpha \in \mathbb{K} \setminus \{0\}). \tag{***}$$

4.1.9 Remark. If $0 < \rho' \leqslant \rho < 1$ then $p_{\rho'} = p^{\frac{\rho'}{\rho}}$, where $p = p_\rho$.

4.1.10. PROPOSITION. *Let S be an absorbing subset of X. Then the ρ-gauge $p = p_S$ has the following properties:*

(i) $p(0) = 0$.

(ii) $p(\lambda x) = \lambda^\rho p(x)$ *if* $\lambda \geqslant 0$.

(iii) $p(\lambda x) = |\lambda|^\rho p(x)$ *for all* $\lambda \in \mathbb{K}$, *provided* S *is balanced.*

(iv) $p(x + y) \leqslant p(x) + p(y)$ *for all* $x, y \in X$ *provided* S *is* ρ*-convex.*

In particular, p *is a* ρ*-seminorm if* S *is absolutely* ρ*-convex (and also absorbing).*

PROOF. (i) Since S is absorbing, $0 \in S$ and hence $0 \in \alpha^{\frac{1}{\rho}} S$ for all $\alpha > 0$, whence $p(0) = 0$

(ii) $x \in \alpha^{\frac{1}{\rho}} S$ iff $\lambda x \in (\lambda^\rho \alpha)^{\frac{1}{\rho}} S (\lambda > 0)$. Hence $p(\lambda x) = \lambda^\rho p(x)$

(iii) Assume first that $|\lambda| = 1$; then $\lambda^{-1} S = S$ since S is balanced. It follows that $\lambda x \in \alpha^{\frac{1}{\rho}} S$ iff $x \in \alpha^{\frac{1}{\rho}} \lambda^{-1} S = \alpha^{\frac{1}{\rho}} S$. Therefore $p(\lambda x) = p(x)$. Next for arbitrary $\lambda \in \mathbb{K}$, $\lambda \neq 0$, write $\lambda = |\lambda| \mu$ where $|\mu| = 1$. Then, by (ii)

$$p(\lambda x) = |\lambda|^\rho p(\mu x) = |\lambda|^\rho p(x) \qquad (\text{since } |\mu| = 1).$$

Finally, the relation clearly holds if $\lambda = 0$.

(iv) Given $\epsilon > 0$, we can choose $\alpha, \beta > 0$ such that $x \in \alpha^{\frac{1}{\rho}} S$, $y \in \beta^{\frac{1}{\rho}} S$; $\alpha < p(x) + \epsilon, \beta < p(y) + \epsilon$. Since, by 4.1.5,

$$x + y \in \alpha^{\frac{1}{\rho}} S + \beta^{\frac{1}{\rho}} S \subseteq (\alpha + \beta)^{\frac{1}{\rho}} S$$

it follows that

$$p(x + y) \leqslant \alpha + \beta < p(x) + p(y) + 2\epsilon.$$

From the arbitrariness of ϵ we conclude that

$$p(x+y) \leqslant p(x) + p(y)$$

proving the first result of (iv). For the second result we remark that it follows from this and (iii), since an absolutely ρ-convex set is both ρ-convex and balanced.

4.1.11. LEMMA. *Let $S_j (j = 1, \cdots, n)$ be absorbing balanced sets and $S = S_1 \bigcap \cdots \bigcap S_n$; then S is absorbing and balanced. If p_j, p are ρ-guages of S_j, S respectively then $p = p_1 \vee \cdots \vee p_n$.* †

PROOF. That S is absorbing and balanced are easy consequences of the definitions (see 2.1.15). Since $S \subseteq S_j$ we have $p_j \leqslant p$. Write $q = p_1 \vee \cdots \vee p_n$. Then $q(x) = \max_j p_j(x) \leqslant p(x)$. Again we have, for some $\epsilon_n \to 0$,

$$\begin{aligned} x \in (p_j(x) + \epsilon_n)^{\frac{1}{\rho}} S_j \quad &\subseteq \quad (q(x) + \epsilon_n)^{\frac{1}{\rho}} S_j \\ &\qquad \text{(since } S_j \text{ is balanced)} \\ &= \quad \lambda S_j \text{ (say)}. \end{aligned}$$

It follows that $\lambda^{-1} x \in S$, so that

$$x \in \lambda S = (q(x) + \epsilon_n)^{\frac{1}{\rho}} S.$$

This implies that $p(x) \leqslant q(x) + \epsilon_n$, whence making $\epsilon_n \to 0$ we get $p(x) \leqslant q(x) \leqslant p(x)$. Thus $q = p$, completing the proof.

4.1.12. PROPOSITION. *Let S be a subset of a LS X (over $\mathbb{K}$), which is absolutely ρ-convex $(0 < \rho \leqslant 1)$ and absorbing. Then:*

(i) *The ρ-gauage $p = p_S$ is a ρ-seminorm such that if*

$$B_1 = \{x \in A : p(x) < 1\}, \qquad \overline{B}_1 = \{x \in A : p(x) \leqslant 1\}$$

then $B_1 \subseteq S \subseteq \overline{B}_1$; in particular, $\ker p \subseteq S$.

(ii) *$S = B_1$ or $\overline{B}_1$ according as S is open or closed in the p-topology.*

† See 3.1.7 (b) for the definition.

(iii) *If A is an algebra and S a (multiplicative) subsemigroup of A then p is sm..*

PROOF. (i) By 4.1.10, p is a ρ-seminorm. If $x \in B_1$ then for some $\epsilon > 0$, $p(x) < \epsilon < 1$, $x \in \epsilon^{\frac{1}{\rho}} S$. Then $\epsilon^{-\frac{1}{\rho}} x \in S$, and since $|\epsilon^{\frac{1}{\rho}}| = |\epsilon|^{\frac{1}{\rho}} < 1$ and S is balanced, $x = \epsilon^{\frac{1}{\rho}}(\epsilon^{-\frac{1}{\rho}} x) \in S$, so that $B_1 \subseteq S$. Again, if $x \in S$ then $x \in 1^{\frac{1}{\rho}} S$, so that $p(x) \leqslant 1$, whence $S \subseteq \overline{B}_1$.

(ii) Assume first that S is open and $x \in S$. Since $(1 - \frac{1}{n})^{\frac{1}{\rho}} x \to x$ and S is open, $(1 - \frac{1}{n})^{\frac{1}{\rho}} x \in S$ for some $n \geqslant 2$, so that $p(x) \leqslant 1 - \frac{1}{n} < 1$, whence $B_1 = S$. Next assume that S is closed. By 3.2.7, $\overline{B}_1$ is the closure of B_1. Therefore, since by (i), $B_1 \subseteq S \subseteq \overline{B}_1$ and S is closed, it follows that $S = \overline{B}_1$.

(iii) By definition of p, since S is balanced, we have for any $\epsilon > 0$ with $0 < \epsilon < 1$,

$$x \in (p(x) + \epsilon)^{\frac{1}{\rho}} S, \qquad y \in (p(y) + \epsilon)^{\frac{1}{\rho}} S.$$

Therefore

$$xy \in [(p(x) + \epsilon)(p(y) + \epsilon)]^{\frac{1}{\rho}} S^2 \subseteq [p(x)p(y) + C\epsilon]^{\frac{1}{\rho}} S,$$

where C is a constant independent ϵ. By making $\epsilon \to 0$ we conclude that $p(xy) \leqslant p(x)p(y)$, i.e., p is sm..

4.1.13. PROPOSITION. *Let p be a ρ-seminorm on A. Then $B_1, \overline{B}_1$ (as defined in 4.1.12) are absolutely ρ-convex and absorbing. Further, B_1 is open, $\overline{B}_1$ is closed in the p-topology and we have*

$$p_{B_1} = p_{\overline{B}_1} = p. \tag{$*$}$$

If p is sm. then $B_1, \overline{B}_1$ are subsemigroups.

PROOF. Using the identity $p(\lambda x) = |\lambda|^\rho p(x)$ it is straightforward to check that $B_1, \overline{B}_1$ are absolutely ρ-convex. Next, for any $x \in A$, set $\lambda_x = 1$ if $p(x) = 0$ and $\lambda_x = (2p(x))^{-\frac{1}{\rho}}$ if $p(x) \neq 0$. Then $p(\lambda_x x) = 0$ or $\frac{1}{2}$, so that $\lambda_\alpha x \in B_1 \subseteq \overline{B}_1$, proving that B_1, $\overline{B}_1$ are absorbing. It is further clear from the definition of the

p-topology that B_1 is open, and $\overline{B}_1$ is closed, being the closure of B_1 (by 3.2.7.). Again it is immediate from the sm. property that $B_1, \overline{B}_1$ are subsemigroups whenever p is sm.. It remains to prove the equalities $(*)$.

Since $B_1 \subseteq \overline{B}_1$ we have

$$p_{\overline{B}_1}(x) \leqslant p_{B_1}(x). \tag{1}$$

Again, if $\lambda > p(x)$ then $x \in \lambda^{\frac{1}{\rho}} B_1$, so that $p_{B_1}(x) \leqslant \lambda$, whence

$$p_{B_1}(x) \leqslant p(x). \tag{2}$$

Finally, $p(x) \leqslant \lambda$ iff $x \in \lambda^{\frac{1}{\rho}} \overline{B}_1$. Hence

$$p_{\overline{B}_1}(x) = \inf\{\lambda : \lambda \geqslant p(x)\} \geqslant p(x). \tag{3}$$

From (1), (2), (3) we obtain $(*)$.

4.1.14. Remark. If p be as in 4.1.12, and $r > 0$, then we can show as above (for the $r = 1$ case) that

$$B_r = \{x \in A : p(x) < r\},\ \overline{B}_r = \{x \in A : p(x) \leqslant r\}$$

are respy. open absolutely ρ-convex and closed absolutely ρ-convex. Also, when p is sm., B_r, $\overline{B}_r$ are subsemigroups for $0 < r \leqslant 1$.

4.1.15. PROPOSITION. *Let p be a ρ-seminorm on X, $0 < \rho \leqslant 1$, $0 < \rho' \leqslant 1$. Then $p' = p^{\frac{\rho'}{\rho}}$ is a ρ'-seminorm iff the unit ball $B_1 = \{x : p(x) < 1\}$ is ρ'-convex. In particular, p' is a ρ'-seminorm for all ρ' with $0 < \rho' < \rho$.*

PROOF. Note first that we have also $B_1 = \{x : p'(x) < 1\}$. If p' is a ρ'-seminorm then, by 4.1.12, B_1 is ρ'-convex. Conversely, let B_1 be ρ' convex. Denote by p'_{β_1} the ρ'-gauge determined by B_1. Then $p'_{B_1}(x) < \lambda \iff \lambda^{-\frac{1}{\rho'}} x \in B_1 \iff p'(\lambda^{-\frac{1}{\rho'}} x) < 1$ iff $p'(x) < \lambda$. Therefor $p' = p'_{B_1}$ and consequently

p' is a ρ'-seminorm. Then second assertion in the proposition is an immediate consequence of the first and 4.1.7.

4.1.16. THEOREM (Rickart[†]). *Let* $A = (A, p)$ *be a real* ρ-*seminormed algebra,* $0 < \rho \leqslant 1$, *and* $\tilde{A}$ *the complexification of the algebra* A. *Then* $\tilde{A}$ *is a* ρ-*seminormed algebra. Further,* $\tilde{A}$ *is unital,* ρ-*normed or* ρ-*Banach according as* A *has the corresponding property. Moreover, we can choose the* ρ-*seminorm* $\tilde{p}$ *of* $\tilde{A}$ *in such a way that the map* $x \mapsto x + i.0$ *of* A *into* $\tilde{A}$ *is an isometry (i.e.* $\tilde{p}(x) = p(x)$), $\tilde{p}(e) = 1$ *whenever* A *has a unity* e, *with* $p(e) = 1$, *and finally* $\tilde{p}$ *is sm. if* p *is sm..*

PROOF (following the method of Bonsall-Duncan [4, p.68]). Write $U = \{x \in A : p(x) < 1\}$, $\tilde{U} =$ the absolute ρ-convex hull of U in $\tilde{A}$. Then $\tilde{U}$ is absorbing in $\tilde{A}$. To see this, suppose that $z = x + iy \neq 0$ $(x, y \in A)$ and

$$\mu^\rho > \max\{p(x), p(y)\}. \tag{1}$$

Then $x/\mu, y/\mu \in U$, so that

$$\frac{1}{2^{1/\rho}\mu} z = \frac{1}{2^{1/\rho}}\left(\frac{x}{\mu}\right) + \frac{i}{2^{1/\rho}}\left(\frac{y}{\mu}\right) \in \tilde{U} \tag{2}$$

proving $\tilde{U}$ is absorbing.

We next prove that $\tilde{U}$ has the property:

$$z = x + iy \in \tilde{U} \Rightarrow x, y \in U. \tag{3}$$

Assume that z has the representation

$$z = \sum_{k=1}^{n} \gamma_k x_k (x_k \in U), \qquad \sum_k |\gamma_k|^\rho \leqslant 1.$$

Writing

$$\gamma_k = \alpha_k + i\beta_k \qquad (\alpha_k, \beta_k \in \mathbb{R})$$

we get

$$z = x + iy, \text{ where } x = \sum_k \alpha_k x_k,\ y = \sum_k \beta_k y_k.$$

[†] Rickart [23, p.8] obtained the results when A is a real normed algebra.

Since $|\alpha_k|, |\beta_k| \leqslant |\gamma_k|$ it follows, by the absolute ρ-convexity of U in A, that $x, y \in U$, whence (3) holds.

Denote by $\tilde{p}$ the gauge of $\tilde{U}$ (in $\tilde{A}$); then $\tilde{p}$ is a ρ-seminorm on $\tilde{A}$. Writing $\tilde{B}_1 = \{z \in \tilde{A} : \tilde{p}(z) < 1\}$, by 4.1.12(i), we get

$$\tilde{B}_1 \subseteq \tilde{U}. \tag{4}$$

On the other hand, if

$$z = \sum_{k=1}^{n} \lambda_k x_k \in \tilde{U} \ (x_k \in U, \lambda_k \in \mathbb{C}), \sum_k |\lambda_k|^\rho \leqslant 1, \text{ we can}$$

choose $\epsilon > 0$ such that $p(x_k) < \epsilon < 1$ for all k. Then $\epsilon^{-\frac{1}{\rho}} x_k \in U$, so that $\epsilon^{-\frac{1}{\rho}} z \in \tilde{U}$, whence $\tilde{p}(z) \leqslant \epsilon < 1$. It follows therefore that

$$\tilde{U} = \tilde{B}_1 \tag{5}$$

and so on particular $\tilde{U}$ is $\tilde{p}$-bounded. From (1),(2) it follows that

$$\tilde{p}(z) \leqslant 2 \max\{p(x), p(y)\}. \tag{6}$$

Again, if $\lambda > 0, \lambda z \in \tilde{U}$ then by (3), $\lambda x, \lambda y \in U$, so that $p(\lambda x), p(\lambda y) < 1$, whence

$$\lambda^\rho \max(p(x), p(y)) < 1. \tag{7}$$

For a given $\epsilon > 0$, we can choose the $\lambda > 0$ such that

$$\frac{1}{\lambda^\rho} < \tilde{p}(z) + \epsilon. \tag{8}$$

We conclude from (7),(8), since ϵ is arbitrary, that

$$\max\{p(x), p(y)\} \leqslant \tilde{p}(z). \tag{9}$$

The inequalities (6),(9) imply that $\tilde{p}$ induces the topology of $\tilde{A}$, and also that $\tilde{p}$ is faithful (respy. complete) when p is faithful (respy. complete).

If $x \in A$ and $\lambda \in \mathbb{R}$ then it is clear from (3) that

$$\lambda x + i0 \in \tilde{U} \text{ iff } \lambda x \in U,$$

whence $p(x) = \tilde{p}(x)$. When A is unital we can, using 3.5.9, assume that $p(e) = 1$, so that then we get $\tilde{p}(e) = p(e) = 1$.

Finally, let p be sm.; then U is a subsemigroup of A. We shall show that $\tilde{U}$ is a subsemigroup of $\tilde{A}$. Suppose now that $z, z' \in \tilde{U}$, then they have representations $z = \sum_k \lambda_k x_k$, $z' = \sum_l \lambda'_l x_l$, where $x_k, x_l \in U$ and $\sum_k |\lambda_k|^\rho, \sum_l |\lambda'_l|^\rho \leqslant 1$. Now

$$zz' = \sum_{k,l} \lambda_k \lambda'_l x_k x_l, \qquad \text{and}$$

$$\sum_{k,l} |\lambda_k \lambda'_l|^\rho = \left(\sum_k |\lambda_k|^\rho \right) \left(\sum_l |\lambda'_l|^\rho \right) \leqslant 1 \cdot 1 = 1.$$

Since U is a semi-group, $x_k x_l \in U$ and consequently the above relations show that $zz' \in \tilde{U}$, so that $\tilde{U}$ is a subsemigroup of $\tilde{A}$, as desired, completing the proof of the theorem.

§ 2. Locally Bounded Algebras

4.2.1. LEMMA. *Let X be a locally bounded TLS and U a bounded balanced nucleus of X. Then we can find a $\rho, 0 < \rho \leqslant 1$ such that*

$$U + U \subseteq 2^{\frac{1}{\rho}} U. \tag{$*$}$$

PROOF. Since U is bounded, by 2.1.25, $\{\frac{1}{n} U : n = 1, 2 \cdots\}$ is a basis of nuclei, so that there is a $N \geqslant 2$ such that

$$\frac{U}{N} + \frac{U}{N} \subseteq U. \tag{$*'$}$$

Write $\rho = \log 2 / \log N$. Then $N = 2^{\frac{1}{\rho}}$ and $(*') \Rightarrow (*)$.

4.2.2. LEMMA (Köthe). *Let S be a balanced subset of a LS X such that*

$$S + S \subseteq 2^{\frac{1}{\rho}} S \tag{1}$$

where $0 < \rho \leqslant 1$. Then

$$2^{-\frac{k_1}{\rho}} S + \cdots + 2^{-\frac{k_n}{\rho}} S \subseteq S \tag{2}$$

for all integers $k_j \geqslant 1$ *such that* $\sum_{j=1}^{n} 2^{-k_j} \leqslant 1$, *and further*

$$|C_\rho|(S) \subseteq 2^{\frac{1}{\rho}} S. \tag{3}$$

PROOF. ([18,p.165]). To prove (2) it is clearly sufficient to prove the special case (2′) of (2) where we have

$$\sum_{j=1}^{n} 2^{-k_j} = 1. \tag{$*$}$$

We call $k = \max_j k_j$, the order of the resolution $(*)$. We prove (2′) by induction on k. If $k = 1$ the relation (2) reduces just to the hypothesis (1) and so (2) holds for this case. Next assume that (2′) holds for order $k \geqslant 1$. To prove that it holds for order $k+1$, observe that each decomposition $(*)$ of order $k+1$ is obtained from a decomposition of order k, by replacing one (or more) of the summands 2^{-k} by $2^{-(k+1)} + 2^{-(k+1)}$, since it follows from $\sum 2^{-k_j} = 1$ that the summands $2^{-(k+1)}$ occurs an even number of times. Thus, to obtain (2′) for order $k+1$ starting from (2′) for order k we have to replace the summand $2^{-\frac{k}{\rho}} S$ by $2^{-\frac{k+1}{\rho}} S + 2^{-\frac{k+1}{\rho}} S$. But this is permissible since $2^{-\frac{1}{\rho}} S + 2^{-\frac{1}{\rho}} S \subseteq S$ (by (1)). This completes the proof of (2′) (by induction).

It remains to prove (3). Suppose that $\sum_{j=1}^{n} |\alpha_j|^\rho \leqslant 1$ and $x_j \in S$. Determine integers k_j such that $2^{-k_j} \leqslant |\alpha_j|^\rho < 2^{-k_j+1} = 2 \cdot 2^{-k_j}$. Then

$$\sum_j |\alpha_j|^\rho < 2 \sum_j 2^{-k_j} \leqslant 2 \cdot \sum |\alpha_j|^\rho \leqslant 2 \cdot 1 = 2$$

since S is balanced,

$$\alpha_j x_j \in \alpha_j S \subseteq |\alpha_j| S \subseteq 2^{\frac{-k_j+1}{\rho}} S,$$

so that

$$\sum_j \alpha_j x_j \in \sum_j 2^{-\frac{k_j+1}{\rho}} S = 2^{\frac{1}{\rho}} \left(\sum_j 2^{-\frac{k_j}{\rho}} S \right) \subseteq 2^{\frac{1}{\rho}} S \text{ (using (2))},$$

proving (3).

4.2.3. PROPOSITION. *A TA A (or more generally a TLS) is ρ-seminormed $(0 < \rho \leqslant 1)$ iff it has a bounded nucleus U such that*

$$U + U \subseteq 2^{\frac{1}{\rho}} U.$$

Moreover, when A is semi-normed we can even choose the U so that it is a subsemigroup of A.

PROOF. If A is a ρ-seminormed algebra (A, p) we may assume that p is sm. (see 3.4.4). By 4.1.13, $U = B_1$ is an open absolutely ρ-convex nucleus which is a subsemigroup. U is, moreover, bounded (by 3.2.13). Finally, if $x, y \in U$ then by ρ-convexity,

$$\frac{x}{2^{\frac{1}{\rho}}} + \frac{y}{2^{\frac{1}{\rho}}} \in U, \text{ so that } U + U \subseteq 2^{\frac{1}{\rho}} U.$$

Conversely, suppose that U is an open nucleus having the stated properties. Then, by 4.2.2,

$$V = |C_\rho|(U) \subseteq 2^{\frac{1}{\rho}} U \tag{$*$}$$

and by 4.1.6, V is an open nucleus. The boundedness of U together with the inclusion $(*)$ implies that V is bounded. Since V is absolutely ρ-convex the gauge $p = p_V$ is a ρ-seminorm (by 4.1.10). By 4.1.12(ii), $V = B_1 = \{x \in A : p(x) < 1\}$ and hence $(*)$ gives:

$$\frac{1}{2^{\frac{1}{\rho}}} V \subseteq U \subseteq V = B_1. \tag{$**$}$$

Since U, V are bounded nuclei it follows from $(**)$ that the topology of A coincides with the p-topology, whence A is ρ-seminormed.

4.2.4. COROLLARY. *Every locally bounded*[†] *TA is a ρ-seminormed algebra for some $\rho, 0 < \rho \leqslant 1$.*

[†] A TA is said to be locally bounded if it is locally bounded as a TLS (i.e. it has a bounded nucleus).

PROOF. This is an immediate consequence of 4.2.1, 4.2.3 .

4.2.5. Remark. Rolewicz [12′] has shown that to each locally bounded TLS X there is a largest number $\rho_0, 0 < \rho_0 \leqslant 1$, such that if $0 < \rho < \rho_0$ then the topology of X can be induced by a ρ-seminorm.

4.2.6. COROLLARY. *A TA A is a semi-normed algebra iff it has a bounded convex nucleus U.*

PROOF. If $A = (A, P)$ is semi-normed we can take $U = B_1 = \{x \in A : p(x) < 1\}$. On the other hand, if A has a bounded convex nucleus U then we have, by 4.1.5,

$$U + U = 1 \cdot U + 1 \cdot U = (1 + 1)U = 2^1 U.$$

By 4.2.3, A is semi-normed.

4.2.7. COROLLARY (Kolmogorov†). *A Hausdorff TA is a normed algebra iff it has a bounded convex nucleus.*

PROOF. This follows from 4.2.6 since the topology induced by a semi-norm p is Hausdorff iff p is a norm.

4.2.8. PROPOSITION. *An infinite direct product of locally bounded Hausdorff TLS's – in particular such a product of TA's – cannot be locally bounded.*

PROOF. Suppose $X = \pi X_\alpha$, where $\{X_\alpha\}$ is an infinite family of locally bounded Hausdorff TLS's. Denote by π_α the factor projection: $X \to X_\alpha$. Then sets of the form

$$U = \pi_{\alpha_1}^{-1}(U_{\alpha_1}) \times \cdots \times \pi_{\alpha_n}^{-1}(U_{\alpha_n})$$

where U_{α_n} run through an open basis of X_{α_j}, form a basis of nuclei for X. Fixing a U, let β be an index with $\beta \neq \alpha_1, \cdots, \alpha_n$ and $x \neq 0$ an element of the space X_β. Then the sequence $(nx)(n = 1, 2, \cdots)$ of elements clearly belong to U (after suitable

† Actually Kolmogorov proved the following result: A Hausdorff TLS is normable iff it has a bounded convex nucleus.

identification †). Since $\frac{1}{n} \cdot nx = x \neq 0$ this sequence does not converge to 0 and consequently U is not bounded. Since U is an arbitrary basis nucleus we conclude that X is not locally bounded.

4.2.9. COROLLARY. *A direct product of ρ-seminormed algebras is ρ-seminormed iff it is a 'finite direct product'.*

PROOF. If $A = A_1 \times \cdots \times A_n$, when $A_j = (A_j, p_j)$ are ρ-seminormed algebras then A is ρ-seminormed under $q : q(x) = \max_j p_j(x)$, where $x \in A, x = (x_1, \cdots, x_n), x_j \in A_j$ (cf. 3.3.11).

Conversely, if a direct product A of ρ-seminormed algebras is ρ-seminormed then A is locally bounded (by 3.2.14) and hence by 4.2.8, A is a finite direct product.

§ 3. Locally Pseudo-convex Spaces

4.3.1. DEFINITION. Following Waelbroeck [29,p.4] we call a subset S of a LS X (over $\mathbb{K}$) *pseudo-convex* if it is ρ-convex for some ρ $(0 < \rho \leqslant 1)$. A TLS X is called *locally pseudo-convex* if it has a basis $\{U_\alpha\}$ of pseudo-convex nuclei U_α; if U_α is ρ_α-convex we also say that X is locally $\{\rho_\alpha\}$-convex. If all the $\rho_\alpha =$ (some) ρ then X is called *locally ρ-convex*, and if $\rho = 1$ then X is called *locally convex.*

A ρ-seminorm p is also called a pseudo-seminorm and ρ is called the homogenity index of the pseudo-seminorm p (cf. 3.2.1).

4.3.2. If $\mathcal{P}$ is a family of pseudo-seminorms on a LS X, the $\mathcal{P}$-topology (see 3.1.7) makes X a locally pseudo-convex space $(X, \mathcal{P})$; if $\mathcal{P} = \{p_\alpha\}$ and $B_r^\alpha = \{x \in X : p_\alpha(x) < r\}$ $(r \in \mathbb{R}, r > 0)$ then the family of all finite intersections of the B_r^α $(\alpha, r$ varying) give a basis of pseudo-convex nuclei for $(X, \mathcal{P})$. Conversely, if X is any locally pseudo-convex space with $\{U_\alpha\}$ as a basis of pseudo-convex nuclei them the gauges p_α associated with the U_α determine a family $\mathcal{P} = \{p_\alpha\}$ of pseudo-seminorms p_α.

---

†. The element $x_\beta \in X_\beta$ is identified with the element $x = (x_\alpha)$ in X such that $x_\alpha = 0$ if $\alpha \neq \beta, x_\alpha = x_\beta$ if $\alpha = \beta$.

4.3.3. Let $\mathfrak{S}^*$ denote the set of all pseudo-seminorms on a LS X. We introduce an order $\prec$ in $\mathfrak{S}^*$ by writing

$$p \prec p' \text{ if } p^{\frac{1}{\rho}} < p'^{\frac{1}{\rho'}} \text{ (i.e. } p(x)^{\frac{1}{\rho}} \leqslant p'(x)^{\frac{1}{\rho'}} \; \forall x \in X) \text{ and } \rho' \leqslant \rho,$$

where $p, p' \in \mathfrak{S}^*$ are respectively a ρ-seminorm and a ρ'-seminorm. Clearly, the relation $\prec$ is reflexive and transitive. It is also anti-symmetric, for, if $p \prec p'$, $p' \prec p$ then $p^{\frac{1}{\rho}} = p'^{\frac{1}{\rho'}}$, $\rho = \rho'$, so that $p = p'$. Thus $\prec$ is a partial order. We also note that if $p \prec p'$ then for $0 < \epsilon \leqslant 1$, $B_\epsilon^{p'} \subseteq B_\epsilon^{p}$ (since if $p'(x) < \epsilon$ then $p(x) \leqslant p'(x)^{\frac{\rho}{\rho'}} < \epsilon$).

4.3.4. LEMMA. *The poset* $\mathfrak{S}^* = (\mathfrak{S}^*, \prec)$ *is closed for all finite lattice sums.*

PROOF. If $p_j \in \mathfrak{S}^*$ and p_j is a ρ-seminorm $(j = 1, \cdots, n)$ set $\rho = \min_j\{\rho_j\}, p'_j = p^{\frac{\rho}{\rho_j}}; q(x) = \max_j p'_j(x)$. By 3.2.9, p'_j is a ρ-seminorm and so by 3.2.3, q is a ρ-seminorm, so that $q \in \mathfrak{S}^*$. It easily follows from the definition of q that $q \succ p_j$, and that if $p' \in \mathfrak{S}^*, p' \succ p_j$ for all j then $p' \succ q$. So q is indeed the lattice sum $p_1 \vee \cdots \vee p_n$.

4.3.5. LEMMA. *If* $p_j \in \mathfrak{S}^*(j = 1, \cdots, n), q = p_1 \vee \cdots \vee p_n$, *and* $0 < \epsilon \leqslant 1$ *then*

$$B_\epsilon^q \subseteq B_\epsilon^{p_1} \bigcap \cdots \bigcap B_\epsilon^{p_n}.$$

PROOF. Since $p_j \prec q$ we have $B_\epsilon^q \subseteq B_\epsilon^{p_0}$ (see 4.3.3). The required result is now clear.

4.3.6. LEMMA. *If* $p_j, p'_j \in \mathfrak{S}^*(j = 1, \cdots, n)$ *and* $p_j \sim p'_j$ *then*

$$p_1 \vee \cdots \vee p_n \sim p'_1 \vee \cdots \vee p'_n.$$

PROOF. By $(**)$ of 3.2.11 we have

$$p_j \leqslant C'_j p_j'^{\frac{\rho_j}{\rho'_j}}, p'_j \leqslant C_j p_j^{\frac{\rho'_j}{\rho_j}}.$$

Set $C' = \max C'_j, \rho = \min \rho_j$. Then

$$\begin{aligned} p_1 \vee \cdots \vee p_n &= \max p_j^{\frac{\rho}{\rho_j}} \leqslant C' \max p_j'^{\frac{\rho}{\rho'_j}} \\ &\leqslant C'(p'_1 \vee \cdots \vee p'_n). \end{aligned}$$

Similarly, $p'_1 \vee \cdots \vee p'_n \leqslant C(p_1 \vee \cdots \vee p_n)$, where $C = \max C_j$. Hence the lemma.

4.3.7. DEFINITION. A family $\mathcal{P}$ of pseudo-seminorms is called *saturated* if it satisfies the condition

$$p_1, \cdots, p_n \in \mathcal{P} \Rightarrow p_1 \vee \cdots \vee p_n \in \mathcal{P}. \qquad (*)$$

For any subset $\mathcal{P}$ of $\mathfrak{S}^*$ we set

$$\overline{\mathcal{P}} = \{q \in \mathfrak{S}^* : q = p_1 \vee \cdots \vee p_n, p_j \in \mathcal{P}\}.$$

The set $\overline{\mathcal{P}}$ clearly satisfies the condition $(*)$ and is called the *saturated closure* of $\mathcal{P}$.

4.3.8. LEMMA. (a) $\mathcal{P}, \overline{\mathcal{P}}$ *induces on* X *the same topology.*

(b) *If* $\mathcal{P}_0$ *is a cofinal*[†] *subset of* $\mathcal{P}$ *(with respect to the order in* 4.3.3) *then* $\mathcal{P}_0$ *induces on* X *the same topology as* $\mathcal{P}$.

(c) *If* $\mathcal{P}$ *is a saturated family then the family* $\{B^p_\epsilon : p \in \mathcal{P}, \epsilon > 0\}$ *is a basis of nuclei.*

PROOF. (a) It is sufficient to observe that $p(x_\alpha - x) \to 0$ for each $p \in \mathcal{P}$ iff $q(x_\alpha - x) \to 0$ for each $q \in \overline{\mathcal{P}}$.

(b) It is enough to note that if $p \prec p_0$ $(p \in \mathcal{P}, p_0 \in \mathcal{P}_0)$ and

$$B = \{x \in X : p(x) < r\}, B_0 = \{x \in X : p_0(x) < r'\}$$

where $r' = r^{\frac{\rho_0}{\rho}}$, ρ_0 (respy. ρ) being the homogenity index of p_0 (respy p.), then $B_0 \subseteq B$.

(c) This follows from 4.3.5 and the simple fact that $B^p_{r_1} \subseteq B^p_{r_2}$ if $0 < r_1 < r_2$.

---

† i.e. given any $p \in \mathcal{P}$ there is a $p_0 \in \mathcal{P}_0$ with $p \prec p_0$.

4.3.9. Let $\mathcal{P}$ be a family of pseudo-seminorms on a LS X. A subset S of X is said to be $\mathcal{P}$*-bounded* if the numerical functions $p(x)(p \in \mathcal{P})$ are bounded on S (i.e. $p|S$ is bounded for each $p \in \mathcal{P}$). The meaning of p-bounded where p is a pseudo-seminorm is clear.

4.3.10. LEMMA. *Let $X = (X, \mathcal{P})$ be a locally pseudo-convex space. A subset S of X is (t.) bounded iff it is $\mathcal{P}$-bounded.*

PROOF. Let S be bounded. If possible let $p_\alpha|S$ be not bounded for some p_α in $\mathcal{P}$. Then we can find a sequence (x_n) in S with $p_\alpha(x_n) \geqslant n$. Writing $y_n = n^{-\frac{1}{\rho_\alpha}} x_n$ we have: $p_\alpha(y_n) \geqslant 1$. It follows that $y_n \not\to 0$ in X, contradicting the boundedness of S.

Conversely, suppose now S is such that $p_\alpha|S$ is bounded for all p_α. Then, for (x_n) in $S, \lambda_n \to 0$ we have $p_\alpha(\lambda_n x_n) = |\lambda_n|^{\rho_\alpha} p_\alpha(x_n) \to 0$, since the sequence is bounded and $\rho_\alpha > 0$. Therefore S is bounded.

4.3.11. PROPOSITION. *Let $X = (X, \mathcal{P}), Y = (Y, \mathcal{Q})$ be locally pseudo-convex spaces, with $\mathcal{P}$ saturated. Then, a linear transformation $T : X \to Y$ is continuous iff for each $q_\beta \in \mathcal{Q}$ there is a $p_\alpha \in \mathcal{P}$ and a constant $C = C_{\alpha\beta} > 0$ such that*

$$q_\beta(Tx) \leqslant C p_\alpha(x)^{\frac{\rho_\beta}{\rho_\alpha}} \quad (x \in X) \tag{$*$}$$

where ρ_α, ρ_β are respectively the homogenity indices of p_α, p_β.

PROOF. The condition $(*)$ clearly implies continuity of T at 0 and and hence also everywhere. Conversely, if T is continuous then (by continuity at 0), given q_β and $\epsilon > 0$ there is a p_α and a $\delta > 0$ such that

$$p_\alpha(x) \leqslant \delta \Rightarrow q_\beta(Tx) \leqslant \epsilon. \tag{$**$}$$

For an x with $p_\alpha(x) \neq 0$, write $x_1 = (\delta p_\alpha(x)^{-1})^{\frac{1}{\rho_\alpha}} x$. Then $p_\alpha(x_1) = \delta$, so that by $(**)$ we get $q_\beta(Tx_1) \leqslant \epsilon$ which reduces to $(*)$ above will $C = \epsilon\delta^{-\frac{\rho_\beta}{\rho_\alpha}}$. If $p_\alpha(x) = 0$, we can argue exactly as in the proof of 3.2.10 and show that $q_\beta(Tx) = 0$, so that $(*)$ holds trivially in this case.

4.3.12. LEMMA. *Let X be a TLS, $Y = (Y, \mathcal{Q})$ a locally pseudo-convex space, $T : X \to Y$ a linear transformation. For each $q_\beta \in \mathcal{Q}$ we set $p_\beta^*(x) = q_\beta(Tx)$. Then p_β^* is a pseudo-seminorm on X with the same homogenity index as q_β. The transformation T is continuous iff all p_β^* are continuous.*

PROOF. We have
$p_\beta^*(x+y) = q_\beta(Tx+Ty) \leqslant p_\beta^*(x) + p_\beta^*(y)$. Further, if ρ is the homogenity index of q_β then:

$$p_\beta^*(\lambda x) = q_\beta(T\lambda x) = q_\beta(\lambda Tx) = |\lambda|^\rho p_\beta^*(x).$$

To prove the criterion for continuity of T it is enough, in view of 2.1.26, 3.1.3, to consider the continuity of both at 0. Now T is continuous at 0 iff: $x_\alpha \to 0$ in $X \Rightarrow p_\beta^*(x_\alpha) = q_\beta(Tx_\alpha) \to 0$ (for all q_β), iff all p_β^* are continuous at 0. This proves the lemma.

4.3.13. LEMMA. *Let $X = (X, \mathcal{P})$ be a locally pseudo-convex space, where $\mathcal{P}$ is saturated. Then:*

(i) *For any continuous linear functional f on X there is a $p_\alpha \in \mathcal{P}$ such that f is p_α-continuous, i.e. f is a continuous functional of the pseudo-seminormed linear space (X, p_α).*

(ii) *If p is any continuous pseudo-seminorm on X then there is a p_α such that*

$p \leqslant C p_\alpha^{\frac{\rho}{\rho_\alpha}}$, where C is some constant > 0 and ρ, ρ_α the homogenity indices of p, p_α.

PROOF. (i) By 2.1.31, there is a p_α and an $\epsilon > 0$ such that

$$p_\alpha(x) < \epsilon \Rightarrow |f(x)| < 1.$$

It follows from 3.5.4 that
$f : (X, p_\alpha) \to \mathbb{K}$ is n. bounded, whence by 3.5.5, it is continuous.

(ii) The identity map : $(X,\mathcal{P}) \to (X,p)$ is continuous by hypothesis on p. So, by 4.3.11, there is a $C > 0$ and a p_α with

$$p(x) \leqslant C p_\alpha(x)^{\frac{\rho}{\rho_\alpha}}$$

as desired.

4.3.14. DEFINITION. Let $X = (X,\mathcal{P})$ be a locally pseudo-convex TLS – or more generally a TLS with its topology induced by a family $\mathcal{P}$ of quarter-norms. A map

$$T : \mathcal{D}_T (\subseteq X) \to X$$

where $\mathcal{D}_T$ is the domain of T, is called a *contraction* if we have for each $p_\alpha \in \mathcal{P}$ a constant ϵ_α with $0 < \epsilon_\alpha < 1$ such that for all $x, y \in \mathcal{D}_T$ we have

$$p_\alpha(Tx - Ty) \leqslant \epsilon_\alpha p_\alpha(x - y).$$

Note that a contraction map is automatically continuous.

4.3.15. THEOREM (Contraction mapping principle). *Let $X = (X,\mathcal{P})$ be a complete (or even sequentially complete) Hausdorff TLS, with $\mathcal{P}$ a family of ρ-seminorms (or more generally quarter-norms). Let $T : S \to S$ be a contraction, where S is a closed subset of X. Then T has a unique fixed point $x^* \in S$ with $x^* = \lim T^n x_0$, where x_0 is an arbitrary point of S.*

PROOF. Define $x_1 = Tx_0, x_n = Tx_{n-1} = T^n x_0$. If $m < n$ then

$$\begin{aligned} p_\alpha(x_m - x_n) &= p_\alpha(T^m x_0 - T^n x_0) \leqslant (\epsilon_\alpha)^m p_\alpha(x_0 - x_{n-m}) \\ &\leqslant (\epsilon_\alpha)^m [p_\alpha(x_0 - x_1) + \cdots + p_\alpha(x_{n-(m-1)} - x_{m-1})] \\ &\leqslant (\epsilon_\alpha)^m p_\alpha(x_0 - x_1)[1 + \epsilon_\alpha + \cdots \epsilon_\alpha^{n-m}] \\ &< (\epsilon_\alpha)^m p_\alpha(x_0 - x_1) \frac{1}{1 - \epsilon_\alpha}. \end{aligned}$$

It follows that $p_\alpha(x_m - x_n) \to 0$ as $m, n \to \infty$ (for each α). By completeness of X the sequence $x_n \to$ some $x^* \in X$; since $x_n \in S$ and S is closed, $x^* \in S$. By continuity of T

we have: $Tx^* = \lim_{n\to\infty} Tx_n$. By the construction of T we have also: $\lim_{n\to\infty} Tx_n = \lim_{n\to\infty} x_{n+1} = x^*$. Therefore, since X is Hausdorff, $Tx^* = x^*$. Thus x^* is a fixed point of T. If $Ty = y$, we have

$$p_\alpha(x^* - y) = p_\alpha(Tx^* - Ty) \leqslant \epsilon_\alpha p_\alpha(x^* - y).$$

Since $\epsilon_\alpha < 1$ we must $p_\alpha(x^* - y) = 0$ for all p_α, whence $x^* = y$ (X being Hausdorff).

4.3.16. LEMMA. *A locally pseudo-convex space* $X = (X, \mathcal{P})$ *is locally connected.*

PROOF. Let $\overline{\mathcal{P}}$ be the saturated closure of $\mathcal{P}$, so that we have also $X = (X, \overline{\mathcal{P}})$. Since the open balls $B_r^\alpha = \{x \in X : p_\alpha(x) < r\}(p_\alpha \in \overline{\mathcal{P}})$ form a base of nuclei the lemma follows since each B_r^α is connected (see proof of 3.2.8).

§ 4. Locally Pseudo-convex Algebras

4.4.1. DEFINITION. A TA A is called *locally pseudo-convex* if it is locally pseudo-convex as a TLS. This means that a locally pseudo-convex algebra A is of the form $A = (A, \mathcal{P})$, where $\mathcal{P}$ is a family of pseudo-seminorms. If each $p_\alpha \in \mathcal{P}$ is sm. then A is called *locally sm. pseudo-convex algebra*. Again, if each p_α is a semi-norm, A is called a *locally*† *sm. convex algebra*.

4.4.2. LEMMA. *If* A *is locally sm. pseudo-convex then it has a basis consisting of pseudo-convex subsemigroup nuclei. Conversely, if* A *is a TA which admits such a basis then it is locally sm. pseudo-convex.*

PROOF. If $A = (A, \mathcal{P})$ with each $p_\alpha \in \mathcal{P}$ sm. then, by 4.1.13, $B_r^\alpha = \{x \in A : p_\alpha(x) < r\}, 0 < r \leqslant 1$, are pseudo-convex subsemigroup nuclei. On the other hand if A has such a bais then the associate gauges p_α are sm. pseudo-convex (by 4.1.12 ((i), (iii))).

† This is the same as locally m. convex algebra in the terminology of Michael [20].

4.4.3. LEMMA. *If $p_1, \cdots p_n$ are sm. ρ_j-seminorms $(j = 1, \cdots, n)$ then $q = p_1 \vee \cdots \vee p_n$ is a sm. ρ-seminorm with $\rho = \min\{\rho_j\}$, where ρ_j is the homogenity index of p_j.*

PROOF. By 4.3.4, $q(x) = \max_j p_j(x)^{\frac{\rho}{\rho_j}}$, and q is a ρ-seminorm. Further, we have

$$q(xy) = \max_j p_j(xy)^{\frac{\rho}{\rho_j}} \leqslant \max_j p_j(x)^{\frac{\rho}{\rho_j}} p_j(y) f^{\frac{\rho}{\rho_j}} = q(x)q(y),$$

whence the lemma.

4.4.4. PROPOSITION. *Let A be an algebra (over $\mathbb{K}$) $\mathcal{P}$ a family of pseudo-seminorms on A, and $\overline{\mathcal{P}}$ the saturated closure of $\mathcal{P}$. Then A is a TA under the $\mathcal{P}$-topology iff for each. $p' \in \mathcal{P}$ there is a $p_0 \in \overline{\mathcal{P}}$ and a $C > 0$ such that*

$$p'(xy) \leqslant C p_0(x)^{\frac{\rho'}{\rho}} p_0(y)^{\frac{\rho'}{\rho}} \quad (x, y \in A) \tag{$*$}$$

where ρ', ρ_0 are respectively the homogenity indices of p', p_0.

PROOF. The proof is somewhat similar to that of 3.4.3.

First suppose that $(*)$ holds for each $p' \in \mathcal{P}$. Assume that in $A, x_\alpha \to x, y_\beta \to y$. Then from the identity

$$x_\alpha y_\beta - xy = (x_\alpha - x)(y_\beta - y) + x(y_\beta - y) + (x_\alpha - x)y$$

we obtain, using $(*)$ and the subadditivity of p', the inequality:

$$p'(x_\alpha y_\beta - xy) \leqslant C \Big\{ p_0(x_\alpha - x)^{\frac{\rho'}{\rho_0}} p_0(y_\beta - y)^{\frac{\rho'}{\rho_0}}$$

$$+ p_0(x)^{\frac{\rho'}{\rho_0}} p_0(y_\beta - y)^{\frac{\rho'}{\rho_0}} + p_0(x_\alpha - x)^{\frac{\rho'}{\rho_0}} p_0(y)^{\frac{\rho'}{\rho_0}} \Big\}.$$

It follows that $p'(x_\alpha y_p - xy) \to 0$, for every p', whence $x_\alpha y_\beta \to xy$, and $(A, \mathcal{P})$ is a TA.

Next suppose that $(A, \mathcal{P})$ is a TA. Since the map $(x, y) \mapsto xy$ is continuous, given $p' \in \mathcal{P}$ and $\epsilon > 0$ then is a $p_0 \in \overline{\mathcal{P}}$ and a $\delta > 0$ such that

$$p_0(x), p_0(y) \leqslant \delta \Rightarrow p'(xy) \leqslant \epsilon. \tag{1}$$

For $x, y \in A$ with $p_0(x), p_0(y) \neq 0$, set $x_1 = \delta^{\frac{1}{\rho}} x / p_0(x)^{\frac{1}{\rho_0}}$, $y_1 = \delta^{\frac{1}{\rho_0}} y / p_0(y)^{\frac{1}{\rho_0}}$ where ρ_0 is the homogenity index of p_0. Then $p_0(x_1) = \delta = p_0(y_1)$, whence by (1), $p'(x_1 y_1) \leqslant \epsilon$ which reduces to

$$p'(xy) \leqslant C p_0(x)^{\frac{\rho'}{\rho_0}} p_0(y)^{\frac{\rho'}{\rho_0}} \tag{2}$$

where $C = \epsilon / \delta^{2\rho'/\rho}$.

It remains to consider the case where $p_0(x)$ or $p_0(y) = 0$. Suppose that $p_0(x) = 0$. Then $p_0(n^2 x) = 0$ for any integer $n \leqslant 1$. For arbitrary $y \in A$ we can choose n (large enough) that

$$p'(\frac{y}{n}) = \frac{p'(y)}{n^{\rho'}} \leqslant \delta.$$

Then, by (1), $p'(nxy) = p'(n^2 x \cdot y/n) \leqslant \epsilon$ which gives $n^{\rho'} p'(xy) \leqslant \epsilon$. Since n is arbitralily large we get $p'(xy) = 0$. Similarly $p'(xy) = 0$ if x is arbitrary and $p_0(y) = 0$. Thus (2) holds for all x, y as required.

4.4.5. COROLLARY. *Let $A = (A, \mathcal{P})$ be a locally pseudo-convex algebra. Then for each $p' \in \mathcal{P}$ we can find a pseudo-seminorm p^* on A such that*

$$p'(xy) \leqslant p^*(x)^{\frac{\rho'}{\rho^*}} p^*(y)^{\frac{\rho'}{\rho^*}} \quad (x, y \in A), \tag{$**$}$$

$p^* \sim p_1 \vee \cdots \vee p_n$ for some $p_j \in \mathcal{P}$.

4.4.6. COROLLARY. *If $\mathcal{P}$ is any family of sm. pseudo-seminorms on an algebra A then under the $\mathcal{P}$-topology A is a locally (sm.) pseudo-convex algebra.*

PROOF. By sm. property of $p_\alpha \in \mathcal{P}$ we have $p_\alpha(xy) \leqslant p_\alpha(x) p_\alpha(y)$ and the condition $(*)$ of 4.4.4 evidently holds.

4.4.7. LEMMA. *Let $A = (A, \mathcal{P})$ be a locally pseudo-convex algebra. If $\mathcal{P} = \{p_\alpha\}$ and $\rho_0 = \inf \rho_\alpha > 0$, when ρ_α is the homogenity index of p_α, then A is a locally ρ_0-pseudoconvex algebra.*

PROOF. By 3.2.9, $q_\alpha = p_\alpha^{\rho_0/\rho_\alpha}$ is a ρ_0-seminorm such that

$q_\alpha \sim p_\alpha$. Thus $A = (A, \mathcal{Q})$ where $\mathcal{Q} = \{q_\alpha\}$ is a family of ρ_0-seminorms, whence A is a locally ρ_0-pseudoconvex algebra.

4.4.8. PROPOSITION. *The locally* $\{\rho_\alpha\}$*-convex algebras* A *are precisely the TA's of the form* $A = (A, \mathcal{P})$, *where* $\mathcal{P}$ *is a family* $\{p_\alpha\}$ *of* ρ_α*-seminorms on* A *satisfying* $(*)$ *of* 4.4.1. *Moreover,* A *is Hausdorff iff* $\ker \mathcal{P} = \{0\}$.

PROOF. The first statement is essentially a restatement of 4.4.4. The second statement follows from 3.1.21 (iii).

4.4.9. DEFINITION. A family $\hat{\mathcal{P}}$ of pseudo-seminorms on an algebra A is called *well-behaved* if $\hat{\mathcal{P}}$ is saturated and further for each $p' \in \hat{\mathcal{P}}$ there is a $p \in \hat{\mathcal{P}}$ such that

$$p'(xy) \leqslant p(x)^{\frac{\rho'}{\rho}} p(y)^{\frac{\rho'}{\rho}} \qquad (**)$$

where p', p are respectively a ρ'-seminorm and a ρ-seminorm.

4.4.10. PROPOSITION. *The locally pseudo-convex algebras are precisely the TA', A of the from* $A = (A, \hat{\mathcal{P}})$, *where* $\hat{\mathcal{P}}$ *is a well-behaved family of pseudo-seminorms.*

PROOF. If $\hat{\mathcal{P}}$ is well-behaved then by $(**)$ of 4.4.5 and $(*)$ of 4.4.4, A is a TA (under the p-topology) which by 4.3.2, is locally pseudo-convex. Next assume that A is a locally pseudo-convex algebra. Then, by 4.4.4, $A = (A, \mathcal{P})$ with $\mathcal{P}$ satisfying $(*)$ of 4.4.4. Let $\overline{\mathcal{P}}$ denote the saturated closure of $\mathcal{P}$ and write

$$\hat{\mathcal{P}} = \{p \in \mathfrak{S}^* : p \sim q \in \overline{\mathcal{P}}\}$$

where $\mathfrak{S}^*$ denotes the set of all pseudo-seminorms on A. Then, by 4.3.6, $\hat{\mathcal{P}}$ is saturated. Further, by 4.4.5, $\hat{\mathcal{P}}$ is well-behaved.

4.4.11. LEMMA. *Let* $A = (A, \mathcal{P})$ *be a locally* ρ*-convex algebra such that:*

(i) $\mathcal{P} = \{p_\alpha\}$ *is a well-behaved family;*

(ii) $\sup_{p_\alpha} p_\alpha(x) < \infty$ *for each* $x \in A$.

Then $p(x) = \sup_{p_\alpha} p_\alpha(x)$ *is a sm.* ρ*-seminorm and* (A,p) *is a* ρ*-seminormed algebra.*

PROOF. Since $\mathcal{P}$ is well-behaved, for each $p_\alpha \in \mathcal{P}$ there is a $p_\beta \in \mathcal{P}$ such that

$$p_\alpha(xy) \leqslant p_\beta(x)p_\beta(y) \leqslant p(x)p(y).$$

Taking the supremum of the first term over all α we get

$$p(xy) \leqslant p(x)p(y).$$

4.4.12. PROPOSITION. *Let* $A = (A, \mathcal{P})$ *be a locally pseudo-convex (respy. locally sm. pseudo-convex) algebra. Then:*

(i) *Any subalgebra* A_0 *of* A *under the relative topology is locally pseudo-convex (respy. locally sm. pseudo-convex).*

(ii) *If* I *is a bi-ideal of* A *and* $A^\# = A/I$*, then* $A^\#$*, under the quotient topology, is locally pseudo-convex (respy. locally sm. pseudo-convex).*

PROOF. (i) We may assume that $\mathcal{P}$ is well-behaved. For $p_\alpha \in \mathcal{P}$ write $p_{\alpha_0} = p_\alpha | A_0$ (restriction to A_0) and $\mathcal{P}_0 = \{p_{\alpha_0}\}$. Then it is clear that $(A_0, \mathcal{P}_0)$ has the required properties.

(ii) Let $p_\alpha^\#$ be defined in terms of p_α as in 3.4.15 and write $\mathcal{P}^\# = \{p_\alpha^\#\}$. Then $(A^\#, \mathcal{P}^\#)$ is the quotient algebra and this has the required properties (the proof that $\mathcal{P}^\#$ induces the quotient topology is similar to that of 3.1.22 (iii)).

4.4.13. PROPOSITION. *The unitization* A_1 *or the complexification* $\tilde{A}$ *of a locally pseudo-convex or locally sm. pseudo-convex algebra* A *is again an algebra of the same type.*

PROOF. If $A = (A, \mathcal{P})$ then for each $p \in \mathcal{P}$ denote by p_1 the canonical extension of p to A_1 as defined in 3.4.11 and by $\tilde{p}$ the extension of p to $\tilde{A}$ as defined in the proof of 4.1.16. Write

$$\mathcal{P}_1 = \{p_1 : p \in \mathcal{P}\}, \quad \tilde{\mathcal{P}} = \{\tilde{p} : p \in \mathcal{P}\}.$$

Then $A_1 = (A_1, \mathcal{P}_1)$, $\tilde{A} = (\tilde{A}, \tilde{\mathcal{P}})$ have the stated properties.

4.4.14. LEMMA. *Let* $A = (A, \mathcal{P})$ *be sequentially complete-in particular complete-Hausdorff locally sm.* ρ*-convex algebra. If for* *an* $x \in A$ *we have*

$$\nu_\alpha(x) < 1 \text{ for all } \nu_\alpha$$

then x *is q. invertible,* $\sum_{n=1}^{\infty}(-1)^n x^n$ *converges in* A *and the q. inverse*

$$x' = \sum_{n=1}^{\infty}(-1)^n x^n.$$

PROOF. As in the proof of 3.3.19, we obtain for each p_α,

$$\sum_{n=1}^{\infty} p_\alpha(x^n) < \infty.$$

So

$$p_\alpha\left(\sum_{n=N}^{N+r}(-1)^n x^n\right) \leqslant \sum_{N}^{N+r} p_\alpha(x^n) \to 0, \text{ as } N, r \to \infty.$$

This means that $\sum_{n=1}^{\infty}(-1)^n x^n$ converges absolutely and so also in A. Exactly as in the proof of 3.3.19 we can show that $\sum_{n=1}^{\infty}(-1)^n x^n = x'$.

4.4.15. PROPOSITION. *A locally sm. pseudo-convex algebra (or more generally a TA whose topology is induced by a family of a. sm. quarter-norms) is a* C *algebra.*

PROOF. Let $A = (A, \mathcal{P})$, where $\mathcal{P} = \{p_\alpha\}$ and each p_α is a. sm.. Then, by 3.3.22, each p_α satisfies the inequality $(*)$ stated therein. It follows that if (each) $p_\alpha(x_\gamma - x) \to 0$ $(x_\alpha, x \in G_q)$ then $p_\alpha(x'_\gamma - x') \to 0$, which proves that the map $x \mapsto x'$ of G_q is continuous, so that A is a C algebra.

4.4.16. LEMMA. *In a locally sm. pseudo-convex algebra* $A = (A, \mathcal{P})$, *where* $\mathcal{P}$ *is saturated, for any continuous pseudo-*

seminorm p *on* A *there is a* $p_\alpha \in \mathcal{P}$ *such that* $\nu_p^{\frac{1}{\rho}} \leqslant \nu_{p_\alpha}^{\frac{1}{\rho_\alpha}}$, *where* ρ, ρ_α *are homogenity induces of* p, p_α *respectively.*

PROOF. By 4.3.13 (ii) there is a p_α with $p \leqslant C p_\alpha^{\frac{\rho}{\rho_\alpha}}$. Therefore

$$\begin{aligned}\nu_p(x) = \lim_{n\to\infty} (p(x^n))^{\frac{1}{n}} &\leqslant \lim_{n\to\infty} C^{\frac{1}{n}} \left[\lim_{n\to\infty} p_\alpha(x^n)^{\frac{1}{n}}\right]^{\frac{\rho}{\rho_\alpha}} \\ &\leqslant \nu_\alpha(x)^{\frac{\rho}{\rho_\alpha}},\end{aligned}$$

so that $\nu_p^{\frac{1}{\rho}} \leqslant \nu_\alpha^{\frac{1}{\rho_\alpha}}$, as required.

§ 5. Projective Limit Decompositions

4.5.1. Let $\{A_\alpha : \alpha \in \mathcal{A}\}$ be a family of TA's with the indexing set $\mathcal{A}$, a poset directed above by $\prec$. Suppose that for each pair (α, β) of elements of $\mathcal{A}$, with $\alpha \prec \beta$, there is given a continuous homomorphism $\varphi_{\alpha\beta} : A_\beta \to A_\alpha$, such that

(i) $\varphi_{\alpha\alpha} = 1_{A_\alpha}$

(ii) $\varphi_{\alpha\beta} \circ \varphi_{\beta\gamma}$ for $\alpha \prec \beta \prec \gamma$.

Write

$A =$ the Cartesian product $\prod_{\alpha \subset a} A_\alpha$.

Under coordinate-wise operations A is an algebra. Set

$$A_0 = \{x = (x_\alpha) \in A : \varphi_{\alpha\beta}(x_\beta) = x_\alpha \text{ whenever } \alpha \prec \beta\}.$$

A_0 is called the *projective limit* of the A_α and we write $A_0 = \varprojlim A_\alpha$, $x = \varprojlim x_\alpha$ $(x \in A_0)$.

The projective limit A_0 has the following *universal mapping property*: for any TA B and continuous homomorphisms $\omega_\alpha : B \to A_\alpha$ such that $\varphi_{\alpha\beta} \circ \omega_\beta = \omega_\alpha$ (whenever $\alpha \prec \beta$), there is a unique continuous homomorphism $\theta : B \to A_0$ such that $\pi_\alpha \circ \theta = \omega_\alpha$ $(\alpha \in \mathcal{A})$, where π_α denotes the natural projection $A \to A_\alpha$. (It is easy to see that θ is given by $\theta(b) = (\omega_\alpha(b))$, for $b \in B$.)

4.5.2. LEMMA. *A_0 is a closed subalgebra of A.*

PROOF. By 2.2.8 (b), A is a TA, and since $\varphi_{\alpha\beta}$ are homo-

4.5.2. LEMMA. *A_0 is a closed subalgebra of A.*

PROOF. By 2.2.8 (b), A is a TA, and since $\varphi_{\alpha\beta}$ are homomorphisms, A_0 is a subalgebra of A. It follows from the continuity of $\varphi_{\alpha\beta}$ that A_0 is closed in A.

4.5.3. THEOREM (Michael[†]). *Every complete Hausdorff locally sm. pseudo-convex algebra A is (t. isomorphic to) a projective limit, $\varprojlim \hat{A}_\alpha$ of pseudo-Banach algebras[†] $\hat{A}_\alpha$. In particular, if A is locally ρ-convex, then $A = \varprojlim \hat{A}_\alpha$, with all $\hat{A}_\alpha$ as ρ-Banach algebras.*

Conversely, a projective limit $B = \varprojlim B_\alpha$ of pseudo-Banach algebras B_α, is a complete Hausdorff locally sm. pseudo-convex algebra.

PROOF. Let $A = (A, \mathcal{P})$, where we may assume that $\mathcal{P} = \{p_\alpha\}$ is saturated. Set $A_\alpha = A/N_\alpha$, where $N_\alpha = \ker p_\alpha$. Then A_α is a ρ_α-normed algebra (ρ_α being the homogenity index of p_x) whose completion $\hat{A}_\alpha$ is, by 3.4.17, a ρ_α-Banach algebra. Make the indexing set $\mathcal{A} = \{\alpha\}$ a directed set (directed above) by defining $\alpha \prec \beta$ if $p_\alpha \prec p_\beta$, where the latter $\prec$ is the order introduced in 4.3.3 (the directedness of the order is a consequence of the family $\mathcal{P}$ being saturated).

If $\alpha \prec \beta$, it is clear that $N_\beta \subseteq N_\alpha$, where $N_\beta = \ker p_\beta$, $N_\alpha = \ker p_\alpha$. Consequently, the map

$$\varphi_{\alpha\beta} : x_\beta = x + N_\beta \mapsto x_\alpha = x + N_\alpha \qquad (x \in A)$$

is well-defined and is a continuous homomorphism of A_β onto A_α, which can be extended, by 3.5.11 (b), to a homomorphism $\hat{\varphi}_{\alpha\beta}$ of $\hat{A}_\beta$ into $\hat{A}_\alpha$. It follows that

$$\{\hat{A}_\alpha : \hat{\varphi}_{\alpha\beta} \qquad (\alpha, \beta \in a,)\ \alpha \prec \beta\}$$

is a projective system of pseudo-Banach algebras which defines the projective limit

† A pseudo-Banach algebra is just a ρ-Banach algebra for some ρ, $0 < \rho \leqslant 1$. This algebra is not to be confused with the pseudo-Banach algebra introduced by Allen-Dales-McClure who use the term for a certain inductive limit of Banach algebras.

$$\hat{A}_0 = \varprojlim \hat{A}_\alpha \subseteq \prod \hat{A}_\alpha.$$

Similarly, $\{A_\alpha : \varphi_{\alpha\beta}\}$ gives rise to the projective limit

$$A_0 = \varprojlim A_\alpha,$$

which is identifiable as a subalgebra of $\hat{A}_0$. The map $\varphi : A \to A_0$ given by $(\varphi(x))_\alpha = x + N_\alpha = x_\alpha$ is a homomorphism, since each of the maps $\varphi_\alpha : x \mapsto x_\alpha (\alpha \in \mathcal{A})$ is a homomorphism. It is 1-1 since $x_\alpha = 0$ for all $\alpha \Rightarrow x \in \bigcap N_\alpha = \{0\}$ (since A is Hausdorff). Further, φ is continuous since each φ_α is continuous.

We now show that φ is "onto". Take any element $y = \varprojlim y_\alpha$ in A_0. For any finite subset S of $\mathcal{A}$, choose a $\beta \in \mathcal{A}$ such that $\beta \succ \alpha$ for all $\alpha \in S$. Take an element $x_S \in A$ such that $(\varphi(x_S))_\beta = y_\beta$; then $(\varphi(\alpha_S))_\alpha = y_\alpha$ for all $x \in S$. Also, given finite subsets S_1, S_2 with $S_1 \bigcap S_2 \neq \emptyset$, we have

$$(\varphi(x_{S_1}))_\alpha = y_\alpha = (\varphi(x_{S_2}))_\alpha \text{ for all } \alpha \in S_1 \bigcap S_2.$$

This relation implies that

$$x_{S_1} - x_{S_2} \in N_\alpha \subseteq U_{\alpha,\epsilon} = \{x \in A : p_\alpha(\alpha) < \epsilon\} \quad (\epsilon > 0).$$

It follows that (x_S) – S varying among all the finite subsets of $\mathcal{A}$ – is a C-net. Since A is complete

$$x_S \to \text{ (some) } x \in A.$$

Since φ is continuous we obtain

$$y_\alpha = (\varphi(x_S))_\alpha \ (S \text{ varying}) \to (\varphi(x))_\alpha = x_\alpha,$$

i.e. $y_\alpha = x_\alpha$, so that $y = \varphi(\alpha)$, proving that φ is "onto".

We next show that φ^{-1} is continuous. Consider a net $(x^{(\lambda)})$ in A such that

$$\varphi(x^{(\lambda)}) \to \varphi(x) \qquad (x \in A).$$

Then by 3.4.16 we have

$$p_\alpha(x^{(\lambda)} - x) = p_\alpha^{\#}(x_\alpha^{(\lambda)} - x_\alpha) \to 0 \text{ for all } \alpha.$$

This means that $x^{(\lambda)} \to x$ in A, proving the continuity of φ^{-1}.

To complete the proof of the first statement of the theorem it remains to show that $A_0 = \hat{A}_0$. Now any non-empty open set $\hat{W}$ of $\hat{A}_0$ contains an open set $\hat{V}$ of the form

$$\hat{V} = \pi_{\alpha_1}^{-1}(\hat{G}_{\alpha_1}) \bigcap \cdots \bigcap \pi_{\alpha_n}^{-1}(\hat{G}_{\alpha_n}) \bigcap \hat{A}_0,$$

where $\hat{G}_{\alpha_j}$ are open sets in $\hat{A}_{\alpha_j}$ and $\pi_{\alpha_j}^{-1}$ the natural projection $\pi \hat{A}_\alpha \to \hat{A}_{\alpha_j}$ $(j = 1, \cdots, n)$ choose an $\alpha_{n+1} \in \mathcal{A}$ with $\alpha_{n+1} \succ \alpha_j (j = 1, \cdots, n)$. Then we have the homomorphisms

$$\hat{\varphi}_{\alpha_j, \alpha_{n+1}} : \hat{A}_{\alpha_{n+1}} \to \hat{A}_{\alpha_j} \qquad (j = 1, \cdots, n).$$

Set

$$\hat{G}_{\alpha_{n+1}} = \hat{\varphi}_{\alpha_1, \alpha_{n+1}}^{-1}(\hat{G}_{\alpha_1}) \bigcap \cdots \bigcap \hat{\varphi}_{\alpha_n, \alpha_{n+1}}^{-1}(\hat{G}_{\alpha_n}).$$

Then

$$\varphi_{\alpha_j, \alpha_{n+1}}(G_{\alpha_{n+1}}) \subseteq \hat{G}_{\alpha_j} \qquad (j = 1, \cdots, n).$$

Set

$$\hat{U} = \hat{V} \bigcap \pi_{\alpha_{n+1}}^{-1}(G_{\alpha_{n+1}}) \subseteq \hat{V}.$$

Since $A_{\alpha_{n+1}}$ is dense in $\hat{A}_{\alpha_{n+1}}$ we can find $x \in A$ so that $x_{\alpha_{n+1}} \in \hat{G}_{\alpha_{n+1}}$. It follows that

$$\varphi(x) \in \hat{U}, \text{ so that } \hat{U} \bigcap A_0 \neq \emptyset.$$

Therefore A_0 is dense in $\hat{A}_0$. But A_0 is complete, being t. isomorphic to the complete space A and consequently closed. Thus, $A_0 = \hat{A}_0$.

The second (or converse) statement follows since every pseudo-Banach algebra is a complete Hausdorff locally sm. pseudo-convex algebra and these properties carry over to direct products and to closed subalgebras.

4.5.4. For descriptive convenience we call a complete Hausdorff locally sm. pseudo-convex algebra A as a *pseudo-Michael algebra*. If it is a locally sm. ρ-convex algebra then it will be referred to as a ρ-Michael algebra. Finally, if $\rho = 1$, A is called a Michael[†] algebra.

If completeness hypothesis is dropped in any of the above definitions then the resulting notions will be indicated by employing pre-Michael instead of Michael in their names

4.5.5. COROLLARY. *Every pseudo-pre-Michael algebra is a dense subalgebra of a projective limit* $\varprojlim \hat{B}_\alpha$ *of pseudo-Banach algebras* $\hat{B}_\alpha$.

PROOF. It suffices to apply 4.5.3 to the completion $B = \hat{A}$ of A.

4.5.6. PROPOSITION (Michael). *Let* $A = (A, \mathcal{P})$ *be a pseudo-Michael algebra with projective limit decomposition* $A = \varprojlim \hat{A}_\alpha$ ($\hat{A}_\alpha$ *a pseudo-Banach algebra).* For $x \in A$, $x = \varprojlim x_\alpha$ $(x_\alpha \in \hat{A}_\alpha)$ *we write* $x = (x_\alpha)$. *Then:*

(i) $x = (x_\alpha)$ *is q. invertible in* A *iff* $x_\alpha \in \hat{A}_\alpha$ *is q. invertible (for each* α*).*

(ii) A *is unital iff all* $\hat{A}_\alpha$ *are unital, and then the unity* $e = (e_\alpha), e_\alpha$ *being unity of* $\hat{A}_\alpha$.

(iii) $u = (u_\alpha)$ *is idempotent in* A *iff for each* α, u_α *is idempotent in* $\hat{A}_\alpha$.

(iv) *If* A *is unital, then an element* $x = (x_\alpha)$ *is invertible iff for each* α, x_α *is invertible in* $\hat{A}_\alpha$.

PROOF. (i) If x is q. invertible then, by 1.1.24, x_α is q. invertible in A_α and so also in $\hat{A}_\alpha \supseteq A_\alpha$. Conversely, if for each α, x_α , has q.i. x'_α, then $(x'_\alpha) \in A$. To see this, note that if $\beta > \alpha$, $\varphi_{\alpha\beta} : \hat{A}_\beta \to \hat{A}_\alpha$ the element $\varphi_{\alpha\beta}(x'_\beta)$ is q.i. of $\varphi_{\alpha\beta}(x_\beta) = x_\alpha$, so that $\varphi_{\alpha\beta}(x'_\beta) = x'_\alpha$. It follows that $(x'_\alpha) \in \varprojlim \hat{A}_\alpha = A$, whence there is an element $x' \in A$ with $(x')_\alpha = x'_\alpha$. Since multiplication and addition are component-wise it is clear that x' is q.i. of x.

(ii), (iii): These follow at once since multiplication in A is component-wise.

[†] Such algebras have also been considered by Arens.

(iv) The proof is similar to that of (i).

4.5.7. COROLLARY. *For* $x = (x_\alpha) \in A$ *we have:*

(1) $\sigma'_A(x) = \bigcup_\alpha \sigma_{\hat{A}_\alpha}(x_\alpha)$;

(2) $r_A(x) = \sup_\alpha r_{\hat{A}_\alpha}(x_\alpha)$;

(3) $\sigma_A(x) = \bigcup_\alpha \sigma_{\hat{A}_1}(x_\alpha)$ whenever A is unital.

PROOF. By 1.7.8 $\lambda(\neq 0) \in \sigma'_A(x)$ iff $-\lambda^{-1}x$ is not q. invertible, i.e. by 4.5.6 (i), iff for some α, $-\lambda^{-1}x_\alpha$ is not q. invertible in $\hat{A}_\alpha$ iff $\lambda \in \sigma'_{\hat{A}_\alpha}(x_\alpha)$. This proves (1), and (2) readily follows from (1).

For (3), assume now that A is unital; then each $\hat{A}_\alpha$ is unital. From (1) and 1.7.21 we obtain

$$\sigma_A(x) \bigcup \{0\} = \bigcup_\alpha \sigma_{\hat{A}_1}(x_\alpha) \bigcup \{0\}.$$

If $0 \notin \sigma_A(x)$ then x is invertible and consequently each x_α is invertible, so that $0 \notin \bigcup_\alpha \sigma_{\hat{A}_\alpha}(x_\alpha)$. On the other hand, if $0 \in \sigma_A(x)$ then x is not invertible and consequently, some x_α is not invertible so that $0 \in \sigma_{\hat{A}_1}(x_\alpha)$. Thus, the equality in (3) holds in both cases.

4.5.8. COROLLARY. $\bigcap_\alpha \varphi_\alpha^{-1}(\sqrt{\hat{A}_\alpha}) \subseteq \sqrt{A} = \bigcap_\alpha \varphi_\alpha^{-1}(\sqrt{A_\alpha})$.

PROOF. By 1.2.17 (i), each $\varphi_\alpha^{-1}(\sqrt{\hat{A}_\alpha})$ or $\varphi_\alpha^{-1}(\sqrt{A_\alpha})$ is a bi-ideal. So

$$J = \bigcap_\alpha \varphi_\alpha^{-1}(\sqrt{\hat{A}_\alpha}),\ I = \bigcap_\alpha \varphi_\alpha^{-1}(\sqrt{A_\alpha})$$

are bi-ideals of A. By 1.2.26, $\varphi(\sqrt{A}) \subseteq \sqrt{A}_\alpha$. So $\sqrt{A} \subseteq I$. On the other hand, if $x \in I$ then $x_\alpha \in \sqrt{A_\alpha}$ and so x_α is q. invertible. Consequently, by 4.5.6 (i), x is q. invertible so that I is q.i. bi-ideal, whence $I \subseteq \sqrt{A}$. Thus $\sqrt{A} = I$. Again, if $x \in J$ then

$x_\alpha \in \sqrt{\hat{A}_\alpha}$, x_α is q. invertible and consequently x is q. invertible. Thus, J is a q. invertible ideal whence $J \subseteq \sqrt{A}$.

§6. Metrizable Locally Pseudo-convex Algebras

4.6.1. PROPOSITION. *Every first countable locally pseudo-convex algebra A is semi-metrizable and has the form $A = (A, \hat{\mathcal{P}})$, where $\hat{\mathcal{P}} = \{p_1, p_2, \cdots\}$ is a countable well-behaved family of pseudo-seminorms such that*

$$p_1 \prec p_2 \prec \cdots \dagger \tag{*}$$

$$p_j(xy) \leqslant p_{j+1}(x)^{\frac{\rho_j}{\rho_{j+1}}} p_{j+1}(y)^{\frac{\rho_i}{\rho_{j+1}}} \tag{**}$$

$(j = 1, 2, \cdots)$, where ρ_j, ρ_{j+1} are respectively the homogenity indices of p_j, p_{j+1}.

PROOF. That first countability implies semi-metrizability follows from 2.1.7. Since A is first countable its topology is induced by a countable family $\mathcal{Q} = \{q_n\}$ of pseudo-seminorms. Set $p'_n = q_1 \vee \cdots \vee q_n$ and $\mathcal{P}' = \{p'_n\}$. Then $p'_1 \prec p'_2 \prec \cdots$. Set $p_1 = p'_1$. By 4.4.4, there an integer $i(1)$ such that

$$p_1(xy) = p'_1(xy) \leqslant C_1 p'_{i(1)}(x)^{\frac{\rho'}{\rho_1}} p'_{i(1)}(y)^{\frac{\rho'}{\rho_1}}$$

where ρ_1, ρ' are homogenity indices of $p_1, p'_{i(1)}$ and C_1 a constant which we may clearly suppose $\geqslant 1$. Set $p_2 = C_1^{\frac{\rho}{2\rho'}} p'_{i(1)}$. Then $p_2 \sim p'_{i(1)}$ and satisfies

$$p_1(xy) \leqslant p_2(x)^{\frac{\rho'}{\rho}} p_2(y)^{\frac{\rho'}{\rho}}$$

For p_2 we can similarly choose $p'_{i(2)}$, with $i(2) \geqslant i(1), 2$, such that

$$p_2(xy) \leqslant C_2 p'_{i(2)}(x)^{\frac{\rho'_2}{\rho_2}} p'_{i(2)}(y)^{\frac{\rho'_2}{\rho_2}}$$

† Recall that $p_j \prec p_{j+1}$ means: $p_j^{\frac{1}{\rho_j}} \leqslant p_j^{\frac{1}{\rho_{j+1}}}$ and $\rho_{j+1} \leqslant \rho_j$.

where we can assume that $C_2 \geqslant C_1$. Set $p_3 = C_2^{\frac{\rho_2}{2\rho_2'}} p'_{i(2)}$ and we have

$$p_2(xy) \leqslant p_3(x)p_3(y).$$

Thus proceeding we get $\hat{\mathcal{P}} = \{p_1, p_2, \cdots\}$ with the desired properties. It is clear from the way we constructed that $\hat{\mathcal{P}}$ is equivalent to a cofinal subset of $\mathcal{P}'$, whence $\hat{\mathcal{P}} \sim \mathcal{P}' \sim \mathcal{Q}$. So $A = (A, \hat{\mathcal{P}})$, completing the proof.

4.6.2. DEFINITION. A first countable Hausdorff locally pseudo-convex algebra is called a *pseudo-pre-Frechet algebra or a pseudo-pre-$\mathfrak{F}$ algebra.* If A is also complete it is called a *pseudo-Frechet* or a *pseudo-$\mathfrak{F}$* algebra.

4.6.3. DEFINITION. A pseudo-pre-$\mathfrak{F}$ algebra A is called a *locally sm.* pseudo-pre-$\mathfrak{F}$ algebra if its topology can be induced by a countable family $\mathcal{Q} = \{q_n\}$ of sm. ρ_n-seminorms for some ρ_n with $0 < \rho_n \leqslant 1$, $n = 1, 2, \ldots$.

4.6.4. LEMMA. *Every pseudo-pre-$\mathfrak{F}$ algebra A is of the form $A = (A, \mathcal{Q})$ where $\mathcal{Q} = \{q_n\}$ is a countable family of pseudo-seminorms with* $\ker \mathcal{Q} = \{0\}$. *If A is locally sm. pre-pseudo-$\mathfrak{F}$ algebra then we can choose $\mathcal{Q}$ such that each $q \in \mathcal{Q}$ is sm.*.

PROOF. This follows from 4.6.1, 3.1.21 (iii).

4.6.5. DEFINITION. If in a pseudo-pre-$\mathfrak{F}$ algebra $A = (A, \mathcal{Q})$, all $q \in \mathcal{Q}$ are ρ-seminorms for some fixed ρ $(0 < \rho \leqslant 1)$ then A is called a ρ-pre-$\mathfrak{F}$ algebra. If in addition each q is sm., then A is called a locally sm. ρ-pre-$\mathfrak{F}$ algebra. The meaning of a ρ-$\mathfrak{F}$ algebra or a locally sm. ρ-$\mathfrak{F}$ algebra is clear. Finally, when $\rho = 1$, we say simply $\mathfrak{F}$ algebra or locally sm. $\mathfrak{F}$ algebra as the case may be.

4.6.6. PROPOSITION. *The locally sm. pseudo-$\mathfrak{F}$ (respy. locally sm. ρ-$\mathfrak{F}$) algebras are precisely projective limits of sequences of pseudo-Banach (respy. ρ-Banach) algebras.*

PROOF. This follows from 4.6.4, 4.5.3.

4.6.7. PROPOSITION. *Every pseudo-pre-$\mathfrak{F}$ (respy. pseudo-$\mathfrak{F}$)*

algebra A is a pre-(F) (respy. (F)) algebra.

PROOF. Let $A = (A, \mathcal{Q})$, $\mathcal{Q} = \{q_n : n = 1, 2, \cdots\}$ and $\ker \mathcal{Q} = \{0\}$. Then

$$|x| = \sum_1^\infty \frac{1}{2^n}\left(\frac{q_n(x)}{1 + q_n(x)}\right)$$

is a (F) norm and $A = (A, |\cdot|)$. Further, $|\cdot|$ is complete whenever $(A, \mathcal{Q})$ is complete.

4.6.8. Examples of metrizable locally pseudo-convex algebras

(i) Let (ρ_n) be a sequence of real numbers such that $0 < \rho_n \leqslant 1$. The set C of all $\mathbb{K}$-valued continuous functions $x = x(t)$ on $\mathbb{R}$ is an algebra (over $\mathbb{K}$) under pointwise operations. Set

$$\|x\|_n = \sup_{|t| \leqslant n} |x(t)|^{\rho_n}; \text{ then } \|\cdot\|_n$$

is a sm. ρ_n-seminorm. It can be seen that $C = (C, \{\|\cdot\|_n : n = 1, 2, \cdots\})$ is a locally sm. pseudo-$\mathfrak{F}$ algebra. If all $\rho_n = \rho$ then C is a locally sm. ρ-$\mathfrak{F}$ algebra. Finally, when $\rho = 1$, we get a locally sm. $\mathfrak{F}$ algebra which we denote by $C(\mathbb{R})$.

(ii) Let $l^\rho = l^\rho(\mathbb{K})$ denote the algebra of all sequences $(x^{(n)})_{n=1}^\infty$ of elements of $\mathbb{K}$ such that $p(x) = \sum_n |x^{(n)}|^\rho < \infty$, under coordinate-wise operations. Consider the Cartesian power

$$A = (l^\rho)^\omega = l^\rho \times l^\rho \times \cdots.$$

If $y = (y_1, \cdots, y_n, \cdots) \in A$, $y_n \in l^\rho$ then set $p_n(y) = p(y_n) = \sum_{m=1}^\infty |y_n^{(m)}|^\rho$, where $y_n = (y_n^{(m)})_{m=1}^\infty$. Then it is easy to check that $(A, \mathcal{P})$ is a ρ-$\mathfrak{F}$ algebra, where $\mathcal{P} = \{p_n\}$. Further, it is locally sm. ρ-$\mathfrak{F}$ algebra since each p_m is sm..

(iii) We have already noted in 3.3.14 (example (iii)) that the algebra $\mathcal{E}$ of all entire functions (under pointwise operations) is a unital (F) algebra. Now set for $f \in \mathcal{E}$,

$$\|f\|_n = \sup_{|z|\leqslant n} |f(z)|.$$

Then $\|\cdot\|_n$'s are clearly sm. seminorms † and the family $\{\|\cdot\| : n = 1, 2, \cdots\}$ induces the same topology as the (F) metric $\|\cdot\|_G$ since convergence with respect to either is uniform convergence over compacta. Thus, $(\mathcal{E}, \{\|\cdot\|_n : n = 1, 2, \cdots\})$ is a locally sm. $\mathfrak{F}$-algebra (over $\mathbb{C}$).

(iv) For $0 < \rho \leqslant 1$, let L^{ω}_{ρ} denote all $\mathbb{K}$-valued measurable functions $x = x(t)$ (or rather equivalence classes of such functions) on the unit interval $[0,1]$ such that

$$\|x\|_n^{(\rho)} = \left(\int_0^1 |x(t)|^{n\rho}\, dt\right)^{\frac{1}{n}} < \infty.$$

Now

$$\begin{aligned} [\|xy\|_n^{(\rho)}]^n &= \int_0^1 |x(t)|^{n\rho} |y(t)|^{n\rho}\, dt \\ &\leqslant \left[\int_0^1 |x(t)|^{2n\rho} dt\right]^{\frac{1}{2}} \left[\int_0^1 |y(t)|^{2n\rho} dt\right]^{\frac{1}{2}}, \end{aligned}$$

whence we conclude that

$$\|xy\|_n^{(\rho)} \leqslant \|x\|_{2n}^{(\rho)} \|y\|_{2n}^{(\rho)}.$$

Again, if $0 < r < s$, then by applying Hölder's inequality for $x^{\rho r}$ and 1, with $p = \frac{s}{r}, \frac{1}{p'} = 1 - \frac{1}{p}$ we obtain

$$\int_0^1 |x(t)|^{\rho r} \cdot 1 dt \leqslant \left(\int_0^1 |x(t)|^{\rho r \cdot \frac{s}{r}} dt\right)^{\frac{r}{s}} \times \left(\int_0^1 1^{p'} dt\right)^{\frac{1}{p'}}$$

which gives

$$\left(\int_0^1 |x(t)|^{\rho r} dt\right)^{\frac{1}{r}} \leqslant \left(\int_0^1 |x(t)|^{\rho s} dt\right)^{\frac{1}{s}}.$$

† Actually they are norms.

It follows that we have

$$\|x\|_1 \leqslant \|x\|_2 \leqslant \cdots \ (\text{ie.} \|x\|_1 \prec \|x\|_2 \prec \cdots)$$

so that $\mathcal{P} = \{\|\cdot\|_n : n = 1, 2, \cdots\}$ is a well-behaved (countable) family of ρ-norms and L^ω_ρ is a ρ-$\mathfrak{F}$ algebra. It is not a C algebra. To see this, consider elements of the algebra given by:

$$g_m(t) = \begin{cases} 1/(1+(m-1)^{\frac{1}{\rho}}) & \text{for } 0 \leqslant t \leqslant \frac{1}{m^{1/\rho}} \\ 1 & \text{for } 1/m^{\frac{1}{\rho}} < t \leqslant 1 \end{cases}$$

Then

$$g_m^{-1}(t) = \begin{cases} 1+(m-1)^{1/\rho} & (0 \leqslant t \leqslant 1/m^{1/\rho}) \\ 1 & (1/m^{1/\rho} < t \leqslant 1). \end{cases}$$

Since

$$\int_0^1 (g_m - 1)^{n\rho} dt = \frac{1}{m} \frac{(m-1)^\rho}{(1+(m-1)^{\frac{1}{\rho}})^{n\rho}} \to 0$$

as $m \to \infty$, for each n, we get $g_m \to 1$. On the other hand

$$\int_0^1 (g_m^{-1} - 1)^{n\rho} dt = \frac{m-1}{m}(m-1)^{n-1} \to \infty \ (\text{if } n > 1)$$

which implies that $g_m^{-1} \not\to 1$. Thus L^ω_ρ is not a C algebra. Therefore, by 4.4.15, it is not locally sm. (ρ-$\mathfrak{F}$) algebra.

The algebra $L^\omega = L^\omega_1$ is a Frechet algebra which may be called *Aren's algebra* since this was introduced and studied first by Arens [1′] who also showed [2′] that it is not a C algebra.

(v) Let T be any locally compact, σ-compact, Hausdorff space. Then the algebra $C = C(T, \mathbb{K})$, of all the $\mathbb{K}$-valued continuous functions under the topology of uniform convergence over compacta, is a locally sm. $\mathfrak{F}$ algebra. (If $T = \bigcup K_n$, K_n compact, $f \in C$, $\|f\|_n = \sup\{|f(t)| : t \in K_n\}$ then $C = (C, \{\|\cdot\|_n\})$. If T is compact C is a Banach algebra.

When T is not compact C may fail to be an I algebra. For example, take $T = \mathbb{R}$, $K_n = [-n, n]$. Then $\|\cdot\|_1 \leqslant \|\cdot\|_2 \leqslant \cdots$. Clearly, if

$$U_{n,\epsilon} = \{f \in C : \|f\|_n < \epsilon\}$$

then $\{U_{n,\epsilon} : n \geqslant 1, \epsilon > 0\}$ is a basis of nuclei, so that

$$\{U_{n,\epsilon}(1)\} = \{1 + U_{n,\epsilon}\}$$

is a basis of neighbourhoods of 1. Since the function $\dfrac{\epsilon x}{2n} \in U_{n,\epsilon} \quad (x \in \mathbb{R})$,

$$f_{n,\epsilon}(x) = 1 + \frac{\epsilon x}{2n} \in U_{n,\epsilon}(1)$$

and $f_{n,\epsilon}$ is not invertible since it vanishes at $x = -\dfrac{2n}{\epsilon}$. It follows that $C(\mathbb{R})$ is not an I algebra.

(vi) Let $C^\infty = C^\infty[0,1]$ be the algebra of all $\mathbb{K}$-valued infinitely differentiable functions on the closed interval $[0,1]$, with pointwise operations. For $f \in C^\infty$ write

$$\|f\|_\infty = \sup_{0 \leqslant \epsilon \leqslant 1} |f(t)|,$$

and set

$$\|f\|_n = \sum_{k=0}^{n} \|f^{(k)}\|_\infty,$$

where $f^{(k)}$ denotes the kth derivative of $f(f^{(0)} = f)$. It follows from the definitions that we have

$$\|f\|_n \leqslant \|f\|_{n+1} \; (n = 0, 1, 2, \cdots).$$

It can be shown by induction, using Leibnitz's theorem, that

$$\|fg\|_n \leqslant 2^{n+1} \|f\|_n \|g\|_n.$$

If we set $\|f\|_n^* = 2^{n+1}\|f\|_n$ then $\|\cdot\|_n^*$ is sm. and $\|\cdot\|_n^* \sim \|\cdot\|_n$. It is easy to see that $(C^\infty, \{\|\cdot\|_n^* : n = 0, 1, 2, \cdots\})$ is a locally sm. Frechet algebra. Further it is an I algebra. To see this consider its group G_i of invertible elements; $G_i = \{f \in C^\infty : f \neq 0 \text{ on } [0,1]\}$. We claim that G_i is open. For, given $f \in G_i$ we have by compactness of $[0,1]$, $m_0 = \inf_{0 \leqslant t \leqslant 1} |f(t)| > 0$ choose an ϵ with $0 < \epsilon < \frac{1}{2} m_0$. If $g \in C^\infty$, $\|g - f\|_0 < \epsilon$ then

$$g(t)| \geqslant |f(t)| - |f(t) - g(t)| \geqslant m_0 - \epsilon > \frac{m_0}{2} > 0,$$

so that $g \in G_i$. Thus G_i is open and C^∞ is an I algebra. Finally, by 4.4.15, it is a C algebra. Thus C^∞ is a locally sm. $\mathfrak{F}$ algebra which is a CI algebra (cf. Example V).

(vii) In 3.6.33 we have seen that the field $\mathbb{C}(X)$ of rational functions carries a topology (Williamson topology) under which it is a CI (division) algebra W. There is another topology on $\mathbb{C}(X)$ making it a pre-$\mathfrak{F}$ algebra which is also an I algebra. This topology as well is due to Williamson (for details, see [14′, pp.731-32] or [31, p.83]).

§ 7. Ample Algebras

4.7.1. DEFINITION. A TLS X is called *ample* if its (continuous) dual X^* separates points of X : given elements $x_1 \neq x_2$ in X there is an $f \in X^*$ with $f(x_1) \neq f(x_2)$. The separation condition is clearly equivalent to the simpler condition: given $x \neq 0$ in X there is an $f \in X^*$ with $f(x) \neq 0$ (it suffices to observe that $f(x_1) \neq f(x_2)$ iff $f(x_1 - x_2) \neq 0$).

A TA A is called ample if it is ample as a TLS.

4.7.2. Remark. Clearly an ample algebra (or TLS) must necessarily be Hausdorff.

4.7.3. PROPOSITION. (a) *Every subalgebra A_0 of an ample algebra A is ample.*

(b) *A direct product or direct sum of ample algebras in ample.*

PROOF. (a) If $y \in A_0$, $y \neq 0$ then by ampleness of A there is an $f \in A^*$ with $f(y) \neq 0$. Write $f_0 = f|A_0$. Then $f_0 \in A_0^*$ and $f_0(y) = f(y) \neq 0$, whence A_0 is ample.

(b) Let $A = \prod A_\alpha$ be a direct product of ample algebras A_α. If $a = (a_\alpha) \neq 0$ in A then $a_{\alpha_0} \neq 0$ for some α_0. Since A_{α_0} is ample we can choose $f_{\alpha_0} \in A_{\alpha_0}^*$ with $f_{\alpha_0}(a_{\alpha_0}) \neq 0$. Define f on A by $f(x) = f_\alpha(\pi_\alpha x)$, where $x = (x_\alpha)$ and π_α is the projection $x \mapsto x_\alpha$. Clearly $f \in A^*$ and $f(a) = f(a_{\alpha_0}) \neq 0$, so that A is ample. Since the direct sum $\sum A_\alpha$ is a subalgebra of A, by (a) it is also ample, completing the proof.

4.7.4. PROPOSITION. (a) *The unitization* A_1 *of an ample algebra* A *is ample.*

(b) *The complexification* $\tilde{A}$ *of an ample real algebra* A *is ample.*

(c) *If an ample real algebra* A *has complex structure then the resulting complexal gebra* $\check{A}$ *is ample.*

PROOF. (a) Let $a_1 = \alpha e_1 + a$ $(a \in A)$ be a non-zero element of A_1. We have to show that there is a $f_1 \in A_1^*$ with $f_1(a_1) \neq 0$. For the construction of f_1 we have to consider two cases.

Case 1: $\alpha = 0, a_1 = a$. Since A is ample there is an $f \in A^*$ with $f(a) \neq 0$. Extend f to f_1 on A_1 by setting for $x_1 = \lambda e_1 + x$ $(x \in A, \lambda \in \mathbb{K})$ $f_1(x_1) = \lambda + f(x)$. Then $f_1 \in A_1^*$ and $f_1(a_1) = f(a) \neq 0$.

Case 2: $\alpha \neq 0$. Set $f_1(\lambda e_1 + x) = \lambda$. Then $f_1 \in A_1^*$ and $f_1(a_1) = \alpha \neq 0$.

(b) Suppose that $z_0 \in \tilde{A}$, $z_0 = x_0 + iy_0 \neq 0$ $(x_0, y_0 \in A)$. Since A is ample, and x_0 or $y_0 \neq 0$, there is an $f_0 \in A^*$ with $f_0(x_0)$ or $f_0(y_0) \neq 0$. Setting

$$\tilde{f}_0(z) = f_0(x) + if_0(y) \qquad (z = x + iy)$$

and $\tilde{f}(z) = \tilde{f}_0(z) - i\tilde{f}_0\,(iz)$, it is easy to check that $\tilde{f}$ is C-linear and $\tilde{f} \in (\tilde{A})^*$. Since $\tilde{f}(z_0) = 2(f_0(x_0) + if_0(y_0)) \neq 0$, $\tilde{A}$ is ample.

(c) For $f \in A^*$ set $\check{f}(x) = f(x) - if(ix)$. Then

$$\begin{aligned}\check{f}((\alpha+i\beta)x) &= f((\alpha+i\beta)x) - if(i(\alpha+i\beta)x)\\ &= \alpha f(x) + \beta f(ix) - i. - \beta f(x) - i\alpha f(ix)\\ &= (\alpha+i\beta)[f(x) - if(ix)] = (\alpha+i\beta)\check{f}(x)\end{aligned}$$

so that $\check{f} \in (\check{A})^*$. If $x_0 \in A = \check{A}$ and $x_0 \neq 0$ then by ampleness of A there is an $f \in A^*$ with $f(x_0) \neq 0$. But then $\check{f} \in (\check{A})^*$ and $\check{f}(x_0) = f(x_0) - if(ix_0) \neq 0$, proving $\check{A}$ is ample.

4.7.5. PROPOSITION. *For a TA which is a division algebra to be ample it is sufficient that* $A^* \neq \{0\}$.

PROOF. Suppose that A is a division TA with $A^* \neq \{0\}$. If $f \in A^*$, $f \neq 0$ there is an $a_0 \in A$ with $f(a_0) \neq 0$. For any a in A with $a \neq 0$ define $f_a(x) = f\ \ (xa^{-1}a_0)$. Since the maps $x \mapsto xa^{-1}a_0$ and $y \mapsto f(y)$ are continuous, f_a is continuous, $f_a \in A^*$. Since $f_a(a) = f(a_0) \neq 0$, A is ample as required.

4.7.6. PROPOSITION. *Every locally convex algebra – in particular a Banach algebra – is ample.*

PROOF. This is an immediate consequence of the Hahn-Banach theorem (see [24, p.59]).

4.7.7. Remark. For $0 < \rho < 1$; there are ρ-Banach algebras which are ample as well as those which are not ample.

For instance the algebra $l^\rho = l^\rho(\mathbb{K})$ (example (v) of 3.4.6) is an ample ρ-Banach algebra. To see this define, for each $n \geqslant 1$, $f_n(x) = x_n$, where $x = (x_1, x_2, \cdots, x_n, \cdots) = (x_n) \in l^\rho$. Then it is easy to see that $f_n \in (l^\rho)^*$. If $a = (a_n) \in l^\rho$, $a \neq 0$ then for some n, $a_n \neq 0$. Then $f_n(a) = a_n \neq 0$, proving l^ρ is ample.

Consider next the space $L^\rho = L^\rho[0,1]$ consisting of (equivalence classes) of $\mathbb{K}$-valued Lebesgue measurable functions f such that

$$\int_0^1 |f(t)|^\rho dt < \infty.$$

With

$$\|f\| = \textstyle\int_0^1 |f(t)|^\rho dt,$$

L^ρ is a ρ-Banach space. It can be made into a ρ-Banach algebra by introducing in it trivial multiplication: for $x, y \in L^\rho$, $xy = 0$.

L^ρ is not ample since it is known that '0' is the only continuous linear functional on L^ρ (see [24, p.36]).

§ 8. Topological Spectral Radius

4.8.1. DEFINITION. Let $A = (A,p)$ be a ρ-seminormed algebra. Then, by 3.4.3, p is a. sm. and so the limit

$$\nu_p(x) = \lim_{n\to\infty} p(x^n)^{\frac{1}{n}} < \infty$$

exists, by 3.3.6. The non-negative real number $\nu_p(x)$ is called the *topological spectral radius* of A, the adjective "topological" being justified by:

4.8.2. LEMMA. *If q is a ρ-seminorm with $q \sim p$ then $\nu_q = \nu_p$.*

PROOF. By 3.4.1, q is also a. sm.. Further, by 3.2.11 (ii) we have constants C_q, C_p such that

$$p \leqslant C_q q, \; q \leqslant C_p p.$$

Therefore,

$$\nu_p(x) = \lim_{n\to\infty} p(x^n)^{\frac{1}{n}} \leqslant \lim_{n\to\infty} C_q^{\frac{1}{n}} q(x^n)^{\frac{1}{n}} \leqslant \nu_q(x),$$

since $\lim\limits_{n\to\infty} C_q^{\frac{1}{n}} = 1$. Similarly, $\nu_q(x) \leqslant \nu_p(x)$. Then $\nu_p(x) \doteq \nu_p(y)$, as desired.

4.8.3. PROPOSITION. *The topological spectral radius $\nu(x) = \nu_p(x)$ possesses, besides the properties (i) – (v) of* 3.3.7, *the following:*

(vi) $\nu(\lambda x) = |\lambda|^\rho \nu(x)$;
(vii) $\nu(x+y) \leqslant \nu(x) + \nu(y)$ provided $x \leftrightarrow y$.

PROOF. The property (vi) has already been established in 3.4.5. So we have only to prove (vii). For this it is convenient to follow the method of Bonsall-Duncan [4, p.71].

We may assume that p is sm.. Choose $\epsilon > 0$ and set

$$u = (\nu(x)+\epsilon)^{-\frac{1}{\rho}} x, \; v = (\nu(y)+\epsilon)^{-\frac{1}{\rho}} y.$$

Since $x \leftrightarrow y$ we have also $u \leftrightarrow v$. Also, by construction $\nu(u), \nu(v) < 1$, so that by 3.3.6, $p(u^n), p(v^n) < 1$ for all sufficiently large n. If follows that the semi-groups $\{u^n : n = 1,2,\cdots\}, \{v^n : n = 1,2,\cdots\}$ are p-bounded. Since $u \leftrightarrow v$ the semi-group S generated by u, v is

$$\{u^k, v^j, u^k v^j : k, j = 1,2,\cdots\}.$$

Since p is sm. we have

$$p(u^k v^j) \leqslant p(u^k)p(v^j).$$

It follows from the above that S is p-bounded. By 3.5.10 we can find a sm. ρ-seminorm q such that $q \sim p$, $q(u), q(v) \leqslant 1$. By 4.8.2 we have $\nu_p = \nu_q = \nu$ (say). Then

$$q(x) = q((\nu(x)+\epsilon)^{\frac{1}{\rho}} u) = (\nu(x)+\epsilon)q(u) \leqslant \nu(x)+\epsilon \text{ (since } q(u) \leqslant 1).$$

Similarly, $q(y) \leqslant \nu(y) + \epsilon$. It follows that

$$\nu(x+y) \leqslant q(x+y) \leqslant q(x) + q(y) \leqslant \nu(x) + \nu(y) + 2\epsilon.$$

From the arbitrariness of ϵ we conclude that $\nu(x+y) \leqslant \nu(x) + \nu(y)$.

4.8.4 COROLLARY. *If A is commutative then ν is a continuous sm. ρ-seminorm on $A = (A, p)$. Further, $\nu_\nu = \nu$.*

PROOF. That ν is a sm. ρ-seminorm follows from 3.3.7 (iii), 4.8.3 ((vi), (vii)). The continuity of ν is a consequence of the inequality in 3.3.7 (i). Finally, $\nu_\nu = \nu$ since $\nu(x^n) = \nu(x)^n$ (by 3.3.7 (ii b)).

4.8.5. LEMMA. $\lim_{n\to\infty} \binom{2n}{n}^{\frac{1}{2n}} = 2.$

PROOF. Write $a_n = n^n (n!)^{-1}$. Then $a_n / a_{n-1} = (1 - \frac{1}{n})/(1 - \frac{1}{n})^n$. Using the well-known result $\lim_{n\to\infty} (1 + \frac{z}{n})^n = \mathbf{e}^z$ † ($z \in \mathbb{C}$) we get

† We denote the classical exponential function by $\mathbf{e}^z$, so that in particular we have $\mathbf{e} = 1 + \frac{1}{1!} + \frac{1}{2!} + \cdots$

$$\lim_{n\to\infty} \frac{a_n}{a_{n-1}} = \frac{1}{\lim\limits_{n\to\infty}(1-\frac{1}{n})^n} = \frac{1}{\mathrm{e}^{-1}} = \mathrm{e}.$$

It follows that

$$\lim_{n\to\infty} a_n^{\frac{1}{n}} = \lim_{n\to\infty} \frac{a_n}{a_{n-1}} = \mathrm{e}. \tag{1}$$

Therefore

$$\begin{aligned}
\lim_{n\to\infty} \binom{2n}{n}^{\frac{1}{2n}} &= \lim_{n\to\infty} \left(\frac{2n!}{(n!)^2}\right)^{\frac{1}{2n}} \\
&= \lim_{n\to\infty} \left\{\frac{2n!}{(2n)^{2n}} \cdot 2^{2n} \left(\frac{n^n}{n!}\right)^2\right\}^{\frac{1}{2n}} \\
&= \lim_{n\to\infty} \left(\frac{2n!}{(2n)^{2n}}\right)^{\frac{1}{2n}} \cdot 2 \cdot \lim_{n\to\infty} \left(\frac{n^n}{n!}\right)^{\frac{1}{n}} \\
&= \mathrm{e}^{-1} \cdot 2 \cdot \mathrm{e}, \text{ where we have used (1).} \\
&= 2.
\end{aligned}$$

4.8.6. THEOREM (Żelazko). *Let A be a commutative algebra (over $\mathbb{K}$, $p(\neq 0)$ a sm. ρ-seminorm on A, $\nu = \nu_p$. Then:*

(i) *The unit ball $\overline{B}_1 = \{x \in A : \nu(x) \leqslant 1\}$ is convex.*

(ii) *$\|x\| = \nu(x)^{\frac{1}{\rho}}$ is the gauge of $\overline{B}_1$ and it is a sm. seminorm.*

PROOF. (i) By 4.8.4, ν is a ρ-seminorm and consequently, by 3.2.7 (i), $\overline{B}_1$ is closed in the ν-topology. To show that $\overline{B}_1$ is convex it is enough to show that it is midpoint convex.[†]

Assume first that $\nu(x), \nu(y) < 1$. Then

$$\nu(\frac{1}{2}(x+y)) = \frac{1}{2^\rho}\nu(x+y) = \frac{1}{2^\rho} \lim_{n\to\infty} p((x+y)^n)^{\frac{1}{n}}. \tag{1}$$

Since $x \leftrightarrow y, (x+y)^n) = \sum\limits_{k=0}^{n} \binom{n}{k} x^k y^{n-k}$, so that by subadditivity of p, we have

$$p((x+y^n) \leqslant \sum_{k=0}^{n} \binom{n}{k}^\rho p(x^k y^{n-k}). \tag{2}$$

Choose $\epsilon > 0$ such that $\nu(x), \nu(y) < \epsilon < 1$.

[†] i.e. $x, y \in \overline{B}_1 \Rightarrow \frac{x+y}{2} \in \overline{B}_1$ (see [15, p.17]).

Since $\nu(x), \nu(y) < 1$ we can find N_1 such that

$$p(x^n)^{\frac{1}{n}}, p(y^n)^{\frac{1}{n}} < \epsilon, \text{ or, } p(x^n), p(y^n) < \epsilon^n < 1$$

for $n \geqslant N_1$. Therefore, by 3.3.4, $p(x^n), p(y^n) \to 0$ and consequently we can find $C > 0$ such that $p(x^n), p(y^n) < C$, for all n.

Since $\epsilon^n \to 0$ we can choose N_2 such that $\epsilon^n C < 1$ for all $n \geqslant N_2$. It follows that for $n \geqslant N = \max(N_1, N_2)$ and arbitrary m

$$p(x^n y^m) \leqslant p(x^n)p(y^m) < \epsilon^n C < 1. \tag{3}$$

Similarly for arbitrary n and $m \leqslant N$,

$$p(x^n y^m) < 1. \tag{4}$$

From (2), (3), (4) we obtain for $n \geqslant N$,

$$p((x+y)^{2n}) \leqslant \sum_{k=0}^{2n} \binom{2n}{k}^{\rho} p(x^k y^{2n-k}) \leqslant \sum_{k=0}^{2n} \binom{2n}{k}^{\rho} \leqslant (2n+1)\binom{2n}{n}^{\rho} \tag{5}$$

since $\binom{2n}{k} \leqslant \binom{2n}{n}$ for all k. From the definition of ν and (5) we get

$$\begin{aligned}
\nu\left(\frac{x+y}{2}\right) &= \frac{1}{2^\rho} \lim_{n\to\infty} p((x+y)^{2n})^{\frac{1}{2n}} \lim_{n\to\infty} \left((2n+1)\binom{2n}{n}^{\rho}\right)^{\frac{1}{2n}} \\
&\leqslant \frac{1}{2^\rho} \lim_{n\to\infty} (2n+1)^{\frac{1}{2n}} \left(\lim_{n\to\infty} \binom{2n}{n}^{\frac{1}{2n}}\right)^{\rho} \\
&\leqslant \frac{1}{2^\rho} \left(\lim_{n\to\infty} \binom{2n}{n}^{\frac{1}{2n}}\right)^{\rho} \text{ (since } \lim_{n\to\infty} (2n+1)^{\frac{1}{2n}} = 1) \\
&\leqslant \frac{1}{2^\rho} \cdot 2^{s} \text{ (using 4.8.5)} \\
&\leqslant 1,
\end{aligned}$$

which proves $\frac{1}{2}(x+y) \in \overline{B}_1$. Next, if $\nu(x)$, $\nu(y) \leqslant 1$, we can choose a sequence λ_n with $0 < \lambda_n < 1$ and $\lambda_n \to 1$. Then $\nu(\lambda_n x)$, $\nu(\lambda_n y) < 1$, so that, by the result above, $z_n = \frac{1}{2}$ $(\lambda_n x + \lambda_n y) \in \overline{B}_1$. It follows that $\frac{1}{2}(x+y) = \lim z_n \in \overline{B}_1$. Thus $\overline{B}_1$ is midpoint convex, completing the proof of (i).

(ii) Since clearly, $\overline{B}_1 = \{x \in A : \|x\| \leqslant 1\}$, it follows from 4.1.13 that $\|\cdot\|$ is the gauge of $\overline{B}_1$. Finally, since p is sm., $\overline{B}_1$ is a subsemigroup and consequently the gauge $\|\cdot\|$ is sm..

4.8.7. COROLLARY. *If the algebra A is a division algebra then $\|\cdot\|$ is a norm.*

PROOF. If $x \in A$, $x \neq 0$ then x is invertible, and so by 3.3.7 (v), $\|x\| > 0$.

4.8.8. DEFINITION. An element x of ρ-seminormed algebra $A = (A, p)$ is called *topologically nilpotent* or *t. nilpotent* if $\nu(x) = \nu_p(x) = 0$. If $A = (A, \{p_\alpha\})$ is a locally pseudo-convex algebra then an element $x \in A$ is called t. nilpotent if *each* $\nu_\alpha(x) = \nu_{p_\alpha}(x) = 0$.

Clearly every nilpotent element is t. nilpotent $(x^n = 0 \Rightarrow p_\alpha(x^n)^{\frac{1}{n}} = 0)$.

4.8.9. LEMMA. *If x is t. nilpotent then*

(i) $x^n \to 0$,

(ii) λx *is t. nilpotent for every* $\lambda \in \mathbb{K}$.

PROOF. (i) Take any ϵ with $0 < \epsilon < 1$. Since $\nu_\alpha(x) = \lim_{n\to\infty} p_\alpha(x^n)^{\frac{1}{n}} = 0$, we have $p_\alpha(x^n) < \epsilon^n < \epsilon$, for $n \geqslant N$. It follows that $p_\alpha(x^n) \to 0$, $x^n \to 0$.

(ii) By 4.8.3 (vii), $\nu_\alpha(\lambda x) = |\lambda|^\rho \nu_\alpha(x) = 0$.

4.8.10. PROPOSITION. *In a commutative ρ-seminormed algebra $A = (A, p)$ the set I of t.nilpotent elements is a closed ideal of A.*

PROOF. By 4.8.4, 3.3.7 (iii) we have:

$$\nu(x+y) \leqslant \nu(x) + \nu(y), \nu(\lambda x) = |\lambda|^\rho \nu(x), \nu(xy) \leqslant \nu(x)\nu(y).$$

Therefore I is an ideal. Moreover, I is closed, as follows from the continuity of ν (see 4.8.4).

4.8.11. PROPOSITION. *Let A be a ρ-Banach algebra and $x \in A$. Then:*

$$r(x) \leqslant \nu(x)^{\frac{1}{\rho}}. \qquad (*)$$

If A is real we have also

$$\tilde{r}(x) \leqslant \nu(x)^{\frac{1}{\rho}}. \qquad (**)$$

PROOF. We may assume the norm $\|\cdot\|$ to be sm. and write $\nu = \nu_{\|\cdot\|}$. If $|\lambda| > \nu(x)^{\frac{1}{\rho}}$ then

$$\nu(-\lambda^{-1}x) = |\lambda|^{-\rho}\nu(x) < 1,$$

so that by 3.3.19, $-\lambda^{-1}x$ is q. invertible. It follows by 1.7.8 that $\sigma'(x) \subseteq \{\lambda \in \mathbb{K} : |\lambda| \leqslant \nu(x)^{\frac{1}{\rho}}\}$ so that $(*)$ holds.

If A is real we apply $(*)$ to $\tilde{A}$ and deduce $(**)$.

4.8.12. COROLLARY. *In any ρ-Banach algebra a t. nilpotent element is q. nilpotent.*

PROOF. Since $r(x) \leqslant \nu(x)^{\frac{1}{\rho}}$, $\nu(x) = 0 \Rightarrow r(x) = 0$ and hence the result.

CHAPTER V

SOME ANALYSIS

§ 1. Vector-valued Differentiability and Analyticity

5.1.1. DEFINITION. Let X be a Hausdorff TLS and X^* its (continuous) dual. Let $G \subseteq \mathbb{K}$ be an open set. A function $f : G \to X$ is called *weakly differentiable* if there is a function $g : G \to X$ such that for each $x^* \in X^*$, the scalar-valued function $x^* f(\lambda) = x^*(f(\lambda))$ has the scalar-valued function $x^* g(\lambda)$ as its derivative in the usual sense; we write $g = f'^{(w)}$ and call $f'^{(w)}$ the *weak derivative* of f.

The function f is called *strongly differentiable* if for each $\lambda \in G$,

$$\lim_{\mu \to \lambda} \frac{f(\mu) - f(\lambda)}{\mu - \lambda}$$

exists in the topology of X. We denote this limit by $f'(\lambda)$ and call f' the *strong derivative* of f.

5.1.2. LEMMA. *A strongly differentiable function f is weakly differentiable and continuous, and moreover, the strong derivative f' coincides with the weak derivative $f'^{(w)}$.*

PROOF. It follows from the definition of the strong derivative that

$$\lim_{\mu \to \lambda} (f(\mu) - f(\lambda)) = \lim_{\mu \to \lambda} (\mu - \lambda) f'(\lambda) = 0,$$

proving f is continuous. Again,

$$\lim_{\mu \to \lambda} \frac{x^* f(\mu) - x^* f(\lambda)}{\mu - \lambda} = \lim_{\mu \to \lambda} x^* \left(\frac{f(\mu) - f(\lambda)}{\mu - \lambda} \right) = x^* f'(\lambda),$$

whence f is weakly differentiable with $f'^{(w)} = f'$.

5.1.3. DEFINITION. A function $f : G \to X$ is called *weakly analytic* if $x^* f(\lambda) = x^*(f(\lambda))$ is an analytic function of λ on G in the usual sense, for each $x^* \in X$. We call f *strongly analytic* on G

if around each point $\lambda_0 \in G \subseteq \mathbb{K}$, there is a neighbourhood $\{\lambda \in \mathbb{K} : |\lambda - \lambda_0| < r_0\}$ in which f has a power series representation †

$$f(\lambda) = \sum_{n=0}^{\infty} x_n(\lambda - \lambda_0)^n,$$

where $x_n \in X$ and the series converges in the topology of X.

If X is a complex TLS we will also use the expression weakly holomorphic (respy. strongly holomorphic) for weak analyticity (respy. strong analyticity).

5.1.4. LEMMA. *A strongly analytic function f is weakly analytic.*

PROOF. Suppose that

$$f(\lambda) = \sum_{n=0}^{\infty} x_n(\lambda - \lambda_0)^n.$$

Then, for each $x^* \in X^*$, we have

$$x^* f(\lambda) = \sum_{n=0}^{\infty} x^*(x_n)(\lambda - \lambda_0)^n,$$

so that f is weakly analytic.

5.1.5. Remark. If X is a complex, Frechet space (i.e. a complete metrizable locally convex space) then for a function $f : G \to X$, weak holomorphy of $f \Rightarrow$ strong holomorphy of f (see [24, p.79]).

5.1.6. Let $X = (X, \|\cdot\|)$ be a ρ-Banach space and

$$f(\lambda) = \sum_{n=0}^{\infty} x_n\lambda^n \quad (x_n \in X, \lambda \in \mathbb{K}) \qquad (*)$$

a series in X. Put

† In the representation we find it convenient to write the scalars on the right side (instead of the customary left side)

$$r_0 = \overline{\lim_{n\to\infty}} \|x_n\|^{\frac{1}{n}}, R_0 = r_0^{-1}, R = R_0^{\frac{1}{\rho}} = r_0^{-\frac{1}{\rho}} \quad (\overline{\lim} = \limsup).$$

R is called the *radius of convergence* and $\{\lambda \in \mathbb{K} : |\lambda| = R\}$ the circle† of convergence of the series $(*)$.

Next let $X = (X, \{p_\alpha\})$ be a complete Hausdorff locally pseudo-convex space. Set

$$\begin{aligned} r_\alpha &= \overline{\lim_{n\to\infty}} p_\alpha(x_n)^{\frac{1}{n}}, \qquad R_\alpha = r_\alpha^{-\frac{1}{\rho_\alpha}}, \\ R &= \inf_\alpha R_\alpha = \inf_\alpha r_\alpha^{-\frac{1}{\rho_\alpha}}. \end{aligned}$$

R again is called the radius of convergence of the series $(*)$ and $\{\lambda \in \mathbb{K} : |\lambda| = R\}$ its circle of convergence.

5.1.7. PROPOSITION. (a) *The series* $(*)$ *converges absolutely (and hence also in* X*) for all* λ *with* $|\lambda| < R$, *and converges uniformly for* $|\lambda| \leqslant r$, *for any* r *with* $0 < r < R$.

(b) *If* $|\lambda| > R$ *the terms of the series are unbounded and consequently the series diverges (i.e. does not converge).*

(c) *The term-wise differentiated series got from* $(*)$

$$\sum_{n=1}^{\infty} n x_n \lambda^{n-1} \qquad (**)$$

has the same radius of convergence R *(as* $(*)$*).*

PROOF. We will first treat the case where X is a ρ-Banach space $(X, \|\cdot\|)$. The proofs are all modelled after those of the corresponding classical results.

(a) The absolute-valued series $\sum_{n=0}^{\infty} \|x_n\|(|\lambda|^\rho)^n$ – which is a numerical series – converges, by the classical Abel's lemma, for $|\lambda|^\rho < R_0$, or equivalently, for $|\lambda| < R$. Further, this series

† In the real case the circle of convergence is to be interpreted as the end-points of an interval.

converges uniformly for $|\lambda|^\rho \leqslant r^\rho$, whence $(*)$ converges uniformly for $|\lambda| \leqslant r$.

(b) If $|\lambda| > R$, choose r so that $|\lambda| > r > R$. Then

$$\frac{1}{r^\rho} < \frac{1}{R^\rho} = \frac{1}{R_0} = \overline{\lim_{n\to\infty}} \|x_n\|^{\frac{1}{n}}$$

and consequently there are arbitrarily large n with

$$\|x_n\|^{\frac{1}{n}} > \frac{1}{r^\rho}, \text{ or } \|x_n\| > \frac{1}{r^{\rho n}}.$$

It follows that

$$\|x_n\lambda^n\| = |\lambda^{n\rho}| \|x_n\| > \left(\frac{|\lambda|}{r}\right)^{\rho n} \to \infty.$$

Thus the series $(*)$ is not bounded (for $|\lambda| > R$) and so not convergent (by 3.2.15).

(c) If we put $r' = \overline{\lim_{n\to\infty}} \|(n+1)x_{n+1}\|^{\frac{1}{n}}$, then

$$r' = \overline{\lim_{n\to\infty}} [\{(n+1)^{\frac{1}{n+1}}\}^\rho \|x_{n+1}\|^{\frac{1}{n+1}}]^{\frac{n+1}{n}} = \overline{\lim_{n\to\infty}} \|x_{n+1}\|^{\frac{1}{n+1}} = r,$$

where we have used the well-known result : $n^{\frac{1}{n}} \to 1$ as $n \to \infty$.

To extend the above proofs to the case where X is locally pseudo-convex, we note that if $|\lambda| < R_\alpha$ (for each α,) so that the proof above for (a) shows that $\sum_{n=0}^{\infty} p_\alpha(x_n)(|\lambda|^{\rho_\alpha})^n < \infty$, for each p_α, where $X = (X, \{p_\alpha\})$. Hence (a) follows for the locally pseudo-convex case. Further, if $|\lambda| > R$ then there is an α such that $|\lambda| > R_\alpha$, whence it can be shown, as above, that $p_\alpha(x_n\lambda^n) \to \infty$, so that the series is not p_α-bounded. Therefore the series is once again divergent, proving (b) for the general case.

Finally, if we set

$$r'_\alpha = \overline{\lim_{n\to\infty}} p_\alpha((n+1)x_{n+1})^{\frac{1}{n+1}}$$

then as in the proof above for ρ-Banach spaces we obtain $r_\alpha = r'_\alpha$, whence $R' = R$, where R' is the radius of convergence of the series (∗∗).

5.1.8. PROPOSITION. *The function*

$$f(\lambda) = \sum_{n=0}^{\infty} x_n \lambda^n \qquad (*)$$

is strongly differentiable inside its circle of convergence and we have

$$f'(\lambda) = \sum_{n=1}^{\infty} x_n n \lambda^{n-1}. \qquad (**)$$

In particularly, $f(\lambda)$ is a continuous function of λ.

PROOF. The proof we give is again modelled after the proof in the classical case (see [25, p.200]). We take $x_n \in X = (X, \{p_\alpha\})$ and assume that the radius of convergence of the series $(*), R > 0$ (if $R = 0$ there is nothing to prove). Choose a $\lambda \in \mathbb{K}$ with $|\lambda| < R$. Take a real number r with $|\lambda| < r < R$. Writing

$$g(\lambda) = \sum_{n=1}^{\infty} x_n n \lambda^{n-1},$$

we have, for $|\mu| < r$,

$$\frac{f(\mu) - f(\lambda)}{\mu - \lambda} - g(\lambda) = \sum_{n=1}^{\infty} x_n \beta_n \qquad (1)$$

where $\beta_n = (\mu^n - \lambda^n)/(\mu - \lambda) - n\lambda^{n-1}$.

Clearly $\beta_1 = 0$, and for $n \geqslant 2$, we get

$$\beta_n = (\mu - \lambda) \sum_{k=1}^{n-1} k \lambda^{k-1} \mu^{n-k-1} \qquad (2)$$

(as can be easily checked).

Since $|\lambda|, |\mu| < r$, and $\sum_{k=1}^{n-1} k \leqslant n^2$, we obtain from (2)

$$|\beta_n| \leqslant |\mu - \lambda| \left(\sum_{k=1}^{n-1} k \right) r^{n-2} \leqslant |\mu - \lambda| n^2 r^{n-2} \ (n \geqslant 2). \qquad (3)$$

Therefore

$$p_\alpha(\sum_{n=1}^{\infty} x_n\beta_n) \leqslant \sum_{n=1}^{\infty} p_\alpha(x_n)\beta_n^{\rho_\alpha} = \sum_{n=2}^{\infty} p_\alpha(x_n)|\beta_n|^{\rho_\alpha}$$

(since $\beta_1 = 0$)

$$\leqslant |\lambda - \mu|^{\rho_\alpha} \sum_{n=2}^{\infty} p_\alpha(x_n) n^{2\rho_\alpha} r^{(n-2)\rho_\alpha}$$

(using (3))

i.e.

$$p_\alpha\left(\sum_{n=1}^{\infty} x_n\beta_n\right) \leqslant |\lambda - \mu|^{\rho_\alpha} \sum_{n=0}^{\infty} p_\alpha(x_{n+2})^{2\rho_\alpha} r^{n\rho_\alpha}. \tag{4}$$

For the series $\sum_{n=0}^{\infty} x_{n+2}(n+2)^2\lambda^n$, we have

$$r'_\alpha = \overline{\lim_{n\to\infty}} p_\alpha(x_{n+2})^{\frac{1}{n}}(n+2)^{\frac{2\rho_\alpha}{n}} = \overline{\lim_{n\to\infty}} p_\alpha(x_n)^{\frac{1}{n}} = r_\alpha.$$

It follows that $R' = R$. Since $r < R = R'$, the series on the RHS of (4) converges, whence by (1), (4), as $\mu \to \lambda$ we obtain

$$f'(\lambda) = g(\lambda)$$

proving $(**)$. The continuity, of f follows from 5.1.2.

§ 2. Exponential and Logarithmic Functions

5.2.1. DEFINITION. Let A be a unital Hausdorff TA with unity e. The exponential function in A is defined by

$$E^{\dagger}(x) = \mathrm{Exp}(x) = \sum_{n=0}^{\infty} \frac{x^n}{n!}, \qquad (x^0 = e,\ 0! = 1) \tag{$*$}$$

whenever the series on the right converges. We denote by $\mathcal{D}_E$ the domain of E and by $\mathcal{R}_E$ its range. Since $0 \in \mathcal{D}_E$, $e \in \mathcal{R}_E$ both

† The Hausdorff assumption ensures that E is well-defined.

$\mathcal{D}_E, \mathcal{R}_E$ are non-empty. The inverse function Log is defined on $\mathcal{R}_E$ by setting $\operatorname{Log} y = x$ if $y = E(x)$. The function Log is in general many-valued.

We denote the classical exponential function by

$$\mathbf{E}(\lambda) = \mathbf{e}^{\lambda} = \mathbf{1} + \lambda + \frac{\lambda^2}{2!} + \cdots + (\lambda \in \mathbb{K}).$$

In particular, we write

$$\mathbf{e} = \mathbf{1} + \mathbf{1} + \frac{1}{2!} + \cdots$$

5.2.2. PROPOSITION. $\mathcal{D}_E = A$ *if* A *is a unital pseudo-Michael algebra – inparticular a unital* ρ*-Banach-algebra; moreover the series converges absolutely for all* $x \in A$.

PROOF. Let $A = (A, \mathcal{P}), \mathcal{P} = \{p_\alpha\}$. Since

$$p_\alpha\left(\frac{x^n}{n!}\right) \leqslant \frac{p_\alpha(x)^n}{(n!)^{\rho_\alpha}} = u_n \qquad \text{(say)}$$

and $\frac{u_n+1}{u_n} = \frac{p_\alpha(x)}{(n+1)^{\rho_\alpha}} \to 0$, the series for $E(x)$ converges absolutely and so also in A for all $x \in A$, by 3.1.24.

5.2.3. PROPOSITION. *Let* A *be a pseudo-Michael algeba – in particular a* ρ*-Banach algebra – with unity* e*. Then:*

(i) $E(0) = e$; $E(e) = \mathbf{e}e$.

(ii) $E(x + y) = E(x)E(y)$ provided $x \leftrightarrow y$.

(iii) $E(-x) = E(x)^{-1}$.

(iv) If A is complex and u an idempotent then $E(2\pi i u) = 0$ $(i = \sqrt{-1})$.

PROOF. (i) Clear

(ii) $E(x + y) = \sum_{n=0}^{\infty} \frac{(x+y)^n}{n!}$. Expanding the summands on the RHS, using the binomial theorem, and regrouping the terms (which is permissible because of absolute convergence) we find that the cofficient of $\frac{y^n}{n!}$ is

$$e + \frac{(n+1)x}{n+1} + \frac{(n+2)(n+1)}{2!} \cdot \frac{x^2}{(n+2)(n+1)} + \cdots$$

$$= e + x + \frac{x^2}{2!} + \cdots = E(x).$$

Thus,

$$E(x+y) = E(x)(e + y + \frac{y^2}{2!} + \cdots) = E(x)E(y).$$

(iii) This follows from (ii) by taking $y = -x$ and using (i).

(iv) $E(2\pi iu) = e + \{2\pi i + \frac{(2\pi i)^2}{2!} + \cdots\}u$. The expression in the bracket above $= \mathbf{e}^{2\pi i} - 1 = 1 - 1 = 0$. So we get $E(2\pi iu) = e$.

5.2.4. COROLLARY. *Let T be a bounded l.o. on a ρ-Banach space X. Then*

$$E(T) = I + T + \frac{T^2}{2!} + \cdots$$

is an invertible bounded l.o. on X.

PROOF. Consider the ρ-Banach algebra $B(X)$ and use 5.2.3 (iii). Note that $E(T)^{-1} = E(-T)$.

5.2.5. LEMMA. *If A is a Banach algebra with unity e and normalized norm $\|\cdot\|$, then*

$$\|E(x)\| \leqslant \mathbf{E}(\|x\|).$$

PROOF. Clear.

5.2.6. If A is a TA without unity we can introduce the *quasi-exponential function* E_q by

$$E_q(x) = x + \frac{x^2}{2!} + \cdots$$

which makes sense (whenever there is convergence).

If A_1 is the unitization of A and E the exponential function in A_1 then we have the obvious relation

$$E(x) = e_1 + E_q(x) \qquad (x \in A),$$

where e_1 is the unity of A_1. Using this relation and 5.2.3 (ii) we can easily deduce the identity

$$E_q(x) \circ E_q(y) = E_q(x+y) \text{ whenever } x \leftrightarrow y.$$

Since $E_q(0) = 0$ we get $E_q(-x) = (E_q(x))'$.

5.2.7. DEFINITION. Let A be a TA and G_q its group of q. invertible elements. The connected component of G_q containing 0 is called the ***principal component*** of G_q and we denote it by G_{q0}. Similarly, if A has unity e, the connected component of the group G_i (of invertible elements) containing e is denoted by G_{ie}; G_{ie} is called the ***principal component*** of G_i.

5.2.8. PROPOSITION. *In a C algebra A, G_{q0} is a closed normal subgroup of G_q. If A has a unity e, G_{ie} is a closed normal subgroup of G_i.*

PROOF. Since A is a C algebra, by 3.6.5 ((a), (b)) G_q and G_i are TG's. The stated results now follow by applying to these groups a standard result [21, p.39] in the theory of topological groups regarding the identity component.

5.2.9. COROLLARY. *If A is a ρ-Banach algebra then G_{qo} is a clopen*† *normal subgroup; if A is, besides unital then G_{ie} is a clopen normal subgroup of G_i. These results hold, more generally, when A is a pseudo-Michael Q algebra.*

PROOF. In view of 5.2.8 we have only to prove G_{qo}, G_{ie} are clopen. By 3.6.23 (b), 3.6.21, 3.6.10 we conclude that G_q, G_i are open when A is a ρ-Banach algebra. Also they are open in the pseudo-Michael algebra case be virtue of the Q algebra hypothesis we have in this case. Finally, in all the cases A is locally connected (see 3.2.8, 4.3.16). Hence the clopen properties of G_{qo}, G_{ie}.

5.2.10. PROPOSITION. *Let $A = (A, \|\cdots\|)$ be a unital ρ-Banach algebra. Then:*

(i) *The series $Log\, x = -\sum_{n=1}^{\infty} \frac{1}{n}(e-x)^n$ converges absolutely for all $x \in A$ with $\nu(e-x) < 1$, where $\nu = \nu_{\|\cdot\|}$.*

---

† clopen=closed and open

(ii) $E(\mathrm{Log} x) = x$.

(iii) *For each* $x \in A$, *the function* $E_x(\lambda) = E(\lambda x)$ *is analytic for all* $\lambda \in \mathbb{K}$. *In particular,* E_x *is entire when* $\mathbb{K} = \mathbb{C}$.

PROOF. Suppose that $\nu(e-x) < \epsilon < 1$, then $\|(e-x)^n\| < \epsilon^n$ for $n \geqslant$ (some) N. It follows that

$$\|\frac{(e-x)^n}{n}\| = \frac{1}{n^\rho}\|(e-x)^n\| < \frac{\epsilon^n}{n^\rho} < \epsilon^n.$$

Since $\sum_{n=N}^{\infty} \epsilon_n$ converges the series for $\mathrm{Log}\, x$ converges absolutely for $\nu(e-x) < 1$.

The identity (ii) can be checked, as in the classical case, by substituting the series for $\mathrm{Log}\, x$ in each summand of the series

$$E(\mathrm{Log} x) = \sum_{n=0}^{\infty} \frac{(\mathrm{Log} x)^n}{n!}$$

and simplifying (the needed steps for the simplication are justified because of absolute convergence).

Finally, we have

$$E_x(\lambda) = \sum_{n=0}^{\infty} \frac{x^n \lambda^n}{n!}.$$

Thus radius of convergence for the above power series in λ (see 5.1.7) by $r_0^{-\frac{1}{\rho}}$, where

$$\begin{aligned} r_0 = \overline{\lim_{n\to\infty}}\|\frac{x^n}{n!}\|^{\frac{1}{n}} &= \lim_{n\to\infty}\frac{\|x^n\|^{\frac{1}{n}}}{(n!)^{\frac{\rho}{n}}} \\ &= \nu(x)(\lim_{n\to\infty}(\frac{1}{n!})^{\frac{1}{n}})^\rho. \end{aligned}$$

Now

$$\lim_{n\to\infty}\left(\frac{1}{n!}\right)^{\frac{1}{n}} = \lim_{n\to\infty}\left(\frac{1}{(n+1)!}/\frac{1}{n!}\right) = \lim \frac{1}{n+1} = 0.$$

So $r_0 = 0, r_0^{-\frac{1}{\rho}} = \frac{1}{0^{\frac{1}{\rho}}} = \infty$, whence the series $E_x(\lambda)$ converges for all λ and (iii) follows.

5.2.11. Remark. Proposition 5.2.10 can be extended to unital pseudo-Michael algebra $A = (A, \{p_\alpha\})$ in the following form: For every x in A with $\nu_\alpha(e - x) < 1$, for all α, the series for $\operatorname{Log} x$ (defined in 5.2.10) converges absolutely and for all such x the identity (ii) of 5.2.10 holds. Further, the function $E_x = E_x(\lambda)$ is analytic for all $\lambda \in \mathbb{K}$, for each $x \in A$.

5.2.12. PROPOSITION. *Let A be a unital ρ-Banach algebra. The range $\mathcal{R}_E$ of the expoential function E generates (algebraically) the principal component G_{ie} (i.e. G_{ie} is the smallest subgroup of G_i containing $\mathcal{R}_E$).*

PROOF. Since A is connected and E continous, $\mathcal{R}_E$ is connected. By 5.2.3 (iii), $\mathcal{R}_E \subseteq G_i$. Since $e \in \mathcal{R}_E$ (by 5.2.3 (i)) and $\mathcal{R}_E$ is connected, $\mathcal{R}_E \subseteq G_{ie}$. Write

$$U = \{x \in A : \|e - x\| < 1\}.$$

Then, by 5.2.10 ((i), (ii)), $U \subseteq \mathcal{R}_E \subseteq G_{ie}$. Since U is an open neighbourhood of e in G_{ie} and G_{ie} is connected, U and hence $\mathcal{R}_E$ generates G_{ie} (see [20, p.37]).

5.2.13. COROLLARY. $\mathcal{R}_E = G_{ie}$ *iff* $\mathcal{R}_E$ *is a subgroup. In particular, if A is commutative,* $\mathcal{R}_E = G_{ie}$.

PROOF. The first statement is immediate. For the second we observe that by 5.2.3 ((i), (iii)), $\mathcal{R}_E$ is a subgroup and hence the result.

5.2.14. COROLLARY. (Theorem of Nagumo [11′]). *If $y \in \mathcal{R}_E$ then y belongs to a connected abelian subgroup of G_i. Conversely, every connected abelian subgroup H of G_i is contained in $\mathcal{R}_E$. Hence $y \in \mathcal{R}_E$ (i.e. y has a logarithm) iff y belongs to a connected abeliean subgroup H of G_i.*

PROOF. If $y \in \mathcal{R}_E$ then $y = E(x)$, and $H = \{E(\lambda x) : -\infty < \lambda < \infty\}$ is clearly a connected abelian subgroup of G_i. Next, by 1.1.9, there is a commutative subalgebra B_0 with $H \subseteq B_0$. Then

$B = \overline{B}_0$ is a commutative unital ρ-Banach algebra with $H \subseteq B$. By 5.2.13, the range $E(B) = G_{ie}(B)$ (the principal component in B). Since $e \in H$ and H is connected we have

$$H \subseteq G_{ie}(B) = E(B) \subseteq E(A) = \mathcal{R}_E.$$

The above two results clearly imply the final assertion in the corollary.

5.2.15. THEOREM (Gleason-Kahane-Żelazko†). *Let A be a complex unital ρ-Banach algeba. A linear functional χ on A is a character iff*

(i) $\chi(e) = 1$.

(ii) $\ker \chi$ *comprises non-invertible elements.*

PROOF. Suppose that χ is a character. Then, by 1.3.10, condition (i) holds. Further, if x is invertible, $\chi(x)\chi(x^{-1}) = \chi(e) = 1$, so that $\chi(x) \neq 0$ and (ii) also holds.

It remains to prove the "if" part of the theorem. So assume that χ is a linear functional satisfying (i), (ii). Write $M = \ker \chi$. Then χ being a linear functional $\neq 0$, M is a subspace of A of codim 1, so that we have $A = \mathbb{C}e + M$. When shall now show that χ is bounded with $\|\chi\| \leqslant 1$. If $x \in M$, $\chi(x) = 0$, so that $0 = |\chi(x)| \leqslant \|x\|$. Next consider an element $y \in A \backslash M$. Then y has the form

$$y = \lambda e - x \qquad (\lambda \in \mathbb{C}, \lambda \neq 0, x \in M).$$

Since $\lambda^{-1}x \in M$, by hypothesis (ii) it is not invertible and so, by 3.3.20 (ii), $\|e - \lambda^{-1}x\|$ †† $\geqslant 1$. Therefore

$$\begin{aligned} |\chi(y)| = |\chi(\lambda e - x)| = |\lambda| \quad &\leqslant \quad |\lambda| \|e - \lambda^{-1}x\|^{\frac{1}{\rho}} = \|\lambda e - x\|^{\frac{1}{\rho}} \\ &= \quad \|y\|^{\frac{1}{\rho}} \text{ (using } \lambda e - x = y). \end{aligned}$$

It follows that $\|\chi\| \leqslant 1$, in particular that χ is continuous.

To prove that χ is a character it is enough by 1.3.19, to show that $x \in M \Rightarrow x^2 \in M$.

Consider now an $x \in M$. We have to show that $x^2 \in M$. We may clearly limit ourselves to the case that $\|x\| = 1$. Set

† These authors have obtained the result for Banach algebras.

†† We assume (without loss of generality) that $\|\cdot\|$ is sm..

$$f(\lambda)=\sum_{n=0}^{\infty}\frac{\chi(x^n)}{n!}\lambda^n \qquad (\lambda\in\mathbb{C}). \tag{1}$$

Since

$$|\chi(x^n)|\leqslant\|\chi\|\,\|x^n\|^{\frac{1}{\rho}}\leqslant\|x\|^{\frac{n}{\rho}}=1 \tag{2}$$

and

$$\sum_{n=0}^{\infty}\frac{|\lambda|^n}{n!}=\mathbf{E}(|\lambda|). \tag{3}$$

it follows that the series for $f(\lambda)$ is absolutely convergent and hence $f(\lambda)$ is an ordinary entire function. Further, since

$$|f(\lambda)|\leqslant\sum_{n=0}^{\infty}\frac{|\lambda|^n}{n!}=\mathbf{E}(|\lambda|),$$

f has order at most 1. By continuity of χ we have

$$f(\lambda)=\chi\left(\sum_{n=0}^{\infty}\frac{(\lambda x)^n}{n!}\right)=\chi(E(\lambda x)) \tag{4}$$

where E is the exponential function in A. Since $E(\lambda x)$ is invertible, $f(\lambda)\neq 0$, by (4) and (ii). We now apply Hadamard's factorization theorem to conclude that

$$f(\lambda)=\mathbf{e}^{\lambda\alpha+\beta}\ \ (\alpha,\beta\in\mathbb{C}) \tag{5}$$

where $\mathbf{e}^{\lambda}$ denotes the classical exponential function. From (1) we find $f(0)=1$ and $f'(0)=\chi(x)=0$ (since $x\in M$). These imply that in (5) $\alpha=\beta=0$, whence $f(\lambda)=1$. Therefore $\chi(x^2)=f''=0$, whence $x^2\in M$, completing the proof.

§ 3. Square Roots and Quasi-square Roots

5.3.1. DEFINITION. Let A be an algebra and $x\in A$. An element y such that $y^2=x$ is called a *square root* or *sq.r.* of x; it may or may not exit. (Since $(-y)^2=y^2=x$ a sq.r. even when it exists is not unique). Similarly, an element y such that $y \circ y = x$ is called a *quasi-square root* or *q.sq.r.* of x.

5.3.2. LEMMA. *If A has a unity e then an element y is a q.sq.r. of an element x iff $y + e$ is a sq.r. of $x + e$.*

PROOF. This follows from the identity $(y+e)^2 = e + y \circ y$

5.3.3. LEMMA. (cf. [4, p.44, Lemma 12]). *Let $A = (A, \mathcal{P} = \{p_\alpha\})$ be a pseudo-Michael algebra, $x, y \in A, x \leftrightarrow y$, and $x \circ x = y \circ y$. If $\nu_\alpha(x+y) < 2^{\rho_\alpha}$ (ρ_α = homogenity index of p_α) for all α then $x = y$.*

PROOF. Write $a = \frac{1}{2}(x+y), b = x - y$. Then $ab = \frac{1}{2}(x^2 - y^2) = -b$ (using $x \circ x = y \circ y$). This implies $a \circ b = a$. Since $\nu_\alpha(a) = \frac{1}{2^{\rho_\alpha}}\nu_\alpha(x+y) < 1$, by 4.4.14, a is q. invertible, so that by 1.1.25 $b = 0, x = y$.

5.3.4. PROPOSITION. (cf. [4, p.44, Proposition 13]). *In a pseudo-Michael algebra $A = (A, \{p_\alpha\})$ every element a with $\nu_\alpha(a) < 2^{\rho_\alpha} - 1$ has a unique q.sq.r.b. with $\nu_\alpha(b) < 2^{\rho_\alpha} - 1$. Further, $b \leftrightarrow a$ and $\nu_\alpha(b) \leqslant \nu_\alpha(a)$. In particular, every t. nilpotent element a has a unique t. nilpotent q.sq.r.b.*

PROOF. Following Bonsall-Duncan [4, p.44], we use the method based on the existence of fixed point for obtaining the results. For each α, fix an η_α such that $\nu_\alpha(a) < \eta_\alpha < 2^{\rho_\alpha} - 1$. Then, by 3.5.10(i), we can choose a sm. ρ_α-seminorm such that $q_\alpha \sim p_\alpha$,

$$\nu_\alpha(a) \leqslant q_\alpha(a) < \eta_\alpha < 2^{\rho_\alpha} - 1 \leqslant^{\dagger} 2^{\rho_\alpha - 1}. \tag{1}$$

Denote by B_a the closed (commutative) subalgebra of A generated by a; B_a is a pseudo-Michael algebra. Set $S = \{x \in B_a : q_\alpha(x) \leqslant \eta_\alpha$ for all $\alpha\}$, where η_α satisfies (i) . Then S is a closed subset of A. Define a mapping T by

$$Tx = \frac{a - x^2}{2}(x \in S).$$

Then we have

---

† $2^\rho - 1 = 2 \cdot 2^{\rho-1} - 1 = 2^{\rho-1} + (2^{\rho-1} - 1) \leqslant 2^{\rho-1}$ (since $\rho < 1$).

$$
\begin{aligned}
q_\alpha(Tx) &= \frac{1}{2^{\rho_\alpha}} q_\alpha(a - x^2) \leqslant \frac{1}{2^{\rho_\alpha}}(\eta_\alpha + \eta_\alpha^2) = \frac{1}{2^{\rho_\alpha}}(1 + \eta_\alpha)\eta_\alpha \\
&\leqslant \frac{1}{2^{\rho_\alpha}} \cdot 2^{\rho_\alpha}\eta_\alpha = \eta_\alpha,
\end{aligned}
$$

so that T is a mapping $S \to S$. Further, if $x, y \in S$,

$$
\begin{aligned}
q_\alpha(Tx - Ty) &= q_\alpha(\frac{x^2 - y^2}{2}) \leqslant \frac{1}{2^{\rho_\alpha}} q_\alpha(x + y) q_\alpha(x - y) \\
&\leqslant \frac{1}{2^{\rho_\alpha}} \cdot 2\eta_\alpha q_\alpha(x - y) = \mu_\alpha q_\alpha(x - y),
\end{aligned}
$$

where $\mu_\alpha = 2^{1-\rho_\alpha}\eta_\alpha < 2^{1-\rho_\alpha} \cdot 2^{\rho_{\alpha-1}} = 1$. Thus T is a contraction mapping and so by 4.3.15, T has a fixed point $b : Tb = b$. It follows that $2b = a - b^2$ or $b \circ b = a$. Since $b \in S \subseteq B_a, b \leftrightarrow a$. Also,

$$\nu_\alpha(b) \leqslant q_\alpha(b) \leqslant \eta_\alpha < 2^{\rho_\alpha} - 1 \leqslant 2^{\rho_\alpha - 1}.$$

To prove the uniqueness of b (under the condition $\nu_\alpha(b) < 2^{\rho_\alpha} - 1$) we suppose that there is a b_1 with $b_1 \circ b_1 = a, \nu_\alpha(b_1) < 2^{\rho_\alpha} - 1 \leqslant 2^{\rho_\alpha - 1}$ (for all α). Then

$$ab_1 = (2b_1 + b_1^2)b_1 = b_1(2b_1 + b_1^2) = b_1 a,$$

i.e. $a \leftrightarrow b_1$. Since $b \in B_a$ it is the limit of a sequence of polynomials in a. Hence $b \leftrightarrow b_1$. Therefore

$$\nu_\alpha(b + b_1) \leqslant \nu_\alpha(b) + \nu_\alpha(b_1) < 2^{\rho_\alpha - 1} + 2^{\rho_\alpha - 1} = 2^{\rho_\alpha}.$$

By 5.3.3, $b = b_1$. Further, the uniqueness assertion implies that b is independent of the particular choice of η_α satisfying $\nu_\alpha(a) < \eta_\alpha < 2^{\rho_\alpha} - 1$. Hence, since $\nu_\alpha(b) \leqslant \eta_\alpha$, we conclude that $\nu_\alpha(b) \leqslant \nu_\alpha(a)$. Finally, the inequalities $\nu_\alpha(b) \leqslant \nu_\alpha(a)$ imply that $\nu_\alpha(b) = 0$ whenever $\nu_\alpha(a) = 0$. This means that when a is t. nilpotent its q.sq.r.b. is also t. nilpotent.

5.3.5. COROLLARY. *In a unital pseudo-Michael algebra* $A = (A, \{p_\alpha\})$ *with unity* e, *every element* a_1 *with* $\nu_\alpha(e - a_1) < 2^{\rho_\alpha} - 1$, *for all* α, *has a unique square root* b_1 *with*

$$\nu_\alpha(e - b_1) < 2^{\rho_\alpha} - 1, b_1 \leftrightarrow a_1; \text{ then } \nu_\alpha(e - b_1) \leqslant \nu_\alpha(e - a_1).$$

PROOF. Write $a_1 - e = a$. Then, by 5.3.4, a has a q.sq.r. b. with $b \leftrightarrow a, \nu(b) \leqslant \nu(a)$. Writing $b_1 = a + b$, by 5.3.2, b_1 is a square root of a_1. Also, the uniqueness of b_1 is an immediate consequence of the uniqueness of b.

5.3.6. COROLLARY. *In a ρ-Banach algebra $A = (A, \|\cdot\|)$ with $\nu = \nu_{\|\cdot\|}$, every element a with $\nu(a) < 2^\rho - 1$ is q. invertible and has a unique q.sq.r. b. with $\nu(b) < 2^\rho - 1$, $b \leftrightarrow a$, $\nu(b) \leqslant \nu(a)$; b is also q. invertible. If a is t. nilpotent so is b.*

PROOF. Since a ρ-Banach algebra is a Michael algebra the existence of b and all the properties stated in the corollary except q. invertibility of a, b are immediate consequences of 5.3.4. The q. invertibility of a or b follows from 3.3.19, since $\nu(a), \nu(b) < 2^\rho - 1, < 1$.

5.3.7. Remark. An element y of an algebra A such that $y^n = x$ $(x \in A)$ is called an n th root. Similarly if we write

$$y^{\circ n} = y \circ y \circ \cdots \circ y(n \text{ factors }) \text{ and } y^{\circ n} = x,$$

then y is called a quasi n th root or q.n th root of x. If A has a unity e we have the relation $e + y^{\circ n} = (e+y)^n$, so that if y is a q.n th root of x then $e + y$ is n th root of $e + x$.

Proposition 5.3.4 can be generalized for q.n th roots as follows: Each element a of a pseudo-Michael algebra A with

$$\nu_\alpha(a) < 2^{\rho_\alpha} - 1/(\frac{2^n}{n} - 1)$$

has a unique q. n th root b with $b \leftrightarrow a, \nu_\alpha(b) \leqslant \nu_\alpha(a)$. In particular every t. nilpotent element a has a unique t. nilpotent q. n th root b. The proof is again by the fixed point theorem method. Here we take $\eta_\alpha < 2^{\rho_\alpha} - 1/(\frac{2^n}{n} - 1)$ and define T by

$$Tx = \frac{1}{n}(a - \binom{n}{2} x^2 \cdots \binom{n}{n-1} x^{n-1} - x^n).$$

(observe: $(e+y)^n - e = \binom{n}{1} y + \cdots + \binom{n}{n-1} y^{n-1} + y^n$ $(y \in A)$.)

5.3.8. PROPOSITION. *Let A be a complex or strictly real,* *commutative unital, pseudo-Michael – in particular* ρ*-Banach-Algebra. For every element* $b \in A$ *with* $\sigma(b) = \{0,1\}$ *there is a unique element* $d \in \sqrt{A}$ *such that* $u = b + d$ *is idempotent.*

PROOF. Since, by 7.2.21, A is t. spectrally Gelfand we have

$$\{0,1\} = \sigma(b) = \{\chi(b) : \chi \in \triangle_c\}.$$

It follows that for every $\chi, \chi(b) = 0$ or 1, so that, $\chi(b^2) = \chi(b), b^2 - b \in \ker \chi$. Therefore

$$c = b^2 - b \in \sqrt[h]{A} = \sqrt{A} \qquad \text{(see 7.2.12)}.$$

Now

$$\sigma(2b - e) = 2\sigma(b) - 1 = \{-1, 1\} \not\ni 0,$$

hence $(2b - e)^{-1}$ exists. It follows that $-4c(2b - e)^{-2} \in \sqrt{A}$ and so it is a t. nilpotent element (see 7.4.10). This element has, by 5.3.4 a t. nilpotent q.sq.r. $f_0 \in A$. By 7.4.10, $f_0 \in \sqrt{A}$. It follows that if $f = e + f_0$ then

$$f^2 = e - 4c(2f - e)^{-2}.$$

Set [†]

$$d = -\frac{2b - e}{2} + \frac{2b - e}{2} f.$$

Then

$$d = \frac{2b - e}{2}(f - e) = \frac{2b - e}{2} f_0 \in \sqrt{A}.$$

Also

$$b + d = \frac{e + (2b - e)f}{2}$$

and a simple computation shows that $(b + d)^2 = b + d$. Thus $u = b + d$, with $d \in \sqrt{A}$, is an idempotent. If $v = b + d_1 (d_1 \in \sqrt{A})$ is also an idempotent then $u - v \in \sqrt{A}$, so that, by 7.2.12, 7.2.21, $\sigma'(u - v) = \{0\}$. If follows from (vii) of 7.2.22, and 7.2.23 that $u = v$, $d = d_1$, proving the uniqueness of the choice of $d \in \sqrt{A}$.

[†] For a motivation of the definition of d see [30, p.48].

5.3.9. COROLLARY. If the algebra A is s.s. then every element $b \in A$, with $\sigma(b) = \{0, 1\}$, is an idempotent.

PROOF. Since $\sqrt{A} = \{0\}$, $d = 0$, and $b = b + d$ is idempotent.

§ 4. Complex Vector-valued Line Integrals and Cauchy's Theorems

5.4.1. Let X be a complex TLS and Γ a rectifiable (continuous) path in the plane $\mathbb{C}$, represented by

$$z = z(t) \ (\alpha \leqslant t \leqslant \beta; \alpha, \beta \in \mathbb{R}).$$

Let $f : \Gamma \to X$ be a (vector-valued) function on Γ. If $\pi = \{t_0 = \alpha < t_1 < t_2 < \cdots < t_n = \beta\}$ be a partition of $[\alpha, \beta]$ we set

$$S(\pi) = S(\pi : z_j') = \sum_{j=1}^{n} f(z_j')(z_j - z_{j-1}) = \sum f(z_j')\Delta z_j,$$

where $z_j = z(t_j), z_j' = z(t_j')$ for some t_j' with $t_{j-1} \leqslant t_j' \leqslant t_j$. We denote $\max_j |t_j - t_{j-1}|$ by $|\pi|$. If $\lim S(\pi)$, as $|\pi| \to 0$, exists we say that the line integral; $\int_\Gamma f dz$ exists and write

$$\int_\Gamma f dz = \int_\Gamma f(z) dz = \lim_{|\pi| \to 0} S(\pi).$$

It is clear that if $\int_\Gamma f dz$, $\int_\Gamma g dz$ exist so does $\int_\Gamma (\alpha f + \beta g) dz$ $(\alpha, \beta \in \mathbb{C})$ and we have

$$\int_\Gamma (\alpha f + \beta g) dz = \alpha \int_\Gamma f dz + \beta \int_\Gamma g dz.$$

5.4.2. LEMMA. *Let X, Y be complex TLS's, $\Gamma \subset \mathbb{C}$ a path, $f = f(z) : \Gamma \to X$ a function such that $\int_\Gamma f dz$ exists. Let T :*

$X \to Y$ *be a continuous linear transformation. Then* $\int T f dz$ *exists and*

$$T \int_\Gamma f dz = \int_\Gamma T f dz.$$

PROOF.

$$\begin{aligned} T \int_\Gamma f dz &= T \lim_{|\Delta|\to 0} \sum f(z_k') \Delta_{z_k} \\ &= \lim_{|\Delta|\to 0} \sum (T f(z_k')) \Delta_{z_k} = \int_\Gamma T f dz. \end{aligned}$$

5.4.3. COROLLARY. *If* $\int_\Gamma f dz$ *exists and* $x^* \in X^*$,[†] *then*

$$x^* \left(\int_\Gamma f dz \right) = \int_\Gamma x^* f(z) dz = \int_\Gamma x^* f dz.$$

PROOF. This follows from 5.4.2 by taking $Y = X^*$ and $T = x^*$.

5.4.4. COROLLARY. *If* $X = A$ *is a TA,* Γ *and* f *as in* 5.4.2, *and* $x \in A$ *then*

(i) $x \left(\int_\Gamma f dz \right) = \int_\Gamma (xf) dz.$

(ii) $\left(\int_\Gamma f dz \right) x = \int_\Gamma (fx) dz.$

PROOF. These follow from 5.4.2, by taking $Y = A$ and $T = l_x$ or r_x.

5.4.5. PROPOSITION. *Let* $X = (X, \|\cdot\|)$ *be a complex Banach space,* Γ *a rectifiable path in* $\mathbb{C}$, *and* $f : \Gamma \to X$ *a continuous function. Then* $\int_\Gamma f dz$ *exists and we have*

[†] X^* denotes the (continuous) dual of X.

$$\|\int_\Gamma f\,dz\| \leqslant M|\Gamma|. \tag{$*$}$$

where $M = \sup\{\|f(\lambda)\| : \lambda \in \Gamma\}$ *and* $|\Gamma|$ *denotes the length of* Γ.

PROOF (following [14, p.62]). Let Γ have parametric representation $z = z(t)(\alpha \leqslant t \leqslant \beta)$. Since f is continuous on Γ, the function $f^*(t) = f(z(t))$ is continuous on $I = [\alpha, \beta]$. Since I is compact, f^* is uniformly continuous on I. Hence, given $\epsilon > 0$, there is a $\delta > 0$ such that

$$\|f^*(t') - f^*(t'')\| < \frac{\epsilon}{2|\Gamma|} \text{ if } \alpha \leqslant t' < t'' \leqslant \beta,$$

$|t' - t''| < \delta$, where $|\Gamma|$ denotes the length of Γ. Let π_1, π_2 be two partitions of I with $|\pi_1|, |\pi_2| < \frac{\rho}{2}$. Let π_3 be a common refinement of π_1, π_2. Then it is easy to see that

$$\|S(\pi_j) - S(\pi_3)\| < \frac{\epsilon}{2|\Gamma|} \cdot |\Gamma| = \frac{\epsilon}{2} \qquad (j = 1, 2).$$

Therefore we have

$$\|S(\pi_1) - S(\pi_2)\| \leqslant \|S(\pi_1) - S(\pi_3)\| + \|S(\pi_3) - S(\pi_2)\| < \frac{\epsilon}{2} + \frac{\epsilon}{2} = \epsilon.$$

It follows from the completeness of X that $\int_\Gamma f\,dz = \lim_{|\pi| \to 0} S(\pi)$ exists. Further, since clearly $\|S(\pi)\| \leqslant M|\Gamma|$, for each π, the inequality $(*)$ follows.

5.4.6. The general existence result for line integrals proved above (5.4.5) is no longer valid if X is a complex ρ-Banach space with $0 < \rho < 1$. However, even for such spaces the line integral exists for a special class of functions (see 5.4.9).

5.4.7. DEFINITION. Let X be a complex ρ-Banach space and Γ a rectifiable path in $\mathbb{C}$. A function $f : \Gamma \to X$ is called ***ρ-admissible*** or ***ρ-admissible on*** Γ if f has a representation

$$f(z) = \sum_{j=1}^{\infty} x_j \varphi_j(z) \tag{1}$$

where $x_j \in X, \varphi_j = \varphi_j(z)$ are bounded complex functions on Γ such that the ordinary line integrals $\int_\Gamma \varphi_j dz$ exists and further

$$\sum_j \|x_j\| \|\varphi_j\|^\rho < \infty \tag{2}$$

where $\|\varphi_j\| = \|\varphi_j\|_\Gamma = \sup_{\lambda \in \Gamma} |\varphi_j(\lambda)|$.

The condition (2) ensures, by the vector analogue of Weierstrass M-test, that the series in (1) converges uniformly on Γ. It follows that the function f will be continuous whenever all the φ_j are continuous.

5.4.8. LEMMA. (a). *If* $f = \sum_{j=1}^{\infty} x_j \varphi_j$, $g = \sum_{j=1}^{\infty} x_j \varphi_j$ *are* ρ*-admissible on* Γ *so is* $f + g$.

(b) *If* $f = \sum_{j=1}^{\infty} x_j \varphi_j$ *is* ρ*-admissible and* ψ *a continuous function on* Γ *then* $\sum_{j=1}^{\infty} x_j (\varphi_j \psi)$ *is* ρ*-admissible on* Γ.

PROOF. (a) Since

$$|\varphi_j(\lambda) + \psi_j(\lambda)|^\rho \leqslant |\varphi_j(\lambda)|^\rho + |\psi_j(\lambda)|^\rho$$

we have

$$\sum_{j=1}^{\infty} \|x_j\| \, \|\varphi_j + \psi_j\|_\Gamma^\rho \leqslant \sum_{j=1}^{\infty} \|x_j\| \, \|\varphi_j\|_\Gamma^\rho + \sum_{j=1}^{\infty} \|x_j\| \, \|\psi_j\|_\Gamma^\rho < \infty.$$

Hence $f + g$ is ρ-admissible.

(b)

$$\|\varphi_j \psi\|_\Gamma^\rho = \sup_\lambda |\varphi_j(\lambda) \; \psi_j(\lambda)|_\Gamma^\rho \leqslant \|\varphi_j\|_\Gamma^\rho \|\psi\|_\Gamma^\rho,$$

so that

$$\sum_{j=1}^{\infty} \|x_j\| \|\varphi_j \psi\|_\Gamma^\rho \leqslant \|\psi\|_\Gamma^\rho (\sum_{j=1}^{\infty} \|x_j\| \|\varphi_j\|_\Gamma^\rho) < \infty,$$

which proves (b).

5.4.9. THEOREM.([31, p.57]). *If* f *is* ρ*-admissible on* Γ *then the line integral* $\int_\Gamma f dz$ *exists and we have*

$$\int_\Gamma f dz = \sum_{j=1}^{\infty} x_j \int_\Gamma \varphi_j dz. \tag{*}$$

PROOF. The series in $(*)$ converges absolutely since

$$\sum_{j=1}^{\infty} \|x_j \int_\Gamma \varphi_j dz\| \leqslant \sum_j \|x_j\|\|\int_\Gamma \varphi_j dz|^\rho \leqslant \sum_j \|x_j\| \, \|\varphi_j\|_\Gamma^\rho \, |\Gamma|^\rho$$
$$\leqslant |\Gamma|^\rho \sum_{j=1}^{\infty} \|x_j\|\|\|\varphi_j\|_\Gamma^\rho < \infty,$$

where $|\Gamma|$ denotes the length of the curve Γ. Therefore, this series converges in X (by 3.1.24).

Let $z = z(t)$ $(\alpha \leqslant t \leqslant \beta)$ be the parametric representation of Γ. Let $\pi = \{\alpha = t_0 < t_1 < \cdots < t_n = \beta\}$ be a decomposition of $[\alpha, \beta], |\pi| = \max |t_j - t_{j-1}|$, and

$$S = S(\pi) = \sum_{k=1}^{n} f(z_k')(z_k - z_{k-1})$$

with $z_k' = z(t_k'), z_k = z(t_k)$ and $t_{k-1} \leqslant t_k' \leqslant t_k$.

We have:

$$\|S - \sum_{j=1}^{\infty} x_j \int_\Gamma \varphi_j dz\| =$$

$$= \left\| \sum_{k=1}^{n} \sum_{j=1}^{\infty} x_j \varphi_j(z_k')(z_k - z_{k-1}) - \sum_{j=1}^{\infty} x_j \int_\Gamma \varphi_j dz \right\|$$

$$= \left\| \sum_{j=1}^{\infty} x_j \left\{ \sum_{k=1}^{n} \varphi_j(z_k')(z_k - z_{k-1}) - \int_\Gamma \varphi_j dz \right\} \right\|$$

$$= \left\| \sum_{j=1}^{\infty} x_j \lambda_j \right\| = \|\Sigma_1 + \Sigma_2\| \tag{1}$$

where

$$\Sigma_1 = \sum_{j=1}^{m} x_j \lambda_j, \ \Sigma_2 = \sum_{j=m+1}^{\infty} x_j \lambda_j$$

and

$$\lambda_j = \sum_{k=1}^{n} \varphi_j(z_k')(z_k - z_{k-1}) - \int_\Gamma \varphi_j dz.$$

Since

$$|\int_\Gamma \varphi_j dz| \leqslant \|\varphi_j\| \ |\Gamma|, \tag{2}$$

where we have written $\|\varphi_j\|$ for $\|\varphi_j\|_\Gamma$, we obtain

$$|\lambda_j| \leqslant \|\varphi_j\| \ |\Gamma| + \|\varphi_j\| \ |\Gamma| = 2\|\varphi_j\| \ |\Gamma|.$$

It follows that

$$\|\Sigma_2\| = \| \sum_{j=m+1}^{\infty} x_j \lambda_j\| \leqslant \sum_{j=m+1}^{\infty} \|x_j\| |\lambda_j|^\rho$$

$$\leqslant 2^\rho |\Gamma|^\rho \sum_{j=m+1}^{\infty} \|x_j\| \|\varphi_j\|^\rho$$

$$\leqslant 2|\Gamma|^\rho \sum_{j=m+1}^{\infty} \|x_j\| \|\varphi_j\|^\rho . \tag{3}$$

From (3), by using (2) of 5.4.7, we can find, for a given $\epsilon > 0$ an integer N such that for $m \geqslant N$ we have

$$\|\Sigma_2\| < \frac{\epsilon}{2}. \tag{4}$$

Next, for any fixed $m \geqslant N$ we can choose a $\delta > 0$ such that for a decomposition π with $|\pi| < \delta$ we have

$$|\lambda_j|^\rho = |\int_\Gamma \varphi_j dz - \sum_{k=1}^{n} \varphi_j(z_k')(z_k - z_{k-1})|^\rho$$

$$\leqslant \frac{\epsilon}{2\sum_{k=1}^{m} \|x_k\|} \quad (j = 1, \cdots, m). \tag{5}$$

It follows that

$$\|\Sigma_1\| \leqslant \sum_{j=1}^{m} \|x_j\| \ |\lambda_j|^\rho < \frac{\epsilon}{2}, \tag{6}$$

so that by the choice of N we get, using (1),(4),(6),

$$\left\| S - \sum_{j=1}^{\infty} x_j \int \varphi_j dz \right\| = \|\Sigma_1 + \Sigma_2\| \leqslant \|\Sigma_1\| + \|\Sigma_2\| < \epsilon,$$

for any $S = S(\pi)$ with $|\pi| < \delta$. Therefore

$$\int_\Gamma f dz = \lim_{|\pi| \to 0} S = \sum_{j=1}^{\infty} x_j \int_\Gamma \varphi_j dz,$$

completing the proof.

5.4.10. COROLLARY. $\| \int_\Gamma f dz \| \leqslant \sum \|x_j\| \; \|\varphi_j\|^\rho |\Gamma|^\rho$.

PROOF. This follows from $(*)$ of 5.4.9, using the inequalities (2) therein.

5.4.11. DEFINITION. Let X be a complex ρ-Banach space, $G \subseteq \mathbb{C}$, an open set, and $f : G \to X$ a function. The function f is called *ρ-admissible holomorphic* on G if f has a representation

$$f = \sum_{j=1}^{\infty} x_j \varphi_j \qquad (*)$$

with $x_j \in X$, φ_j an ordinary holomorphic function on G, and further for every compact set $K \subseteq G$, we have

$$\sum_{j=1}^{\infty} \|x_j\| \; \|\varphi_j\|_K^\rho < \infty \qquad (**)$$

where $\|\varphi_j\|_K = \sup_{\lambda \in K} |\varphi_j(\lambda)|$.

The condition $(**)$ implies that the series $\sum_j x_j \varphi_j(\lambda)$ converges absolutely for every $\lambda \in K$. For any $\lambda \in G$, by taking $K = \{\lambda\}$ we conclude that the series $\sum_j x_j \varphi_j(\lambda)$ converges absolutely and we have $f(\lambda) = \sum_{j=1}^{\infty} x_j \varphi_j(\lambda)$.

5.4.12. DEFINITION. A function $f : G \to X$ is said to be *locally* ρ-admissible holomorphic at $\lambda_0 \in G$ if there is an open set G_0 such that $\lambda_0 \in G_0 \subseteq G$ and $f|G_0 \to X$ is ρ-admissible holomorphic. If f is locally ρ-admissible holomorphic at every point $\lambda \in G$ we call f locally ρ-admissible holomorphic on G. Clearly "ρ-admissible holomorphic on G" $\Rightarrow$ "locally ρ-admissible holomorphic on G", but not vice-versa (see 6.2.9).

5.4.13. LEMMA. (a) *If* $f = \sum_{j=1}^{\infty} x_j\varphi_j$, $g = \sum_{j=1}^{\infty} x_j\psi_j$ *are locally ρ-admissible (respy. ρ-admissible) holomorphic on G then so is $f+g$.*

(b) *If* $f = \sum_{j=1}^{\infty} x_j\varphi_j$ *is locally ρ-admissible (respy. ρ-admissible) holomorphic on G and ψ a holomorphic function on G then*

$$\psi f = f\psi = \sum_{j=1}^{\infty} x_j(\varphi_j\psi)$$

is locally ρ-admissible (respy. ρ-admissible) holomorphic on G.

PROOF. Similar to that of 5.4.8.

5.4.14. PROPOSITION. *A ρ-admissible holomorphic function $f : G \to X$ is strongly holomorphic and its strong derivative f' is ρ-admissible holomorphic. Moreover, we have*

$$f'(\lambda) = \sum_{j=1}^{\infty} x_j\varphi_j'(\lambda) \tag{$*$}$$

and in general,

$$f^{(n)}(\lambda) = \sum_{j=1}^{\infty} x_j\varphi_j^{(n)}(\lambda) \qquad (n = 1, 2, \cdots) \tag{$**$}$$

where $f^{(n)}$ is the nth strong derivative of f and $\varphi_j^{(n)}$ the nth ordinary derivative of φ_j.

PROOF. We have, for $\lambda, \lambda + \mu \in G$,

$$\|\frac{f(\lambda+\mu)-f(\lambda)}{\mu}-\sum_j x_j\varphi_j'(\lambda)\|$$

$$= \quad \|\sum_j x_j\{\frac{\varphi_j(\lambda+\mu)-\varphi_j(\lambda)}{\mu}-\varphi_j'(\lambda)\}\|$$

$$\leqslant \quad \sum_j \|x_j\| |\frac{\varphi_j(\lambda+\mu)-\varphi_j(\lambda)}{\mu}-\varphi_j'(\lambda)|^\rho. \qquad (1)$$

Let D be a closed disc centre λ and radius r such that $D \subseteq G$, and write $C = \partial D = \{\mu \in \mathbb{C} : |\mu| = r\}$. Using the classical Cauchy integral representation theorem we have

$$\frac{\varphi_j(\lambda+\mu)-\varphi_j(\lambda)}{\mu}-\varphi_j'(\lambda) =$$

$$= \frac{1}{2\pi i}\int_C \varphi_j(z)\left[\frac{1}{\mu}\left\{\frac{1}{z-\lambda-\mu}-\frac{1}{z-\lambda}\right\}-\frac{1}{(z-\lambda)^2}\right]dz.$$

Now choose μ such that $|\mu| < \frac{r}{2}$, so that for $z \in \mathbb{C}$ we have $|z-\lambda-\mu| \geqslant |z-\lambda|-|\mu| = r-|\mu| > \frac{r}{2}$. Then

$$\left|\frac{\varphi_j(\lambda+\mu)-\varphi_j(\lambda)}{\mu}-\varphi_j'(\lambda)\right| \leqslant \frac{|\mu|}{2\pi}\|\varphi_j\|_C\cdot\frac{1}{r^2}\cdot\frac{1}{\frac{r}{2}}\cdot 2\pi r = \frac{2\|\varphi_j\|_C}{r^2}|\mu|. \qquad (2)$$

From (1),(2) we get

$$\|\frac{f(\lambda+\mu)-f(\lambda)}{\mu}-\sum_{j=1}^\infty x_j\varphi_j'(\lambda)\|$$

$$\leqslant \sum_j \|x_j\|(\frac{2}{r^2})^\rho\|\varphi_j\|_C^\rho|\mu|^\rho = (\frac{2}{r^2})^\rho|\mu|^\rho\sum_{j=1}^\infty\|x_j\|\|\varphi_j\|_C^\rho. \qquad (3)$$

Since f is ρ-admissible the last sum $\sum$ in (3) is finite. By making $\mu \to 0$ in (3) the $R.H.S \to 0$ and we get $(*)$.

It remains to prove that f' is also ρ-admissible. Let $K \subseteq G$ be a compact set. Since G is Hausdorff locally compact we can find an open set H with

$$K \subseteq H \subseteq K_1 = \overline{H} \subseteq G \text{ and } K_1 \text{ compact .}$$

Choose an open covering $\{D_\alpha\}$ of K by open discs $D_\alpha \subseteq H$. Let δ be the Lebegue number of this covering. Then for any $\lambda \in K$, we have (the disc) $D = D(\lambda, \frac{\delta}{3}) \subseteq$ some $D_\alpha \subseteq H \subseteq K_1$. By Cauchy's estimates we have:

$$|\varphi_j'(\lambda)| \leqslant \frac{\|\varphi_j\|_D}{\delta/3} \leqslant \frac{3\|\varphi_j\|_{K_1}}{\delta}$$

so that $\|\varphi_j'\|_K \leqslant \frac{3\|\varphi_j\|_{K_1}}{\delta}$. It follows that

$$\sum_{j=1}^{\infty} \|x_j\| \|\varphi_j'\|_K^\rho \leqslant \left(\frac{3}{\delta}\right)^\rho \sum_{j=1}^{\infty} \|x_j\| \|\varphi_j\|_{K_1}^\rho < \infty,$$

whence f' is ρ-admissible. Finally, the representation $(**)$ is obtained by iteration (using $(*)$).

5.4.15. We proceed to obtain a version of the Cauchy integral formula for vector-valued functions. First we recall some classical concepts. We call a closed rectifiable path in the plane $\mathbb{C}$ a *contour*. The *index* of a point $\lambda \in \mathbb{C}\backslash\Gamma$ with respect to a contour Γ is the integer $\mathbf{I}$ given by the integral

$$\mathbf{I} = \mathbf{I}(\lambda, \Gamma) = \frac{1}{2\pi i} \int_\Gamma \frac{dz}{z - \lambda}.$$

If Γ is contained in an open set $G \subseteq \mathbb{C}$, we call Γ homologous to zero (in G), in symbols, $\Gamma \sim 0$, if

$$\mathbf{I}(\lambda, \Gamma) = 0 \text{ for all } \lambda \in \mathbb{C}\backslash G.$$

Again, we call Γ homotopic to 0 (in G), in symbols, $\Gamma \approx 0$, if Γ is homotopic to a constant curve (i.e. a point) in G in the usual topological sense.

It is known† that:

$$\Gamma \approx 0 \Rightarrow \Gamma \sim 0.$$

Let $G \subseteq \mathbb{C}$ be an open set and $S \subseteq G$ a subset. Following [24, p.241] we say that a *contour* $\Gamma \subseteq G$ *surrounds* S in G if the following conditions are satisfied: $S \bigcap \Gamma = \emptyset$,

† See, for example [8, p.93, 6.10].

$$\mathbf{I}(\lambda, \Gamma) = \begin{cases} 1 & \text{if } \lambda \in S \\ 0 & \text{if } \lambda \in \mathbb{C}\backslash G. \end{cases}$$

5.4.16. THEOREM (Cauchy integral theorems). *Let X be a complex ρ-Banach space $(0 < \rho \leqslant 1)$ and $f = f(\lambda)$ a X-valued (complex) strongly differentiable function on an open set $G \subseteq \mathbb{C}$. If $0 < \rho < 1$ we also assume that f is locally ρ-admissible holomorphic on G. Let Γ be contour in G with $\Gamma \sim 0$. Then we have:*

(a) *Integral theorem:* $\int_\Gamma f(z)dz = 0$.

(b) *Integral formulae: For every $\lambda \in G\backslash\Gamma$ with $\mathbf{I}(\lambda, \Gamma) = 1$,*

$$f(\lambda) = \frac{1}{2\pi i}\int_\Gamma \frac{f(z)}{z-\lambda}dz \tag{$*$}$$

and more generally,

$$f^{(n)}(\lambda) = \frac{n!}{2\pi i}\int_\Gamma \frac{f(z)}{(z-\lambda)^{n+1}}dz \quad (n = 1, 2, \cdots) \tag{$**$}$$

where $f^{(n)}$ denotes the nth strong derivative of f.

PROOF. (a) First assume that X is a Banach space $(\rho = 1)$. Then by 5.1.2, 5.4.5 $\int_\Gamma f(z)dz$ exists. If $x^* \in X$ then by 5.4.3

$$x^*\left(\int_\Gamma f(z)dz\right) = \int_\Gamma x^* f(z)dz = 0,$$

by the classical Cauchy theorem since $x^* f$ is holomorphic. Since the above conclusion holds for all $x^* \in X^*$ and X^* separates points (X being ample by 4.7.6) we obtain $\int_\Gamma f(z)dz = 0$, completing the proof of (a) for a Banach space X. Now assume that X is a ρ-Banach space and f is ρ-admissible holomorphic on G. Then we have

$$f(\lambda) = \sum_j x_j \varphi_j(\lambda) \; (\lambda \in G)$$

when φ_j are holomorphic. By 5.4.9,

$$\int_\Gamma f(z)dz = \sum_j x_j \int_\Gamma \varphi_j(z)dz = 0,$$

since each $\int_\Gamma \varphi_j dz = 0$, by the classical Cauchy theorem. Next consider the case where f is only locally ρ-admissible holomorphic on G. Since Γ is compact we find open sets

$$G_j (j = 1, \cdots, n) \subseteq G$$

such that $f_j = f|G_j \to X$ is ρ-admissible holomorphic on G_j. It is not hard† to see that we can find contours Γ_j such that $\Gamma_j \subseteq G_j$ and

$$\Gamma \sim \Gamma_1 + \cdots + \Gamma_n.$$

It follows that

$$\int_\Gamma f(z)dz = \sum_{j=1}^{n} \int_{\Gamma_j} f(z)dz = 0.$$

(b) In the case X is a Banach space the integral formula $(*)$ is obtained by applying $x^* \in X^*$, exactly as in the above proof of (a). Thus

$$x^*(\frac{1}{2\pi i}\int_\Gamma \frac{f(z)}{z-\lambda}dz) = \frac{1}{2\pi i}\int_\Gamma \frac{x^* f(z)}{z-\lambda}dz = x^* f(z)$$

for every x^*, whence $(*)$ follows. The proof of $(**)$ (for this case) is similar.

Next consider the case where X is a ρ-Banach space and f ρ-admissible holomorphic. Write

$$f(z) = \sum_{j=1}^{\infty} x_j \varphi_j(z) \ (\varphi_j \text{ holomorphic }) \tag{1}$$

$$g(z) = \frac{f(z)}{(z-\lambda)^{n+1}}, \ \psi_j(z) = \frac{\varphi_j(z)}{(z-\lambda)^{n+1}}.$$

Then

$$g(z) = \sum_{j=1}^{\infty} x_j \psi_j(z). \tag{2}$$

---

† See [31, pp.62-63 (diagrams)].

First assume that f is ρ-admissible on G. Then we see that each ψ_j is holomorphic on $G_0 = G\backslash\{\lambda\}$. Let $K \subseteq G_0$, be compact. By compactness of K there is an $r > 0$ such that

$$|z - \lambda| > r \text{ for all } z \in K.$$

Therefore

$$\|\psi_j\|_K = \sup_{z\in K} \frac{|\varphi_j(z)|}{|z-\lambda|^{n+1}} \leqslant \frac{\|\varphi_j\|_K}{r^{n+1}},$$

so that

$$\begin{aligned}\sum_{j=1}^{\infty} \|x_j\| \|\psi_j\|_K^{\rho} &\leqslant \sum_j \|x_j\| \frac{\|\varphi_j\|_K^{\rho}}{r^{(n+1)\rho}} \\ &\leqslant \frac{1}{r^{(n+1)\rho}} \sum_{j=1}^{\infty} \|x_j\| \|\varphi_j\|_K^{\rho} \\ &< \infty.\end{aligned}$$

It follows that $g(z)$ is ρ-admissible holomorphic on G_j, so that by $(*)$ of 5.4.9 we obtain from (2)

$$\begin{aligned}\frac{1}{2\pi i}\int_{\Gamma} g(z)dz &= \sum_{j=1}^{\infty} x_j \frac{1}{2\pi i}\int_{\Gamma} \frac{\varphi_j(z)}{(z-\lambda)^{n+1}}dz \\ &= \sum_j x_j \frac{\varphi_j^n(\lambda)}{n!} \\ &= \frac{1}{n!} f^{(n)}(\lambda) \quad (\text{by } (**) \text{ of } 5.4.14). \qquad (3)\end{aligned}$$

Again from (1) we get

$$\frac{1}{2\pi i}\int_{\Gamma} g(z)dz = \frac{1}{2\pi i}\int_{\Gamma} \frac{f(z)}{(z-\lambda)^{n+1}}dz. \qquad (4)$$

From (3),(4) we conclude that

$$\frac{n!}{2\pi i}\int_{\Gamma} \frac{f(z)}{(z-\lambda)^{n+1}}dz = f^{(n)}(\lambda).$$

The extension of the proof to the case where f is locally ρ-holomorphic on G is carried out as in the proof of (a)

5.4.17. COROLLARY. *If* $f(\lambda) = \sum_j x_j\varphi_j(\lambda)$ *then*

$$f^{(n)}(\lambda) = \sum_j x_j\varphi_j^{(n)}(\lambda).$$

PROOF. We have

$$\begin{aligned} f^{(n)}(\lambda) &= \frac{n!}{2\pi i}\int_\Gamma \frac{f(z)}{(z-\lambda)^{n+1}}dz = \sum_j x_j\frac{n!}{2\pi i}\int_\Gamma \frac{\varphi_j(z)}{(z-\lambda)^{n+1}}dz \\ &= \sum_j x_j\varphi_j^{(n)}(\lambda) \end{aligned}$$

(using the classical Cauchy representation theorem for φ_j).

5.4.18. COROLLARY(Cauchy's estimates). *Suppose that* $\lambda \in G \subseteq \mathbb{C}$ *and* Γ_0 *a circle centre* λ, *radius* r, *such that* $\Gamma_0 \subseteq G$. *Then:*

(i) *If* X *is a complex Banach space and* $f : G \to X$ *strongly differentiable (strongly holomorphic) then*

$$\|f^{(n)}(\lambda)\| \leqslant \frac{Mn!}{r^n}$$

where $M = \sup_{|z-\lambda|=r} |f(z)|$.

(ii) *If* X *is a complex* ρ*-Banach space and* $f : G \to X$ *admissible* ρ*-holomorphic, then*

$$\|f^{(n)}(\lambda)\| \leqslant (\frac{n!}{r^n})^\rho \sum_{j=1}^{\infty} \|x_j\| \|\varphi_j\|_{\Gamma_0}^\rho. \tag{$*$}$$

PROOF. (i) We have from the integral formula

$$\begin{aligned} \|f^{(n)}(\lambda)\| &= \|\frac{n!}{2\pi i}\int_\Gamma \frac{f(z)}{(z-\lambda)^{n+1}}dz\| \\ &\leqslant \frac{n!}{2\pi}\cdot\frac{M}{r^{n+1}}\cdot 2\pi r(\text{ using } (*) \text{ of } 5.4.5) \\ &= \frac{Mn!}{r^n}. \end{aligned}$$

(ii) Here we have

$$\|f^{(n)}(\lambda)\| \leqslant \sum_j \|x_j\| |\varphi_j^{(n)}(\lambda)|^\rho \leqslant \sum_{j=1}^{\infty} \|x_j\| (\frac{n!}{r^n})^\rho \|\varphi_j\|_{\Gamma_0}^\rho,$$

using the classical Cauchy estimates. Thus we get $(*)$.

5.4.19. COROLLARY. *A (strongly) entire (respy. ρ-admissible entire) function on a complex, Banach space (respy. ρ-Banach space) is a constant.*

PROOF. This follows, as in the classical case, from the Cauchy estimate for $n = 1$, by making $r \to \infty$.

§ 5. Power Series Operating in TA's

5.5.1. Let $f(\lambda) = \sum_{n=0}^{\infty} \gamma_n \lambda^n$ ($\gamma_n \in \mathbb{K}$) be a power series. Let A be a unital TA (with unity e) and $x \in A$. We say that f *operates* on x if the series $\sum_{n=0}^{\infty} \gamma_n x^n$ (recall $x^0 = e$) converges in A. Even when A has no unity it is meaningful to consider the operation of a power series f with $f(0) = \gamma_0 = 0$ on an element x : $\sum_{n=1}^{\infty} \gamma_n x^n$. (A power series f with $f(0)$ is called a power series vanishing at 0.)

Let now $A = (A, \{p_\alpha\})$ be a locally pseudo-convex algebra. Then we can speak of $f = \sum_n \gamma_n x^n$ operating absolutely on x if $\sum_n p_\alpha(x^n)$ converges for each p_α. If A is complete (or even sequentially complete) and f operates absolutely on x then f also operates on x (because of 3.1.24).

An entire[†] function f with $f(0) = 0$ is said to operate on

† entire in the real case means real analytic everywhere, i.e. is represented by a power series which converges everywhere in $\mathbb{R}$.

an element x of a TA A if its power series expansion $\sum_n \gamma_n \lambda^n$ operates on x; if A is unital we can consider the operation of *any* entire function on x. If a power series f operates on every $x \in A$ we say that f *operates on* A.

5.5.2. Remark. Every entire function f with $f(0) = 0$ (respy. every entire function f) clearly operates on any nilpotent element a of a TA A (respy. unital TA A). Every such f operates also on an idempotent element u $\left(\sum_{n=0}^{\infty} \gamma_n u^n = \gamma_0 e + \left(\sum_{n=1}^{\infty} \gamma_n\right) u\right)$.

5.5.3. PROPOSITION. *Let $A = (A, \{p_\alpha\})$ be a complete (or even sequentially complete) locally pseudo-convex TA. Then all entire functions vanishing at 0 operate on an element x iff*

$$\nu_\alpha(x) = \overline{\lim_n} p_\alpha(x^n)^{\frac{1}{n}} < \infty,$$

or equivalently

$$\nu_\alpha^*(x) = \sup_n p_\alpha(x^n)^{\frac{1}{n}} < \infty,$$

for every p_α. In particular, entire functions vanishing at 0 operate on any t. nilpotent element.

PROOF. First suppose that $\nu_\alpha^*(x) < \infty$ for all α. Choose a $C = C_\alpha > 0$ such that $\nu_\alpha^* \leqslant C$. Then $p_\alpha(x^n) \leqslant C^n$ $(n = 1, 2, \cdots)$ Let $f(\lambda) = \sum_{n=1}^{\infty} \gamma_n \lambda^n$ be an entire function (vanishing at 0). Then $\lim_{n\to\infty} |\gamma_n|^{\frac{1}{n}} = 0$. It follows that, given an ϵ with $0 < \epsilon < 1$, there is an integer $N = N_\alpha > 0$ such that for $n \geqslant N, |\gamma_n|^{\frac{1}{n}} < (\epsilon C^{-1})^{\frac{1}{\rho_\alpha}}$. Therefore

$$\begin{aligned}\sum_{n=N}^{\infty} p_\alpha(\gamma_n x^n) \quad &= \quad \sum_n |\gamma_n|^{\rho_\alpha} p_\alpha(x^n) \leqslant \sum_n (\epsilon C^{-1})^n C^n \\ &\leqslant \quad \sum_{n=N}^{\infty} \epsilon^n < \infty.\end{aligned}$$

Thus $\sum_n \gamma_n x^n$ converges absolutely and so also in A. This means that f operates on x, as required .

Next suppose that $\nu_\alpha(x) = \infty$ for some α. Let $\rho = \rho_\alpha$ be the homogenity index of p_α. Since $\nu_\alpha(x) = \infty$, there is a sequence $0 < n_1 < n_2 < \cdots$ of integers such that

$$p_\alpha(x^{n_k})^{\frac{1}{n_k}} > k \quad (k = 1, 2, \cdots).$$

But then the entire function

$$f(\lambda) = \sum_{k=1}^{\infty} \frac{\lambda^{n_k}}{k^{n_k/\rho}}$$

which vanishes at 0, does not operate on x, since $p_\alpha(x^{n_k})/k^{n_k} > 1$ (so that) $\dfrac{x^{n_k}}{k^{\frac{n_k}{\rho}}} \not\to 0$ and the series for $f(x)$ does not converge (by 2.1.32).

For the particular case statement in the proposition it is enough to observe that if x is t. nilpotent then, by definition, $\nu_\alpha(x) = 0$ for all α.

5.5.4. COROLLARY. *All entire functions vanishing at 0 operate on A iff*

$$\nu_\alpha^*(x) < \infty \text{ for all } x \text{ and all } \alpha.$$

5.5.5. COROLLARY. *All entire functions vanishing at 0 operate on any complete locally sm. pseudo-convex algebra A.*

PROOF. If $A = (A, \{p_\alpha\})$ then since p_α is sm., by 3.3.6, $\nu_\alpha(x)(\leqslant p_\alpha(x)) < \infty$, and consequently $\nu_\alpha^*(x) < \infty$. The corollary now follows from 5.5.4.

5.5.6. PROPOSITION. *Let $A = (A, \{p_j\})$ be a pseudo-$\mathfrak{F}$-algebra with $p_1 \prec p_2 \prec \cdots$. If there are constants $c_{j,n}$ $(j, n = 1, 2, \cdots)$ such that*

(i) $p_j(x_1 x_2 \cdots x_n) \leqslant c_{j,n} p_{j+1}(x_1) \cdots p_{j+1}(x_n)$ for $x_1, \cdots, x_n \in A$

(ii) $\sup_n \sqrt[n]{c_{j,n}} = c_j < \infty$,

then there exists a family $\{q_j : j \geqslant 2\}$ *of sm.* ρ_j *-seminorms with* $\{q_j : j \geqslant 2\} \sim \{p_j : j \geqslant 1\}$, *so that* A *is a locally sm. pseudo-* $\mathfrak{F}$ *-algebra.*

PROOF. Write $B_j(r) = \{x \in A : p_j(x) < r\}$. By conditions (i),(ii) we have

$$p_j(x_1 x_2 \cdots x_n) \leqslant c_j^n p_j(x_1) \cdots p_j(x_n). \tag{1}$$

It follows that the Cartesian n th power

$$B_{j+1}^n(c_j^{-1}) = (B_{j+1}(c_j^{-1}))^n \subseteq B_j(1). \tag{2}$$

Set

$$U_{j+1} = C_{\rho_{j+1}}\left(\bigcup_{n=1}^{\infty} B_{j+1}^n(c_j^{-1})\right) \quad (j = 1, 2, \cdots) \tag{3}$$

where $C_{\rho_{j+1}}$ denotes the ρ_{j+1}-convex hull. Since $\rho_{j+1} \leqslant \rho_j$, $B_j(1)$ – which is absolutely ρ_j-convex – is also, by 4.1.7, absolutely ρ_{j+1}-convex, so that (2) implies that $U_{j+1} \subseteq B_j(1)$. Also, from the definition of U_{j+1} we have $B_{j+1}(c_j^{-1}) \subseteq U_{j+1}$. Thus

$$B_{j+1}(c_j^{-1}) \subseteq U_{j+1} \subseteq B_{j+1}(1). \tag{4}$$

By 4.1.10, U_{j+1} determines a ρ_{j+1}-gauge q_{j+1} which is sm. (since U_{j+1} is a subsemigroup as can be easily seen). Taking the ρ_{j+1}-gauges of the sets in (4) and using the relations $(*), (***)$ of 4.1.8 and Remark 4.1.9 we get

$$p_j^{\frac{\rho_{j+1}}{\rho_j}} \leqslant q_{j+1} \leqslant c_j^{-1} p_{j+1}. \tag{5}$$

It follows from (5) that $\{q_n : n \geqslant 2\} \sim \{p_n : n \geqslant 1\}$, completing the proof.

5.5.7. LEMMA. *Let* A *be a commutative algebra. For* $x_1, \cdots, x_n \in A$ *set*

$$W_k^{(n)}(x_1, \cdots, x_n) = \sum_n (x_{i_1} + \cdots + x_{i_k})^n$$

where the summation is over all $i_1, \cdots, i_k$ *with* $1 \leqslant i_1 < i_2 \cdots i_k \leqslant n$. *Then*

$$x_1 x_2 \cdots x_n = \frac{(-1)^n}{n!} \sum_{k=1}^{n} (-1)^k W_k^{(n)}(x_1, \cdots, x_n). \qquad (*)$$

PROOF. Using the multinomial theorem the coefficient of the typical term $x_{i_1}^{n_1} \cdots x_{i_r}^{n_r}$ $(1 \leqslant r \leqslant k)$ in the development of $W_k^{(n)}$ is $\dfrac{n!}{(n_1!) \cdots (n_r!)} \dbinom{n-r}{k-r}$, since $\dbinom{n-r}{k-r}$ of the subsets of $\{1, \cdots, n\}$ which have k elements contain $i_1, \cdots, i_r$ and each of these subsets yield the typical term with the coefficient $\dfrac{n!}{(n_1!) \cdots (n_r!)}$. Therefore the coefficient of $x_{i_1}^{n_1} \cdots x_{i_r}^{n_r}$ in $\sum_{k=1}^{n} (-1)^k W_k^{(n)}$, is equal to

$$\frac{n!}{(n_1!) \cdots (n_r!)} \sum_{k=r}^{n} (-1)^k \binom{n-r}{k-r} = \frac{n!}{(n_1!) \cdots (n_r!)} (1-1)^{n-r} = 0$$

when $r \neq n$, and is equal to $n!(-1)^n$ when $r = n$ This completes the proof.

5.5.8. THEOREM. *Let* $A = (A, \{p_j\})$ *be a commutative* $\mathfrak{F}$*-algebra, where we may assume that* $\{p_j\}$ *satisfy the condition*

$$p_1 \prec p_2 \prec \cdots; \qquad (*)$$

$$p_j(xy) \leqslant p_{j+1}(x) p_{j+1}(y) \qquad (**)$$

(*see* 4.6.1). *If there exist constants* $c_{j,n}^*$ *such that*

(i) $p_j(x^n) \leqslant c_{j,n}^* \, p_{j+1}(x)^n \quad (n, j = 1, 2 \cdots)$

(ii) $\sup_n \sqrt[n]{c_{j,n}^*} = c_j^* < \infty \quad (j = 1, 2, \cdots)$

then A *is a locally sm.* $\mathfrak{F}$*-algebra.*

PROOF. Suppose that $x_1, x_2, \cdots, x_n \in A$. Assume first that $p_{j+1}(x_k) \neq 0$ $(k = 1, \cdots, n)$. Write $y_k = x_k / p_{j+1}(x_k)$; then $p_{j+1}(y_k) = 1$ $(k = 1, \cdots, n)$. It follows that

$$\begin{aligned}
p_j(x_1x_2\cdots x_n) &= \\
&= p_{j+1}(x_1)\cdots p_{j+1}(x_n)p_j(y_1y_2\cdots y_n) \\
&= p_{j+1}(x_1)\cdots p_{j+1}(x_n)p_j(\frac{(-1)^n}{n!}\sum_k W_k^{(n)}(y_1,\cdots,y_n)) \\
&\leqslant \frac{1}{n!}p_{j+1}(x_1)\cdots p_{j+1}(x_n)\sum_k p_j(W_k^{(n)}(y_1,\cdots,y_n)) \qquad (1)
\end{aligned}$$

where we have used indentity $(*)$ of 5.5.7 and the subadditvity of p_j. Using the definition of $W_k^{(n)}$ we obtain the inequality:

$$\begin{aligned}
p_j(W_k^{(n)}(y_1,\cdots,y_n)) &\leqslant \sum_{1\leqslant i_1\leqslant i_2\cdots\leqslant i_k\leqslant n} p_j((y_{i_1}+\cdots+y_{i_k})^n) \\
&\leqslant \sum_{1\leqslant i_1\cdots\leqslant i_k\leqslant n} c^*_{j,n}p_{j+1}(y_{i_1}+\cdots+y_{i_k})^n \\
&\leqslant \sum_{i_1,\cdots,i_k} c^*_{j,n}[p_{j+1}(y_{i_1})+\cdots+p_{j+1}(y_{i_k})]^n \\
&\leqslant \sum c^*_{j,n}k^n \leqslant \sum c^*_{j,n}n^n \\
&\leqslant c^*_{j,n}\binom{n}{k}n^n. \qquad (2)
\end{aligned}$$

From (1),(2) we get

$$\begin{aligned}
p_j(x_1x_2\cdots x_n) &\leqslant (\frac{1}{n!}\sum_k c^*_{j,n}\binom{n}{k}n^n)p_{j+1}(x_1)\cdots p_{j+1}(x_n) \\
&\leqslant c_{j,n}p_{j+1}(x_1)\cdots p_{j+1}(x_n) \qquad (3)
\end{aligned}$$

where

$$c_{j,n} = \frac{1}{n!}c^*_{j,n}(\sum_k \binom{n}{k})n^n = \frac{2^n n^n}{n!}c^*_{j,n}. \qquad (4)$$

If for some $k, p_j(x_k) = 0$, then by condition $(**)$ above we have

$$\begin{aligned}
p_j(x_1x_2\cdots x_n) &= p_j(x_1x_2\cdots x_{k-1}\ x_{k+1}\cdots x_nx_k) \\
&\leqslant p_{j+1}(x_1x_2\cdots x_{k-1}\ x_{k+1}\cdots x_n)p_j(x_k) = 0
\end{aligned}$$

so that (3) is satisfied trivially. Thus (3) holds without any restriction on x_k. This means that condition (i) of 5.5.6 holds. We shall next show condition (ii) of 5.5.6 also holds. Now

$$(\frac{2^n n^n}{n!})^{\frac{1}{n}} = \frac{2n}{\sqrt[n]{n!}}$$

so that

$$\lim_{n\to\infty} \frac{2n}{\sqrt[n]{n!}} = 2 \lim_{n\to\infty} \frac{n}{\sqrt[n]{n!}} = 2\mathrm{e} \tag{5}$$

where we have used for the last equality the result (1) in the proof of 4.8.5. It follows from (4),(5) that

$$\lim_{n\to\infty} (c_{j,n})^{\frac{1}{n}} = 2\mathrm{e} \lim_{n\to\infty} (c^*_{j,n})^{\frac{1}{n}} = 2\mathrm{e}c^*_j.$$

The theorem now follows from 5.5.6

5.5.9. Remark. One does not know whether the result in 5.5.8 is valid for ρ-$\mathfrak{F}$-algebras $(0 < \rho < 1)$. The above method of proof, however, breaks down in the ρ-$\mathfrak{F}$ case, since corresponding to the limit in (4) we now have $\lim 2n/(n!)^{\frac{\rho}{n}}$ which turns out to be infinite

$$(\lim_{n\to\infty} \frac{n}{(n!)^{\frac{\rho}{n}}} = \lim_{n\to\infty} \left[\frac{n}{(n!)^{\frac{1}{n}}}\right]^{\rho} \cdot n^{1-\rho} \to \mathrm{e}^{\rho} \cdot \infty = \infty).$$

5.5.10. THEOREM (Żelazko [31, p.140]). *Let* $A = (A, \{p_\alpha\})$ *be a commutative* $\mathfrak{F}$*-algebra, where we assume that the sequence* $\{p_\alpha\}$ *satisfy the conditions* $(*), (**)$ *in* 5.5.8. *If for every* $x \in A$ *and every* j *we have*

$$\nu^*_j(x) = \sup_n p_j(x^n)^{\frac{1}{n}} < \infty$$

then A *is a locally sm.* $\mathfrak{F}$ *algebra.*

PROOF. Observe first that $\nu^*_j(x) = \lim_n f_n(x)$, where

$$f_n(x) = \max_{1\leqslant k\leqslant n} p_j(x^k)^{\frac{1}{k}}.$$

Since the p_j are continuous so are f_n. Hence $\nu^*_j(x)$ is the pointwise limit of continuous functions of x on a complete metric space. Therefore ν_j^* belongs to the first Baire class of functions and consequently ν_j^* has at least one point of continuity x_0 in A. This means that there are constants $C, \delta > 0$ and an integer k' such that

$$p_{k'}(x - x_0) \leqslant \delta \Rightarrow \nu_j^*(x) \leqslant C.$$

If $k = \max(j, k')$ then we have

$$p_k(x - x_0) \leqslant \delta \Rightarrow \nu_j^*(x) \leqslant C,$$

since by hypothesis $p_j \leqslant p_{j+1}$ for every j. So we get from the definition of ν_j that

$$p_k(x - x_0) \leqslant \delta \Rightarrow p_j(x^n) \leqslant C^n \ (n = 1, 2, \cdots). \tag{1}$$

In particular, we have $p_j(x_0^n) \leqslant C^n$.

Writing $y = x - x_0$, we have for any x satisfying the condition in (1),

$$\begin{aligned} p_{j-1}(y^n) &= p_{j-1}((x - x_0)^n) = p_{j-1}\left(\sum_{k=0}^{n} \pm \binom{n}{k} x^k x_0^{n-k}\right) \\ &\leqslant \sum_{k=0}^{n} \binom{n}{k} p_j(x^k) p_j(x_0^{n-k}) \leqslant \sum_{k=0}^{n} \binom{n}{k} C^k C^{n-k} \\ &\leqslant 2^n C^n = (2C)^n. \end{aligned} \tag{2}$$

Let now y be an element of A with $p_k(y) \neq 0$. Writing $y_1 = \delta y / p_k(y)$ we have $p_k(y_1) = \delta$, so that by (2) we obtain

$$\begin{aligned} p_{j-1}(y_1^n) &\leqslant (2C)^n, \text{ which gives} \\ p_{j-1}(y^n) &\leqslant (\frac{2C}{\delta})^n p_k(y)^n. \end{aligned} \tag{3}$$

If $p_k(y) = 0$ then since $k \geqslant j$ we have

$$p_{j-1}(y^n) \leqslant p_j(y) p_j(y^{n-1}) \leqslant p_k(y) p_k(y^{n-1}) = 0.$$

This means that (3) holds in this case as well.

The integer k figuring above depends on j, so we shall write $k = k(j)$. Set

$$n_1 = 1,\ n_2 = k(2), \cdots,\ n_j = k(n_{j-1} + 1).$$

If we set $q_j = \{p_{n_j}\}$ then $\{q_j\} \sim \{p_j\}$ and by virtue of inequality (3), the conditions (i),(ii) of 5.5.8 are satisfied and the theorem follows from 5.5.8.

5.5.11. COROLLARY (Mitagin-Rolewicz-Żelazko [10′ *p*.295]). *A commutative $\mathfrak{F}$-algebra on which all entire functions vanishing at 0 operate is a locally sm. $\mathfrak{F}$-algebra.*

PROOF. This follows from 5.5.4, 5.5.10.

CHAPTER VI

SPECTRAL ANALYSIS IN TA's

§ 1. Spectral Properties

6.1.1. PROPOSITION. *Let A be a Q algebra (in particular a ρ-Banach algebra) and $x \in A$. Then the quasi-resolvent*† *set $\rho'(x)$ is an open subset and the quasi-spectrum*† *$\sigma'(x)$ a compact subset of $\mathbb{K}$.*

PROOF. If $\lambda_0 \in \rho'(x)$ then $\lambda_0 \neq 0$ (by 1.7.6) and $-\lambda_0^{-1}x$ is q. invertible. Since A is a Q algebra there is an open neighbourhood U of $-\lambda_0^{-1}x$, comprising q. invertible elements. Since the map $\lambda \mapsto -\lambda^{-1}x$ is continuous, there is an open neighbourhood N of λ_0 such that for $\lambda \in N$, $-\lambda^{-1}x \in U$, so that $-\lambda^{-1}x$ is q. invertible. Thus $N \subseteq \rho'(x)$ and $\rho'(x)$ is open, its complement $\sigma'(x)$ is closed.

Next let $U(0)$ be an open nucleus comprising q. invertible elements. Since $-\lambda^{-1}x \to 0$ as $|\lambda| \to \infty$ and $U(0)$ is open there is a constant C such that $-\lambda^{-1}x \in U(0)$ for $|\lambda| > C$. Hence, for such λ, $-\lambda^{-1}x$ is q. invertible and $\lambda \in \rho'(x)$. This implies that if $\lambda \in \sigma'(x)$ then $|\lambda| \leqslant C$, whence $\sigma'(x)$ is bounded. Being both bounded and closed $\sigma'(x)$ is compact, as required.

6.1.2. Corollary. *In an I algebra (in particular a unital ρ-Banach algebra) A, $\rho(x)$ is open, $\sigma(x)$ is compact, and $\sigma(x) \neq \mathbb{K}$.*

PROOF. By 3.6.13, A is a Q algebra and so by 6.1.1, $\sigma(x) \bigcup \{0\} = \sigma'(x)$ is compact. If $0 \notin \sigma(x)$ then x is invertible. Since A is an I algebra there is an open neighbourhood U of x, comprising invertible elements. Since $x - \lambda e \to x$ as $|\lambda| \to 0$, there is an $\epsilon > 0$ such that for $|\lambda| < \epsilon$, $x - \lambda e \in U$ and hence invertible, so that $\lambda \in \rho(x)$. It follows that $0 \notin \overline{\sigma(x)}$ (the closure of $\sigma(x)$). On the other hand, since $\sigma(x) \subseteq \sigma'(x)$ and

---

† For definition see 1.7.5.

$\sigma'(x)$ is closed we have $\overline{\sigma(x)} \subseteq \sigma'(x) = \sigma(x) \bigcup \{0\}$. Hence $\overline{\sigma(x)} = \sigma(x), \sigma(x)$ is compact and $\rho(x)$ is open. Finally, since $\sigma(x)$ is bounded, $\sigma(x) \neq \mathbb{K}$.

6.1.3. PROPOSITION. *If A is a real Q algebra and $x \in A$ then the extended quasi-spectrum $\tilde{\sigma}'(x)$ is compact, in particular closed, and $r(x) \leqslant \tilde{r}(x) < \infty$.*

PROOF. Let $\lambda = \alpha + i\beta \neq 0$, be a complex number. Then

$$\frac{x^2 - 2\alpha x}{\alpha^2 + \beta^2} = \frac{x^2 - 2\alpha x}{|\lambda|^2} \to 0, \text{ as } |\lambda| \to \infty$$

(note $|\alpha|/|\lambda|^2 \leqslant \frac{1}{|\lambda|} \to 0$). It follows that if U is a nucleus of A, comprising q. invertible elements then there is a $C > 0$ such that

$$y = \frac{x^2 - 2\alpha x}{\alpha^2 + \beta^2} \in U,$$

and hence y q. invertible, if $|\lambda| > C$. This implies, by 1.8.5, that for $|\lambda| > C$, $\lambda \notin \tilde{\sigma}'(x)$. This means that $\tilde{\sigma}'(x)$ is bounded. To complete the proof of its compactness it remains to show that $\tilde{\sigma}'(x)$ is closed, or equivalently, $\tilde{\rho}'(x)$ is open. Using 1.8.5 and the fact that

$$\frac{x^2 - 2\alpha x}{|\lambda|^2} = \frac{x^2 - 2(\operatorname{Re} \lambda)x}{|\lambda|^2}$$

is a continuous function of λ (for $\lambda \neq 0$), it can be shown, exactly as in the proof in 6.1.1 for $\rho'(x)$ being open, that $\tilde{\rho}'(x)$ is open, completing the proof of the proposition.

6.1.4. COROLLARY. *If A is a unital Q algebra then $\tilde{\sigma}(x)$ is compact.*

PROOF. It is clearly sufficient to prove that $\tilde{\sigma}(x)$ is closed in $\tilde{\sigma}'(x) = \tilde{\sigma}(x) \bigcup \{0\}$ and we carry out the proof as in 6.1.2. We can assume that $0 \notin \tilde{\sigma}(x)$, so that $0 \in \tilde{\rho}(x)$. Then x, and hence x^2, is invertible. Therefore, A being an I algebra we can choose an open neighbourhood U of x^2, comprising invertible elements. Since

$$x_{\alpha,\beta} = (x - \alpha e)^2 + \beta^2 e \to x^2, \text{ as } \alpha, \beta \to 0,$$

we can find an $\epsilon > 0$ such that for $|\alpha|, |\beta| < \epsilon$, $x_{\alpha,\beta} \in U$. If

$$z_{\alpha,\beta} = x - (\alpha + i\beta)e$$

then

$$z_{\alpha,\beta}\overline{z}_{\alpha,\beta} = x_{\alpha,\beta} \in U \text{ for } |\alpha|, |\beta| < \epsilon.$$

From the choice of U, $x_{\alpha,\beta}$ and hence $z_{\alpha,\beta}$, is invertible, whence

$$\lambda = \alpha + i\beta \in \tilde{\rho}(x) \text{ for } |\alpha|, |\beta| < \epsilon.$$

This implies that $0 \notin \overline{\tilde{\sigma}(x)}$, whence $\tilde{\sigma}(x)$ is closed in $\tilde{\sigma}'(x)$, as desired.

§ 2. The Resolvent Function

6.2.1. DEFINITION. Let A be a unital algebra (over a field F), $x \in A$ and $\rho(x)$ the resolvent set of x. We assume that $\rho(x) \neq \emptyset$. The vector-valued function

$$R_x : \lambda \mapsto x_\lambda = x(\lambda) = (x - \lambda e)^{-1}$$

on $\rho(x)$ is called the *resolvent function* of x; x_λ is called a *resolvent* of x.

6.2.2. PROPOSITION. *Let A be a unital algebra and $x \in A$. Then:*

$$x_\lambda - x_\mu = (\lambda - \mu)x_\lambda x_\mu \ (\lambda, \mu \in \rho(x)) \quad \text{(Hilbert relation)}; \qquad (*)$$

$$(e - \lambda x)^{-1} - (e - \mu x)^{-1} = (\lambda - \mu)(e - \lambda x)^{-1}x(e - \mu x)^{-1}$$
$$(\lambda = 0 \text{ or } \lambda^{-1} \in \rho(x), \mu = 0 \text{ or } \mu^{-1} \in \rho(x)). \qquad (**)$$

PROOF.

$$(x - \lambda e)(x - \mu e)^{-1} = [(x - \mu e) + (\mu - \lambda)e]\,(x - \mu e)^{-1} = e + (\mu - \lambda)x_\mu.$$

Multiplying both sides of the above equation, on the left, by x_λ we get

$$(x - \mu e)^{-1} = x_\lambda + (\mu - \lambda)x_\lambda x_\mu$$

which gives

$$x_\mu - x_\lambda = (\mu - \lambda)x_\lambda x_\mu$$

changing signs on both sides we get relation $(*)$. The relation $(**)$ can be obtained by noting that

$$\begin{aligned}(e-\lambda x)(e-\mu x)^{-1} &= [(e-\mu x)+(\mu-\lambda)x](e-\mu x)^{-1}\\ &= e+(\mu-\lambda)x(e-\mu x)^{-1}.\end{aligned}$$

6.2.3. COROLLARY. *Any two resolvents x_λ, x_μ of x commute: $x_\lambda \leftrightarrow x_\mu$.*

PROOF. Interchanging λ, μ in $(*)$ of 6.2.2 we get

$$x_\mu - x_\lambda = (\mu - \lambda)x_\mu x_\lambda \qquad (*')$$

Changing the sign in $(*')$ and comparing it with $(*)$ we get

$$(\lambda - \mu)x_\lambda x_\mu = (\lambda - \mu)x_\mu x_\lambda.$$

Since we may assume that $\lambda \neq \mu$ we conclude that $x_\lambda \leftrightarrow x_\mu$.

6.2.4. PROPOSITION. *Let A be a unital Hausdorff TA and $x \in A$. Assume that*

(i) *$\rho(x)$ is open;*

(ii) *$\lambda \mapsto x_\lambda$ $(\lambda \in \rho(x))$ is continuous.*

Then the resolvent function x_λ is a strongly infinitely differentiable function with

$$\frac{dx_\lambda}{d\lambda} = x_\lambda^2 \qquad (*)$$

and in general,

$$\frac{d^n x_\lambda}{d\lambda^n} = n! x_\lambda^{n+1}. \qquad (**)$$

Further, if

(iii) *$\sigma(x) \neq \emptyset$ and compact and $\rho^{\#}(x) = (\rho(x))\backslash\{0\})^{-1} \bigcup \{0\}$* †

† $(\rho(x)\backslash\{0\})^{-1} = \{\lambda^{-1} : \lambda \in \rho(x)\backslash\{0\}\}$

then $\rho^{\#}(x)$ *is open and on it* $(e-\lambda x)^{-1}$ *is infinitely differentiable with*

$$\frac{d^n}{d\lambda^n}[(e-\lambda x)^{-1}] = n!x^n[(e-\lambda x)^{-1}]^n \ (\lambda \in \rho^{\#}(x)). \qquad (***)$$

PROOF. If $\lambda \in \rho(x)$ then $\lambda+\mu \in \rho(x)$ for sufficiently small μ since $\rho(x)$ is open. Using the Hilbert relation we get

$$\frac{x_{\lambda+\mu} - x_\lambda}{\mu} = x_{\lambda+\mu}x_\lambda.$$

Making $\mu \to 0$, and using condition (ii) above we get $(*)$. The formulae $(**)$ is obtained by induction. Assume thus

$$\frac{d^{n-1}x_\lambda}{d\lambda^{n-1}} = (n-1)!x_\lambda^n.$$

Then $\dfrac{d^n x_\lambda}{d\lambda^n} = (n-1)!\dfrac{d}{d\lambda}(x_\lambda^n)$. Now

$$\frac{d}{d\lambda}(x_\lambda^n) = L_{\mu\to 0}\frac{1}{\mu}(x_{\lambda+\mu}^n - x_\lambda^n) = L_{\mu\to 0}\frac{x_{\lambda+\mu} - x_\lambda}{\mu}\sum_{0\leqslant j+k=n-1} x_{\lambda+\mu}^j x_\lambda^k$$

$$= \frac{dx_\lambda}{d\lambda}\cdot nx_\lambda^{n-1} = nx_\lambda^{n+1}(\text{ using } (*))$$

where we have written L for lim. Therefore $\frac{d^n x_\lambda}{d\lambda^n} = (n!)x_\lambda^{n+1}$. (The factorization of $x_{\lambda+\mu}^n - x_\lambda^n$ used in the above calculation is based on the property $x_{\lambda+\mu} \leftrightarrow x_\lambda$ which is assured by 6.2.3.)

It remains to prove the assertions on $\rho^{\#}(x)$. Since $\rho(x)$ is open, $(\rho(x)/\{0\})^{-1}$ is also open. To prove that $\rho^{\#}(x)$ is open it is enough to show that 0 is an interior point of $\rho^{\#}(x)$. Since $\sigma(x)$ is compact we have $0 \leqslant r(x) < \infty$. If $r(x) = 0$, then $\sigma(x) = \{0\}$ and so $\rho^{\#}(x) = \mathbb{K}$ is open. Next assume that $0 < r(x) < \infty$. If $0 < |\lambda| < \frac{1}{r(x)}$ then $|\lambda^{-1}| > r(x)$, so that $\lambda^{-1} \in \rho(x), \lambda \in \rho^{\#}(x)$. Therefore 0 is an interior point of $\rho^{\#}(x)$ and $\rho^{\#}(x)$ is open as required.

Finally, the relation $(***)$ is established by using $(**)$ of 6.2.2 and induction in exactly the same way as for the proof of

(**) above.

6.2.5. COROLLARY. *Let A^* be the continuous dual of A and $x^* \in A^*$. Then*

$$F_{x^*}(\lambda) = x^*(x_\lambda)$$

is a $\mathbb{K}$-valued infinitely differentiable function on $\rho(x)$ with

$$\frac{d^n F_{x^*}(\lambda)}{d\lambda^n} = x^*\left(\frac{d^n x_\lambda}{d\lambda^n}\right) \quad (n = 1, 2 \cdots).$$

When $\mathbb{K} = \mathbb{C}, F_{x^*}$ is a holomorphic function and x_λ a weakly holomorphic function on $\rho(x)$.

PROOF. This follows form 5.1.2.

6.2.6. COROLLARY. *The conclusions and formulae $(*), (**), (***)$ of 6.2.4 hold in any Hausdorff CI algebra (in particular, a unital ρ-Banach algebra or more generally a unital sm. (F) algebra).*

PROOF. It is enough to prove that the hypotheses (i),(ii) of 6.2.4 are satisfied. By 3.6.22, 3.6.23 a unital (F) algebra A is a CI algebra and so, by 6.1.2, hypothesis (i) of 6.2.4 is satisfied. Again the continuity of the maps $\lambda \mapsto x - \lambda e$, $y \mapsto y^{-1}$ (A being a C algebra) implies that their composite map $\lambda \mapsto x_\lambda$ is continuous (which is hypotheses (ii) of 6.2.4).

6.2.7. LEMMA. *In a CI – in particular a ρ-Banach – algebra $A, x_\lambda \to 0$ as $|\lambda| \to \infty$ $(\lambda \in \rho(x))$.*

PROOF. As $|\lambda| \to \infty, |\lambda^{-1}| \to 0$, and A being a CI algebra we have

$$x_\lambda = -\lambda^{-1}(e - \lambda^{-1}x)^{-1} \to 0 \cdot e = 0.$$

6.2.8. PROPOSITION. *Let $A = (A, \|\cdot\|)$ be a unital ρ-Banach algebra and $x \in A$. Then we have*

(i) *If $|\lambda|^\rho > \nu(x) = \lim_n \|x^n\|^{\frac{1}{n}}$ then $\lambda \in \rho(x)$ and the resolvent x_λ is given by*

$$x_\lambda = (x - \lambda e)^{-1} = -\sum_{n=0}^{\infty} \frac{x^n}{\lambda^{n+1}}, \qquad (*)$$

with the series converging absolutely.

(ii) *If* $\lambda \in \rho(x)$ *and* $|\lambda - \mu| < \|x_\lambda\|^{-\frac{1}{\rho}}$ *then* $\mu \in \rho(x)$ *and*

$$x_\mu = \sum_{n=0}^{\infty} (\mu - \lambda)^n x_\lambda^{n+1}, \qquad (**)$$

with the series converging absolutely.

PROOF. Write $\Lambda = \lambda^{-1}$. Since $\nu(\Lambda x) = |\Lambda|^\rho \nu(x) < 1$ we have, by 3.3.20 (i),

$$x_\lambda = (x - \lambda e)^{-1} = -\Lambda(e - \Lambda x)^{-1} = -\sum_{n=0}^{\infty} \Lambda^{n+1} x^n = -\sum_{n=0}^{\infty} \frac{x^n}{\lambda^{n+1}}$$

which is $(*)$. To prove $(**)$, we first note that we can write

$$x - \mu e = (x - \lambda e)[e + (\lambda - \mu)x_\lambda]. \qquad (1)$$

Since

$$\|(\lambda - \mu)x_\lambda\| = |\lambda - \mu|^\rho \|x_\lambda\| < \frac{1}{\|x_\lambda\|} \cdot \|x_\lambda\| = 1,$$

by 3.3.20, $e + (\lambda - \mu)x_\lambda$ is invertible and so by (1)

$$\begin{aligned} x_\mu &= [e + (\lambda - \mu)x_\lambda]^{-1} x_\lambda = [e - (\mu - \lambda)x_\lambda]^{-1} x_\lambda \\ &= \sum_{n=0}^{\infty} (\mu - \lambda)^n x_\lambda^n \cdot x_\lambda = \sum_{n=0}^{\infty} (\mu - \lambda)^n x_\lambda^{n+1}, \end{aligned}$$

proving $(**)$.

6.2.9. THEOREM. *Let* A *be a complex unital* ρ*-Banach algebra and* $x \in A$. *Then the resolvent function* x_λ *is locally* ρ*-admissible holomorphic on* $\rho(x) \bigcup \{\infty\}$.

PROOF. Let $A = (A, \|\cdot\|)$ where we may assume that $\|\cdot\|$ is sm.. By 6.2.8(ii), we have for $\lambda_0 \in \rho(x)$ and $|\lambda - \lambda_0| < \|x_{\lambda_0}\|^{-\frac{1}{\rho}}$,

$$x_\lambda = \sum_{n=0}^{\infty} (\lambda - \lambda_0)^n x_{\lambda_0}^{n+1}.$$

Write $\varphi_n(\lambda) = (\lambda - \lambda_0)^n, x_n = x_{\lambda_0}^{n+1}$ and $D_0 = \{\lambda \in \mathbb{C} : |\lambda - \lambda_0| < \|x_{\lambda_0}\|^{-\frac{1}{\rho}}\}$. Clearly φ_n is holomorphic on D_0. If $K \subseteq D_0$ and K is compact it is clear that we can enclose[†] K in a closed disc $\overline{D}(\lambda_0, r_0)$ with $r_0 < \|x_{\lambda_0}\|^{-\frac{1}{\rho}}$. So

$$\|\varphi_n\|_K \leqslant \sup_\lambda\{|\lambda - \lambda_0|^n, \lambda \in D_0\} \leqslant r_0^n.$$

It follows that

$$\begin{aligned}\sum_{n=0}^{\infty} \|x_n\| \|\varphi_n\|_K^\rho &\leqslant \sum_n \|x_{\lambda_0}^{n+1}\| r_0^{n\rho} \leqslant \sum_n \|x_{\lambda_0}\|^{n+1} r_0^{n\rho} \\ &= \|x_{\lambda_0}\| \sum_{n=0}^{\infty} \|x_{\lambda_0} r_0\|^n < \infty \text{ (since } \|x_{\lambda_0} r_0\| < 1)\end{aligned}$$

proving that x_λ is locally ρ-admissible at ρ_0. Write $G_x = \{\lambda \in \mathbb{C} : |\lambda| > \nu(x)^{\frac{1}{\rho}}\}$.

By 6.2.8 (i), we have

$$x_\lambda = (x - \lambda e)^{-1} = -\sum_{n=1}^{\infty} \frac{x^{n-1}}{\lambda^n}, \tag{$*$}$$

the series on the right converging absolutely for $|\lambda| > \nu(x)^{\frac{1}{\rho}}$. Writing $x_n = x^{n-1}, \varphi_n(\lambda) = -1/\lambda^n$, we get for x_λ the representation

$$x_\lambda = \sum_n x_n \varphi_n.$$

Each φ_n is clearly holomorphic on G_x. Any compact set in G_x can be enclosed in an annulus

$$K = \{\lambda \in \mathbb{K} : r_1 \leqslant |\lambda| \leqslant r_2\}$$

where $\nu(x)^{\frac{1}{\rho}} < r_1 < r_2$. Now we have for $\lambda \in K$,

---

[†] If $C = \sup\{|\lambda - \lambda_0| : \lambda \in K\}$, then by compactness of K there is a $\lambda_1 \in K$ with $C = |\lambda_1 - \lambda_0| < \|x_{\lambda_0}\|^{-\frac{1}{\rho}}$. We can choose for the radius of $\overline{D}$ any r_0 will $C < r_0 < \|x_{\lambda_0}\|^{-\frac{1}{\rho}}$.

$$|\varphi_n(\lambda)| \leqslant \frac{1}{r_1^n}, \text{ so that } \|\varphi_n\|_K \leqslant \frac{1}{r_1^n}.$$

It follows that

$$\sum_{n=1}^{\infty} \|x^{n-1}\| \|\varphi_n\|_K^{\rho} \leqslant \frac{\|x^{n-1}\|}{r_1^{n\rho}} = \sum_{n=1}^{\infty} \|\frac{x^{n-1}}{r_1^n}\| < \infty,$$

where for the last inequality we have used the absolute convergence property of the series in $(*)$. This proves the ρ-admissible holomorphy of x_λ on G_x. By interpreting G_x as an open neighbourhood of ∞ we see that what we have just proved amounts to x_λ being locally ρ-admissible holomorphic at ∞. This completes the proof of the theorem.

6.2.10. PROPOSITION. *Let A be a complex unital ρ-Banach algebra and $x \in A$. Let Γ be a contour in $\mathbb{C}$ surrounding $\sigma(x) \bigcup \{0\}$ (in $\mathbb{C}$). Then:*

$$\int_\Gamma \lambda^m x_\lambda d\lambda = -2\pi i x^m \ (m = 0, 1, 2, \cdots; x^0 = e). \qquad (*)$$

In particular, if $r > \nu(x)^{\frac{1}{\rho}}$ and C_r is the circle $|\lambda| = r$ ($\lambda \in \mathbb{C}$) we have

$$\int_{C_r} \lambda^m x_\lambda d\lambda = -2\pi i x^m. \qquad (**)$$

PROOF. Since $\Gamma \bigcap \varphi(x) = \emptyset$, $\Gamma \subseteq \rho(x)$ and so by 5.4.9, we have

$$\int_\Gamma \lambda^m x_\lambda d\lambda = -\sum_{n=0}^{\infty} x^n \int_\Gamma \frac{\lambda^m}{\lambda^{n+1}} d\lambda. \qquad (1)$$

We can choose r sufficiently large that $C_r \subseteq \rho(x)$ and $C_r \sim \Gamma$ in $\rho(x)$. Then we have

$$\frac{1}{2\pi i} \int_\Gamma \frac{\lambda^m}{\lambda^{n+1}} d\lambda = \frac{1}{2\pi i} \int_{C_r} \frac{\lambda^m}{\lambda^{n+1}} d\lambda = \begin{cases} 0 & (n \neq m) \\ 1 & (n = m) \end{cases} \qquad (2)$$

by the classical results. The formulae $(*), (**)$ now ready follow from (1), (2).

6.2.11. THEOREM (Beurling-Gelfand spectral radius formula). *In a complex Banach algebra A we have for any $x \in A$,*

$$r(x) = \nu(x) = \lim \|x^n\|^{\frac{1}{n}}. \tag{$*$}$$

If the norm $\|\cdot\|$ is sm. then we have also

$$r(x) = \nu(x) \leqslant \|x\|. \tag{$**$}$$

PROOF. Our proof makes use of the formula $(*)$ of 6.2.10 and is essentially on the lines in the exposition in [24, p.236].

We assume (as we may) that A is unital. By 4.8.11, $r(x) \leqslant \nu(x)$ (since here $\rho = 1$). So to prove $(*)$ it is enough to show that $\nu(x) \leqslant r(x)$. If $r > r(x)$, then by 6.2.10 $(**)$, we have

$$-x^m = \frac{1}{2\pi i}\int_{C_r} \lambda^m x_\lambda d\lambda. \tag{1}$$

Since x_λ is a continuous function of λ and C_r is compact we have

$$M(r) = \sup\{\|x_\lambda\| : |\lambda| = r\} < \infty \tag{2}$$

using $(*)$ of 5.4.5 we obtain from (1),(2)

$$\|x^m\| \leqslant \frac{1}{2\pi} r^m M(r) \cdot 2\pi r = r^{m+1} M(r).$$

Therefore

$$\|x^m\|^{\frac{1}{m}} \leqslant r(rM(r))^{\frac{1}{m}}$$

whence

$$\nu(x) \leqslant r.$$

It follows that $\nu(x) \leqslant r(x)$, completing the proof.

6.2.12. COROLLARY. *If A is a real Banach algebra then*

$\tilde{r}(x) = \nu(x)$, *where* $\tilde{r}(x)$ *is the extended spectral radius. In particular, if* A *is strictly real then* $r(x) = \nu(x)$.

PROOF. Apply 6.2.11 to the complexification $\tilde{A}$.

6.2.13. Remark. We shall obtain later (see 7.4.6) extensions of the above results for ρ-Banach algebras.

6.2.14. PROPOSITION. *Let* $A = (A, \|\cdot\|)$, *with* $\|\cdot\|$ *sm., be a unital* ρ*-Banach algebra and* $x \in A$. *Then:*

(i) $\sigma(x)$ *is an upper semi-continuous function of* x, *i.e., given an open set* $G \subseteq \mathbb{K}$ *with* $\sigma(x) \subseteq G$, *there is a* $\delta > 0$ *such that* $\sigma(x+y) \subseteq G$ *for every* $y \in A$ *with* $\|y\| < \delta$.

(ii) *For* $\lambda \in \rho(x)$ *if* $d(\lambda)$ *denotes the distance of* λ *from* $\sigma(x)$ *then* $\|x_\lambda\| \geqslant 1/d(\lambda)^\rho$; *whence* $\|x_\lambda\| \to \infty$ *as* $d(\lambda) \to 0$.

PROOF. By 6.2.6, x_λ is a continuous function of λ; by 6.2.7, $x_\lambda \to 0$ as $|\lambda| \to \infty$. It follows that there are numbers $C_1, r > 0$ such that $\|x_\lambda\| < C_1$ for all λ with $|\lambda| > r$, Write $\overline{D_r} = \{\lambda \in \mathbb{K} : |\lambda| \leqslant r\}$. Then $K = (\mathbb{K}\backslash G) \bigcap \overline{D_r}$ is compact, so that there is a $C_2 > 0$ such that $\|x_\lambda\| < C_2$ for $\lambda \in K$. It follows that if $\lambda \in \mathbb{K}\backslash G$ then $\|x_\lambda\| < C = \max\{C_1, C_2\}$. If y in A with $\|y\| < \delta = C^{-1}$ and $\lambda \in \mathbb{K}\backslash G$ then

$$x + y - \lambda e = (x - \lambda e)[e + x_\lambda y]$$

is invertible since $\lambda \in \rho(x)$ and

$$\|x_\lambda y\| \leqslant \|x_\lambda\| \|y\| < CC^{-1} = 1 \text{ (see 3.3.20)}.$$

This means that $\lambda \notin \sigma(x+y)$, whence $\sigma(x+y) \subseteq G$, proving (i).

If $\lambda \in \rho(x)$ and $\mu \in \sigma(x)$ then by 6.2.8 (ii) we have $|\lambda - \mu| \geqslant \|x_\lambda\|^{-\frac{1}{\rho}}$, so that

$$\|x_\lambda\| \geqslant 1/|\lambda - \mu|^\rho, \|x_\lambda\| \geqslant 1/d(\lambda)^\rho,$$

proving (ii).

6.2.15. COROLLARY. $r(x) \to 0$ *as* $x \to 0$. *Thus* $r(x)$ *is continuous at* $x = 0$.

PROOF. Take $G = \{\lambda \in \mathbb{K} : |\lambda| < \epsilon\}$. Then $G \supseteq \sigma(0) = 0$, whence by upper semicontinuity at 0 we have $\sigma(x) \subseteq G$, $r(x) \leqslant \epsilon$ for $\|x\| < \delta$. This shows that $r(x) \to 0$ as $x \to 0$.

6.2.16. COROLLARY. *If A is a ρ-Banach algebra and $x \in A$ then $\sigma'(x)$ is upper semi-continuous.*

PROOF. It suffices to observe that $\sigma'(x) = \sigma_{A_1}(x)$ where A_1 is the unitization of A.

6.2.17. PROPOSITION (Rickart† [23, p.36]). *Let x be an element of a unital ρ-Banach algebra $A = (A, \|\cdot\|)$ which is either complex or strictly real. Let V be an open neighbourhood of 0 in $\mathbb{K}$. Then there is a $\delta > 0$ such that for every $y \in A$ with $\|x - y\| < \delta$ and $xy = yx$ we have:*

$$\sigma(y) \subseteq \sigma(x) + V; \tag{$*$}$$

$$\sigma(x) \subseteq \sigma(y) + V. \tag{$**$}$$

PROOF. Since $\sigma(x) + V$ is an open set containing $\sigma(x)$, the relation $(*)$ follows from 6.2.14(i). To prove $(**)$ we will assume that is is false and show that this leads to a contradiction

Without loss of generality we can take $V = \{\lambda \in \mathbb{K} : |\lambda| < 2\epsilon\}$. By our assumption we can find a sequence $x_n \to x$ with $x_n \leftrightarrow x$, such that

$$\sigma(x) \nsubseteq G_n = \sigma(x_n) + V \text{ for all } n. \tag{1}$$

Choose $\lambda_n \in \sigma(x) \backslash G_n$. If $\mu \in \sigma(x_n)$ then

$$\lambda_n - \mu \notin V, \text{ so that } |\lambda_n - \mu| \geqslant 2\epsilon. \tag{2}$$

Since $\lambda_n \in \sigma(x)$ and $\sigma(x)$ is compact we can, by passing to a subsequence if necessary, assume that

$$\lambda_n \to \lambda_0 \in \sigma(x). \tag{3}$$

From (2),(3) we get:

$$|\lambda_0 - \mu| \geqslant 2\epsilon > \epsilon \text{ for every } \mu \in \sigma(x_n) \tag{4}$$

so that $\lambda_0 \notin \sigma(x_n)$ for every n. For obtaining the contradiction we have to consider separately the cases $\lambda_0 = 0, \lambda_0 \neq 0$.

---

† He obtained the result for complex Banach algebras.

Case 1. $\lambda_0 = 0$. Then $0 \notin \sigma(x_n)$ and x_n is invertible. Since $|\mu| > \epsilon$ for $\mu \in \sigma(x_n)$ we get $r(x_n^{-1}) = \sup|\mu^{-1}| \leqslant \epsilon^{-1}$. So, by 7.4.6, $\nu(x_n^{-1}) = r(x_n^{-1})^\rho \leqslant \epsilon^{-\rho}$, whence, by 3.7.29(i), $x = \lim x_n$ is invertible contradicting $0 = \lambda_0 \in \sigma(x)$.

Case 2. $\lambda_0 \neq 0$. Since $\lambda_0 \notin \sigma(x_n)$, $y_n = -\lambda_0^{-1}x_n$ is q. invertible for all n. By 1.7.12

$$\sigma'(y_n') = \left\{-\frac{\lambda}{1+\lambda} : \lambda \in \sigma(x_n)\right\}. \tag{5}$$

Since $\lambda \in \sigma(y_n)$ iff $-\lambda\lambda_0 \in \sigma(x_n)$, and by (4) $|\lambda_0 + \lambda\lambda_0| > \epsilon$ for $-\lambda\lambda_0 \in \sigma(x_n)$, we get

$$|-\frac{\lambda}{1+\lambda}| = |\frac{-\lambda\lambda_0}{\lambda_0 + \lambda_0\lambda}| \leqslant \frac{\|x_n\|^{\frac{1}{\rho}}}{\epsilon} \leqslant \frac{C}{\epsilon} \tag{6}$$

for some $C > 0$ (since the sequence x_n being convergent is bounded). From (5),(6) we conclude that $r(y_n')$, and hence by GB formula (see 7.4.6.) $\nu(y_n')$ is bounded. So, by 3.7.29(ii), $\lim y_n = -\lambda_0^{-1}x$ is q. invertible, contradicting the fact that $\lambda_0 \in \sigma'(x)$.

6.2.18. COROLLARY. *If the algebra A of 6.2.17 is not unital then we have inclusion relations analogous to $(*)(**)$ obtained by replacing σ by σ' (the quasi-spectrum).*

PROOF. We have only to apply 6.2.17 to the unitization A_1 of A.

6.2.19. COROLLARY. *If A is commutative, $r(x)$ is continuous everywhere.*

PROOF. This follows from $(*), (**)$ of 6.2.17.

6.2.20. PROPOSITION. *The completion $\hat{A}$ of a commutative strictly real ρ-normed algebra A is strictly real.*

PROOF. Consider an element $x \in \hat{A}$. If possible let $\tilde{\sigma_{\hat{A}}}(x)$ contain a complex number $\lambda_0 = \alpha_0 + i\beta_0$, with $\beta_0 \neq 0$. Write $\eta = \frac{1}{2}|\beta_0| > 0$. Choose a sequence (x_n) in A with $x_n \to x$. Since A is strictly real, $\tilde{\sigma_A}(x_n) \subseteq \mathbb{R}$. Since $A \subseteq \hat{A}$ we have also $\tilde{A} \subsetneq \tilde{\hat{A}}$, and so

$$\tilde{\sigma_{\hat{A}}}(x_n) \subseteq \tilde{\sigma_A}(x_n) \subseteq \mathbb{R}. \tag{1}$$

Write

$$V = \{\lambda \in \mathbb{C} : \lambda = \alpha + i\beta \text{ with } \alpha \in \mathbb{R}, |\beta| < \eta\}.$$

By (**) of 6.2.17, there is a $\delta > 0$ such that

$$\sigma(x) \subseteq \sigma(y) + V \text{ whenever } \|x - y\| < \delta.$$

Choose n sufficiently large that $\|x_n - x\| < \delta$. Then

$$\tilde{\sigma_{\hat{A}}}(x) \subseteq \tilde{\sigma_{\hat{A}}}(x_n) + V \subseteq \mathbb{R} + V = V.$$

It follows that $\lambda_0 \in V, |\beta_0| < \eta = \frac{1}{2}|\beta_0|$, whence $\beta_0 = 0$ – a contradiction. Therefore $\tilde{\sigma_{\hat{A}}}(x) \subseteq \mathbb{R}$ and $\hat{A}$ is strictly real.

§ 3. Pseudo-Resolvent Function

6.3.1. DEFINITION. Let A be an algebra (over a field F) and $x \in A$. Set

$$\rho^p(x) = \{\lambda \in F : (\lambda x)' \text{ exists}\}; \ \rho^p(x)$$

is called the *pseudo-resolvent set of* x. Since $(0x)' = 0' = 0$, $0 \in \rho^p(x)$ and we have always $\rho^p(x) \neq \emptyset$. Write

$$x'_\lambda = (\lambda x)' \qquad (\lambda \in \rho^p(x))$$

and call x'_λ a *pseudo-resolvent* or a p-resolvent of x; $\lambda \mapsto x'_\lambda$ is called a *pseudo-resolvent function.*†

6.3.2. Remark. The map $\lambda \mapsto -\lambda^{-1}$ is a bijection of $\rho'(x) = \rho'(x)\backslash\{0\}$ onto $\rho^p(x)\backslash\{0\}$, where $\rho'(x)$ denotes the quasi-resolvent sent of x. If $F = \mathbb{K}$ the map is also a homeomrohpism.

If x is q. nilpotent then $\sigma'(x) = \{0\}$, $\rho'(x) = F\backslash\{0\}$. Therefore,

† This function has been considered implicity by Kaplansky [10″, p.400, Lemma 3.2].

$$\rho^p(x) = -\{F\backslash\{0\}\}^{-1}\bigcup\{0\} = \{F\backslash\{0\}\}\bigcup\{0\} = F.$$

6.3.3. LEMMA. (a) $x'_\lambda \leftrightarrow x'_\mu \quad (\lambda, \mu \in \rho^p(x))$.

(b) $\dfrac{x'_\mu}{\mu} - \dfrac{x'_\lambda}{\lambda} = \mu - \lambda \quad (\lambda, \mu \in \rho^p\backslash\{0\})$.

PROOF. (a) Since $\lambda x \leftrightarrow \mu x$, by applying 1.1.18 twice we get $x'_\lambda \leftrightarrow x'_\mu$.

(b) We have

$$\lambda x + (\lambda x)' + (\lambda x)(\lambda x)' = 0 \tag{1}$$

$$\mu x + (\mu x)' + (\mu x)(\mu x)' = 0 \tag{2}$$

Denoting the expressions in the equations (1), (2) above also by (1), (2) we get

$$\mu \times (1) - \lambda \times (2) = \mu \cdot 0 - \lambda \cdot 0 = 0,$$

which gives

$$\mu(\lambda x)' - \lambda(\mu x)' = \lambda\mu x[(\mu x)' - (\lambda x)']. \tag{3}$$

Again, $(1) \times \dfrac{(\mu x)'}{\lambda}$ and $\dfrac{(\lambda x)'}{\mu} \times (2)$ give

$$x(\mu x)' + \frac{(\lambda x)'(\mu x)'}{\lambda} + x(\lambda x)'(\mu x)' = 0. \tag{4}$$

$$x(\lambda x)' + \frac{(\lambda x)'(\mu x)'}{\mu} + x(\lambda x)'(\mu x)' = 0. \tag{5}$$

From (4), (5) we obtain by subtraction

$$x[(\mu x)' - (\lambda x)'] = \left(\frac{1}{\mu} - \frac{1}{\lambda}\right)(\lambda x)'(\mu x)' \tag{6}$$

and from (3), (6) we get

$$\frac{\mu(\lambda x)' - \lambda(\mu x)'}{\lambda\mu} = \frac{(\lambda - \mu)}{\lambda\mu}(\lambda x)'(\mu x)',$$

which reduces, after a change of sign, to the equation in (b).

6.3.4. LEMMA. *In a Q algebra A, for each $x \in A$, $\rho^p(x)$ is an open set containing* 0.

PROOF. For $\lambda \in \rho^p(x)$, let U be an open neighbourhood of λx comprising q. invertible elements. By continuity of scalar multiplication there is an open neighbourhood N of λ such that $Nx \subseteq U$. If follows that $N \subseteq \rho^p(x)$ and $\rho^p(x)$ is open.

6.3.5. PROPOSITION. (a) *Let A be a Hausdorff C algebra, and x an element of A such that $\rho^p(x)$ is open in $\mathbb{K}$.*

Then x'_λ is a strongly differentiable function of λ on $\rho^p(x)$, with

$$\frac{dx'_\lambda}{d\lambda} = -x(1 + x'_\lambda)^2, \tag{$*$}$$

$$\frac{d^n x'_\lambda}{d\lambda^n} = (-1)^n n! x^n (1 + x'_\lambda)^{n+1}. \tag{$**$}$$

We have also the relation

$$\frac{d^n}{d\lambda^n}\left(\frac{x'_\lambda}{\lambda}\right) = n!\left(\frac{x'_\lambda}{\lambda}\right)^{n+1} \quad (\lambda \neq 0). \tag{$***$}$$

(b) *In a Hausdorff CQ algebra A, the formulae $(*)$, $(**)$, $(***)$ are valid for every element x in A.*

PROOF. (a) From the equation (1) in the proof of 6.3.3 we obtain, for $\lambda \in \rho_q(x)\backslash\{0\}$,

$$\frac{x'_\lambda}{\lambda} = \frac{(\lambda x)'}{\lambda} = -x(1 + x'_\lambda). \tag{1}$$

Making $\lambda \to 0$ we obtain

$$\left[\frac{dx'_\lambda}{d\lambda}\right]_0 = -x. \tag{2}$$

If $\lambda, \mu \in \rho^p(x)\backslash\{0\}$, by 6.3.3 (b) we have

$$\frac{x'_\mu}{\mu} - \frac{x'_\lambda}{\lambda} = (\mu - \lambda)\frac{x'_\lambda x'_\mu}{\lambda\mu}.$$

Making $\mu \to \lambda$ we obtain

$$\frac{d}{d\lambda}\left(\frac{x'_\lambda}{\lambda}\right) = \frac{x'^2_\lambda}{\lambda^2}, \tag{3}$$

which reduces to

$$-\frac{x'_\lambda}{\lambda^2} + \frac{1}{\lambda}\frac{dx'_\lambda}{d\lambda} = \frac{x'^2_\lambda}{\lambda^2},$$

whence we get

$$\frac{dx'_\lambda}{d\lambda} = \frac{x'_\lambda}{\lambda}(1 + x'_\lambda) = -x(1 + x'_\lambda)^2, \tag{4}$$

where we have used (1). Thus we have proved $(*)$ when $\lambda \neq 0$. On the other hand (2) shows that the formula in (4) (or $(*)$) is valid for $\lambda = 0$, completing the proof of $(*)$ for all λ.

To prove $(**)$, assume by induction that

$$\frac{d^{n-1}x'_\lambda}{d\lambda^{n-1}} = (-1)^{n-1}(n-1)!x^{n-1}(1 + x'_\lambda)^n.$$

Differentiating both sides of the above equation we get

$$\begin{aligned}\frac{d^n x_\lambda}{d\lambda^n} &= (-1)^{n-1}(n-1)!x^{n-1} \cdot n(1 + x'_\lambda)^{n-1}\frac{dx'_\lambda}{d\lambda} \\ &= (-1)^{n-1}n!x^{n-1}(1 + x'_\lambda)^{n-1} \cdot -x(1 + x'_\lambda)^2 \text{ (using (4))} \\ &= (-1)^n n!x^n(1 + x'_\lambda)^{n+1}\end{aligned}$$

which is $(**)$. Finally, by successive differentiation we obtain $(***)$ from (3).

(b) This follows from (a) and 6.3.4.

6.3.6. PROPOSITION. *Let A be a ρ-Banach algebra and $x \in A$. Then x'_λ exists for all λ such that $\nu(\lambda x) < 1$, or equivalently, $|\lambda| < \nu(x)^{-\frac{1}{\rho}}$, and we have*

$$x'_\lambda = \sum_{n=1}^{\infty}(-1)^n(\lambda x)^n \tag{*}$$

with the series on the right converging absolutely. In particular, x'_λ is analytic for $|\lambda| < \nu(x)^{-\frac{1}{\rho}}$.

PROOF. Consider the unitization A_1 of A. Then we have

$$(e_1 + \lambda x)^{-1} = e_1 + (\lambda x)'$$

where e_1 is the unity of A_1. By $(*)$ of 3.3.20 we have

$$(e_1 + \lambda x)^{-1} = \sum_{n=0}^{\infty} (-1)^n (\lambda x)^n,$$

the series converging absolutely. It follows that we have

$$x'_\lambda = (\lambda x)' = \sum_{n=1}^{\infty} (-1)^n (\lambda x)^n$$

(with the series converging absolutely).

6.3.7. COROLLARY. *If A is a complex ρ-Banach algebra then x'_λ is ρ-admissible holomorphic on $\{\lambda \in \mathbb{C} : |\lambda| < \nu(x)^{-\frac{1}{\rho}}\}$.*

PROOF. Set

$$\psi_n(\lambda) = (-1)^n \lambda^n; \ x_n = x^n.$$

Then $(*)$ of 6.3.6 can be rewritten as

$$x'_\lambda = \sum_{n=1}^{\infty} x_n \psi_n(\lambda).$$

For any r_0 with $0 \leqslant r_0 < \nu(x)^{-\frac{1}{\rho}}$, set

$$K = K_{r_0} = \{\lambda \in \mathbb{C} : |\lambda| \leqslant r_0\}.$$

Then

$$\|\psi_n\|_K \leqslant r_0^n.$$

So

$$\sum_{n=1}^{\infty} \|x_n\| \, \|\psi_n\|_K^\rho \leqslant \sum_{n=1}^{\infty} \|x^n\| r_0^{n\rho} = \sum_{n=1}^{\infty} \|(r_0 x)^n\| < \infty,$$

since $\nu(r_0 x) = r_0^\rho \nu(x) < 1$. Hence the corollary.

6.3.8. PROPOSITION. *Let A be a complex ample C-algebra. If $x \in A\backslash\{0\}$ is q. nilpotent then $\{x'_\lambda : \lambda \in \mathbb{C}\}$ is unbounded.*

PROOF. First note that by 6.3.1 and definition of q. nilpotent, $\rho^p(x) = \mathbb{C}$. Since A is a Hausdorff C-algebra, by 6.3.5 (a),

$x'_\lambda = (\lambda x)'$ is strongly differentiable, and hence for $x^* \in A^*$ $F'_{x^*}(\lambda) = x^*(x'_\lambda)$ is an entire function. If possible let x'_λ be a bounded function of λ. Then $F'_{x^*}(\lambda)$ is bounded and so by classical Liouville's theorem $F'_{x^*}(\lambda)$ is a constant. Since $F'_{x^*}(0) = 0$ we must have

$$F'_{x^*}(\lambda) = 0 \qquad \text{for all } \lambda \in \mathbb{C}.$$

Putting $\lambda = 1$, we get $x^*(x') = 0$. Since A is ample we conclude that $x' = 0$ which implies $x = 0$, contradicting the choice of x. Therefore $\{x'_n\}$ is unbounded as required.

6.3.9. COROLLARY. (Kaplansky). *In a complex normed algebra A, if $x \neq 0$ is q. nilpotent then $\{x'_\lambda\}$ is unbounded.*

PROOF. Since a normed algebra is an ample C-algebra, the corollary follows from 6.3.8 .

6.3.10. LEMMA. *Let A be a normed algebra and $x \in A$ be q. nilpotent. If $\lambda_n \in \mathbb{K}$ are such that the sequence $((\lambda_n x)')$ is unbounded then $|\lambda_n| \to \infty$.*

PROOF. First note that since x is q. nilpotent $(\lambda x)'$ exists for all $\lambda \in \mathbb{K}$. Assume now that, to the contrary, $|\lambda_n| \leqslant C$ for all n. Since $((\lambda_n x)')$ is unbounded there is, by 2.1.23, a nucleus U and a subsequence $((\lambda_{n'} x)')$ such that

$$\frac{(\lambda_{n'} x)'}{n'} \notin U \quad \text{for all } n'. \tag{$*$}$$

Since $|\lambda_{n'}| \leqslant C$, for all n', we can choose a subsequence $(\lambda_{n''})$ of $(\lambda_{n'})$ with $\lambda_{n''} \to$ (some) λ_0. Since A is a C-algebra we get $(\lambda_{n''} x)' \to (\lambda_0 x)'$, whence the sequence $((\lambda_{n''} x)')$ is bounded.

But by $(*)$,

$$\frac{(\lambda_{n''} x)'}{n''} \notin U,$$

contradicting boundedness of $((\lambda_{n''x})')$. Hence the Lemma.

6.3.11. PROPOSITION. (Kaplansky). *In a complex normed (or more generally, ample ρ-normed) algebra A every q. nilpotent element x is s.t.z.d.. In particular, every element of the radical*

$\sqrt{A}$ *is a s.t.z.d.*.

PROOF. We may assume that $x \neq 0$. By 6.3.8 there is a sequence (λ_n) in $\mathbb{C}$ with

$$\|(\lambda_n x)'\| \to \infty.$$

By 6.3.10, $|\lambda_n| \to \infty$. Write

$$x_n = \lambda_n x, \qquad y_n = x'_n / \|x'_n\|^{\frac{1}{\rho}}$$

(so that $\|y_n\| = 1$). Then since

$$x_n x'_n = -x_n - x'_n$$

we get

$$x_n y_n = -x_n \|x'_n\|^{-\frac{1}{\rho}} - y_n. \tag{1}$$

Since $x_n = \lambda_n x$ the equation (1) becomes

$$x y_n = -x \|x'_n\|^{-\frac{1}{\rho}} - \lambda_n^{-1} y_n. \tag{2}$$

Since $\|x'_n\|, |\lambda_n| \to \infty$ as $n \to \infty$, and $\|y_n\| = 1$ we get from (2) $xy_n \to 0$ (as $n \to \infty$). In exactly similar manner we also get $y_n x \to 0$. Thus, x is a s.t.z.d., as desired.

6.3.12. COROLLARY. *In a real normed – or more generally, ample ρ-normed – algebra A every ext. q. nilpotent element x is a s.t.z.d.*.

PROOF. By definition x is a q. nilpotent element of the complexification $\tilde{A}$. By 4.7.4. (b), $\tilde{A}$ is also ample. So, by 6.3.11, x is a s.t.z.d. of $\tilde{A}$ and hence by 3.7.16 (ii), a s.t.z.d. of A.

6.3.13. DEFINITION. A TA A is called a *topological integral domain* or TID if it has no non-zero t.z.d..

6.3.14. PROPOSITION. *A complex ample ρ-normed algebra A which is a TID is q.s.s., in particular s.s.*.

PROOF. This is an immediate consequence of 6.3.11 and the definition of a TID.

§ 4. Spectral Algebras

6.4.1. DEFINITION. A unital algebra over a field is called *spectral* if for every x in A, $\sigma(x) \neq \emptyset$.

6.4.2. LEMMA. *Every subunital algebra B of a spectral algebra A is spectral.*

PROOF. By 1.7.19 (**), $\sigma_B(x) \supseteq \sigma_A(x) \neq \emptyset$.

6.4.3. THEOREM. *Every complex ample Hausdorff CI algebra is spectral.*

PROOF. If possible let $\sigma(x) = \emptyset$, for some x in A, so that $\rho(x) = \mathbb{C}$. By 6.2.5, $F_{x^*}(\lambda) = x^*(x_\lambda)$ $(x^* \in A^*)$ is holomorphic on $\rho(x) = \mathbb{C}$. Also, F_{x^*} is bounded since by virtue of 6.2.7, $x^*(x_\lambda) \to 0$ as $|\lambda| \to \infty$. By Liouville's theorem, F_{x^*} is a constant which must by 0 (since $F_{x^*}(\lambda) \to 0$ as $|\lambda| \to \infty$). Thus $F_{x^*}(\lambda) = 0$ for every $x^* \in A^*$, whence by ampleness of $A, x_\lambda = 0$. It follows that $e = (x - \lambda e)x_\lambda = 0$ – a contradiction. Thus $\sigma(x) \neq \emptyset$, as required.

6.4.4. COROLLARY. *Every complex Hausdorff locally convex CI algebra A is spectral.*

PROOF. By 4.7.6, A is ample and so the result follows from 6.4.3.

6.4.5. COROLLARY. *Every complex Hausdorff locally sm. convex I algebra A is spectral. In particular, every complex Hausdorff locally sm. convex division algebra D is spectral.*

PROOF. By 4.4.15, 3.6.21, A is a CI algebra and hence the first statement follows from 6.4.4. The second statement follows from the first since D being a division algebra is an I algebra ($G_i = D\backslash\{0\}$ is open).

6.4.6. COROLLARY. *A complex, ample ρ-Banach algebra – in particular, Banach algebra – is spectral.*

PROOF. This follows from 6.4.3 since every Banach algebra is a CI algebra (by 3.6.23 (b)) and is also Hausdorff.

6.4.7. THEOREM (Żelazko). *Every complex unital ρ-Banach algebra A is spectral.*

PROOF. This theorem goes beyond 6.4.6 since it covers also those algebras which are not ample.

If possible let there be an element $x \in A$ with $\sigma(x) = \emptyset$. Then $\rho(x) = \mathbb{C}$ and by 6.2.9, x_λ is a locally ρ-admissible entire function. If Γ is any circle in $\mathbb{C}$ then $\Gamma \sim 0$ and hence by 5.14.16 (a)

$$\int_\Gamma x_\lambda d\lambda = 0.$$

On the other hand, by taking $m = 0$ in the formula $(*)$ of 6.2.10, we obtain

$$\int_\Gamma x_\lambda d\lambda = -2\pi i e,$$

where e is the unity of A. This contradiction proves that $\sigma(x) \neq \emptyset$, for every $x \in A$, and A is spectral.

6.4.8. COROLLARY. *Every strictly real unital ρ-Banach algebra A is spectral.*

PROOF. Let $\tilde{A}$ be the complexification of A. Since A is strictly real we have

$$\sigma_A(x) = \sigma_{\tilde{A}}(x) \neq 0 \text{ (by 6.4.7)}.$$

6.4.9. COROLLARY. *Every complex or strictly real unital pseudo-Michael algebra A is spectral.*

PROOF. By 4.5.3, A has a projective limit decomposition $A = \varprojlim \hat{A}_\alpha$, where each $\hat{A}_\alpha$ is a unital ρ_α-Banach algebra. Again, by 4.5.7 (3),

$$\sigma_A(x) = \bigcup_\alpha \sigma_{\hat{A}_\alpha}(x_\alpha), \qquad x = (x_\alpha). \tag{$*$}$$

First let A be complex. Then $\hat{A}_\alpha$ is complex and so by 6.4.7, $\sigma_{\hat{A}_\alpha}(x_\alpha) \neq \emptyset$. It follows by $(*)$ that $\sigma_A(x) \neq \emptyset$, and A is spectral.

Next let A be strictly real. In view of 1.7.26, 1.9.11 we may assume that A is commutative. By 1.9.8, A_α is commutative strictly real and so by 6.2.20, $\hat{A}_\alpha$ is strictly real. Hence by 6.4.8 and $(*)$, A is spectral.

6.4.10. Remark. A complex unital commutative complete locally convex algebra may fail to be spectral. To overcome this deficiency Waelbroeck [13′] has given a modified definition of spectrum with respect to which these algebras are spectral. His modified spectrum which we shall denote by $sp(x)$ is the complement (in $\mathbb{K}$) of the set of all $\lambda_0 \in \mathbb{K}$ such that $(x - \lambda e)^{-1}$ exists and is (t). bounded in a neighbourhood of λ_0. Using the properties of his spectrum $sp(x)$, Waelbroeck has obtained interesting results concerning locally convex algebras.

§ 5. Gelfand-Mazur and Other Similar Theorems

6.5.1. LEMMA. *If a division algebra A over a field F is spectral then A is of the form $A = Fe$, where e is the unity of A. Moreover, if $x = \lambda_x e$, the map $\omega : x \mapsto \lambda_x$ is an isomorphism of A onto F.*

PROOF. If $x \in A$ and $\lambda_x \in \sigma(x)$ then $x - \lambda_x e$ is non-invertible and so $x - \lambda_x e = 0$ since A is a division algebra. Thus, $x = \lambda_x e$, $A = Fe$ and ω is clearly an isomorphism.

6.5.2. COROLLARY. *If a spectral division algebra A over $\mathbb{K}$ is a Hausdorff TA then the map $\omega : x \mapsto \lambda_x$ is a t. isomorphism.*

PROOF. This is because of the result that an isomorphism between two finite-dimensional TLS's is automatically a homeomorphism (here $\dim A = \dim \mathbb{K} = 1$) (see 2.1.12).

6.5.3. THEOREM. *A complex ample CI division algebra A is of the form $A = \mathbb{C}e$, with the map $\omega : x \mapsto \lambda_x$ $(x = \lambda_x e)$ a t. isomorphism.*

PROOF. Since $A \backslash \{0\} = G_i$ is open, A is Hausdorff. Hence, by 6.4.3, A is spectral. The theorem now follows from 6.5.1, 6.5.2.

6.5.4. COROLLARY (Arens). *A complex Hausdorff locally sm. convex division algebra A is of the form $A = \mathbb{C}e$ (with ω a t. isomorphism).*

PROOF. Since A is a Hausdorff division algebra, by 3.6.10, A is an I algebra. Further, by 4.4.15, A is a C algebra, and by 4.7.6. it is ample. The required conclusion now follows from 6.5.3.

6.5.5. THEOREM (Gelfand-Mazur). *A complex normed division algebra A is of the form $A = \mathbb{C}e$ and $\omega : x \mapsto \lambda_x$ is a t. isomorphism. If $\|e\| = 1$ then ω is also an isometry.*

PROOF. Without loss of generality we may suppose that the norm of A is sm.. Then A is locally sm. convex, so that by 6.5.4, $A = \mathbb{C}e$ and ω a t. isomorphism.

If $\|e\| = 1$, then $\|x\| = \|\lambda_x e\| = |\lambda_x|$ and ω is an isometry.

6.5.6. COROLLARY (Żelazko). *A complex ρ-seminormed division algebra $A = (A, p$ [†]$)$ has the form $A = \mathbb{C}e$. Consequently, every locally bounded complex TA B which is a division algebra has the form $B = \mathbb{C}e$.*

PROOF. We may assume that p is sm.. Let a be an element of A and A_m a maximal commutative subalgebra containing a (see 1.1.9.); by 1.1.19, A_m is a division algebra. For $x \in A_m$, write $\|x\| = \nu^{\frac{1}{\rho}}(x)$, where $\nu(x) = \lim_{n\to\infty} p(x^n)^{\frac{1}{n}}$. By 4.8.6, 4.8.7, the above defined functional $\|\cdot\|$ is a norm on A_m, so that $(A_m, \|\cdot\|)$ is a complex normed division algebra. By 6.5.5, $A_m = \mathbb{C}e$, so that $a = \lambda e \quad (\lambda \in \mathbb{C})$. It follows that $A = \mathbb{C}e$, proving the first assertion. For the second assertion it suffices to observe that, by 4.2.4, B is ρ-seminormed (so that it follows from the first).

6.5.7. PROPOSITION. *A strictly real, ample CI division algebra A is of the form $A = \mathbb{R}e$ (with $\omega : x \to \lambda_x$ a t. isomorphism). In particular, such an algebra is commutative.*

---

† As always we assume $p \neq 0$

PROOF. As in 6.5.6, for any element $a \neq 0$ in A, take a maximal commutative subalgebra A_m with $a \in A_m$; A_m is a division algebra and so in particular inverse-closed in A. By 4.7.3 (a), 3.6.27, A_m is a commutative ample CI algebra. Further, by 1.9.11, A_m is strictly real, and consequently, by 1.9.14, it is also formally real. By 1.6.20, its complexification $\tilde{A}_m$ is a division algebra which is moreover ample (see 4.7.4 (b)). By 6.5.3, $\tilde{A}_m = \mathbb{C}e$, whence $A_m = \mathbb{R}e$, so that $a = \lambda e$ $(\lambda \in \mathbb{R})$. Since a is an arbitrary element of A we conclude that $A = \mathbb{R}e$, as desired.

6.5.8. COROLLARY. *If A is a strictly real, Hausdorff locally sm. convex division algebra then $A = \mathbb{R}e$.*

PROOF. The same argument as in the proof of 6.5.4 shows that A is an ample CI algebra. The required result is now an immediate consequence of 6.5.7.

6.5.9. PROPOSITION. *A commutative real ample CI (in particular, a Hausdorff locally sm. convex) division algebra A is of the form $A = \mathbb{R}e$ or $\mathbb{C}e$ $(A \simeq \mathbb{R}$ or $\mathbb{C})$ according as A is formally real or not.*

PROOF. If A is formally real then, by 1.9.14, it is strictly real and so $A = \mathbb{R}e$ (by 6.5.7). It remains to consider the case where A is not formally real. By 1.6.20 (b), A has a complex structure. Since $ix = jx$, with $j \in A$, multiplication by i is continuous so that $\check{A}$ is a complex TA which is moreover, ample by 4.7.4 (a). By 6.5.3 (& 6.5.4) we conclude that $A = \check{A} = \mathbb{C}e$.

6.5.10. THEOREM. *Every real ample CI division algebra A is t. isomorphism to $\mathbb{R}$, $\mathbb{C}$, or $\mathbb{H}$ (the Hamilton quarternions).*

PROOF. In view of 6.5.9, we may assume that A is not commutative. Let Z denote the centre of A. By 1.1.18, Z is a commutative division algebra. For any $x \neq 0$ in A the algebra $Z(x)$ generated by Z, x, x^{-1} is a commutative division algebra which is moreover, ample (by 4.7.3 (a)) and CI (by 3.6.24). Applying 6.5.9 to $Z(x)$ we get

$$Z(x) = \mathbb{R}e \text{ or } \mathbb{C}e.$$

Thus,

$$x = (\alpha + i\beta)e \ (\alpha, \beta \in \mathbb{R}, \text{ with } \beta = 0 \text{ if } Z(x) = \mathbb{R}e).$$

It follows that

$$(x - (\alpha + i\beta)e)(x - (\alpha - i\beta)e) = 0$$

or,

$$x^2 - 2\alpha x + (\alpha^2 + \beta^2)e = 0,$$

which implies that A is algebraic of degree 2 at most. By a theorem of Jacobson † A is finite-dimensional, and so by the classical Frobenius theorem, $A \simeq \mathbb{R}$, $\mathbb{C}$ or $\mathbb{H}$; the isomorphism is topological since A is finite-dimensional Hausdorff.

6.5.11. COROLLARY (Arens). *Every real Hausdorff locally sm. convex division algebra A is t. isomorphic to $\mathbb{R}, \mathbb{C}$ or $\mathbb{H}$.*

PROOF. As in 6.5.4 we see that A is ample and CI, and the required conclusion follows from 6.5.10.

6.5.12. Theorem (Żelazko). *Every real ρ-normed division algebra $A = (A, p)$ is t. isomorphic to $\mathbb{R}$, $\mathbb{C}$ or $\mathbb{H}$ according as it is formally real, commutative but not formally real, or not commutative. If ω denotes the t. isomorphism then*

$$|\omega(x)| = \nu(x) = \lim_n p(x^n)^{\frac{1}{n}},$$

where $|\cdot|$ denotes the standard norm †† *on $\mathbb{R}$, $\mathbb{C}$ or $\mathbb{H}$.*

PROOF. Since we do not know *à priori* that A is ample we cannot deduce this theorem from 6.5.10. However, we can prove it by directly considering separately the three cases that arise.

Case 1. A is formally real. For any non-zero x is A consider a maximal commutative subalgebra A_m with $x \in A_m$. As in 6.5.7, A_m is a formally real division algebra, so that A_m is

---

† Every algebraic division algebra of bounded degree over a perfect field is finite-dimensional ("Structure theory for algebraic algebras, Annals of Mathematics 46 (1945), p.701").

†† $|t| = (t\bar{t})^{\frac{1}{2}}$ ($t \in \mathbb{R}, \mathbb{C}$ or $\mathbb{H}$ and $\bar{t}$ the conjugate of t).

now a commutative formally real ρ-normed division algebra. Its complexification $\tilde{A}_m$ is a ρ-normed division algebra, whence by 6.5.6, $\tilde{A}_m = \mathbb{C}e$, $A_m = \mathbb{R}e$, $x = \lambda e$, $A = \mathbb{R}e$.

Case 2. A is commutative but not formally real. Then A has a complex structure and as in the proof of 6.5.9, we get $A = \check{A} = \mathbb{C}e$.

Case 3. A is not commutative. Let Z be the centre of A. For $x \neq 0$ in A we can form, as in the proof of 6.5.10, the algebra $Z(x)$ which is now a real commutative ρ-normed division algebra. By the conclusions in cases 1, 2 we have $Z(x) = \mathbb{R}e$ or $\mathbb{C}e$. This implies, as in 6.5.10, that A is finite-dimensional, and hence by Frobenius, $A \simeq \mathbb{R}$, $\mathbb{C}$ or $\mathbb{H}$.

It remain to prove $(*)$. Set $\|x\| = |\omega(x)|$. Then $\|\cdot\|$ is a norm on A with $\|xy\| = \|x\|\,\|y\|$. Since A is finite-dimensional, $\|\cdot\| \sim p$. By 4.8.2

$$\nu(x) = \nu_p(x) = \nu_{\|\cdot\|}(x) = \lim \|x^n\|^{\frac{1}{n}} = \lim \|x\| = \|x\|.$$

6.5.13. Remark. If A is a formally real normed division algebra whose norm $\|\cdot\|$ is normalized, then the t. isomorphism $\omega : A \to \mathbb{R}$ is an isometry: if $x = \lambda_x e$, $\|x\| = |\lambda_x| = |\omega(x)|$.

6.5.14. LEMMA. *Let A be a unital ρ-normed algebra which is a TID. Let $\hat{A}$ be the completion of A. Then every non-zero element x of A has an inverse in $\hat{A}$.*

PROOF. Suppose that $x \in A$ has no inverse in $\hat{A}$. Then $0 \in \sigma_{\hat{A}}(x)$, so that $\sigma_{\hat{A}}(x) \neq \emptyset$. Also, by 6.1.2, $\sigma_{\hat{A}}(x)$ is closed and $\sigma_{\hat{A}}(x) \neq \mathbb{K}$. It follows that $\partial\sigma_{\hat{A}}(x) \neq \emptyset$. If $\lambda \in \sigma_{\hat{A}}(x)$ then, by 3.7.28, $x - \lambda e$ is a s.t.z.d. of $\hat{A}$ and so by 3.7.14, it is a t.z.d. of A. By the hypothesis on A, $x = \lambda e$ and since x is not invertible, $x = 0$.

6.5.15. COROLLARY. *If a unital ρ-Banach algebra A is a TID then it is a division algebra.*

6.5.16. Remark. (cf. Żelazko [31, p.112]). The result in 6.5.15 does not hold for a locally sm. $\mathfrak{F}$ algebra. For example, consider the algebra $\mathcal{E}$ of entire functions (see 4.6.8 (iii)). $\mathcal{E}$ is not a division algebra, since for instance, $z \in \mathcal{E}$ has no inverse. But $\mathcal{E}$ is a TID. To see this, suppose that in $\mathcal{E}$ we have $g \neq 0$, and

$$f_k g \to 0 \quad \text{as } k \to \infty. \tag{$*$}$$

Since the zeros of g are isolated we can find a sequence r_n of reals with $0 < r_n \to \infty$ and such that

$$\inf_{|z|=r_n} |g(z)| > 0 \ (n = 1, 2, \cdots). \tag{$**$}$$

If we write

$$\|f_n\|^* = \sup_{|z|=r_n} |f(z)|,$$

the family $\{\|\cdot\|_n^*\}$ of norms is easily seen to be equivalent to the family $\{\|\cdot\|_n\}$ defining the topology in $\mathcal{E}$ (see 4.6.8 (iii)). The condition $(*)$ implies that $\|f_k g|_n^* \to 0$ as $k \to \infty$, for all n. But this result along with $(**)$ gives: $\|f_k\|^* \to 0$. This means that g is not a t.z.d., proving $\mathcal{E}$ is a TID.

6.5.17. THEOREM. (cf. [31, pp.30-32.]) *Let A be a unital ρ-normed algebra which is a TID. Then:*
$A = \mathbb{C}e$, $A \simeq \mathbb{C}$, *if A is a complex algebra*
and
$A \simeq \mathbb{R}, \mathbb{C}$ or $\mathbb{H}$ *if A is a real algebra.*

PROOF. Let $\hat{A}$ be the completion of A. By 6.5.14, every $x \neq 0$ in A has an inverse x^{-1} in $\hat{A}$. Consider the unital subalgebras $A(x)$ of $\hat{A}$, consisting of all rational functions of x, over $\mathbb{K}$, i.e. all elements of the form $f(x)/g(x) = f(x)g(x)^{-1}$ $(g(x) \neq 0)$, where f, g are polynomials over $\mathbb{K}$. $A(x)$ is a division algebra containing x, which being a subalgebra of $\hat{A}$ is ρ-normed. If A is complex, it follows from 6.5.6 that $A(x) = \mathbb{C}e$, $x = \lambda_x e$, $A = \mathbb{C}e$.

It remains to consider the case where A is real. By 6.5.12, $A(x) \simeq \mathbb{R}$, $\mathbb{C}$ or $\mathbb{H}$. It follows that x satisfies a relation of the form

$$\alpha x^2 + \beta x + \gamma e = 0, \text{ where } \alpha, \beta, \gamma \in \mathbb{R}, \gamma \neq 0.$$

It follows that $x^{-1} = -\dfrac{\alpha}{\gamma}x - \dfrac{\beta}{\gamma}e \in A$, showing that A is a division algebra. Now we apply 6.5.12 to A and conclude that $A \simeq \mathbb{R}$,

$\mathbb{C}$ or $\mathbb{H}$.

6.5.18. Remark. Theorem 6.5.17 has been extended by Żelazko to locally sm. convex algebras in the following form: If a unital locally sm. convex algebra A has no nonzero g.t.z.d. then $A \simeq \mathbb{C}$ if A is complex, and $\simeq \mathbb{R}$, $\mathbb{C}$ or $\mathbb{H}$ if A is real (see [31, pp.112-14]).

6.5.19. PROPOSITION. *Suppose that the norm of a unital ρ-normed algebra A satisfies the condition*

$$C\|x\| \; \|y\| \leqslant \|xy\| \tag{$*$}$$

for some $C > 0$ and all $x, y \in A$. Then A is t. isomorphic to $\mathbb{C}$ if A is complex and to $\mathbb{R}$, $\mathbb{C}$ or $\mathbb{H}$ if A is real.

PROOF. In view of 6.5.17 it is sufficient to show that A is a TID. Suppose that $x, y_n \in A$, $\|y_n\| = 1$, $xy_n \to 0$. Then, by $(*)$ we have

$$C\|x\| = C\|x\| \; \|y_n\| \leqslant \|xy_n\| \to 0.$$

This means that $\|x\| = 0$, $x = 0$, so that A has no non-zero l.t.z.d.; similarly it has no non-zero r.t.z.d.. Thus A is a TID, completing the proof.

6.5.20. COROLLARY (Arens-Shilov). *A unital normed algebra A satisfying condition $(*)$ above is t. isomorphic to $\mathbb{C}$ if A is complex and to $\mathbb{R}$, $\mathbb{C}$ or $\mathbb{H}$ if A is real.*

6.5.21. Remark. A special case of 6.5.20 was proved earlier by Lorch and Mazur (see [14, p.127]) in the form: A complex unital Banach algebra A whose norm satisfies the condition $\|xy\| = \|x\| \; \|y\|$ for all x, y in A is (t.) isomorphic to $\mathbb{C}$.

§ 6. Turpin's Theorem on Locally Convex Algebras

6.6.1. Let A be a Hausdorff complete locally convex algebra. Denote by $\mathcal{P}_C$ the set of all continuous semi-norms on A and by A^* (the continuous) dual of A.

Set

$$\begin{aligned} r_1(x) &= \sup_{p_\alpha \in \mathcal{P}_c} \overline{\lim_{n\to\infty}}\, p_\alpha(x^n)^{\frac{1}{n}}, \\ r_2(x) &= \sup_{f\in A^*} \overline{\lim_{n\to\infty}} |f(x^n)|^{\frac{1}{n}}, \\ r_3(x) &= \inf\{r : (\alpha_n)_1^\infty,\ \alpha_n \in \mathbb{K},\ \text{the series } \sum_{n=1}^\infty \alpha_n \lambda^n \\ &\qquad \text{has radius of convergence} > r \\ &\qquad \Rightarrow \sum_{n=1}^\infty \alpha_n x^n \text{ converges in } A\}. \end{aligned}$$

We have clearly $0 \leqslant r_j(x) \leqslant \infty$ $(j = 1, 2, 3)$.

6.6.2. LEMMA. $r_2(x) \leqslant r_1(x)$.

PROOF. If $f \in A^*$, by 4.3.13 (i) there is a p_α such that f is p_α-continuous, and so p_α-bounded. Therefore we have $|f(x^n)| \leqslant \|f\| p_\alpha(x^n)$ so that

$$\overline{\lim_{n\to\infty}} |f(x^n)|^{\frac{1}{n}} \leqslant \overline{\lim_{n\to\infty}} (\|f\|^{\frac{1}{n}} p_\alpha(x^n)^{\frac{1}{n}}) \leqslant 1. \overline{\lim_{n\to\infty}} p_\alpha(x^n)^{\frac{1}{n}} \leqslant r_1(x).$$

By taking the sup over $f \in A^*$ we get $r_2(x) \leqslant r_1(x)$.

6.6.3. LEMMA. *If* $r_3(x) < \infty$ *and* $|\mu| > r_3(x)$, *then:*

(i) *the series* $\displaystyle\sum_{n=1}^\infty \left(\frac{x}{\mu}\right)^n$ *converges absolutely;*

(ii) $\mu \in \rho'(x)$;

(iii) $r(x) \leqslant r_3(x)$, *where* $r(x)$ *denotes the spectral radius of* x.

PROOF. By definition of $r_3(x)$ there is an r_0 with $|\mu| > r_0 > r_3(x)$ such that, if $\sum_{n=1}^\infty \alpha_n \lambda^n$ is a series over $\mathbb{K}$ with radius of convergence $\geqslant r_0$, the series $\sum_{n=1}^\infty \alpha_n x^n$ converges absolutely.

Now the numerical series $\sum_{n=1}^{\infty}\left(\frac{\lambda}{\mu}\right)^n$ has radius of convergence $= 1/\overline{\lim}|\frac{1}{\mu^n}|^{\frac{1}{n}} = |\mu| > r_0$, so that by choice of r_0, $\sum_{n=1}^{\infty}\left(\frac{x}{\mu}\right)^n$ converges absolutely. This implies, by 3.1.24, that $\sum_{n=1}^{\infty}\left(\frac{x}{\mu}\right)^n$ converges in A, whence, by 2.2.17, $\left(-\frac{x}{\mu}\right)'$ exists, so that $\mu \in \rho'(x)$. Therefore, if $\lambda \in \sigma'(x)$ then $\lambda \notin \rho'(x)$ and so we must have $|\lambda| \leqslant r_3(x)$. It follows that $r(x) \leqslant r_3(x)$, which completes the proof of the lemma.

6.6.4. PROPOSITION. *Let A be a complex Hausdorff complete locally convex algebra which is CQ.*

Then we have

$$r_1(x) = r_2(x) = r_3(x) = r(x). \qquad (*)$$

PROOF. By 6.1.1, $\sigma'(x)$ is compact. So $0 \leqslant r(x) < \infty$. Suppose that

$$0 < |\lambda| < \frac{1}{r(x)}.$$

Then $r(x) < |\lambda|^{-1} = |-\lambda^{-1}|$, so that $-\lambda^{-1} \in \rho'(x)$ and consequently $\lambda \in \rho^p(x)$ (see 6.3.2). Since A is a CQ algebra, by 6.3.5, x'_λ is a strongly holomorphic function with

$$\frac{d^n x'_\lambda}{d\lambda^n} = (-1)^n n! x^n (1 + x'_\lambda)^{n+1}.$$

If $f \in A^*$, the $F(\lambda) = f(x'_\lambda)$ is holomorphic on $\rho^p(x)$ and in particular, on $|\lambda| < \dfrac{1}{r(x)}$. Since

$$F^{(n)}(0) = f((-1)^n n! x^n) = (-1)^n n! f(x^n)$$

the Taylor expansion for F is given by

$$F(\lambda) = \sum_{n=1}^{\infty}(-1)^n f(x^n)\lambda^n \quad (\text{note } F(0) = f(0) = 0)$$

with the series converging for $|\lambda| < 1/r(x)$.

It follows that

$$(-1)^n f(x^n)\lambda^n = f((-1)^n \lambda^n x^n) \to 0, \text{ as } n \to \infty$$

and this is true for every $f \in A^*$. This implies that the sequence $((-1)^n \lambda^n x^n)$ is weakly bounded. Since A is locally convex it is also strongly bounded †, i.e. we have a constant $M_\alpha > 0$ (depending on λ) such that

$$p_\alpha((-1)^n \lambda^n x^n) = p_\alpha(\lambda^n x^n) < M_\alpha \text{ for all } n \geqslant 1.$$

It follows that

$$\overline{\lim_{n\to\infty}} \sqrt[n]{p_\alpha(x^n)} \leqslant \overline{\lim_{n\to\infty}} (M_\alpha)^{\frac{1}{n}} |\lambda|^{-1} = |\lambda|^{-1}.$$

This being true for every p_α we obtain, by taking sup over p_α,

$$r_1(x) \leqslant |\lambda|^{-1} \text{ (for any } \lambda \text{ with } |\lambda|^{-1} > r(x)).$$

By taking the infimum over λ we get

$$r_1(x) \leqslant r(x).$$

Combining the above inequality with that in 6.6.2 we get

$$r_2(x) \leqslant r_1(x) \leqslant r(x) \quad (< \infty). \tag{1}$$

Next choose r, r' with

$$r_2(x) < r' < r.$$

By definition of r_2, we have for any $f \in A^*$.

$\sqrt[n]{|f(x^n)|} < r' < r$ for sufficiently large n. It follows that $\left|f\left(\frac{x^n}{r'^n}\right)\right| < 1$ ($n \geqslant$ some N). Therefore $\left(\frac{x^n}{r'^n}\right)$ is weakly bounded and consequently strongly bounded. This means that there is an $M_\alpha > 0$ such that

$$p_\alpha\left(\frac{x^n}{r'^n}\right) \leqslant M_\alpha.$$

We have therefore $p_\alpha\left(\frac{x^n}{r^n}\right) \leqslant M_\alpha\left(\frac{r'}{r}\right)^n$, $\frac{r'}{r} < 1$, so that by the

† See [24, p.68].

Weierstrass M-test $\sum_{n=1}^{\infty} \frac{x^n}{r^n}$ converges absolutely in A. Let $\sum_{n=1}^{\infty} \alpha_n \lambda^n$ be a power series with radius convergence $> r$. Then $|\alpha_n| \leqslant r^{-n}$ (for sufficiently large n) so that

$$p_\alpha(\alpha_n x^n) = |\alpha_n| p_\alpha(x^n) \leqslant \frac{p_\alpha(x^n)}{r^n} \leqslant M_\alpha \left(\frac{r'}{r}\right)^n .$$

It follows that the series $\sum_{n=1}^{\infty} \alpha_n x^n$ converges absolutely in A. This means that $r \geqslant r_3(x)$ whence

$$r_2(x) \geqslant r_3(x). \tag{2}$$

From (1), (2) we get

$$r_3(x) \leqslant r_2(x) \leqslant r_1(x) \leqslant r(x). \tag{3}$$

But by 6.6.3 (iii), $r(x) \leqslant r_3(x)$. This together with (3) gives the relation $(*)$.

6.6.5. THEOREM (Turpin). *A commutative complex $\mathfrak{F}$ algebra which is a Q algebra is locally sm.*.

PROOF. By 3.6.18, 4.6.7, A is CQ. Since A is a locally convex Q algebra we can find an open absolutely convex nucleus U comprising q. invertible elements. Since U is balanced, by 3.6.8 (i), we have $r(x) \leqslant 1$ for $x \in U$. Set $V = U/2$. Then V is open and also absolutely convex (see 4.1.2 (c)); since U is balanced, $V \subseteq U$. If $y \in V$ then $y = x/2$ $(x \in U)$, so that

$$r(y) = r\left(\frac{x}{2}\right) = \frac{1}{2} r(x) \leqslant \frac{1}{2} \cdot 1 = \frac{1}{2} < 1.$$

By 6.6.4, $r_3(y) = r(y) < 1$. So, by 6.6.3 (i), $\sum_{n=1}^{\infty} y^n$ converges absolutely in A.

Let $\varphi(\lambda) = \sum_{n=1}^{\infty} \alpha_n \lambda^n$ be an entire function vanishing at 0. Then

$$|\alpha_n \lambda^n| < 1 \text{ for } n \geqslant n(\lambda). \tag{$*$}$$

If $x \in A$, then using the fact that U as a nucleus is absorbing we find $\lambda_0 \neq 0$ such that $x = \lambda_0 y$ $(y \in U)$. Then, using $(*)$,

$$p_\alpha(\alpha_n x^n) = p_\alpha(\alpha_n \lambda_0^n y^n) = |\alpha_n \lambda_0^n| p_\alpha(y^n) < p_\alpha(y^n)$$

for $n \geqslant n_0 = n(\lambda_0)$. Since $\sum_n y^n$ converges absolutely we have

$$\sum_{n=n_0}^{\infty} p_\alpha(\alpha_n x^n) \leqslant \sum_{n=n_0}^{\infty} p_\alpha(y^n) < \infty,$$

whence $\sum_{n=1}^{\infty} \alpha_n x^n$ converges absolutely in A. Thus, all entire functions vanishing at 0 operate on A and the theorem follows from 5.5.11.

CHAPTER VII

GELFAND REPRESENTATION THEORY

§ 1. Ideals of Topological Algebras

7.1.1. LEMMA. (cf. [20, p.70]). *Let A be dense subalgebra of a TA $\overline{A}$. Then:*

(i) *The closure $\overline{I}$ in $\overline{A}$ of an (l. or r.) ideal I of A is an ideal of $\overline{A}$ of the same type.*

(ii) *If the ideal I is regular with a (l. or r.) relative unity u then $\overline{I}$ is also regular with u as a relative unity.*

(iii) *If $I \neq A$ is a closed regular ideal of A then $\overline{I} \neq \overline{A}$ and $\overline{I}$ is a closed regular ideal of $\overline{A}$.*

PROOF. It is enough to prove the results when I is a l. ideal. Clearly $\overline{I}$ is a subspace of $\overline{A}$. Further if $\overline{x} \in \overline{A}$, $\overline{a} \in \overline{I}$, $x_\alpha \in A$, $a_\beta \in \overline{I}$ and $x_\alpha \to \overline{x}$, $a_\beta \to \overline{a}$ then $x_\alpha a_\beta \to \overline{x}\,\overline{a}$. Since each $x_\alpha a_\beta \in I$ it follows that $\overline{x}\ a \in \overline{I}$, proving $\overline{I}$ is a l. ideal. Hence (i).

Next let I be a regular l. ideal with relative r. unity u. If $\overline{x} \in \overline{A}$, $x_\alpha \in A$ and $x_\alpha \to \overline{x}$ then

$$\overline{x}u - \overline{x} = \lim_\alpha (x_\alpha u - x_\alpha) \in \overline{I} \quad (\text{since } x_\alpha u - x_\alpha \in I)$$

and $\overline{I}$ is regular with u as a relative r. unity, proving (ii).

Finally, if u is a relative unity for closed ideal $I \neq A$, then $u \notin I$ and hence $u \notin \overline{I}$ (since $A \bigcap \overline{I} = I$) and therefore $\overline{I} \neq \overline{A}$, proving (iii).

7.1.2. COROLLARY. *Let A be a TA. Then the closure $\overline{I}$ of a l. ideal, a r. ideal, or a bi-ideal I of A is an ideal of A of the same type.*

PROOF. Apply 7.1.1 with $\overline{A} = A$.

7.1.3. COROLLARY. *Every maximal ideal M of A is either closed or dense.*

PROOF. By maximality of M, the closure $\overline{M} = M$ or A.

7.1.4. DEFINITION. Following Michael [20] we call a TA A *normal* if every closed regular l. (respy. r.) ideal $I \neq A$ is contained in some closed maximal regular l. (respy. r.) ideal of A. Further, if every such I is contained in some closed *hypermaximal* ideal then A will be called *hypernormal*. Trivially, every hypernormal ideal is normal. Finally, we call a TA A *hyponormal* if every maximal regular l. or r. ideal is closed.

7.1.5. Remark. Since a radical algebra A has no regular ideal $\neq A$, such an algebra is vacuously hypernormal and hyponormal.

7.1.6. LEMMA. (a) *Every closed maximal regular l. or r. ideal M of a hypernormal TA A is hypermaximal.*

(b) *Every hyponormal TA A is normal.*

(c) *Every hyponormal TA A is functionally*† *continuous.*

PROOF. (a) By hypernormality there is a closed hypermaximal ideal M_1 with $M \subseteq M_1$. The maximality of $M \Rightarrow M = M_1$.

(b) By Krull (1.2.10) if I is a closed regular ideal $\neq A$ there is maximal regular ideal M with $I \subseteq M$. Since A is hyponormal, M is closed and so A is normal.

(c) Since A is hyponormal every hypermaximal ideal is closed and so by 1.3.9, 2.1.30 A is functionally continuous.

7.1.7. We give below examples to show that neither of the properties hypernormality, hypornormality need imply the other.

Example 1. Let A be any radical algebra $(A = \sqrt{A})$ and A_1 its unitization. Then A_1 is a TA under the indiscrete topology. The ideal A is hypermaximal in A_1. Since A is not closed in A_1, A_1 is not hyponormal. On the other hand, since no ideal $I \neq A_1$ is closed it is vacuously hypernormal.

Example 2. The Williamson algebra $\mathcal{W}$. We have seen in

---

† For definition see 2.2.19.

3.6.33 that $\mathcal{W}$ is a commutative Hausdorff division algebra over $\mathbb{C}$. Since $\{0\}$ is the only maximal ideal of $\mathcal{W}$ and it is closed, $\mathcal{W}$ is hyponormal. But since $\{0\}$ is not hypermaximal, $\mathcal{W}$ is not hypernormal.

7.1.8. PROPOSITION. *In a Q algebra A if $I \neq A$ is a regular l or r. ideal then its closure $\overline{I} \neq A$.*

PROOF. Since A is a Q algebra, G_q is open. If u is a relative (r. or l.) unity of I the same is true of $u + a$ for any $a \in I$ (see 1.2.8(a)). Therefore, by 1.2.9(ii), $-u - a \notin G_q$, so that $(-u + I) \bigcap G_q = \emptyset$ and hence also $\overline{-u + I} \bigcap G_q = \emptyset$ (since G_q is open). It follows that

$$-u + \overline{I} = \overline{-u + I} \neq A, \text{ so that } \overline{I} \neq u + A = A.$$

7.1.9. COROLLARY. *Every maximal regular (l. r. or bi-) ideal M of A is closed. Hence, every Q algebra – in particular (see 3.6.23) a pseudo-Banach or (more generally) a sm. (F) algebra – is hypernormal.*

7.1.10. LEMMA. *Let $A = (A, |\cdot|)$ be a unital sm. (F) algebra with unity e. Then for any ideal $I \neq A$ we have:*

$$1 \leqslant d(e, I)^{\dagger} \leqslant |e|; \qquad (*)$$

in particular when $|e| = 1$, $d(e, I) = 1$.

PROOF. Since $0 \in I$, $d(e, I) \leqslant d(e, 0) = |e|$.

If $d(e, x) = |e - x| < 1$ then, by 3.3.18 (ii), x is invertible and hence $x \notin I$ (since $I \neq A$). It follows that, for $x \in I$, we have $d(e, x) \geqslant 1$. Hence the inequality $(*)$.

7.1.11. LEMMA. *In a TA A, if I_l (respy. I_r) is a closed l. (respy. r.) ideal of A then $(I_l : A)$ (respy. $(I_r : A)$) is closed. Hence the primitive ideal P associated with a closed maximal regular l. (respy. r.) ideal M_l (respy. M_r) is closed.*

PROOF. If $x_\alpha \in (I_l : A)$ and $x_\alpha \to x \in A$, then, for any $y \in A$, $x_\alpha y \in I_l$ and consequently $xy \in I_l$ (since $x_\alpha y \to xy$ and I_l is closed). It follows that $x \in (I_l : A)$, proving that $(I_l : A)$

---

† $d(e, I) = \inf\{|e - a| : a \in I\}$.

is closed. Similarly, $(I_r : A)$ is closed. Finally, since $P = (M_l : A)$ (respy. $(M_r : A)$), P is closed.

7.1.12. PROPOSITION. *A primitive ideal P as well as the radicals $\sqrt{A}, \sqrt[s]{A}, \sqrt[h]{A}$ of a hyponormal algebra – in particular of a Q or of a pseudo-Banach algebra – A, are closed.*

PROOF. By hyponormality of A and 7.1.11 every primitive ideal is closed. Also, every maximal regular bi-ideal being primitive (by 1.5.10) is also closed. Therefore $\sqrt{A}$ † (respy. $\sqrt[s]{A}$) as the intersection of primitive ideals (respy. maximal regular bi-ideals) is closed. Finally, since each hypermaximal ideal of A is closed (see 7.1.6. (c)) their intersection $\sqrt[h]{A}$ is also closed.

7.1.13. PROPOSITION. *Every pseudo-Michael algebra A is normal.*

PROOF. Let $A = (A, \mathcal{P})$, where $\mathcal{P}$ is saturated, and $A = \varprojlim \hat{A}_\alpha$ the projective limit decomposition, with $\hat{A}_\alpha$ pseudo-Banach algebras.

Let $I \neq A$ be a closed regular (l. or r.) ideal with relative (r. or l.) unity u. Since $u \notin I$ and I is closed there is a $p_\alpha \in \mathcal{P}$ and an $\epsilon > 0$ such that $B_\epsilon = \{x \in A : p_\alpha(x - u) < \epsilon\}$ is disjoint with I. It follows that

$$p_\alpha(a - u) \geqslant \epsilon \ \ (a \in I). \tag{1}$$

Using the notations in (the proof of) 4.5.3, 3.4.15 we can write

$$a^\# = a + N_\alpha = a_\alpha;\ \varphi_\alpha : x \mapsto x_\alpha = x + N_\alpha = x^\# \ \ (N_\alpha = \ker p_\alpha).$$

Then we have

$$p_\alpha^\#(a_\alpha - u_\alpha) = p_\alpha^\#(a^\# - u^\#) = p_\alpha(a - u) \geqslant \epsilon(a \in I). \tag{2}$$

Clearly, $I_\alpha = \varphi_\alpha(I)$ is a regular ideal of A_α, with relaive unity u_α; $u_\alpha \notin I_\alpha$ because of (2). By 7.1.1, $J_\alpha = \overline{I}_\alpha$ is a closed regular ideal (with relative unity u_α) $\neq \hat{A}_\alpha$. By Krull J_α is

† That $\sqrt{A}$ is closed can also be deduced from the fact that it is the intersection of all maximal regular l. ideals.

contained in some regular maximal ideal M. Since $\hat{A}_\alpha$ is pseudo-Banach, by 7.1.9, M_α is closed. By continuity of φ_α and 1.2.17 (iii) $M = \varphi_\alpha^{-1}(M_\alpha)$ is a closed regular ideal of A which is maximal, and clearly $I \subseteq M$. This proves that A is normal.

7.1.14. COROLLARY. *A complex or strictly real commutative pseudo-Michael algebra A is hypernormal.*

PROOF. The ideal M_α in the proof above (of 7.1.13) is now closed hypermaximal since $\hat{A}_\alpha$ is Gelfand (by 7.2.17, 7.2.19)†. Let χ_α be the character determined by M_α; χ_α is continuous. Then, if we set $\chi = \chi_\alpha \circ \varphi_\alpha$, $\chi(u) = \chi_\alpha(u_\alpha) = 1$, so that χ is a continuous character of A. If $M = \ker \chi$ then M is a closed hypermaximal ideal. Also, if $x \in I$ then $\varphi_\alpha(x) \in J_\alpha \subseteq M_\alpha$, $\chi(x) = 0$, so that $I \subseteq \ker \chi = M$. This proves A is hypernormal.

7.1.15. Remark. There are commutative Michael algebras which are not functionally continuous (and so not hyponormal) (see [20, p.49]).

7.1.16. Remark. There are hypernormal TA's having dense regular maximal ideals. For an example, consider the algebra $\mathsf{C} = \mathsf{C}(\mathbb{R}, \mathbb{K})$ of all $\mathbb{K}$-valued continuous functions on $\mathbb{R}$ topologized by the family $\{\|\cdot\|_n\}$ of sm. seminorms, where

$$\|f\|_n = \sup_{|t| \leqslant n} |f(t)| \quad (f \in \mathsf{C}) \qquad \text{(cf. Example of 4.6.8)}$$

C is a unital commutative locally sm. $\mathfrak{F}$ algebra. Note that C is strictly real when $\mathbb{K} = \mathbb{R}$. By 7.1.14, C is hypernormal. C has, however, dense maximal ideals. For instance, if

$$I = \{f \in \mathsf{C} : f(t) : 0 \text{ for } t > \text{(some) } t_0(f)\}$$

I is clearly an ideal. We claim that I is dense in A. To see this define for an $f \in \mathsf{C}$,

$$f_{n,\epsilon}(t) = \begin{cases} f(t) & \text{if } t < n \\ f(t)\frac{n+\epsilon-t}{\epsilon} & \text{if } n \leqslant t \leqslant n+\epsilon \\ 0 & \text{if } t > n+\epsilon. \end{cases}$$

† That $\hat{A}_\alpha \neq \sqrt{\hat{A}_\alpha}$ holds since $\hat{A}_\alpha$ admits the regular ideal J_α.

Then $f_{n,\epsilon} \in I$, and $\|f_{n,\epsilon} - f\|_n = 0 < \epsilon$, proving that I is dense in C. By Krull's lemma there is a maximal ideal $M \supseteq I$ and then M is a dense maximal ideal.

7.1.17. PROPOSITION (Rickart[†]). *Let A be a hyponormal algebra-in particular a ρ-Banach algebra or (more generally) a Q algebra. Let A^* be an arbitrary algebra and $\varphi : A \to A^*$ an epimorphism. Then $\varphi(\overline{\ker\varphi}) \subseteq \sqrt{A^*}$ where bar denotes closure. In particular, when A^* is s.s., $\ker\varphi$ is closed and $\sqrt{A} \subseteq \ker\varphi$.*

PROOF. Let M_l^* be a maximal regular l. ideal of A^*. By 1.2.17 (iii), $M_l = \varphi^{-1}(M_l^*)$ is a maximal regular l. ideal of A and $\ker\varphi \subseteq M_l$. Since A is hyponormal M_l is closed. So $\overline{\ker\varphi} \subseteq M_l$. It follows that

$$\varphi(\overline{\ker\varphi}) \subseteq \bigcap M_l^* = \sqrt{A^*}.$$

When $\sqrt{A^*} = \{0\}$, we obtain $\varphi(\overline{\ker\varphi}) = \{0\}$, so that $\overline{\ker\varphi} = \ker\varphi$. Further, By 1.2.26, $\varphi(\sqrt{A}) \subseteq \sqrt{A^*} = \{0\}$, whence $\sqrt{A} \subseteq \ker\varphi$, completing the proof.

7.1.18. DEFINITION. A bi-ideal I of an algebra A is called *primary* or a primary ideal if it is contained in a unique maximal regular bi-ideal M. An *algebra* A is called *primary* if the ideal $\{0\}$ is primary - which is the same as A having a unique maximal regular ideal.

7.1.19. Remark. Every maximal regular bi-ideal is primary. Moreover, in a Q algebra – in particular in a pseudo-Banach algebra – the closure $\overline{I}$ of a primary ideal I is primary (if $I \subseteq M$, where M is a maximal regular bi-ideal, then since M is closed, $\overline{I} \subseteq M$ and $\overline{I}$ is primary).

7.1.20. Examples of primary ideals

(i) Denote by $\mathsf{C}^{(1)} = \mathsf{C}^{(1)}(\mathbb{R}, \mathbb{K})$ the algebra of $\mathbb{K}$-valued continuously differentiable functions on $\mathbb{R}$. $\mathsf{C}^{(1)}$ is a locally sm. $\mathfrak{F}$ algebra under the family $\{\|\cdot\|_n^{(1)}\}$ of seminorm, where

$$\|f\|_n^{(1)} = \sup\{|f(t)| : |t| \leqslant n\} + \sup\{|f^{(1)}(t)| : |t| \leqslant n\}$$

[†] He has proved this result for Banach algebras [23, p.74].

where $f^{(1)}$ denotes the derivative of f. Define in $\mathsf{C}^{(1)}$ the ideals: $M_0 = \{f \in \mathsf{C}^{(1)} : f(0) = 0\}$, $I_1 = \{f \in \mathsf{C}^{(1)} : f(0) = f^{(1)}(0) = 0\}$. We have $I_1 \subset M_0$, and M_0 is maximal (since for $f \in \mathsf{C}^{(1)}$, $f - f(0) \in M_0$). We claim I_1 is primary. To see this, suppose that I is an ideal with $I \supset I_1$, $I \not\subseteq M_0$. Then there is an $f \in I$ with $f(0) = \alpha \neq 0$. Then if $\beta = f^{(1)}(0)$, $g(t) = f(t) - \alpha - \beta t$, then $g \in I_1 \subset I$. Therefore $\alpha + \beta t \in I$, whence $\alpha t + \beta t^2 = (\alpha + \beta t)t \in I$. Since clearly $\beta t^2 \in I_1 \subset I$ we get $\alpha t \in I$, $t \in I$. Therefore $\alpha = (\alpha + \beta t) - \beta t \in I$ and $I = A$. It follows that I_1 is not contained in any other maximal ideal, so that I_1 is primary.

(ii) The algebra $\mathsf{C}^{(1)}([0,1], \mathbb{K})$ is a Banach algebra under the norm.

$$|f|^{(1)} = \|f\|_\infty + \|f^{(1)}\|_\infty, \quad \text{where } \|\cdot\|_\infty$$

denotes the sup norm. If I_1 and M_0 are defined analogouly as above then I_1 is a primary ideal with M_0 as the containing maximal ideal.

(iii) Denote by $\mathsf{C}^{(n)} = \mathsf{C}^{(n)}([0,1], \mathbb{K})$ the algebra of all $\mathbb{K}$-valued functions f whose nth derivative exists and is continuous. $\mathsf{C}^{(n)}$ is a Banach algebra with respect to the norm:

$$\|f\|^{(n)} = \sum_{r=0}^{n} \frac{1}{r!} \|f^{(r)}\|_\infty \;\dagger\; .$$

Set $I_r = \{f \in \mathsf{C}^{(n)} : f(0) = f^{(1)}(0) = \cdots = f^{(r)}(0) = 0\}$. Then it can be shown that I_r $(1 \leqslant r \leqslant n)$ are primary ideals with $I_n \subset I_{n-1} \subset \cdots \subset I_1 \subset M_0$. Also, I_n is the smallest primary ideal $\subset M_0$ (see [10, p.205]).

(iv) In the (F) algebra $\mathcal{E}$ of entire functions (defined in 3.3.14 (iii)) if we set

$$I_r = \{f \in \mathcal{E} : f^{(j)}(0) = 0 \; (0 \leqslant j \leqslant r)\}$$

then I_r an primary ideals such that

$$I_n \subset I_{n-1} \; (n = 1, 2, \cdots)$$

† The weighted sum has been taken to make the norm sm..

and $I_0 = M_0 = \{f \in \mathcal{E} : f(0) = 0\}$ is maximal regular. Clearly $\bigcap I_n = \{0\}$. Since $\{0\}$ is obviously not primary, there does not exist a smallest primary ideal contained in M_0.

7.1.21. Remark. For information regarding primary ideals and other related matters in Banach algebras, see [23, p.92] and references cited therein, as also [22, pp.238-99].

§ 2. Gelfand Algebras

7.2.1. DEFINITION. A TA A is called *Gelfand algebra* if (i) $A \neq \sqrt{A}$ (ii) every maximal regular l. or r. ideal is hypermaximal and closed.

Note that when A is unital, condition (i) is automatically satisfied (by 1.2.24 (d), $e \notin \sqrt{A}$).

7.2.2 PROPOSITION. *Let A be a Gelfand algebra. Then:*

(i) *The radical $\sqrt{A}$ is the intersection of all hypermaximal ideals of A, i.e. $\sqrt{A} = \sqrt[h]{A}$.*

(ii) *Every character of A is continuous, i.e. A is functionally continuous; $\Delta = \Delta_c \neq \emptyset$.*

(iii) *A is both hypernormal and hyponormal.*

(iv) *If $x \in A$ then $\sigma'(x) = \{\chi(x) : \chi \in \Delta\} \bigcup \{0\}$.*

(v) *If A is unital and $x \in A$ then $\sigma(x) = \{\chi(x) : \chi \in \Delta\}$.*

PROOF. (i) This follows from 7.2.1 (ii), 1.2.22, 1.2.24 (b).

(ii) The ideal $\ker \chi$ is hypermaximal by 1.3.9, so closed by 7.2.1 (ii) and hence χ is continuous by 2.1.30. The condition 7.2.1 (i) ensures the existence of a maximal regular l. ideal M; M is closed hypermaximal (by 7.2.1 (ii)) and so determines a continuous character, so that $\Delta_c \neq \emptyset$.

(iii) The condition 7.2.1 (ii) implies that A is hyponormal. Also, the same condition together with Krull's lemma (1.2.10) shows that A is hypernormal.

(iv) Since always $0 \in \sigma'(x)$ it is enough to consider non-zero elements on both sides of the equation. If $\lambda \neq 0$ in $\sigma'(x)$ then $-\lambda^{-1}x$ is not q. invertible and hence, by 1.2.21 (b), 7.2.21 (ii) there is a hypermaximal ideal $M = M_\chi$ for which $\lambda^{-1}x$ is a relative unity. So, by 1.3.10, $\chi(\lambda^{-1}x) = 1$, $\chi(x) = \lambda$. Thus $\sigma'(x) \subseteq$ R.H.S. of the equation in (iv). The reverse inclusion follows from 1.7.11., completing the proof.

(v) In view of 1.7.24, it is enough to prove that if $\lambda \in \sigma(x)$ then $\lambda = \chi(x)$ for some χ. Since $x - \lambda e$ is not invertible, by 1.2.14, 7.2.1 (ii) there is a hypermaximal ideal M with $x - \lambda e \in M$. If χ is the character determined by M then $\chi(x - \lambda e) = 0$, $\chi(x) = \lambda$.

7.2.3. PROPOSITION. *The unitization A_1 of an algebra $A \neq \sqrt{A}$ is a Gelfand algebra iff A is a Gelfand algebra.*

PROOF. By 1.4.9 (b), there is a bijection between the hypermaximal ideals $M_1(\neq A)$ of A_1 and the hypermaximal ideals M of A, given by:

$$M_1 \mapsto M = A \bigcap M_1 \qquad (*)$$

$$M_1 = \check{M} = \{x_1 \in A_1 : x_1 u \in M\} = \{x_1 \in A_1 : u x_1 \in M\} \qquad (**)$$

where u is the relative unity of M.

It follows from $(*), (**)$ that M_1 is closed iff M is closed. These results along with those in 1.4.8 (connecting the maximal l. or r, ideals of A_1 and maximal regular l. or r. ideals of A) prove the proposition.

7.2.4. Remark. The restriction $A \neq \sqrt{A}$ in the statement of 7.2.3 cannot be omitted. For example, any radical TA A is not Gelfand (since $A = \sqrt{A}$) though its unitization A_1 is Gelfand (by 1.4.11, its sole maximal ideals A is hypermaximal (see 1.1.12) and closed (see 2.2.9)).

7.2.5. DEFINITION. Let A be an algebra without unity. Then A is called *spectrally Gelfand* if it satisfies the condition

$$\sigma'(x) = \{\chi(x) : \chi \in \Delta\} \bigcup \{0\} \; (x \in A) \qquad (*)$$

where Δ denotes the set of characters of A.

If A is TA then A is called *t.* (=topologically) *spectrally Gelfand* if

$$\sigma'(x) = \{\chi(x) : \chi \in \Delta_c\} \bigcup \{0\} \ (x \in A) \qquad (**)$$

where Δ_c denotes the set of continuous characters of A.

7.2.6. DEFINITION. A *unital* algebra A is called spectrally Gelfand if

$$\sigma(x) = \{\chi(x) : \chi \in \Delta\} \ (x \in A). \qquad (*1)$$

Similarly, unital TA A is called t. spectrally Gelfand if

$$\sigma(x) = \{\chi(x) : \chi \in \Delta_c\} \qquad (x \in A). \qquad (**1)$$

7.2.7. Remark. Since $\sigma'(x) = \sigma(x) \bigcup \{0\}$, $(*1) \Rightarrow (*)$ and $(**1) \Rightarrow (**)$.

7.2.8. Proposition. (a) *A t. spectrally Gelfand algebra is also spectrally Gelfand; the two concepts coincide when A is functionally continuous – in particular when A is ρ-Banach or even hyponormal.*

(b) *Every Gelfand algebra is t. spectrally Gelfand.*

(c) *The unitizsation A_1 of an algebra (respy. TA) A is spectrally Gelfand (respy. t. spectrally Gelfand). iff A is spectrally Gelfand (respy. t. spectrally Gelfand).*

PROOF. (a) By 1.7.11 (respy. 1.7.24)

$$\chi(x) \in \sigma'(x) \ (\text{respy.}\ \chi(x) \in \sigma(x)) \ \text{ for all} \chi \in \Delta.$$

When A is functionally continuous, $\Delta = \Delta_c$. Also, ρ-Banach $\Rightarrow$ hyponormal $\Rightarrow$ functionally continuous. Hence (a).

(b) This follows from 7.2.2 ((ii), (iv), (v)).

(c) If $x_1 \in A_1$, $x_1 = \mu e_1 + x$ $(x \in A)$ then $\Delta_{A_1}(x_1) = \mu + \sigma'_A(x)$. Also, $\Delta_1 = \Delta \bigcup \{\chi_0\}$, $\Delta_{1c} = \Delta_c^1 \bigcup \{\chi_0\}$, where Δ_1 (respy. Δ_{1c} denotes the set of characters (respy. continuous characters) of A_1, Δ^1 (respy. Δ_c^1) the set of extensions of characters (respy. continuous characters) of A to A_1, and χ_0 the distinguished character of A_1 $(\ker \chi_0 = A)$. The assertions (c) of the proposition are easy consequences of the above relations.

7.2.9. PROPOSITION. *Let A be a spectrally Gelfand algebra and $x \in A$. Then x is q. invertible iff $\chi(x) \neq -1$ for every*

$\chi \in \Delta$. If A is a t. spectrally Gelfand algebra and $\chi(x) \neq -1$ for every $\chi \in \Delta_c$ then x is q. invertible.

PROOF. If x is q. invertible then by 1.1.24, $\chi(x)$ is q. invertible, so that by 1.1.26 (ii), $\chi(x) \neq -1$. If x is not q. invertible then $-1 \in \sigma'(x)$. Since A is spectrally Gelfand there is a $\chi \in A$ with $\chi(x) = -1$. This completes the proof of the first assertion.

For the second, assume now that A is t. spectrally Gelfand, $x \in A$ and $\chi(x) \neq 1$ for all $\chi \in \Delta_c$. If possible let x be not q. invertible; then $-1 \in \sigma'(x)$ and so by hypothesis on A there is a $\chi \in \Delta_c$ with $\chi(x) = -1$, which contradicts our assumption on χ. This contradiction shows that x is q. invertible as required.

7.2.10. COROLLARY. *Every unital spectrally Gelfand (respy. t. spectrally Gelfand) algebra A has the Wiener property: $x \in A$ is invertible iff $\chi(x) \neq 0$ for every $\chi \in \Delta$ (respy. every $\chi \in \Delta_c$).*

PROOF. An element $x \in A$ is invertible iff $x-e$ is q. invertible iff (by 7.2.9) $\chi(x - e) \neq -1$, i.e. $\chi(x) \neq 1 - 1 = 0$, for every $\chi \in \Delta$ (respy. Δ_c).

7.2.11. Remark. The commutative unital ρ-Banach algebra W^ρ considered in 3.4.10 has the Wiener property by 7.2.10, since it is spectrally Gelfand by 7.2.17, 7.2.8 (b).

7.2.12. PROPOSITION. *In a spectrally Gelfand algebra we have:*

$$\sqrt{A} = \sqrt[s]{A}^{\dagger} = A^{qn\dagger} = \sqrt[h]{A}^{\dagger}. \tag{*}$$

Further, in a t. spectrally Gelfand algebra we have also

$$\sqrt[h]{A} = \sqrt[ch]{A}^{\dagger} \tag{**}$$

so that $\sqrt{A}$ is closed.

PROOF. By 1.5.12,

$$\sqrt{A} \subseteq \sqrt[s]{A} \subseteq \sqrt[h]{A}. \tag{1}$$

Also, by 1.7.15 (iii),

---

† For definitions see 1.5.11, 1.7.13, 1.3.7, 2.2.18.

$$\sqrt{A} \subseteq A^{qn} \subseteq \sqrt[h]{A}. \tag{2}$$

If $x \in \sqrt[h]{A}$ it is clear that $x \in M_\chi$, $\chi(x) = 0 \neq -1$ for every $\chi \in \Delta$, whence by 7.2.9, x is q. invertible. Thus $\sqrt[h]{A}$ is a q. invertible ideal, whence by 1.2.24 (b), $\sqrt[h]{A} \subseteq \sqrt{A}$. This together with (1), (2) proves $(*)$.

Now let A be t. spectrally Gelfand. Then, by $(*)$ of 2.2.18, $\sqrt{A} \subseteq \sqrt[h]{A} \subseteq \sqrt[ch]{A}$.

Therefore, to prove $(**)$ it is enough to show that $\sqrt[ch]{A} \subseteq \sqrt{A}$ and this is done exactly as above for $\sqrt[h]{A}$ (with Δ_c replacing Δ).

7.2.13. COROLLARY. *A t. spectrally Gelfand algebra – in particular a Gelfand algebra – is s.s. iff it is q.s.s. iff it is h.s.s..*

PROOF. This is an immediate consequence of $(*)$ of 7.2.12.

7.2.14. PROPOSITION. *A t. spectrally Gelfand ρ-seminormed s.s. algebra A is ρ-normed.*

PROOF. Let us take $A = (A, p)$. Then, if $p(x) = 0$ we have: $x, x, \cdots \to 0$. If $\chi \in \Delta_c$ then $\chi(x) \to 0$, i.e. $\chi(x) = 0$. So

$$\ker p \subseteq \sqrt[ch]{A} = \sqrt{A} = \{0\},$$

whence p is faithful and p is a ρ-norm.

7.2.15. PROPOSITION. *Let $A \neq \sqrt{A}$ be either a complex or formally real commutative TA. Then A is a Gelfand algebra if it satisfies either of the two conditions:*

A is ρ-seminormed and hyponormal. $(*)$

A is locally sm. convex and CQ. $(**)$

PROOF. Let M be a maximal regular ideal with relative unity u. Then the quotient $A^\# = A/M$ has $u^\#$ as unity and $A^\#$ is a division algebra (see 1.2.5).

Let now A satisfy $(**)$. Then M is closed in A (by hyponormality) and so $A^\#$ is ρ-normed. By 6.5.6, 6.5.12, $A^\# = \mathbb{K}e$ and M is hypermaximal.

Let next A satisfy $(*)$. Then M is closed (by 7.1.9) so that $A^\#$ is a Hausdorff division algebra. Further, $A^\#$ is locally sm. convex (by 4.4.12 (ii)) and CQ (by 3.6.27). Also, by 1.6.21, $A^\#$

is formally real whenever A is formally real. It now follows, by 6.5.4, 6.5.8, that $A^{\#} = \mathbb{K}e$, M is hypermaximal.

Thus we have shown that when either of the conditions $(*), (**)$ is satisfied M is closed hypermaximal, so that A is Gelfand.

7.2.16. COROLLARY. *Every complex or formally real unital commutative algebra satisfying either the condition* $(*)$ *above or the condition*

$$A \text{ is locally sm. convex and } CI \qquad (**1)$$

is a Gelfand algebra.

PROOF. Since A is unital, $A \neq \sqrt{A}$ and further, by 3.6.22 the hypothesis "A is CI" is equivalent to "A is CQ". The corollary now follows from the proposition.

7.2.17. COROLLARY. *Every complex or formally real commutative* ρ*-Banach algebra which is either unital or at least has* $A \neq \sqrt{A}$ *is a Gelfand algebra.*

PROOF. By 7.1.9, A is hyponormal. The desired conclusion now follows from 7.2.15 since A satisfies the condition $(*)$ therein.

7.2.18. PROPOSITION. *Let* A *be a strictly real algebra and* $\tilde{A}$ *its complexification. Then* $\tilde{A}$ *is a Gelfand algebra iff* A *is a Gelfand algebra.*

PROOF. This follows from 1.9.16 (ii), the result (by virtue of 1.9.18) that $M + iM$ is hypermaximal iff M is hypermaximal, and the elementary observation that $M + iM$ is closed in $\tilde{A}$ iff M is closed in A.

7.2.19. COROLLARY. *Every commutative strictly real* ρ*-Banach algebra* A *with* $A \neq \sqrt{A}$ *is a Gelfand algebra.*

PROOF. By 1.9.17 the complexification $\tilde{A}$ of A satisfies the condition $\tilde{A} \neq \sqrt{\tilde{A}}$. Therefore, by 7.2.17, $\tilde{A}$ is Gelfand and so, by 7.2.18, A is Gelfand.

7.2.20. Remark. $\mathbb{C}$ is a 2-dimensional real Banach algebra $\mathbb{C}^{[\mathbb{R}]}$ which is a division algebra. $\mathbb{C}^{[\mathbb{R}]}$ is not a Gelfand algebra since the ideal $\{0\}$ which is (regular) maximal ($\mathbb{C}^{[\mathbb{R}]}$ being a division algebra) is not hypermaximal (since codim $\{0\} = 2$).

Note that $\mathbb{C}^{[\mathbb{R}]}$ is neither formally real (since $i^2 + i^2 = 0$) nor strictly real (since $i^2 = -1$ is not q. invertible). This example therefore shows that a commutative real Banach algebra can fail to be Gelfand if it is neither formally real nor strictly real.

The algebra $\mathbb{H}^{[\mathbb{R}]}$ gives a 4-dim non-commutative real division algebra which again is not Gelfand (here codim $\{0\} = 4$).

7.2.21. THEOREM. *Let A be a complex or strictly real commutative pseudo-Michael algebra with $A \neq \sqrt{A}$. Then:*
(i) $\Delta_c \neq \emptyset$ *(ii) A is t. spectrally Gelfand.*

PROOF. (i) Let $A = \varprojlim \hat{A}_\alpha$ be the projective limit decomposition, where $\hat{A}_\alpha$ are ρ_α-Banach algebras. Since $A \neq \sqrt{A}$ there is an element $x \in A$ which is not q. invertible. By 4.5.6 (i) there is an α such that $x_\alpha \in \hat{A}_\alpha$ is not q. invertible. It follows that $\sqrt{\hat{A}_\alpha} \neq \hat{A}_\alpha$, and by 7.2.17, 7.2.19, $\hat{A}_\alpha$ is a Gelfand algebra. By 7.2.2 (ii), $\hat{A}_\alpha$ has a continuous character $\hat{\chi}_0$ (say). Then $\chi_\alpha = \hat{\chi}_0 \circ \varphi_\alpha$ is a continuous character of A, so that $\Delta_c \neq \emptyset$.

(ii) Suppose that $\lambda \in \sigma'_A(x)$, $\lambda \neq 0$. Then, by 4.5.7 (i), $\lambda \in \sigma'_{\hat{A}_\alpha}(x_\alpha)$ for some α. But $\hat{A}_\alpha$ being Gelfand is t. spectrally Gelfand and there is a (continuous) character $\hat{\chi}_\alpha \in \hat{\Delta}_\alpha = \Delta(\hat{A}_\alpha)$, with $\hat{\chi}_\alpha(x_\alpha) = \lambda$. Then $\chi = \hat{\chi}_\alpha \circ \varphi_\alpha \in \Delta_c$ and $\chi(x) = \lambda$. Further, when A is unital each $\hat{A}_\alpha$ is also unital (by 4.5.6 (ii)). In this case if $0 \in \sigma(x)$ then x is not invertible, so that some x_α is not invertible in $\hat{A}_\alpha$, whence $0 \in \hat{\sigma}_\alpha(x_\alpha)$. Since $\hat{A}_\alpha$ is Gelfand, by 7.2.2 (v) there is a $\hat{\chi}_\alpha \in \hat{\Delta}_\alpha$ with $\hat{\chi}_\alpha(x_\alpha) = 0$. Then $\chi = \hat{\chi} \circ \varphi_\alpha \in \Delta_c$ and $\chi(x) = \hat{\chi}_\alpha(x_\alpha) = 0$. On the other hand if $\lambda \in \sigma(x)$, $\lambda \neq 0$ then $\lambda \in \sigma'(x)$ and so by what has been proved above there is a $\chi \in \Delta_c$ with $\chi(x) = \lambda$. This completes the proof.

7.2.22. PROPOSITION. *Let A be spectrally Gelfand – in particular Gelfand. Then we have:*

(1) *For any $x, y \in A$,*

(i) $\sigma'(x + y) \subseteq \sigma'(x) + \sigma'(y)$ (ii) $\sigma'(xy) \subseteq \sigma'(x)\sigma'(y)$;

(iii) $\sigma(x+y) \subseteq \sigma(x)+\sigma(y)$ (iv) $\sigma(xy) \subseteq \sigma(x)\sigma(y)$
(*here we assume A is unital*);

(v) $r(x+y) \leqslant r(x)+r(y)$ (vi) $r(xy) \leqslant r(x)r(y)$.

(2) *For any two idempotents $u, v (u \neq v)$ of A such that $u \leftrightarrow v$,*

(vii) $\{0\} \subset \sigma'(u-v) = \{0, \pm 1\}$; $r(u-v) = 1$.

If A is unital and $u, v \neq 0$ or e then

(viii) $\{0\} \subset \sigma(u-v) \subseteq \{0, \pm 1\}$.

PROOF. If $\chi \in \Delta$ then

$$\chi(x+y) = \chi(x) + \chi(y) \text{ and } \chi(xy) = \chi(x)\chi(y).$$

From these relations and the definition of spectrally Gelfand the relations (i)-(iv) follow. Also, results (v), (vi) follow from (i), (ii) respectively.

It remains to consider (vii), (viii). By 1.7.9 we have: $\sigma'(u), \sigma'(v) \subseteq \{0,1\}$. Therefore, by (i),

$$\sigma'(u-v) \subseteq \sigma'(u) + \sigma'(-v) = \sigma'(u) - \sigma'(v) \subseteq \{0, \pm 1\}.$$

If we now show that $\sigma'(u-v) \neq \{0\}$ then (vii) will follow.

Suppose to the contrary $\sigma'(u-v) = \{0\}$. Then

$$\sigma'(u-uv) = \sigma'(u(u-v)) \subseteq \sigma'(u)\sigma'(u-v) \subseteq \sigma'(u)\{0\} = \{0\}.$$

Therefore $u - uv \in A^{qn} = \sqrt{A}$ (see 7.2.12). Since $u \leftrightarrow v$, $u - uv$ is easily checked to be an idempotent. Since it is in $\sqrt{A}$, by 1.2.24 (d), $u - uv = 0$. Similarly $v - uv = 0$. This, $u = v$ – a contradiction, proving (vii). The result (viii) is proved by similar arguments.

7.2.23. COROLLARY. (Michael[†]). *Let A be a complex or strictly real pseudo-Michael algebra. Then for $x, y \in A$ with $x \leftrightarrow y$, results* (i) - (vi) *of* 7.2.22 *hold. Further, results* (vii), (viii) *of* 7.2.22 *also hold for* A.

PROOF. We may assume that $A \neq \sqrt{A}$ (if $A = \sqrt{A}$ the results hold either trivially or vacuously). Let A_m be a maximal

---

[†] He considered and proved the results only for complex Michael algebras.

commutative subalgebra containing x, y. Then we know by 1.7.26 that for $a \in A_m$

$$\sigma'_{A_m}(a) = \sigma'_A(a);\ \sigma_{A_m}(a) = \sigma_A(a)$$

(when A is unital). So we may assume that A is commutative. But then, by 7.2.21, A is spectrally Gelfand and so the desired results follow from 7.2.22.

§3. The Gelfand Representation

7.3.1 Let A be a TA and Δ (respy. Δ_c) the set of all (respy. all continuous) characters of A. As remarked earlier Δ or Δ_c may be empty. In the sequel we assume that they are non-empty. We topologize Δ by equipping it with the weak topologize (or topology of simple convergence): thus a net $\chi_\alpha \to \chi$ iff $\chi_\alpha(x) \to \chi(x)$ for every $x \in A$ $(\chi_\alpha, \chi \in \Delta)$. Δ_c as a subset of Δ inherits the relative topology. Let $\mathcal{M}$ (respy. $\mathcal{M}_c$) denote the set of hypermaximal (respy. closed hypermxaximal) ideals of A. Since there is a bijection between Δ (respy. Δ_c and $\mathcal{M}$ (respy. $\mathcal{M}_c$) (see 1.3.9, 2.1.30) the weak topology on Δ (respy. Δ_c can be transferred to $\mathcal{M}$ (respy. $\mathcal{M}_c$; $\mathcal{M}_c$ is a subspace of $\mathcal{M}$.

We refer to Δ or Δ_c as *Gelfand space* or *spectrum* of A; we refer to $\mathcal{M}$ or $\mathcal{M}_c$ as the *maximal ideal spectrum* of A.

If A_1 is the unitization of A we denote the corresponding spectra by Δ_1, Δ_{1c}, $\mathcal{M}_1, \mathcal{M}_{1c}$.

7.3.2. LEMMA. (a) $\Delta \simeq \Delta_1 \backslash \{\chi_0\}, \Delta_c \simeq \Delta_{1c} \backslash \{\chi_0\}$, *where* χ_0 *is the distinguished character of* A_1 *and* $\simeq$ *denotes homeomorphism.*

(b) Δ *is homeomorphic to a subspace of the Cartesian product* $K = \prod K_x$ $(x \in A)$, *where each* $K_x = \mathbb{K}$. *In particular,* Δ, Δ_c *are Hausdorff completely regular spaces.*

PROOF. (a) The homeomorphims are given by $\chi \mapsto \chi_1$, where χ_1 denotes the unique extension of χ (see 1.3.17, 1.3.9, 1.4.9).

(b) The map

$$\chi \mapsto (\cdots, \chi(x)(\in K_x), \cdots)$$

of Δ into K is clearly bijective. Moreover, the weak topology of Δ can be indentified with the relative topology induced by the product topology of K on the image of Δ under the above map. This means that the above map is a homeomorphism. Since each $K_x = \mathbb{R}$ or $\mathbb{C}$ is Hausdorff completely regular it follows that Δ has also the same properties (since these properties are preserved by products and subspaces).

7.3.3. PROPOSITION. *Let A be a strictly real TA and $\tilde{A}$ its complexification. Then the map*

$$\Lambda : \chi \in \Delta(A) \mapsto \tilde{\chi} \in \Delta(\tilde{A}) \text{ (as defined in 1.9.17)}$$

is a homeomorphism. Similarly, if $\Lambda_c = \Lambda|\Delta_c$ (restriction of Λ to the subspace Δ_c) then

$$\Lambda_c : \Delta_c(A) \to \Delta_c(\tilde{A})$$

is a homeomorphism.

PROOF. We have already seen that in 1.9.18 that Λ is a bijection. Since

$$\tilde{\chi}(z) = \chi(x) + i\chi(y) \ (z = x + iy,\ z \in \tilde{A}; x, y \in A)$$

$\tilde{\chi}$ is continuous iff χ is continuous. So Λ_c is also a bijection. Finally, it is clear that

$$\tilde{\chi}_\alpha \to \tilde{\chi} \text{ iff } \chi_\alpha \to \chi.$$

Thus, Λ and Λ_c are homeomorphisms.

7.3.4. Denote by $C(\Delta)$ the algebra of $\mathbb{K}$-valued continuous functions on Δ; these functions can also be regarded as functions on $\mathcal{M}$ (since $\mathcal{M}$ can be identified with Δ). For each $x \in \Delta$, define $\hat{x}$ on $\Delta(= \mathcal{M})$ by

$$\hat{x}(\chi) = \chi(x) = \hat{x}(M) = x(M) \ (M = \ker \chi).$$

It is clear from the definition of the weak topology that $\hat{x} \in C(\Delta) = C(\mathcal{M})$; $\hat{x}$ is called the *Gelfand transform* of x. We write $\hat{A} = \{\hat{x} : x \in A\}$ and call $\hat{A}$ the *transform algebra* of A. We can consider $\hat{x}$ operating on Δ or $\mathcal{M}$ according to our contextual needs. We can also clearly make each $\hat{x}$ operate on Δ_c or $\mathcal{M}_c$. The algebra $C(\Delta)$ is a TA under the weak toplogy. This means that if $f_\alpha, f \in C(\Delta)$ then $f_\alpha \to f$ iff $f_\alpha(\chi) \to f(\chi)$ for each $\chi \in \Delta$. Similarly the algebra $C(\Delta_c)$ is a TA under its weak topology.

7.3.5. PROPOSITION. (i) *The map $\mathcal{G} : x \mapsto \hat{x}$ is a homomorphism of A into $C(\Delta)$ and the map $\mathcal{G}_c : x \mapsto \hat{x}$ is a continuous homomorphism of A into $C(\Delta)$.*

(ii) $\ker \mathcal{G} = \sqrt[h]{A}$ *($= \sqrt{A}$ if A is sepctrally Gelfand).*

(iii) *$\mathcal{G}$ is injective iff A is h.s.s. (=s.s. if A is spectrally Gelfand).*

(iv) *If A is functionally continuous then $\mathcal{G}$ $(= \mathcal{G}_c)$ is a continuous homomorphism.*

PROOF. (i) That $\mathcal{G}$, $\mathcal{G}_c$ are homomorphisms follow from the fact that each $\chi \in \Delta$ or Δ_c is a homomorphism of A. Further, if $x_\alpha \to x$ in A and $\chi \in \Delta_c$ then

$$\hat{x}_\alpha(\chi) = \chi(x_\alpha) \to \chi(x) = \hat{x}(\chi).$$

Hence $\mathcal{G}_c$ is continuous.

(ii) $\ker \mathcal{G} = \bigcap \mathcal{M}_\chi (\chi \in \Delta) = \sqrt[h]{A} = \sqrt{A}$ (when A is spectrally Gelfand, see 7.2.12).

(iii) This is immediate consequence of (ii).

(iv) When A is funtionally continuous we have $\Delta = \Delta_c$, so that $\mathcal{G} = \mathcal{G}_c$ is continuous.

7.3.6. COROLLARY. *Every real h.s.s. TA (or s.s. t. spectrally Gelfand algebra) A is formally real.*

PROOF. By 7.3.5 ((i), (iii)), A is (algebraically) isomorphic to $\hat{A}$. Since elements of $\hat{A}$ are real-valued functions, by 1.6.18, $\hat{A}$ and hence A is formally real.

7.3.7. COROLLARY. *A commutative strictly real s.s., pseudo-Michael algebra (in particular, ρ-Banach algebra) A is formally real.*

PROOF. By 7.2.21, A is t. spectrally Gelfand and so the result follows from 7.3.6.

7.3.8. PROPOSITION. *If $A = (A,p)$ is a ρ-seminormed algebra, with p sm., then*

$$|\chi(x)| \leqslant \nu(x)^{\frac{1}{\rho}} \leqslant p(x)^{\frac{1}{\rho}} \ (x \in A, \chi \in \Delta_c). \qquad (*)$$

Hence $\|\chi\| \leqslant 1$; *if A is unital with $p(e) = 1$, then* $\|\chi\| = 1$.

PROOF. If $M = \ker\chi$ then $A_\chi = A/M \simeq \mathbb{K}$. Since χ is continuous, M is closed and consequently the identity coset $E = u+M$ (where u is an element of A such that $\chi(u) = 1$) is closed. Since $E^2 = E$ we have for any $y \in E$, $y^n \in E$ $(n = 1,2,\cdots)$.

Since E is disjoint with $M, 0 \notin E$. It follows that $y^n \not\to 0$ (since $y^n \in E$ and E is closed). Therefore, by 3.3.4, we have: $p(y^n) \geqslant 1$. If $x \in A$ and $\chi(x) = \lambda \neq 0$, then $\lambda^{-1}x \in E$. So

$$p(\lambda^{-n}x^n) = p((\lambda^{-1}x)^n) \geqslant 1,$$

which gives

$$p(x^n) \geqslant |\lambda|^{n\rho} = |\chi(x)|^{n\rho},$$

so that we have

$$|\chi(x)| \leqslant (p(x^n)^{\frac{1}{n}})^{\frac{1}{\rho}}.$$

Making $n \to \infty$, we get $|\chi(x)| \leqslant \nu(x)^{\frac{1}{\rho}} \leqslant p(x)^{\frac{1}{\rho}}$, which is $(*)$. Futher, Definition 3.5.1. and $(*)$ imply that $\|\chi\| \leqslant 1$, and $\|\chi\| = 1$ when A is unital with $p(e) = 1$ (since $\chi(e)|p(e)^{\frac{1}{\rho}} = 1$).

7.3.9. COROLLARY. *Δ_c is an equicontinuous family of $\mathbb{K}$-valued functions on A.*

PROOF. For $\chi \in \Delta_c$ we have

$$|\chi(x) - \chi(y)| = |\chi(x-y)| \leqslant p(x-y)^{\frac{1}{\rho}} \ (x, y \in A). \qquad (*)$$

The equicontinuity of Δ_c is therefore a consequence of the continuity of p.

7.3.10. COROLLARY. *If $N = \ker p$, $A^\# = A/N$ then $\Delta_c(A)$ is canonically homeomorphic to $\Delta_c(A^\#)$.*

PROOF. For $\chi \in \Delta(A)$ define $\chi^\# \in \Delta(A^\#)$ by $\chi^\#(x+N) =$

$\chi(x)$. This is well-defined since χ vanishes on N (see 3.1.21 (v)). The map $\chi \mapsto \chi^{\#}$ is clearly a homeomorphism.

7.3.11. PROPOSITION. *Let $A = (A, p)$ be a ρ-seminormed algebra and $\hat{A} = (\hat{A}, \hat{p})$ be its completion. Then each $\chi \in \Delta_c(A)$ has a unique extension $\hat{\chi} \in \Delta_c(\hat{A})$ and the map $\chi \mapsto \hat{\chi}$ is a homeomorphism of $\Delta_c(A)$ onto $\Delta_c(\hat{A})$.*

PROOF. The existence of the unique extension $\hat{\chi}$ of χ follows from 3.5.11 (b). Again, if $\hat{\chi} \in \Delta_c(\hat{A})$ then $\ker \hat{\chi} \neq \hat{A}$ and being closed we have $A \not\subseteq \ker \hat{\chi}$, whence $\hat{\chi}|A \in \Delta_c(A)$. Thus, the map $\chi \mapsto \hat{\chi}$ ($\hat{\chi} =$ extension of χ) is a bijection. It remains to show that the map is bi-continuous. Suppose that $\chi_\alpha \to \chi$. If $\hat{x} \in \hat{A}$ then $\hat{x} = \lim_{n\to\infty} x_n$, $x_n \in A$. Given $\epsilon > 0, 0 < \epsilon < 1$, we can find a n such that

$$\hat{p}(\hat{x} - x_n) < \frac{\epsilon}{3}. \tag{1}$$

Since $\chi_\alpha(x_n) \to \chi(x_n)$, there is an α_0 such that for $\alpha \succ \alpha_0$,

$$|\chi_\alpha(x_n) - \chi(x_n)| < \frac{\epsilon}{3}. \tag{2}$$

Using the inequality $(*)$ of 7.3.9 for $\hat{p}$, we have

$$|\hat{\chi}_\alpha(\hat{x}) - \hat{\chi}_\alpha(x_n)| < \hat{p}(\hat{x} - x_n)^{\frac{1}{\rho}} < \left(\frac{\epsilon}{3}\right)^{\frac{1}{\rho}} < \frac{\epsilon}{3}, \tag{3}$$

$$|\hat{\chi}(x_n) - \hat{\chi}(\hat{x})| < \frac{\epsilon}{3}. \tag{4}$$

From (2), (3), (4) we obtain, for $\alpha > \alpha_0$,

$$\begin{aligned} |\hat{\chi}_\alpha(\hat{x}) - \hat{\chi}(\hat{x})| &\leqslant |\hat{\chi}_\alpha(\hat{x}) - \hat{\chi}_\alpha(x_n)| + |\chi_\alpha(x_n) - \chi(x_n)| + \\ &\qquad |\hat{\chi}(x_n) - \hat{\chi}(\hat{x})| \\ &< \frac{\epsilon}{3} + \frac{\epsilon}{3} + \frac{\epsilon}{3} = \epsilon. \end{aligned}$$

This implies that $\hat{\chi}_\alpha \to \hat{\chi}$. Thus we have shown that $\chi_\alpha \to \chi \Rightarrow \hat{\chi}_\alpha \to \hat{\chi}$. On the other hand, the reverse implication is trivial since $\chi_\alpha = \hat{\chi}_\alpha|A$, $\chi = \hat{\chi}|A$. Thus, the map $\chi \mapsto \hat{\chi}$ is bicontinuous, as we wished to show.

7.3.12. PROPOSITION. *The spectrum* Δ_c *of a* ρ*-seminormed algebra* A *is locally compact Hausdorff. If* A *is unital, or more generally, if* $\sqrt[ch]{A}$ † *is regular then* Δ_c *is compact Hausdorff.*

PROOF. By 7.3.2, we can identify Δ_c with a subspace of the carterian product $K = \prod K_x \quad (x \in A)$. Write $S_x = \{\lambda \in K_x : |\lambda| \leqslant p(x)^{\frac{1}{\rho}}\}$ and $S = \prod S_x \quad (x \in A)$. Then S is a compact subset of K. By the inequality $(*)$ of 7.3.8, we have: $\Delta_c \subseteq S \subseteq K$. Denote by $\overline{\Delta}_c$ the closure of Δ_c in $S(\text{or}K)$. Take a point $\Lambda = (\lambda_x) \in \overline{\Delta}_c$ and write $\Lambda(x) = \lambda_x$. Then $\Lambda(x) = \lim \chi_\alpha(x)$ for some net (χ_α) in Δ_c. Since each χ_α is a homomophism so is Λ. Also, since

$$|\chi_\alpha(x)| \leqslant p(x)^{\frac{1}{\rho}}$$

we get

$$|\Lambda(x)| \leqslant p(x)^{\frac{1}{\rho}}.$$

Thus, Λ is bounded and so, by 3.5.5, continuous. By 1.3.3, Λ is either the zero homomorphism h_0 or $\Lambda \in \Delta_c$. Therefore

$$\overline{\Delta}_c = \Delta_c \text{ or } \overline{\Delta}_c = \Delta_c \bigcup \{h_0\}.$$

Since S is compact and $\overline{\Delta}_c$ is closed in S, $\overline{\Delta}_c$ is compact. Thus Δ_c is either compact or locally compact (with $\overline{\Delta}_c$ as its 1-point compactification). Also, by 7.3.2, Δ_c is Hausdorff.

Finally, if $\sqrt[ch]{A}$ is regular, with relative unity u, then $xu - x \in \sqrt[ch]{A} \subseteq \ker \chi \ (\chi \in \Delta_c)$, whence

$$\chi(x)\chi(u) = \chi(x), \chi(u) = 1.$$

It follows that if $\chi_\alpha \to \Lambda$ then $\Lambda(u) = \lim \chi_\alpha(u) = 1$, so that $\Lambda \in \Delta_c$ and $\Delta_c = \overline{\Delta_c}$ is compact.

7.3.13. COROLLARY. *Any subset* Δ_ϵ *of* Δ_c *of the form* $\Delta_\epsilon = \{\chi \in \Delta_c : |\chi(x_0)| \geqslant \epsilon\}$, *for some fixed* $x_0 \in A$ *and* $\epsilon > 0$, *is compact.*

PROOF. Clearly Δ_ϵ is closed in Δ. Let $\Delta_\infty = \Delta \bigcup \{h_0\}$ be the 1- point compactification of Δ. Since $h_0(x_0) = 0$, while

† For defintion see 2.2.18.

$|\chi(x_0)| \geqslant \epsilon$ for $\chi \in \Delta_\epsilon$ it follows that h_0 is not a limit point of Δ_ϵ. Therefore Δ_ϵ is closed in Δ_∞ and hence is compact.

7.3.14. LEMMA. Let a ρ-seminormed algebra A be a direct sum $A = A_1 \oplus A_2$, with $A_j (j = 1, 2)$ closed bi-ideals of A. Then the spectrum $\Delta_c = \Delta_c(A)$ has a decoposition

$$\Delta_c = \Delta_{1c}^0 \bigcup \Delta_{2c}^0$$

where $\Delta_{j,c}^0$ are (disjoint) closed subspaces such that

$$\{\chi | A_j : \chi \in \Delta_{j,c}^0\} = \Delta_{j,c} = \Delta_c(A_j) \quad (j = 1, 2).$$

PROOF. The above decomposition into subspaces follows from 1.3.18. It remains to show that they are closed. But this is quite easy. For, if $\chi \in \Delta_c$ then $\chi \in \Delta_{1c}^0$ iff $\chi | A_2 = 0$. If $\chi_\alpha | A_2 = 0$ and $\chi_\alpha \to \chi$ then for $y \in A_2$

$$\chi(y) = \lim \chi_\alpha(y) = 0, \text{ so that}$$

$\chi \in \Delta_{1c}^0$, proving Δ_{1c}^0 is closed. Similarly, Δ_{2c}^0 is closed.

7.3.15. DEFINITION. A subset S of a TA A which has no unity is said to *topologically generate* or *t. generate* A if A is the smallest closed subslgebra containing S; S is called a *t. generating set* of A. If A is *unital*, with unity e, then the condition is modified by requiring that A is the smallest closed subalgebra containing S and e.

A TA is said to be *finitely* or *countably t. generated* according as it has a finite or a countable t. generating set of elements.

7.3.16. PROPOSITION. *Let Δ_c be the spectrum of a ρ-seminormed algebra $A = (A, p)$ and S a t. generating set of A. If $(\chi_\alpha), \chi_0 \in \Delta_c$ then for $\chi_\alpha \to \chi_0$ in Δ_c it is sufficient if*

$$\chi_\alpha(y) \to \chi_0(y) \text{ for each } y \in S. \tag{$*$}$$

PROOF. Suppose that $(*)$ holds. Then given $x \in A$ and $0 < \epsilon < 1$, by definition of t. generating set, there is a polynomial P in, say, n (possibly) non-commuting variables and elements $y_1, \cdots, y_n$ in S such that

$$p(x - P(y_1, \cdots, y_n)) < \frac{\epsilon}{3}. \tag{1}$$

Further we have

$$|\chi_\alpha(x) - \chi_0(x)| \leqslant |\chi_\alpha(x) - \chi_\alpha(P(y_1, \cdots, y_n))| + |\chi_\alpha(P(y_1, \cdots, y_2)) - \chi_0(P(y_1, \cdots, y_n))| + |\chi_0(P(y_1, \cdots, y_n)) - \chi_0(x)|. \tag{2}$$

Now

$$\begin{aligned}|\chi_\alpha(x) - \chi_\alpha(P(y_1, \cdots, y_n))| &\leqslant |\chi_\alpha(x - P(y_1, \cdots, y_n))| \\ &\leqslant p(x - P(y_1, \cdots, y_n))^{\frac{1}{\rho}} \\ &\qquad (\text{ by } (*) \text{ of } 7.3.8) \\ &\leqslant \left(\frac{\epsilon}{3}\right)^{\frac{1}{\rho}} < \frac{\epsilon}{3} (\text{ since } 1/\rho \geqslant 1). \end{aligned} \tag{3}$$

Similarly

$$|\chi_0(P(y_1, \cdots, y_n)) - \chi_0(x)| < \frac{\epsilon}{3}. \tag{4}$$

By our hypothesis

$$\chi(y_k) \to \chi_0(y_k) \qquad (k = 1, \cdots, n)$$

so that, since P is a polynomial, we have

$$\begin{aligned}\chi_\alpha(P(y_1, \cdots, y_n)) &= P(\chi_\alpha(y_1), \cdots, \chi_\alpha(y_n)) \\ &\to P(\chi_0(y_1), \cdots, \chi_0(y_n)) = \chi_0(P(y_1, \cdots, y_n)).\end{aligned}$$

So we can choose α_0 such that for $\alpha \succ \alpha_0$ we have

$$|\chi_\alpha(P(y_1, \cdots, y_n)) - \chi_0(P(y_1, \cdots, y_n))| < \frac{\epsilon}{3}. \tag{5}$$

From the inequalities (2)-(5) we obtain, for $\alpha \succ \alpha_0$,

$$|\chi_\alpha(x) - \chi_0(x)| < \frac{\epsilon}{3} + \frac{\epsilon}{3} + \frac{\epsilon}{3} = \epsilon,$$

whence $\chi_\alpha(x) \to \chi(x)$, completing the proof.

7.3.17. COROLLARY. *The spectrum Δ_c of a separable ρ-seminormed algebra A is metrizable.*

PROOF. Let $\{x_n\}$ be a countable dense subset of A. Set, for $\chi_1, \chi_2 \in \Delta_c$

$$d(\chi_1, \chi_2) = \sum_{n=1}^{\infty} \frac{1}{2^n} \frac{|\chi_1(x_n) - \chi_2(x_n)|}{1 + |\chi_1(x_n) - \chi_2(x_n)|}. \tag{$*$}$$

Then d is a metric. For, if $d(\chi_1, \chi_2) = 0$ then $\chi_1(x_n) = \chi_2(x_n)$, for all x_n, so that by density of $\{x_n\}$ and continuity of χ_1, χ_2 it follows that $\chi_1 = \chi_2$. The symmetry property of d is immediate from the definition of d. Finally, for the triangle inequality property of d it is clearly enough to prove that each of the summands in $(*)$ satisfies the triangle inequality. But this can be established on the same lines as in the proof of 3.1.10.

Now it is clear that a net $\chi_\alpha \to \chi$, under d, iff

$$\chi_\alpha(x_n) \to \chi(x_n) \qquad (n = 1, 2, \cdots). \tag{$*$}$$

The set $\{x_n\}$, being a dense subset of A, is clearly a t. generating set of A. So, by 7.3.16, condition $(*)$ is equivalent to the convergence of the net $\chi_\alpha \to \chi$ under the topology of Δ_c. Thus, the metric topology coalesces with the topology of Δ_c, completing the proof.

7.3.18. COROLLARY. *The spectrum Δ_c of a countably t. generated ρ-seminormed algebra A is separably metrizable.*

PROOF. If S is a countable set of t. generators of A then the set S_1 of all finite products of elements of S, is countable. Denote by S_2 the set of all rational linear combination of elements from S_1 (where a rational linear combination in the complex case means the coefficients of the combination have rational real and imaginary parts). S_2 is clearly countable and dense in A, whence the required result follows from 7.3.17.

7.3.19. THEOREM. *Let A be a pseudo-Michael algebra with projective limit decomposition*

$$A = \varprojlim \hat{A}_\alpha \ (\hat{A}_\alpha \ \textit{pseudo-Banach algebras}).$$

Let $\Delta_c, \hat{\Delta}_\alpha$ *denote the sets of continuous characters of* $A, \hat{A}_\alpha$ *respy.. Let* Δ_α *denote the subset of* Δ_c *comprising characters which are* p_α*-continuous.*[†] *Then:*

(i) $\Delta_c = \bigcup_\alpha \Delta_\alpha$;

(ii) Δ_α *homeomorphic to* $\hat{\Delta}_\alpha$;

(iii) *If we define* $\psi_{\alpha\beta} : \hat{\Delta}_\alpha \to \hat{\Delta}_\beta$ *by* $\hat{\chi}_\beta = \psi_{\alpha\beta}(\hat{\chi}_\alpha)$ *with* $\hat{\chi}_\beta(\hat{x}_\beta) = \hat{\chi}_\alpha(\hat{\varphi}_{\alpha\beta}(\hat{x}_\beta))$, *where* $\hat{\varphi}_{\alpha\beta}$ *are the maps associated with the projective decomposition:* $\hat{x}_\alpha \in \hat{A}_\alpha, \hat{x}_\beta \in \hat{A}_\beta, \hat{\varphi}_{\alpha\beta}(\hat{x}_\beta) = \hat{x}_\alpha$, *then* $\{\hat{\Delta}_\alpha; \{\varphi_{\alpha\beta}\}\}$ *is an inductive*[††] *system of topological spaces and continuous maps. Write* $\hat{\Delta}_c =$ *(the direct limit)*$\varinjlim \hat{\Delta}_\alpha$. *Then there is a bijective continuous map* $\Theta :$[‡] $\hat{\Delta}_c \to \Delta_c$ *given by* $\Theta(\hat{\chi})(x) = \chi_\alpha(x) = \hat{\chi}_\alpha(x_\alpha)$ *(for any* α*) where* $\hat{\chi} = \varinjlim \hat{\chi}_\alpha$.

PROOF. The relation (i) is an immediate consequence of 4.3.13. By 7.3.10, 7.3.11 we have $\Delta_\alpha \simeq \Delta(A_\alpha) \simeq \hat{\Delta}_\alpha$ which is (ii).

If $\Theta(\hat{\chi}_1(x)) = \chi_{1\alpha}(x_\alpha) = \Theta(\hat{\chi}_2(x)) = \chi_{2\alpha}(x_\alpha)$ for all α then $\hat{\chi}_1 = \hat{\chi}_2$, so that Θ is injective. It is also surjective. For, if $\chi \in \Delta_c, \chi \in \Delta_\alpha$ for some α and $\chi = \hat{\chi}_\alpha \circ \varphi_\alpha$ for some $\hat{\chi}_\alpha \in \hat{\Delta}_\alpha$. It follows that $\chi = \Theta(\hat{\chi_\alpha})$. Finally, the continuity of Θ is clear from its definition.

7.3.20. COROLLARY. *If* A *is unital, each* Δ_α *is compact.*

PROOF. By 4.5.6(ii) each $\hat{\Delta}_\alpha$ is unital and consequently, by 7.3.12, $\hat{\Delta}_\alpha$ is compact, whence $\Delta_\alpha \simeq \hat{\Delta}_\alpha$ is also compact.

7.3.21. DEFINITION. A TA A which is t. generated by a single element is called *monogenic*.

Evidently a monogenic Hausdorff TA is commutative.

[†] i.e. is a continuous character of the algebra (A, p_α).

[††] For definition see [12, pp.184-5].

[‡] In general Θ is not a homeomorphism (see [12,p.161, Remark]).

7.3.22. Examples of monogenic TA's.

(i) The Banach algebra $\mathsf{C}[0,1]$. This has $f_0(t) = t \ (t \in [0,1])$ as t. generator. That f_0 is a t. generator is a consequence of the Weieritrass approximation theorem.

(ii) (cf. [10, p.33]). The algebra $\mathfrak{A}$ of complex valued continuous function on unit disc $|z| \leqslant 1$ which are holomorphic on $|z| < 1$ is a Banach algebra under the sup norm. This algebra has $f_0(z) = z$ as a t. generator. To see this observe that if $f_\epsilon(z) = f(z/1+\epsilon) \ (\epsilon > 0)$ then f can be uniformly approximated by f_ϵ in $|z| \leqslant 1$. Also, each f_ϵ being holomorphic on $|z| < 1+\epsilon$ can be uniformly approximated by polynomials in z in $|z| \leqslant 1$, whence f_0 is a t. generator.

(iii) The Banach algebra $\mathsf{L}^1 = \mathsf{L}^1[0,1]$ (see 3.4.6 (vi)). If f_1 denotes the constant function $1, f_1(t) = 1 \ (t \in [0,1])$, then it is easy to see that if $f_1^n = f_1 * \cdots * f_1$ (n factors), $f_1^n(s) = s^{n-1}/(n-1)$. It follows from Weirstrass approximation theorem and the fact that $\mathsf{C}[0,1]$ is dense in L^1 that f_1 is a t. generator of L_1.

(iv) The subalgebra W_+^ρ of W^s (defined in 3.4.10) consisting of all elements $f = \sum\limits_{n\in\mathbb{Z}} \hat{f}(n)e^{int}$ with $\hat{f}(n) = 0$ for $n < 0$, is closed and so a ρ-Banach algebra. It has e^{it} as a t. generator as can be easily seen.

7.3.23. PROPOSITION. *Let A be a monogenic unital ρ-Banach algebra which is either complex, or when real is strictly real or formally real. Let $\Delta = \Delta_c$ be the spectrum of A and a a t. generator of A. Then the map $\Lambda : \chi : \chi(a)$ is a homeomorphism of Δ onto the spectrum $\sigma(a)$.*

PROOF. By 7.2.16, 7.2.19,7.2.17 A is Gelfand and consequently, by 7.2.8(b), Λ is surjective. Λ is also injective. For, suppose that $\chi_1(a) = \chi_2(a)$. If $x \in A$ then $x = \lim_n P_n(a)$ ($P_n \in \mathbb{K}[X]$), so that

$$\begin{aligned} \chi_1(x) &= \lim_n \chi_1(P_n(a)) = \lim_n P_n(\chi_1(a)) = \lim_n P_n(\chi_2(a)) \\ &= \lim_n \chi_2(P_n(a)) = \chi_2(x), \end{aligned}$$

whence $\chi_1 = \chi_2$. That Λ is continuous follows from the definition of the topology of Δ (the weak topology). By 7.3.12, Δ is compact, and $\sigma(a)$ is clearly Hausdorff. It follows that Λ is a homeomorphism, completing the proof.

7.3.24. PROPOSITION (Shilov). *If A is a complex monogenic unital ρ-Banach algebra, with a t. generator $a, \rho(a) = \mathbb{C}\backslash\sigma(a)$ is connected.*

PROOF. Since $\sigma(a)$ is compact we can enclose it in a closed disc D. Then $\mathbb{C}\backslash D$ is connected unbounded and $\mathbb{C}\backslash D \subseteq \rho(a)$. If $\rho(a)$ is not connected it has a bounded component G_0 which is open (since $\rho(a)$ is open and A is locally connected). It follows that $\partial G_0 \subseteq \sigma(a)$. If $\lambda_0 \in G_0$ then by the maximum modulus principle applied to G_0 we obtain, for any complex polynomial P,

$$\begin{aligned} |P(\lambda_0)| &\leqslant \sup_{\lambda\in\partial G_0} |P(\lambda)| \leqslant \sup_{\chi\in\sigma(a)} |P(\lambda)| = \sup_{\chi\in\Delta} |P(\chi(a))| \\ &= \sup_{\chi\in\Delta} |\chi(P(a))| \leqslant \|P(a)\|^{\frac{1}{\rho}}. \end{aligned}$$

Let A_0 be the subalgebra of A, generated by a; A_0 is dense in A. If $x \in A_0$ then $x = P(a)$ for some $P \in \mathbb{C}[X]$. The map $\chi_0 : P(a) \mapsto P(\lambda_0)$ is a bounded homomorphism. By 3.5.11 (b) this can be extended to all of A, yielding a (continuous) character χ_0 of A. Then $\lambda_0 = \chi_0(a) \in \sigma(a)$ (by 1.7.24) while by choice of $\lambda_0, \lambda_0 \in G_0 \subseteq \rho(a)$. This contradiction proves the desired result.

7.3.25. Remark. If K is a compact subset of $\mathbb{C}$ such that $\mathbb{C}\backslash K$ is connected then it has been shown by Shilov that there then is a monogenic unital Banach algebra A such that K is homeomorphic will $\sigma(a)$, where a is a t. generator of A (see [10, p.72]).

7.3.26. PROPOSITION. *Let A be a complex monogenic unital ρ-Banach algebra with a t. generator a. Then the transform algebra $\hat{A}$ can be identified with the algebra of all continuous complex functions on $\sigma(a)$ which are holomorphic on its interior $\sigma(a)^o$.*

PROOF(cf.[28, p.412]). Since Δ is homeomorphic to $\sigma(a)$ we can regard $\hat{x}$ as a function on $\sigma(a)$: if $\lambda \in \sigma(a), \lambda = \chi(a)$ then

$\hat{x}(\lambda) = \hat{x}(\chi)(x \in A)$. Moreover, $\hat{x}$ is continuous. Thus, $\hat{A}$ can be identified with a subalgebra of the algebra $\mathsf{C} = \mathsf{C}(\sigma(a))$ of all complex continuous functions on $\sigma(a)$. If $x \in A$ and $x = \lim_{n\to\infty} P_n(a)$ then

$$\|\hat{x} - \widehat{P_n(a)}\|_\infty \leqslant \|x - P_n(a)\|^{\frac{1}{\rho}}$$

so that $\hat{x}$ is a uniform limit on $\sigma(a)$ of polynomials on $\sigma(a)$. Hence $\hat{x} = \hat{x}(\lambda)$ is holomorphic on $\sigma(a)^o$. On the other hand, if $f \in \mathsf{C}$ is holomorphic on $\sigma(a)^\circ$, then since, by 7.3.24, $\rho(a)$ is connected we can apply a theorem† of Mergelyan to conclude that f is a uniform limit on $\sigma(a)$ of polynomials, so that $f \in \hat{A}$. This completes the proof.

7.3.27. PROPOSITION. *Let $A = (A, p)$ be a ρ-seminormed algebra, with p.sm., and $C_0(\Delta_c)$ the algebra of continuous $\mathbb{K}$-valued functions on the (locally compact) spectrum Δ_c vanishing at ∞. Then:*

(i) *$\mathcal{G}_c : x \mapsto \hat{x}$ is a continuous homomorphism of A into $C_0(\Delta_c)$ such that*

$$\|\hat{x}\|_\infty \leqslant \nu(x)^{\frac{1}{\rho}} \leqslant p(x)^{\frac{1}{\rho}}. \qquad (*)$$

(ii) *If A is also t. spectrally Gelfand (in particular, Gelfand) then we have*

$$r(x) = \|\hat{x}\|_\infty < \infty, \qquad (**)$$

which implies in particular that $\sigma'(x)$ is bounded.

PROOF. We first observe that $\hat{x} \in C_0(\Delta_c)$ since $\hat{x}(h_0) = h_0(x) = 0$ and $\Delta_c \bigcup \{h_0\}$ is the 1-point compactification of Δ_c. In view of 7.3.5(i), to prove statement (i) we have only to verify the inequality $(*)$. But this is an immediate consequence of $(*)$ of 7.3.8 (since $\|\hat{x}\|_\infty = \sup_\chi |\chi(x)|$).

When A is t. spectrally Gelfand we have $\sigma'(x) = \{\chi(x) : \chi \in \Delta_c\} \bigcup \{0\}$, whence

$$\|\hat{x}\|_\infty = \sup_\chi |\chi(x)| = r(x), \text{ proving } (**).$$

7.3.28. COROLLARY. *If $\mathcal{G}_c$ is a t. isomorphism then there is a positive constant C such that*

† See [25, p.385].

$$p(x)^2 \leqslant Cp(x^2) \text{ for all } x \in A. \qquad (*)$$

PROOF. Since $\mathcal{G}_c^{-1}$ is continuous there is a constant C_0 such that $p(x)^{\frac{1}{\rho}} \leqslant C_0\|\hat{x}\|_\infty$. Therefore

$$p(x)^{\frac{2}{\rho}} \leqslant C_0^2\|\hat{x}\|_\infty^2 = C_0\|\hat{x}^2\|_\infty \leqslant C_0 p(x^2)^{\frac{1}{\rho}},$$

where in obtaining the last inequality we have used $(*)$ of 7.3.27. Thus, we obtain $p(x)^2 \leqslant Cp(x^2)$ $(C = C_0^\rho)$.

7.3.29. PROPOSITION. *Let A be a pseudo-Michael algebra with projective limit decomposition $A = \varprojlim \hat{A}_\alpha$. Then we have*

$$r(x) \leqslant \sup_\alpha \nu_\alpha(x)^{\frac{1}{\rho_\alpha}}. \qquad (*)$$

PROOF. By 4.5.5 (2), we have

$$r(x) = r_A(x) = \sup_\alpha r_{\hat{A}_\alpha}(x_\alpha). \qquad (1)$$

Since $\hat{A}_\alpha$ is a pseudo-Banach algebra the inequalities $(*), (**)$ of 7.3.27 give

$$r_{\hat{A}_\alpha}(x) \leqslant \hat{\nu}_\alpha(x_\alpha)^{\frac{1}{\rho_\alpha}} = \nu_\alpha(x)^{\frac{1}{\rho_\alpha}}. \qquad (2)$$

The desired inequality $(*)$ now follows from (1),(2).

7.3.30. COROLLARY. *A t. nilpotent*[†] *element x of a pseudo-Michael algebra A is q. nilpotent.*

PROOF. Since all $\nu_\alpha(x) = 0$ we conclude from inequality $(*)$ of 7.3.29 that $r(x) = 0$ and x is q. nilpotent, as required.

§ 4. GB Algebras

7.4.1. DEFINITION. A ρ-seminormed algebra $A = (A, p)$ with p sm., is called a *Gelfand-Beurling algebra* or a *GB algebra* if it satisfies the condition

[†] For definition see 4.8.8.

$$r(x)^\rho = \nu(x) = \lim_{n\to\infty} p(x^n)^{\frac{1}{n}} \quad (\leqslant p(x)) \tag{$*$}$$

for any $x \in A$. A ρ-Banach algebra A satisfying $(*)$ is called a GB ρ-Banach algebra.

7.4.2. Remark. By 6.2.11, 6.2.12 every complex or strictly real Banach algebra is a GB algebra.

7.4.3. LEMMA. *Let A be a ρ-seminormed GB algebra. Then we have:*

(i) *For any $x \in A$, $\sigma'(x)$ is bounded.*

(ii) *If $x, y \in A$ and $x \leftrightarrow y$ then*

$$r(x+y) \leqslant (r(x)^\rho + r(y)^\rho)^{\frac{1}{\rho}} \tag{$*$}$$

$$r(xy) \leqslant r(x)r(y). \tag{$**$}$$

PROOF. (i) Since $r(x) = \nu(x)^{\frac{1}{\rho}} < \infty$, $\sigma'(x)$ is bounded.

(ii) By 4.8.3, we have $\nu(x+y) \leqslant \nu(x)+\nu(y)$, $\nu(xy) \leqslant \nu(x)\nu(y)$. It follows that

$$\begin{aligned} r(x+y) &= \nu(x+y)^{\frac{1}{\rho}} \leqslant (\nu(x)+\nu(y))^{\frac{1}{\rho}} = (r(x)^\rho + r(y)^\rho)^{\frac{1}{\rho}} \\ r(xy) &= \nu(xy)^{\frac{1}{\rho}} \leqslant (\nu(x)\nu(y))^{\frac{1}{\rho}} = \nu(x)^{\frac{1}{\rho}}\nu(y)^{\frac{1}{\rho}} = r(x)r(y). \end{aligned}$$

7.4.4. PROPOSITION. *A unital ρ-seminormed GB algebra A is spectral.*

PROOF. We have $r(x)^\rho = \nu(x)$. If $\nu(x) > 0$ then $r(x) > 0$, so that there is a $\lambda \neq 0$ in $\sigma'(x)$. It follows that $\lambda \in \sigma(x)$ and $\sigma(x) \neq \emptyset$. If $\nu(x) = 0$, then by 3.3.7(v), x is not invertible. So $0 \in \sigma(x)$ and $\sigma(x) \neq \emptyset$.

7.4.5. PROPOSITION. *In a spectrally Gelfand GB ρ-seminormed algebra we have*

$$\|\hat{x}\|_\infty = r(x) = \nu(x)^{\frac{1}{\rho}} \quad (x \in A). \tag{$*$}$$

Hence, the concepts "essentially nilpotent", "q. nilpotent" and "t. nilpotent" coalesce and we have

$$\sqrt{A} = \{x \in A : \nu(x) = 0\} = \{x \in A : r(x) = 0\}. \tag{$**$}$$

PROOF. The relation (∗) above results by combining (∗∗) of 7.3.27 and condition (∗) of 7.4.1. The equivalence of "q. nilpotent" and "t. nilpotent" follows readily from (∗) above; that of "q. nilpotent" "essentially nilpotent" from (∗) of 7.2.12.

7.4.6. THEOREM (Beurling-Gelfand-Żelazko). *A complex or strictly real ρ-Banach algebra A is a GB algebra.*

PROOF. We have to prove (see (∗) of 7.4.1.) that

$$r(x) = \nu(x)^{\frac{1}{\rho}} \quad (x \in A). \tag{$*$}$$

Since both sides of (∗) remain the same when x is considered an element of the unitization of A, we may assume that A is unital.

Consider first the case where A is commutative. By 7.2.17,7.2.19, A is a Gelfand algebra. So, by 7.3.27, we have

$$r(x) = \|\hat{x}\|_\infty \leqslant \nu(x)^{\frac{1}{\rho}}. \tag{1}$$

Write

$$\|x\|_* = \nu(x)^{\frac{1}{\rho}}. \tag{2}$$

By 4.8.6(ii), $\|\cdot\|_*$ is a sm. semi-norm. Write $I = \ker\|\cdot\|_*$. Then $A^\# = A/I$ is a normed algebra with

$$\|x^\#\|_*^\# = \|x + I\|_*^\# = \|x\|_* \text{ (see 3.4.15)}. \tag{3}$$

The completion B of $A^\#$ is a commutative complex or strictly real (see 6.2.20) Banach algebra. Since B, A are Gelfand algebras, by 7.2.2(v) and (1) (above) we have

$$r(x^\#) = \sup_{\chi\in\Delta(B)} |\chi(x^\#)| = \sup_{\chi\in\Delta(A)} |\chi(x)| = \|\hat{x}\|_\infty = r(x). \tag{4}$$

By 7.4.2, B is a GB algebra, so that we have

$$r(x^\#) = \nu_{\|\cdot\|_*}(x^\#). \tag{5}$$

Now

$$\begin{aligned}
\nu_{\|\cdot\|_*^\#}(x^\#) &= \lim_{n\to\infty}(\|x^{\#n}\|_*^\#)^{\frac{1}{n}} = \lim_{n\to\infty}(\|x^n\|_*)^{\frac{1}{n}} \\
&= \lim_{n\to\infty} \nu(x^n)^{\frac{1}{n\rho}} \text{ (where we have used (3), (2))} \\
&= \lim_{n\to\infty} \nu(x)^{\frac{1}{\rho}} = \nu(x)^{\frac{1}{\rho}} \text{ (using 3.3.7}(ii)\text{(6))}.
\end{aligned}$$

Thus

$$\nu_{\|\cdot\|^{\#}_{*}}(x^{\#}) = \nu(x)^{\frac{1}{\rho}}. \tag{6}$$

From (4),(5),(6) we conclude that $r(x) = \nu(x)^{\frac{1}{\rho}}$, completing the proof in the commutative case.

When A is not commutative, consider a maximal commutative subalgebra A_0 with $x \in A_0$. Then A_0 is a ρ-Banach algebra which is again complex or strictly real (see 3.7.32). By virtue of $(***)$ of 1.7.25 we have

$$r_{A_0}(x) = r_A(x) = r(x).$$

Since A_0 is commutative, by what we have just proved, $r_{A_0}(x) = \nu(x)^{\frac{1}{\rho}}$. Therefore we obtain

$$r(x) = r_{A_0}(x) = \nu(x)^{\frac{1}{\rho}}.$$

completing the proof in the general case.

7.4.7. COROLLARY. *In a ρ-normed algebra A we have for $x \in A$*

$$\nu(x) \leqslant \begin{cases} r(x)^{\rho} & \textit{if } A \textit{ is complex} \\ \tilde{r}(x)^{\rho} & \textit{if } A \textit{ is real} \end{cases} \tag{$*$}$$

If A is strictly real we have $\nu(x) \leqslant r(x)^{\rho}$.

PROOF. First assume that A is complex and let $\hat{A}$ be its completion. Then by applying 7.4.6 to $\hat{A}$ we get

$$\nu(x) = r_{\hat{A}}(x)^{\rho} \leqslant r_A(x)^{\rho} \text{ (by virtue of 1.7.20).} \tag{$*'$}$$

Next, if A is real, let $\tilde{A}$ be its complexification. By considering x as an element of $\tilde{A}$ and applying $(*')$ we get $\nu(x) \leqslant \tilde{r}(x)^{\rho}$.

If A is strictly real then $\tilde{\sigma}(x) = \sigma(x)$ and so we then have

$$\nu(x) \leqslant \tilde{r}(x)^{\rho} = r(x)^{\rho}.$$

7.4.8. THEOREM (Michael†). *Let A be a pseudo-Michael algebra which is either complex or strictly real and $A = \varprojlim \hat{A}_{\alpha}$*

† He considered only the case of complex locally convex algebras.

its projective limit decomposition. Then for $x \in A$, $x = (x_\alpha)(x_\alpha \in \hat{A}_\alpha)$ *we have the GB formula*

$$r_A(x) = \sup_\alpha \nu_\alpha(x)^{\frac{1}{\rho\alpha}}, \text{ where } \nu_\alpha(x) = \lim_{n\to\infty} p_\alpha(x^n)^{\frac{1}{n}}$$

PROOF. By 4.5.7(2)

$$r_A(x) = \sup_\alpha r_{\hat{A}_\alpha}(x_\alpha).$$

Since by 7.4.6, $\hat{A}_\alpha$ is a GB algebra we have also

$$r_{\hat{A}_\alpha}(x_\alpha) = \nu_\alpha^{\#}(x_\alpha)^{\frac{1}{\rho\alpha}} = \nu_\alpha(x)^{\frac{1}{\rho\alpha}} \text{ (since } p_\alpha(x) = p_\alpha^{\#}(x_\alpha)).$$

Combining these two results we get the above GB formula.

7.4.9. PROPOSITION. *Let A be a* ρ*-normed algebra. If A is complex, every q. nilpotent element x of A is t. nilpotent; if A is real every ext. q. nilpotent element is t. nilpotent; if A is strictly real every q. nilpotent element is t. nilpotent. The radical* $\sqrt{A}$ *of every* ρ*-normed algebra A is a topologically nil*[†] *(bi-) ideal.*

PROOF. The first three assertions are immediate consequences of $(*)$ of 7.4.7 and the fact that when A is strictly real, $\tilde{r}(x) = r(x)$. For the final assertion we have to show that every element of $\sqrt{A}$ is t. nilpotent. If $x \in \sqrt{A}$, then by 1.7.15(ii), x is q. nilpotent. Therefore x is t. nilpotent when A is complex (as just seen above). To obtain this conclusion when A is real it is enough to show that x is ext. q. nilpotent. If $\lambda \neq 0$ in $\mathbb{C}$ then $\frac{x^2 - 2\alpha x}{|\lambda|^2} \in \sqrt{A}$ $(\alpha = \text{Re}\,\lambda)$ and so this element is q. invertible. It follows from 1.8.5 that $\lambda \notin \tilde{\sigma}'(x)$, so that $\tilde{\sigma}'(x) = \{0\}$ and x is ext.q. nilpotent as required.

7.4.10. PROPOSITION. *Let* $A \neq \sqrt{A}$ *be a commutative pseudo-Michael algebra which is either complex or strictly real. Then all the three concepts-essentially nilpotent, q. nilpotent and t. nilpotent-coincide.*

† i.e. every element of the ideal is t. nilpotent.

PROOF. By 7.2.21 (ii), A is t. spectrally Gelfand. So, by 7.2.12, "essentially nilpotent" is the same as "q. nilpotent". Also, by virtue of GB formula of 7.4.8, "q. nilpotent" is the same as "t. nilpotent". Hence the proposition.

7.4.11. LEMMA. *Let $A = (A, p)$ be a GB algebra. Then the following two conditions are equivalent:*

(i) *There is a constant $C_1 > 0$ such that*

$$C_1 p(x)^2 \leqslant p(x^2) \text{ for all } x \in A. \tag{$*$}$$

(ii) *There is a constant $C_2 > 0$ such that*

$$C_2 p(x) \leqslant r(x)^\rho \text{ for all } x \in A. \tag{$**$}$$

PROOF. Assume that $(*)$ holds and apply the $(*)$ inequality successively to the elements $x^2, x^4, \cdots, x^n$, where n is of the form $n = 2^k$, to obtain

$$C_1^{n-1} p(x)^n \leqslant p(x^n),$$

from which follows that

$$C_1^{\frac{n-1}{n}} p(x) \leqslant p(x^n)^{\frac{1}{n}}.$$

Making, $n \to \infty$, we get

$$C_1 p(x) \leqslant \nu_p(x) = r^\rho(x) \quad (\text{ since } A \text{ is GB})$$

which is $(**)$ (with $C_2 = C_1$). Next assume that $(**)$ holds. Then

$$C_2^2 p(x)^2 \leqslant (r(x))^{2\rho} = \nu_p(x)^2 = \nu_p(x^2) \leqslant p(x^2),$$

which gives $(*)$ (with $C_1 = C_2^2$). This completes the proof.

7.4.12. PROPOSITION. *Let $A = (A, \|\cdot\|)$ be a Gelfand ρ-normed algebra which is also a GB algebra. Then the map $\mathcal{G}$: $x \mapsto \hat{x}$ is a t. isomorphism iff there is a constant $C > 0$ such that*

$$\|x\|^2 \leqslant C\|x^2\| \text{ for all } x \in A. \tag{$*$}$$

In particular, the above conclusion regarding $\mathcal{G}$ holds when A is a complex or strictly real commutative ρ-Banach algebra with $A \neq \sqrt{A}$.

PROOF. In view of 7.3.28, we have only to prove the "if" part. It follows from $(*)$ that

$$\begin{aligned} \|x\| &\leqslant C^{\frac{1}{2}}\|x^2\|^{\frac{1}{2}} \leqslant C^{\frac{1}{2}}C^{\frac{1}{4}}\|x^4\|^{\frac{1}{4}} \cdots \leqslant C^{(\frac{1}{2}+\frac{1}{2^2}+\cdots\frac{1}{2^n})}\|x^{2^n}\|^{\frac{1}{2^n}} \\ &\leqslant C\nu(x) = Cr(x)^{\rho} = C\|\hat{x}\|_{\infty}^{\rho}, \end{aligned}$$

so that we have $\|x\| \leqslant C\|\hat{x}\|_{\infty}^{\rho}$. It follows that $\hat{x} = 0 \Rightarrow x = 0$, which means $\mathcal{G}$ is $1-1$. Again it follows from the above inequality that $\mathcal{G}^{-1}$ is continuous. Since A is Gelfand, $\Delta = \Delta_c$ and $\mathcal{G} = \mathcal{G}_c$ is continuous (see 7.3.5). Thus, $\mathcal{G}$ is a t. isomorphism, completing the proof.

7.4.13. COROLLARY. $\mathcal{G}$ is an isometry iff $\rho = 1$ and

$$\|x\|^2 = \|x^2\| \text{ for all } x \in A. \qquad (*)$$

PROOF. If $\mathcal{G}$ is an isometry then $\|x\| = \|\hat{x}\|_{\infty} \leqslant \|x\|^{\frac{1}{\rho}}$ (by $(*)$ of 7.3.27). This implies that $\rho = 1$ (for, if $0 < \rho < 1$ and $\|x\| < 1$ then $\|x\|^{\frac{1}{\rho}} < \|x\|$ since $\frac{1}{\rho} > 1$).

Further we have,

$$\|x\| = \|\hat{x}\|_{\infty}^2 = \|\widehat{x^2}\|_{\infty} = \|x^2\| \text{ which is relation } (*).$$

Conversely, suppose $(*)$ holds and $\rho = 1$. Then by 7.4.11 (with $C_1 = 1$) we get (see proof therein)

$$\|x\| \leqslant r(x)^{\rho} = r(x) = \|\hat{x}\|_{\infty} \text{ (see } (**) \text{ of 7.3.27).}$$

On the other hand, by $(*)$ of 7.3.27, $\|\hat{x}\|_{\infty} \leqslant \|x\|$. Thus $\|x\| = \|\hat{x}\|_{\infty}$ and $\mathcal{G}$ is an isometry.

7.4.14. LEMMA. *Let $A = (A, \|\cdot\|)$ be an ample complex ρ-Banach, with $\|\cdot\|$ sm., and $x, y \in A$. If there is a constant C such that*

$$\|E(\lambda x)yE(-\lambda x)\| \leqslant C \text{ for all } \lambda \in \mathbb{C},$$

then $xy = yx$.

PROOF. Write $F(\lambda) = E(\lambda x)yE(-\lambda x)$. Then we have

$$F(\lambda) = [e_1 + \lambda x + \frac{(\lambda x)^2}{2!} + \cdots]y[e_1 - \lambda x + \frac{(\lambda x)^2}{2!} + \cdots]$$

where e_1 denotes the unity element of the unitization A_1 of A. By continuity of multiplication in A_1 we get

$$F(\lambda) = [e_1 + \lambda x + \frac{(\lambda x)^2}{2!} + \cdots][y - \lambda yx + \frac{y(\lambda x)^2}{2!} + \cdots].$$

The first series on the right above is absolutely convergent (by 5.2.2) and the second series also is absolutely convergent (as can be checked by ratio test). So we can multiply (as in the classical case) the two series term by term to obtain

$$F(\lambda) = y + \lambda(xy - yx) + \lambda^2 \left(\frac{x^2 y}{2} - xyx + \frac{yx^2}{2} \right) + \cdots . \quad (1)$$

Since the series in (1) converges absolutely for all $\lambda \in \mathbb{C}$, $F(\lambda)$ is a strongly entire function. If $\varphi \in A^{*\dagger}$ then $\varphi \circ F(\lambda) = \varphi(F(\lambda))$ is an ordinary entire function. Further, it is bounded since

$$|\varphi(F(\lambda))| \leqslant \|\varphi\| \|F(\lambda)\|^{\frac{1}{p}} \leqslant \|\varphi\| C,$$

for some constant C. By Liouville's theorem $\varphi \circ F$ is constant, so that

$$\varphi(F(\lambda)) = \varphi(F(0)) = \varphi(y).$$

Since A is ample, $F(\lambda) = y$ for all λ. Differentiating the series in λ in (1) we get

$$0 = F'(0) = xy - yx$$

so that $xy - yx = 0$, as desired.

7.4.15. THEOREM (La Page-Hirschfeld- Żelazko).† *An ample*

† A^* = continuous dual of A.

complex ρ-Banach algebra $A = (A, \|\cdot\|)$, *with* $\|\cdot\|$ *sm., which satisfies the condition*

$$C_1\|x\|^2 \leqslant \|x^2\| \text{ for all } x \in A \tag{$*$}$$

is necessarily commutative.

PROOF. Let A_1 be the unitization of A, with unity e_1. For $x, y \in A,\ \lambda \in \mathbb{C}$ set

$$z = F(\lambda) = E(\lambda x)yE(-\lambda x). \tag{1}$$

Clearly

$$\mu e_1 = E(\lambda x)\mu e_1 E(-\lambda x) \tag{2}$$

(where we have used 5.2.3 (iii)). It follows that

$$z - \mu e_1 = E(\lambda x)(y - \mu e_1)E(-\lambda x). \tag{3}$$

The equation (3) shows that $z - \mu e_1$ is invertible iff $y - \mu e_1$ is invertible. Therefore we have

$$\sigma'(z) = \sigma'(y), \quad r(z) = r(y).$$

It follows from $(*)$ and 7.4.11 that there is a constant $C_2 > 0$ with

$$C_2\|z\| \leqslant r(z)^\rho = r(y)^\rho.$$

Therefore

$$\|F(\lambda)\| = \|z\| \leqslant \frac{1}{C_2} r(y)^\rho.$$

Applying 7.4.14, we conclude that $xy = yx$, completing the proof.

7.4.16. LEMMA. *A complex or strictly real ρ-Banach algebra* $A = (A, \|\cdot\|)$, *with* $\|\cdot\|$ *sm., satisfying the condition*

$$C_1\|x\|^2 \leqslant \|x^2\| \text{ for all } x \in A \tag{$*$}$$

is s.s..

PROOF. The condition $(*)$ together with 7.4.11 and the fact A is GB (see 7.4.6) gives

$$C_1\|x\| \leqslant \nu(x) = r(x)^\rho.$$

Therefore, $r(x) = 0 \Rightarrow x = 0$, whence A is q.s.s. and so s.s. (see 1.7.18).

7.4.17. THEOREM (Kaplansky). *Every strictly real s.s. ρ-normed algebra A is commutative.*

PROOF. First suppose that A is primitive. By 1.9.15, A is a division algebra and so by 6.5.12, A is isomorphic to $\mathbb{R}, \mathbb{C}$, or $\mathbb{H}$. The strict reality of A rules out $\mathbb{C}$ or $\mathbb{H}$. Thus $A \simeq \mathbb{R}$. Next let A be s.s. and P be a primitive ideal. The quotient $A_P = A/P$ is primitive and by 1.9.8, A_P is strictly real and so $A_P \simeq \mathbb{R}$. Since A is s.s., $\bigcap P = \{0\}$, so that A is isomorphic to a subalgebra of the direct product $\prod A_P$ (of isomorphic copies of $\mathbb{R}$) and so commutative.

7.4.18. COROLLARY. *Every strictly real primitive ρ-normed algebra is isomorphic to $\mathbb{R}$.*

7.4.19. COROLLARY. *Every strictly real ρ-Banach algebra satisfying condition $(*)$ of 7.4.16 is commutative.*

PROOF. By 7.4.16, A is s.s. and so by 7.4.17, A is commutative.

7.4.20. PROPOSITION (Yood††) *Let $A = (A, \|\cdot\|)$ be a ρ-normed algebra with $\|\cdot\|$ sm. Then the following statements are equivalent:*

(i) *A is a Q algebra.*

(ii) $\tilde{r}(x)^\rho = \nu(x) = \lim_{n\to\infty} \|x^n\|^{\frac{1}{n}} \quad (x \in A)$.

(iii) $\tilde{r}(x)^\rho \leqslant \|x\| \quad (x \in A)$.

(Here $\tilde{r}$ denotes the spectral radius in $\tilde{A}$, where $\tilde{A}$ denotes A itself when A is complex and the complexification of A when it is real.)

---

† Yood proved the proposition for a normed algebra (i.e. $\rho = 1$).

†† These authors obtained the result for Banach algebras (i.e. $\rho = 1$).

PROOF. Assume (i). Then there is an $\eta > 0$ such that for any x with $\|x\| < \eta$, x is q. invertible. Set $\epsilon = \eta^{-1}$ if A is complex, and $= [(1+\eta)^{\frac{1}{2}} - 1]^{-1}$ if A is real. For a given $x \in A$ take a $\lambda = \alpha + i\beta \neq 0$ in $\mathbb{C}$ with $|\lambda|^\rho > \epsilon\|x\|$. When A is real we have:

$$\begin{aligned}\||\lambda|^{-2}(x^2 - 2\alpha x)\| &\leqslant |\lambda|^{-2\rho}(\|x\|^2 + 2^\rho|\alpha|^\rho\|x\|) \\ &\leqslant |\lambda|^{-2\rho}(\|x\|^2 + 2|\lambda|^\rho\|x\|) \\ &\qquad\qquad (\text{since } \rho \leqslant 1, |\alpha| \leqslant |\lambda|) \\ &= (\|\lambda^{-1}x\| + 1)^2 - 1 < (\epsilon^{-1} + 1)^2 - 1 \\ &\leqslant 1 + \eta - 1 = \eta \quad (\text{ by the choice of } \eta). \qquad (1)\end{aligned}$$

When A is complex we have

$$\| - \lambda^{-1}x\| = |\lambda|^{-\rho}\|x\| < \eta. \qquad (1')$$

Using 1.8.5, 1.7.8 we conclude from (1), (1′) that $\lambda \notin \tilde{\sigma}(x)$, whence

$$\tilde{r}(x)^\rho \leqslant \epsilon\|x\|. \qquad (2)$$

It follows that

$$\begin{aligned}\tilde{r}(x)^\rho = [\tilde{r}(x^n)^\rho]^{\frac{1}{n}} &\leqslant \epsilon^{\frac{1}{n}}\|x^n\|^{\frac{1}{n}} \\ &\leqslant \nu(x). \qquad (3)\end{aligned}$$

Now the completion $\hat{\tilde{r}}$ of $\tilde{A}$ is a GB algebra (see 7.4.6). So we have

$$\nu(x) = \hat{\tilde{r}}(x)^\rho \leqslant \tilde{r}(x)^\rho \quad (\text{ since } \tilde{A} \subseteq \hat{\tilde{r}}). \qquad (4)$$

From (3),(4) we obtain $\nu(x) = \tilde{r}(x)^\rho$, proving (i) $\Rightarrow$ (ii). Again, since $\|\cdot\|$ is sm., (ii) $\Rightarrow$ (iii). Finally, to prove (iii) $\Rightarrow$ (i), assume to the contrary. Thus, suppose A satisfies (iii) but not (i). Then we can find an $x \in A$ such that $\|x\| < 1$ and x is not q. invertible. It follows from 1.7.8 that $-1 \in \sigma'(x)$, so that $|-1| = 1 \leqslant \tilde{r}(x)^\rho \leqslant \|x\| < 1$ a contradiction whence (iii) $\Rightarrow$ (i), completing the proof.

§ 5. Holomorphic Functional Calculus for a Single Algebra Element †

7.5.1. DEFINITION. Let A be a complex unital ρ-Banach algebra (with unity e) and $G \subseteq \mathbb{C}$, a nonempty open set. Write

$$A_G = \{x \in A : \sigma(x) \subseteq G\}.$$

If $\lambda \in G$ then $\sigma(\lambda e) = \{\lambda\}$, $\lambda e \in A_G$, so that A_G is nonempty. Moreover, by 6.2,14, A_G is an open set. Denote by $\mathsf{H}(G)$ the set of (ordinary) holomorphic functions on G.

7.5.2. LEMMA. *With respect to pointwise operations,* $\mathsf{H}(G)$ *is a commutative unital algebra over* $\mathbb{C}$. $\mathsf{H}(G)$ is,*moreover, a locally sm.* $\mathcal{F}$ *algebra, the metric convergence in it being the same as uniform convergence on compacta.*

PROOF. It suffices to prove the second statement. Since G is an open subset of $\mathbb{C}$ it is locally compact Hausdorff as well as σ-compact. It follows †† that we can find an increasing sequence K_n of compact sets such that (i) $\bigcup K_n = G$ (ii) if $K \subseteq G$ and K is compact then $K \subseteq$ some K_n. For $f \in \mathsf{H}(G)$, set

$$p_n(f) = \sup_{\lambda \in K_n} |f(\lambda)|; |f| = \sum_{n=1}^{\infty} \frac{1}{2^n} \frac{p_n(f)}{1 + p_n(f)}.$$

Then p_n are sm. seminorms, $|\cdot|$ an F-metric, and $(\mathsf{H}(G), \{p_n\})$ a sm. $\mathcal{F}$ algebra. Further, it is clear from the definition of $|\cdot|$ that metric convergence is the same as uniform convergence over compacta.

7.5.3. DEFINITION. For $f \in \mathsf{H}(G)$ set

$$\tilde{f}(x) = \int_{\Gamma} (\lambda e - x)^{-1} f(\lambda) d\lambda \quad (x \in A_G) \qquad (*)$$

where Γ is any contour surrounding $\sigma(x)$ in G. If we write $\dot{x}_\lambda = -x_\lambda = -(x - \lambda e)^{-1}$ then we can rewrite $(*)$ as

---

† The exposition follows closely that of Rudin [24, pp.240-45]

†† See for example [9, p.241].

$$\tilde{f}(x) = \int_\Gamma \dot{x}_\lambda f(\lambda) d\lambda. \qquad (**)$$

7.5.4. LEMMA. *The A-valued integral in* $(**)$ *exists and is independent of the particular contour* Γ *surrounding* $\sigma(x)$ *in* G.

PROOF. By 6.2.9, 5.4.13(b) the integrand in $(**)$ is locally ρ-admissible holomorphic on $G\backslash\sigma(x)$. The integral exists by 5.4.9 and the independence of the integral on the contour Γ is a consequence of the Cauchy integral theorem (5.4.16 (a)).

7.5.5. LEMMA. *Let* A *be a complex unital* ρ*-Banach algebra and suppose that* $x \in A$ *and* $\alpha \in \rho(x)$. *Let* Γ *be a contour in* $G = \mathbb{C}\backslash\{\alpha\}$ *such that* Γ *surrounds* $\sigma(x)$ *in* G. *Then*

$$\frac{1}{2\pi i}\int_\Gamma (\lambda - \alpha)^n (\lambda e - x)^{-1} d\lambda = (x - \alpha e)^n \quad (n = 0, \pm 1, \pm 2, \cdots). \; (*)$$

PROOF. Set

$$I_n = \frac{1}{2\pi i}\int_\Gamma (\lambda - \alpha)^n x_\lambda \, d\lambda.$$

By Hilbert relation (6.2.2, $(*)$) we have $x_\lambda = x_\alpha + (\lambda - \alpha) x_\lambda x_\alpha$, from which it follows that

$$I_n = \frac{1}{2\pi i}\int_\Gamma (\lambda - \alpha)^n x_\alpha d\lambda + \frac{1}{2\pi i}\int_\Gamma (\lambda - \alpha)^{n+1} x_\lambda x_\alpha d\lambda.$$

For $n \neq -1$, the evaluation of the first indefinite integral gives $\frac{(\lambda-\alpha)^{n+1}}{n+1} x_\alpha$ so that the corresponding definite integral vanishes since Γ is closed. On the other hand, for $n = -1$, the first integral, vanishes since the index $I(\Gamma, \alpha) = 0$, by hypothesis. Thus we obtain the recurrence relation.

$$I_{n+1} = I_n x_\alpha^{-1} = I_n(x - \alpha e). \qquad (1)$$

If $\dot{I}_n$ denotes the integral in $(*)$ we have $\dot{I}_n = -I_n$, so that (1) gives

$$\dot{I}_{n+1} = \dot{I}_n(x - \alpha e) \quad (n = 0, \pm 1, \pm 2, \cdots).$$

To complete the proof it is clearly enough to show that $\dot{I}_0 = e$. By 6.2.10 $(*)$, with $m = 0$, we obtain $I_0 = -e$, so that $\dot{I}_0 = e$ as required.

7.5.6. PROPOSITION. *Let A be a complex unital ρ-Banach algebra and suppose that $x \in A$ and $R = R(\lambda)$ a rational function over $\mathbb{C}$ whose poles do not lie in $\sigma(x)$. Further, let*

$$R(\lambda) = P(\lambda) + \sum_{r,s} c_{rs}(\lambda - \alpha_r)^{-s} \quad (1 \leqslant r \leqslant m)(1 \leqslant s \leqslant k_r)$$

be the representation of R (obtained by division algorithm and partial fraction development) where P is a polynomial (which may be 0), α_r the poles of R, and k_r the multiplicities of α_r. Set

$$R(x) = P(x) + \sum_{r,s} c_{rs}(x - \alpha_r e)^{-s}.$$

Then

$$R(x) = \tilde{R}(x) = \frac{1}{2\pi i}\int_\Gamma R(\lambda)(\lambda e - x)^{-1} d\lambda. \qquad (*)$$

PROOF. Write $R_1(\lambda) = \sum_{r,s} c_{rs}(\lambda - \alpha_r)^{-s}$. By $(*)$ of 7.5.5 and linearity we obtain

$$R_1(x) = \frac{1}{2\pi i}\int_\Gamma R_1(\lambda)(\lambda e - x)^{-1} d\lambda. \qquad (1)$$

Again, by $(*)$ of 6.2.10 and linearity we get

$$P(x) = \frac{1}{2\pi i}\int_\Gamma P(\lambda)(\lambda e - x)^{-1} d\lambda. \qquad (2)$$

Adding (1),(2) we get the representation $(*)$.

7.5.7. LEMMA. *Suppose that $\alpha \in G$ and $f \in \mathsf{H}(G)$. Then*

$$\tilde{f}(\alpha e) = f(\alpha)e.$$

PROOF.

$$\begin{aligned} \tilde{f}(\alpha e) &= \frac{1}{2\pi i}\int_\Gamma f(\lambda)(\lambda - \alpha)^{-1} e\, d\lambda \\ &= \Big[\frac{1}{2\pi i}\int_\Gamma f(\lambda)(\lambda - \alpha)^{-1} d\lambda\Big] e \\ &= f(\alpha)e \text{ (by classical Cauchy theorem).} \end{aligned}$$

7.5.8. Set $\tilde{H} = \tilde{H}(A_G) = \{\tilde{f} : f \in H(G)\}$, where each $\tilde{f}$ is a A-valued function on A_G. Under pointwise operations $\tilde{H}$ is an algebra over $\mathbb{C}$.

7.5.9. THEOREM. *$\tilde{\mathsf{H}}(A_G)$ is a complex Hausdorff TA under the weak topology:*

$$\tilde{f}_\alpha \to \tilde{f} \text{ iff } \tilde{f}_\alpha(x) \to \tilde{f}(x) \text{ for each } x \in A_c.$$

The map $\Lambda : f \mapsto \tilde{f}$ is a continuous isomorphism of $\mathsf{H}(G)$ onto $\tilde{\mathsf{H}}(A_G)$. In particular, $\tilde{H}(A_G)$ is commutative.

PROOF. It is clear that the map Λ is linear. If $\tilde{f} = 0$ then, by 7.5.7,

$$f(\lambda)e = \tilde{f}(\lambda e) = 0 \qquad (\lambda \in G)$$

so that $f = 0$, proving Λ is $1-1$.

Since $(\lambda e - x)^{-1} = -x_\lambda$ is locally ρ-admissible holomorphic on $\rho(x)$, it has a local representation

$$(\lambda e - x)^{-1} = \sum_j \varphi_j(\lambda) x_j \text{ with } \varphi_j$$

ordinary holomorphic for $\lambda \in G_x = \{\lambda \in \mathbb{C} : \lambda > \nu(x)^{\frac{1}{\rho}}\}$ (see proof of 6.2.9). It follows that, for $f_\alpha, f \in \mathsf{H}(G)$ we have

$$\begin{aligned} \tilde{f}_\alpha(x) - \tilde{f}(x) &= \frac{1}{2\pi i}\int_\Gamma (f_\alpha(\lambda) - f(\lambda))(\lambda e - x)^{-1} \\ &= \frac{1}{2\pi i}\sum_j x_j \int_\Gamma (f_\alpha(\lambda) - f(\lambda))\varphi_j(\lambda) d\lambda \end{aligned}$$

so that

$$\begin{aligned} \|\tilde{f}_\alpha(x) - \tilde{f}(x)\| &\le \frac{1}{(2\pi)^\rho}\sum_j \|x_j\| \, |\int_\Gamma (f_\alpha(\lambda) - f(\lambda))\varphi_j(\lambda) d\lambda|^\rho \\ &\le \frac{1}{(2\pi)^\rho}\sum_j \|x_j\| \, \|f_\alpha - f\|_\Gamma^\rho \|\varphi_j\|_\Gamma^\rho |\Gamma|^\rho \\ &\le \|f_\alpha - f\|_\Gamma^\rho M, \end{aligned}$$

where $M = \dfrac{|\Gamma|^\rho}{(2\pi)^\rho}\sum_j \|x_j\| \, \|\varphi_j\|_\Gamma^\rho < \infty$. It follows that $f_\alpha \to f$ in $\mathsf{H}(G) \Rightarrow \tilde{f}_\alpha \to \tilde{f}$ in $\mathsf{H}(A_G)$, whence Λ is continuous. To complete the proof of the theorem it remains to show that Λ is multiplicative: $\Lambda(fg) = \Lambda(f)\Lambda(g)$. Thus if $h = fg$ we have to show that

$$\tilde{h}(x) = \tilde{f}(x)\tilde{g}(x) \quad (x \in A_G).$$

If f, g are rational functions without poles in $\sigma(x)$, then by 7.5.6,

$$\tilde{h}(x) = h(x) = fg(x) = f(x)g(x) = \tilde{f}(x)\tilde{g}(x).$$

In the general case, we make use of Runge's approximation theorem to find rational functions f_n, g_n such that

$$f_n \to f,\ g_n \to g \text{ in } \mathsf{H}(G).$$

Then $f_n g_n \to fg = h$. It follows that

$$\begin{aligned}\tilde{h} = \tilde{fg} &= \Lambda(fg) = \Lambda(\lim_{n\to\infty} f_n g_n) = \lim_{n\to\infty} \Lambda(f_n g_n) \\ &= \lim_{n\to\infty} \Lambda(f_n)\Lambda(g_n) = \tilde{f}\tilde{g},\end{aligned}$$

completing the proof of the multiplicativity. Finally, the commutativity of $\mathsf{H}(A_G)$ is an immediate consequence of that of $\mathsf{H}(G)$ and the isomorphism.

7.5.10. COROLLARY. $x \leftrightarrow \tilde{f}(x) \quad (x \in A_G)$.

PROOF. Define $j \in \mathsf{H}(G)$ by $j(\lambda) = \lambda$. Then, since A_G is commutative, $\tilde{j} \leftrightarrow \tilde{f}$ which clearly implies: $\tilde{j}(x) = x \leftrightarrow \tilde{f}(x)$.

7.5.11. COROLLARY. (cf. [15,p 206]). *If* $f \in \mathsf{H}(G)$ *has a power-series representation* $f(\lambda) = \sum_{n=0}^{\infty} \alpha_n \lambda^n$ $(\alpha_n \in \mathbb{C})$ *throughout* G *then*

$$\tilde{f}(x) = \sum_{n=0}^{\infty} \alpha_n x^n \quad (x \in A_G).$$

PROOF. Write $P_N(\lambda) = \sum_{n=0}^{N} \alpha_n \lambda^n$. Then $P_N \to f$ in $\mathsf{H}(G)$. So, by construction of map Λ, we have

$$\tilde{f}(x) = \lim_N \tilde{P}_N(x) = \lim_N P_N(x) = \sum_{n=0}^{\infty} \alpha_n x^n.$$

7.5.12. LEMMA. *If* $f \in \mathsf{H}(G)$ *and* $x \in A_G$ *then* $\tilde{f}(x)$ *is invertible in* A *iff* f *has no zero in* $\sigma(x)$.

PROOF. Suppose that $f \neq 0$ on $\sigma(x)$. By continuity of f there is an open set G_1 such that

$$\sigma(x) \subseteq G_1 \subseteq G \text{ and } f \neq 0 \text{ on } G_1.$$

It follows that $g = 1/f$ is holomorphic on G_1. Since $fg = 1$ on G_1, by applying 7.5.9 (with G_1 replacing G) we obtain $\tilde{f}(x)\tilde{g}(x) = e$ (using 7.5.7), so that $\tilde{f}(x)$ is invertible. Next suppose that $f(\alpha) = 0$ for some $\alpha \in \sigma(x)$. Since f is holomorphic we can write

$$f(\lambda) = (\lambda - \alpha)g(\lambda), \text{ with } g \in \mathsf{H}(G).$$

It follows that

$$\tilde{f}(x) = (x - \alpha e)\tilde{g}(x) = \tilde{g}(x)(x - \alpha e).$$

Since $(x - \alpha e)$ is not invertible (since $\alpha \in \sigma(x)$), by 1.1.30, $\tilde{f}(x)$ is not invertible.

7.5.13. THEOREM (Spectral Theorem). *Suppose that* A *is a complex unital* ρ*-Banach algebra,* $G \subsetneq \mathbb{C}$ *an open set. If* $x \in A_G$ *and* $f \in \mathsf{H}(G)$ *then*

$$\sigma(\tilde{f}(x)) = f(\sigma(x)).$$

PROOF. Now $\alpha \in \sigma(\tilde{f}(x))$ iff $\tilde{f}(x) - \alpha e$ is not invertible. By 7.5.12, this is equivalent to $f(\lambda) - \alpha$ vanishing in $\sigma(x)$, i.e. $\alpha \in f(\sigma(x))$.

7.5.14. COROLLARY. *An n.a.s.c*† *for a complex unital* ρ*-Banach algebra* A *to have an idempotent* $u \neq 0, e$ *is that there is an element* $a \in A$ *with* $\sigma(a)$ *disconnected.*

PROOF. The condition is necessary since, if u exists, we have by 1.7.9, $\sigma(u) = \{0, 1\}$, so that $\sigma(u)$ is dis connected.

To prove that the condition is sufficient assume that there is

---

† n.a.s.c = necessary and sufficient condition.

an element $a \in A$ with $\sigma(a)$ disconnected. Since $\sigma(a)$ is closed we have a disjoint decomposition

$$\sigma(a) = F_1 \dot{\bigcup} F_2$$

where $F_j \neq \emptyset$, F_j closed in $\mathbb{C}$ $(j = 1, 2)$. It follows that we can find open sets $G_j \supseteq F_j (j = 1, 2)$ such that $G_1 \bigcap G_2 = \emptyset$. Set $G = G_1 \bigcup G_2$. Then $G_j \bigcap \sigma(a) \neq \emptyset$. Define $f \in \mathsf{H}(G)$ by

$$f(\lambda) = \begin{cases} 0 & \text{if } \lambda \in G_1 \\ 1 & \text{if } \lambda \in G_2 \end{cases}$$

Then f is holomorphic and $f^2 = f$. If $u = \tilde{f}(a)$ then $u^2 = u$. Also,

$$\sigma(u) = \sigma(\tilde{f}(a)) = f(\sigma(a)) = \{0, 1\}.$$

Since $\sigma(0) = 0$, $\sigma(e) = 1$, we have $u \neq 0, e$.

7.5.15. PROPOSITION. *Let A be a complex unital ρ-Banach algebra and $\hat{A}$ its transform algebra. Holomorphic functions operate on $\hat{A}$ in the sense that if $x \in A$, $f = f(\lambda)$ holomorphic in an open neighbourhood of $\sigma(x)$ there is a $y \in A$ such that $y \leftrightarrow x$ and*

$$\hat{y} = f \circ \hat{x}, \text{ i.e. } \hat{y}(\chi) = \chi(y) = f(\chi(x)) \quad (\chi \in \Delta).$$

The element y is uniquely determined whenever A is h.s.s. $(\sqrt[h]{A} = \{0\})$.

PROOF. Set

$$y = \frac{1}{2\pi i} \int_\Gamma (\lambda e - x)^{-1} f(\lambda) d\lambda$$

where Γ is a contour surrounding $\sigma(x)$. By 7.5.10, $y \leftrightarrow x$. Since χ is a continuous homomorphism we have

$$\begin{aligned} \chi(y) &= \frac{1}{2\pi i} \int_\Gamma \chi((\lambda e - x)^{-1} f(\lambda)) d\lambda \\ &= \frac{1}{2\pi i} \int_\Gamma (\lambda - \chi(x))^{-1} f(\lambda) d\lambda \\ &= f(\chi(x)) \text{ (by the classical Cauchy theorem).} \end{aligned}$$

The uniqueness statement for y follows since, when A is h.s.s., $\chi(y_1) = \chi(y)$ for all $\chi \in \Delta \Rightarrow y_1 = y$.

7.5.16. THEOREM (a) (Wiener- Żelazko). *If* $f = f(t) \in W^\rho$ † *and* $f(t) \neq 0$ *for any* t *then* $1/f \in W^\rho$.
(b) (Wiener-Levi-Żelazko). *If* $f \in W^\rho$ *and* F *is holomorphic in an open set* G *containing the range of* f *then* $F \circ f \in W^\rho$.

PROOF. (a) Note first that $\chi_{t_1} : f \mapsto f(t_1), t_1 \in \mathbb{R}$, is a character of W^ρ. Let next χ be any character of W^ρ. Since W^ρ is Gelfand (by 7.2.17), χ is continuous. So, if $f \in W^\rho$, $f = \sum_{-\infty}^{\infty} \hat{f}(n)e^{int}$ then

$$\chi(f) = \sum_{-\infty}^{\infty} \hat{f}(n)\chi(e^{it})^n.$$

Now $|\chi(e^{it})| \leqslant r(e^{it}) \leq \|e^{it}\|_\rho^{\frac{1}{\rho}} = 1$. Similarly, $|\chi(e^{-it})| \leqslant 1$, so that

$$|\chi(e^{it})|^{-1} = |\chi(e^{-it})| \leqslant 1.$$

Therefore

$$|\chi(e^{it})| = 1, \text{ so that } \chi(e^{it}) = e^{it_1},$$

for some $t_1 \in \mathbb{R}$. It follows that

$$\chi(f) = \sum_{-\infty}^{\infty} \hat{f}(n)\chi(e^{it_1})^n = f(t_1),$$

so that $\chi = \chi_{t_1}$. The Wiener-Zèlazko theorem is now an immediate consequence of 7.2.10 (since "$\chi(f) \neq 0\ \forall f$" $\Leftrightarrow$ "$f(t) \neq 0, \forall t$").

(b) Since the range of f is $\sigma(f)$, the holomorphic function F satisfies the hypothesis (for f) of 7.5.15. Also $\hat{W}^\rho$ can be identified †† with W^ρ. Hence, by 7.5.15

$$F \circ f = F \circ \hat{f} = F \hat{\circ} f \in \hat{W}^\rho = W^\rho.$$

† See 3.4.10 for Definition of W^ρ.

†† This is possible since $A = W^\rho$ is s.s. ($f \in \sqrt{A} \Rightarrow \chi_t(f) = f(t) = 0,\ \forall\, t \in \mathbb{R} \Rightarrow f = 0$).

7.5.17. THEOREM (Michael).[†] *Let A be a complex unital pseudo-Michael algebra. Given $x \in A$ and a function f holomorphic in an open neighbourhood of $\sigma(x)$, then is a $y \in A$ such that $y \leftrightarrow x$ and*

$$\hat{y}(\chi) = f \circ \hat{x}(\chi) \quad (\chi \in \Delta_c).$$

The element y is uniquely determined if A is h.s.s. (in particular if A is commutative s.s.).

PROOF. Let $A = (A, \mathcal{P})$, $\mathcal{P}$ saturated, and $A = \varprojlim \hat{A}_\alpha$ be the projective limit decomposition of A. For $x = (x_\alpha) \in A$ denote by Γ_α a contour surrounding $\sigma_\alpha(x_\alpha) = \sigma_{\hat{A}_\alpha}(x_\alpha)$. (Note that by 4.5.7(3)), $\sigma_A(x) = \bigcup_\alpha \sigma_\alpha(x_\alpha)$.) Since $\hat{A}_\alpha$ is a ρ_α-Banach algebra, by 7.5.3, 7.5.4,
we can set

$$y_\alpha = \int_{\Gamma_\alpha} (\lambda e_\alpha - x_\alpha)^{-1} f(\lambda) d\lambda$$

and the definition of y_α is independent of the particular choice of Γ_α. In the notation of the proof of 4.5.3, $\hat{\varphi}_{\alpha p}$ is a continuous homomorphism of $\hat{A}_\beta$ onto $\hat{A}_\alpha$, when $\alpha \prec \beta$. Then

$$\begin{aligned} \hat{\varphi}_{\alpha\beta}(y_\beta) &= \hat{\varphi}_{\alpha\beta} \int_{\Gamma_\beta} (\lambda e_\beta - x_\beta)^{-1} f(\lambda) d\lambda \\ &= \int_{\Gamma_\beta} (\lambda e_\alpha - x_\alpha)^{-1} f(\lambda) d\lambda = \int_{\Gamma_\alpha} (\lambda e_\alpha - x_\alpha)^{-1} f(\lambda) d\lambda \end{aligned}$$

(since, by 1.7.19, $\sigma_\alpha(x_\alpha) \subseteq \sigma_\beta(x_\beta)$), so that $\hat{\varphi}_{\alpha\beta}(y_\beta) = y_\alpha$. It follows that $y = (y_\alpha) \in A$. If $\chi \in \Delta_c = \Delta_c(A)$, then by 7.3.19, $\chi \in \Delta_\alpha$ for some α, and χ determines $\chi_\alpha \in \hat{\Delta}_\alpha$ with $\chi_\alpha(x_\alpha) = \chi(x)$. By 7.5.15 applied to $\hat{A}_\alpha$ there is a $y_\alpha \in \hat{A}_\alpha$ such that $\chi_\alpha(y_\alpha) = f(\chi_\alpha(x_\alpha))$. Hence

$$\chi(y) = f(\chi(x)) \text{ or } \hat{y} = f(\hat{x}).$$

[†] He obtained the result for the case of locally sm. convex algebras.

The uniqueness of y, when A is h.s.s., is clear. Finally, we have $y \leftrightarrow x$, since $y_\alpha \leftrightarrow x_\alpha$ for each α.

7.5.18. Remark. The result in 7.5.16(a) can be restated in the following form: if $f = f(t)$ is a continuous complex-valued 2π-periodic function on $\mathbb{R}$ such that

(i) $\sum_{n=-\infty}^{\infty} |\hat{f}(n)|^\rho < \infty$ ($\hat{f}(n), n \in \mathbb{Z}$, denoting the n th Fourier coefficient of f)

(ii) f does not vanish anywhere on $\mathbb{R}$, then

$$g = \frac{1}{f} \text{ has the property } : \sum_{n=-\infty}^{\infty} |\hat{g}(n)|^\rho < \infty.$$

For $\rho = 1$ the above result reduces to a celebrated lemma of Wiener which he uses for proving his general Tauberian theorem.

§ 6. Automorphisms and Derivations

7.6.1. An isomorphism of an algebra A onto itself is called an *automorphism.* If A is unital then every invertible element a in A determines an automorphism $I_a = \ell_a r_{a^{-1}}; I_a(x) = axa^{-1}$ $(a \in G_i, x \in A)$. (Note that the identities $I_a I_{a-1} = I_{a-1} I_a = I$ guarantee that I_a is bijective; also it is clear that I_a preserves algebraic operation). The automorphism I_a is called an *inner automorphism* of A.

7.6.2. LEMMA. *$I_a = I$ iff $a \in Z$ (the centre of A). In particular, if A is commutative every $I_a = I$.*

PROOF. Clear.

7.6.3. In a TA we have automorphisms which are continuous as well as automorphisms which are also homeomorphisms, i.e. t. automorphisms. In general, not every automorphisms is continuous nor every continuous automorphism a t. automorphisms.

7.6.4. PROPOSITION. *Let A be a TA, with unity e. Then:*

(i) *Every inner automorphism of* A *is a t. automorphism.*

(ii) *If* A *is a* (F) *algebra – inparticular a* $\mathfrak{F}$ *algebra – then every continuous automorphism is a t. automorphism.*

(iii) *In a semi-simple* ρ*-Banach algebra every automorphism is a t. automorphism.*

PROOF. (i) This follows since $I_a = l_a r_{a-1}, l_a, r_{a-1}$ are continuous, and $I_a^{-1} = I_{a^{-1}}$.

(ii) This is a consequence of the open mapping theorem (see 3.1.15).

(iii) This is proved in chapter 9 (see 9.4.7).

7.6.5 DEFINITION. Let A be an algebra (over F). A linear map $D : A \to A$ is called a *derivation* if $D(xy) = (Dx)y + xDy$ for all $x, y \in A$.

Let $\mathcal{P}$ the note the algebra of polynomials $f(x)$ in a single variable x over F. For $f(x) = \gamma_0 + \ldots + \gamma_n x^n \, (\gamma_j \in F)$ define $f'(x) = \gamma_1 + 2\gamma_2 x + \ldots + n\gamma_n x^{n-1}$. Then $f \mapsto f'$ is a derivation of the algebra $\mathcal{P}$.

7.6.6. Remarks. If $\mathcal{D} = \mathcal{D}(A)$ denotes the set of all derivation of A it is clear that it is a linear space (over F). Moreover, if we write for $D_1, D_2 \in \mathcal{D}, [D_1, D_2] = D_1 D_2 - D_2 D_1$, then it is easy to see that $[D_1, D_2] \in \mathcal{D}; [D_1, D_2]$ is called the Lie product of D_1, D_2. $\mathcal{D}$ is in fact a Lie algebra i.e. a non-associative algebra (= an algebra without the associative law for multiplication) whose multiplication satisfies: $[D, D] = 0, [D_1, [D_2, D_3]] + [D_2, [D_3, D_1]] + [D_3, [D_1, D_2]] = 0$. $(D, D_1, D_2, D_3 \in \mathcal{D})$.

7.6.7. LEMMA. *Let* D *be a derivation of* A *and* u *an idempotents. Then:*

(i) $u(Du)u = 0$.

(ii) *If* $u \leftrightarrow Du$ *– in particular if* u *lies on the centre of* A *– then* $Du = 0$.

(iii) *If* A *has a unity* e *then* $De = 0$, *hence also* $D(\lambda e) = 0$ *for all* $\lambda \in F$.

PROOF. (i) $Du = Du^2 = (Du)u + uDu$ (1)
so $uDu = u(Du)u + uDu$, whence

$$u(Du)u = 0. \tag{2}$$

(ii) If $u \leftrightarrow Du$ then equation (1) gives $Du = 2uDu = 2u^2Du = 2u(Du)u = 0$ (by (2)).

(iii) This follows from (ii).

7.6.8. LEMMA (Leibniz rule).

$$D^n(xy) = \sum_{r=0}^{n} \binom{n}{r} D^{n-r}x(D^r y) \ (n = 1, 2, \dots ; x, y \in A).$$

PROOF. By induction, making use of the well-known relation

$$\binom{n}{r} + \binom{n}{r-1} = \binom{n+1}{r}.$$

7.6.9. COROLLARY. *Suppose that* $D^2x = 0$. *Then*:

(i) $D^m x = 0$ $(m \geqslant 2)$,

(ii) $D(Dx)^m = 0$ $(m \geqslant 1)$,

(iii) $D^n(x^n) = n!(Dx)^n$ $(n \geqslant 1)$.

PROOF. (i) $D^m x = D^{m-2}(D^2x) = 0$ $(m \geqslant 2)$. (1)

(ii) For $m = 1$, we have $D(Dx) = D^2x = 0$.

Assume now that $D(Dx)^{m-1} = 0$ $(m \geqslant 2)$. Then

$$\begin{aligned} D((Dx)^m) &= D(Dx(Dx)^{m-1}) = D^2x(Dx)^{m-1} + Dx.D(Dx)^{m-1} \\ &= 0.(Dx)^{m-1} + Dx.0 = 0. \end{aligned}$$

(iii) Assume that

$$D^{n-1}(x^{n-1}) = (n-1)!(Dx)^{n-1}. \tag{2}$$

Then

$$D^n(x^{n-1}) = D((n-1)!(Dx)^{n-1}) = (n-1)!D(Dx)^{n-1} = 0 \tag{3}$$

where we have used (ii). By Leibniz rule

$$D^n(x^n) = D^n(x^{n-1}.x)$$

$$= D^n x^{n-1}.x + \binom{n}{1} D^{n-1}x^{n-1}.Dx + \sum_{r=2}^{n} \binom{n}{r} D^{n-r}x^{n-1}.D^r x.$$

Using (1),(2),(3) the above RHS reduces to $0 + n.(n-1)!(Dx)^{n-1}Dx + 0 = n!(Dx)^n$, completing the proof (by induction).

7.6.10. PROPOSITION. *Let D be a derivation of an algebra A and x an element of A such that $Dx \leftrightarrow x$. Then:*

(i) $Dx^n = nx^{n-1}Dx$ $(n \geqslant 1)$,

(ii) $D^m x^n = x^{n-m}a_m$ *for some element* $a_m \in A$ $(1 \leqslant m < n, n \geqslant 2)$,

(iii) $D^n x^n = n!(Dx)^n + xb_n$ *for some element* $b_n \in A$ $(n \geqslant 1)$.

PROOF. We prove the results by induction.

(i) Assume that $Dx^{n-1} = (n-1)x^{x-2}Dx$. Then

$$\begin{aligned} Dx^n &= Dx^{n-1}.x + x^{n-1}Dx = (n-1)x^{x-2}Dx.x + x^{n-1}Dx \\ &= (n-1)x^{n-1}Dx + x^{n-1}Dx = nx^{n-1}Dx. \end{aligned}$$

(ii) Assume that $D^{m-1}x^n = x^{n-m+1}a_{m-1}$. Then

$$\begin{aligned} D^m x^n &= D(D^{m-1}x^n) = D(x^{n-m+1}a_{m-1}) \\ &= Dx^{n-m+1}.a_{m-1} + x^{n-m+1}Da_{m-1} \\ &= (n-m+1)x^{n-m}Dx.a_{m-1} + x^{n-m+1}Da_{m-1} \\ &= x^{n-m}a_m(\text{ with } a_m = (n-m+1)Dx + xDa_{m-1}). \end{aligned}$$

(iii) Assume that

$$D^{n-1}x^{n-1} = (n-1)!(Dx)^{n-1} + xb_{n-1}. \tag{1}$$

Then

$$\begin{aligned} D^n x^n &= D^{n-1}D(x^n) = D^{n-1}(nx^{n-1}Dx) = nD^{n-1}(x^{n-1}Dx) \\ &= nD^{n-1}x^{n-1}.Dx + n\sum_{r=1}^{n-1}\binom{n-1}{r} D^{n-1-r}x^{n-1}D^r Dx \\ &= n[(n-1)!(Dx)^{n-1} + xb_{n-1}]Dx + n\sum_{r=1}^{n-1}\binom{n-1}{r} x^r a_{n+r}D^{r+1}x. \\ &= n!(Dx)^n + xb_n, \end{aligned}$$

for some element b_n, where we have used result (ii) for evaluating each of summands is the second line.

7.6.11. Let A be an algebra (over F). For $a \in A$, set

$$D_a x = (\ell_a - r_a)x = ax - xa.$$

Then D_a is a derivation (see below) called an inner derivation. Further we have clearly the relations.

$$D_{a+b} = D_a + D_b; D_{\lambda a} = \lambda D_a \ (\lambda \in F).$$

7.6.12. LEMMA. (i) *D_a is a derivation*

(ii) *$D_a = 0$ iff $a \in Z$ (the centre of A); $D_a = 0$ for all a if A is commutative.*

(iii) *If A is a TA then D_a is continuous.*

(iv) *If $A = (A, \|\cdot\|)$ is a ρ-Banach algebra, with $\|\cdot\|$sm., then $\|D_a\| \leqslant 2\|a\|$.*

PROOF. (i) Since $D_a = \ell_a - r_a$ it is linear. Further

$$\begin{aligned} D_a(xy) &= axy - xya = (ax - xa)y + x(ay - ya) \\ &= D_a x.y + xD_a y. \end{aligned}$$

(ii) Clear.

(iii) The continuity of D_a follows from that of ℓ_a, r_a.

(iv) $\|D_a(x)\| = \|ax - ax\| \leqslant 2\|a\| \ \|x\|$.

7.6.13. Remark. The inner derivations form on ideal[†] $\mathcal{D}_i$ of the Lie algebra $\mathcal{D}$ of all derivations. To see this it is enough to observe that for any derivation $D, [D, D_a] = D_b$, where $b = Da$, (this can be easily checked).

7.6.14. LEMMA. $(D_a)^n(x) = \sum_{r=0}^{n}(-1)^r \binom{n}{r} a^{n-r} x a^r$

PROOF. The proof is by induction. Assume that

[†] Since a Lie algebra is anti-commutative ($[x, y] = -[y, x]$) every ideal is a bi-ideal.

$$(D_a)^{n-1}(x) = \sum_{r=0}^{n-1}(-1)^r \binom{n-1}{r} a^{n-1-r}xa^r.$$

Then $(D_a)^n(x) = D_a(D_a)^{n-1}(x) = D_a \sum_{r=0}^{n-1}$

$$= \sum_{r=0}^{n-1}(-1)^r\binom{n-1}{r}a^{n-r}xa^r - \sum_{r=0}^{n-1}(-1)^r\binom{n-1}{r}a^{n-r-1}xa^{r+1}$$

$$= a^n x + \sum_{r=1}^{n-1}(-1)^r\left[\binom{n-1}{r}+\binom{n-1}{r-1}\right]a^{n-r}xa^r - (-1)^{n-1}xa^n$$

$$= \sum_{r=0}^{n}(-1)^r\binom{n}{r}a^{n-r}xa^r.$$

7.6.15. PROPOSITION. *Any derivation D of an algebra A can be uniquely extended to its unitization A_1.*

PROOF. Define $D_1(\lambda e + x) = Dx$, where $x \in A, e_1$ is the unity of A_1 and $\lambda \in F$. It is straightforward to check that D_1 is a derivation of A_1. The uniqueness assertion follows from the fact that for any derivation D' of $A_1, D'(\lambda e) = 0$, by 7.6.7 (iii).

7.6.16. PROPOSITION. *Let A be a real algebra, D a derivation of A and $\tilde{A}$ the complexification of A. The (unique) linear extension $\tilde{D}$ of D to $\tilde{A}$ is a derivation of $\tilde{A}$.*

PROOF. $\tilde{D}(x+iy) = Dx+iDy$. It is straightforward to verify that $\tilde{D}$ is a derivation of $\tilde{A}$.

7.6.17. LEMMA. *Let A, A^* be algebras and $\varphi : A \to A^*$ an epimorphism. If D is a derivation of A such that $D(\ker\varphi) \subseteq \ker\varphi$ then D induces a derivation D^* of A^*.*

PROOF. If $x^* \in A^*, x^* = \varphi(x)$ set $D^*x^* = \varphi(Dx)$. Then D^* is a well-define map on A^* since if $\varphi(x) = \varphi(y), \varphi(x-y) = 0$ and $\varphi(Dx) - \varphi(Dy) = \varphi(D(x-y)) = 0$ (since $x - y \in \ker\varphi, D(\ker\varphi) \subseteq \ker\varphi$). Further, D^* is clearly linear and

$$\begin{aligned}D^*(x^*y^*) &= \varphi(Dxy) = \varphi(Dx.y + x.Dy) = \varphi(Dx)\varphi(y) + \varphi(x)\varphi(xy)\\ &= D^*x^*.y^* + x^*.D^*y^*.\end{aligned}$$

7.6.18. PROPOSITION. *Let A be a ρ-Banach algebra and D a continuous derivation of A. Then $E(D) = \operatorname{Exp} D$ is a continuous automorphism of A.*

PROOF. By 5.2.4, E(D) is an invertible, bounded – hence continuous – linear operator. So to prove the proposition we have only to show that $E(D)$ preserves multiplication. Now

$$\begin{aligned} E(D)(xy) &= \sum_{n=0}^{\infty} \frac{1}{n!} D^n(xy) = \sum_{n=0}^{\infty} \frac{1}{n!} \sum_{r=0}^{\infty} \binom{n}{r} (D^{n-r}x)(D^r y_x) \\ &= \sum_{n=0}^{\infty} \sum_{r=0}^{n} \left(\frac{1}{(n-r)!} D^{n-r} x \right) \left(\frac{1}{r!} D^r y \right) \\ &= {}^{\dagger} \left(\sum_{n=0}^{\infty} \frac{1}{n!} D^n x \right) \left(\sum_{n=0}^{\infty} \frac{1}{n!} D^n y \right) \\ &= (E(D)x)(E(D)y), \end{aligned}$$

completing the proof.

7.6.19. PROPOSITION. *In a unital ρ-Banach algebra A, $E(D_a) = I_{E(a)} (a \in A)$.*

PROOF. Using 7.6.14 we get

$$\begin{aligned} E(D_a)(x) &= \sum_{n=0}^{\infty} \frac{D_a^n x}{n!} = \sum_{n=0}^{\infty} \frac{1}{n!} \left(\sum_{r=0}^{n} (-1)^r \binom{n}{r} a^{n-r} x a^r \right) \\ &= \sum_{n=0}^{\infty} \sum_{r=0}^{n} \frac{1}{(n-r)!} a^{n-r} x (-1)^r \frac{a^r}{r!} \\ &= \sum_{n=0}^{\infty} \frac{a^n}{n!} x \sum_{n=0}^{\infty} (-1)^n \frac{a^n}{n!} \\ &= E(a) x E(-a) = I_{E(a)}(x). \end{aligned}$$

† Note that the expression on the LHS of the equality gives the Cauchy product of the two series on the RHS.

7.6.20. THEOREM (Singer-Wermer)†. *Let A be commutative ρ-Banach algebra which is either complex or strictly real. If D is any continuous derivation of A then $D(A) \subseteq \sqrt{A}$.*

PROOF (following Sinclair). First assume that A is complex and also that it is unital. For $a \in A$, $\chi \in \Delta = \Delta_c$, write $x = a - \chi(a)e$. Then $x \in \ker \chi$ and $Dx = Da$. By 7.6.10 (iii),

$$D^n x^n = n!(Dx)^n + xb_n.$$

Therefore $\chi(D^n x^n) = n!(\chi(Dx))^n$ (since $\chi(x) = 0$). So $\chi(Dx) = (n!)^{-\frac{1}{n}}\chi(D^n x^n)^{\frac{1}{n}}$, so that

$$\begin{aligned} |\chi(Dx)| &\leqslant (n!)^{-\frac{1}{n}} \mid \chi(D^n x^n)^{\frac{1}{n}} \\ &\leqslant (n!)^{-\frac{1}{n}} \|D^n(x^n\|^{\frac{1}{n\rho}} \text{ (by 7.3.8)} \\ &\leqslant \mid (n!)^{-\frac{1}{n}}(\|D^n\|\|x^n\|)^{\frac{1}{n\rho}}, \end{aligned}$$

where $\|D^n\|$ denote the bound of D^n (for definition see 3.5.1). The above inequality, gives

$$|\chi(Dx)| \leqslant (n!)^{-\frac{1}{n}}\|D\|^{\frac{1}{\rho}}(\|x^n\|\frac{1}{n})^{\frac{1}{\rho}}.$$

Making $n \to \infty$ we conclude (since $(n!)^{-\frac{1}{n}} \to 0$) that $|\chi(Dx)| = 0, Dx \in \ker \chi$.

$Dx \in \sqrt[h]{A} = \sqrt{A}$ (since A is Gelfand).

If A has no unity, consider its unitization A_1. By 7.6.15, D can be extended to D_1 on A_1. By what we have proved above we get for $x \in A$,

$$Dx = D_1 x \in \sqrt[h]{A_1} \bigcap A = \sqrt[h]{A} \text{ (by 1.4.9(c))}.$$

Now we may assume that $A \neq \sqrt{A}$ (if $A = \sqrt{A}$, trivially $Dx \in \sqrt{A}$). Then by 7.2.17, 7.2.8, and 7.2.12 $\sqrt[h]{A} = \sqrt{A}$. Thus, $Dx \in \sqrt{A}$.

Finally, if A is strictly real consider the complexification $\tilde{A}$. By 7.6.16 D has an extension $\tilde{D}$ to $\tilde{A}$. Then $DA \subseteq \tilde{D}\tilde{A} \subseteq \sqrt{\tilde{A}}$. Also $DA \subseteq A$. So

---

† Singer-Wermer proved the result for commutative complex Banach algebras.

$$DA \subseteq A \bigcap \sqrt{\tilde{A}} = \sqrt{A} \text{ (by 1.9.17).}$$

7.6.21. COROLLARY. *Let D be a continuous derivation of a complex or strictly real, commutative pseudo-Michael algebra $A = (A, \{p_\alpha\})$ such that each $\ker p_\alpha$ is invariant under $D : D(\ker p_\alpha) \subseteq \ker p_\alpha$. Then $D(A) \subseteq \sqrt{A}$.*

PROOF. Let $A = \varprojlim \hat{A}_\alpha$ be the projective limit decomposition. The continuous derivation D induces, by 7.6.17, a derivation D_α on A_α which can be extended to a derivation $\hat{D}_\alpha$ on $\hat{A}_\alpha$. Since $\hat{A}_\alpha$ is a ρ-Banach algebra we have by 7.6.20, for any $x \in A$, $\hat{D}_\alpha(x_\alpha) \subseteq \sqrt{\hat{A}_\alpha}$. It follows that

$$\varphi_\alpha(Dx) = \hat{D}_\alpha(x_\alpha) \subseteq \sqrt{\hat{A}_\alpha},$$

and hence

$$Dx \in \bigcap_\alpha \varphi_\alpha^{-1} \sqrt{\hat{A}_\alpha} \subseteq \sqrt{A} \quad \text{(see 4.5.8).}$$

7.6.22. Remark. It follows from 7.6.20 that in a complex or strictly real commutative s.s. ρ-Banach algebra, 0 is the only continuous derivation. Johnson has shown that in the case of a complex commutative s.s. Banach algebra, 0 is also the only derivation (continuous or not). Further, Johnson and Sinclair have proved that all derivations on any s.s. Banach algebra are continuous (see [4, pp.93, 95]).

CHAPTER VIII

COMMUTATIVE[†] TOPOLOGICAL ALGEBRAS

§ 1. Function Algebras[††]

8.1.1. DEFINITION. An algebra A over $\mathbb{K}$ is called a *function algebra* if there is a set S and every element of A is a $\mathbb{K}$-valued funtion f on S and the algebra operations are point-wise; we call A a function algebra *over* S and write $A = \mathfrak{F}(S) = \mathfrak{F} = \{f : f \in A\}$. If $\mathfrak{F}$ is ρ-normed (respy. ρ-Banach) we call it a ρ-normed (respy. ρ-Banach) function algebra. $\mathfrak{F}$ is called a *canonically* ρ-normed (respy. *canonically* ρ*-Banach*) funtion algebra iff all $f \in \mathfrak{F}$ are bounded and the ρ-norm of A is given by $\|f\|_\infty^\rho$, where $\|f\|_\infty = \sup_{s\in S} |f(s)|$.

8.1.2. Examples.

(i) If X is a locally compact Hausdorff space, $C_0(X)$ the algebra of $\mathbb{K}$-valued continuous functions on X vanishing at ∞, then under the sup norm it is a (canonically) Banach function algebra.

(ii) If $\mathfrak{F} = \mathfrak{F}(S)$ is a ρ-seminormed Gelfand algebra then $\hat{\mathfrak{F}}$ is a canonically normed function algebra.

(iii) The Banach algebra A of holomorphic functions (defined in 7.3.22 (iii)) is a canonically Banach function algebra.

(iv) The algebra $\mathsf{C}(\mathbb{R})$ (see 4.6.8 (i)) is a locally sm. $\mathfrak{F}$ function algebra.

[†] Some of the theory developed in this chapter, especially in §3, §4 deals with non-commutative algebras (=algebras not necessarily commutative).

[††] The treatment here is very limited. For more information on the topic consult the books of *A.* Browder (Introduction to function algebras, Benjamen, 1969) and T.W. Gamelin (Uniform algberas, Prentice Hall, 1969).

8.1.3. Remark. Every function algebra is evidently commutative.

8.1.4. DEFINITION. Let $\mathfrak{F} = \mathfrak{F}(S)$ be a funtion algebra. For a subset $\mathfrak{F}_0$ of $\mathfrak{F}$ we define

$$\ker \mathfrak{F}_0 = \{s \in S : f(s) = 0, \quad \forall\, f \in \mathfrak{F}_0\}.$$

8.1.5. LEMMA. *Let $\mathfrak{F} = \mathfrak{F}(S)$ be a function algebra. Then for an $s \in S \setminus \ker \mathfrak{F}$, $\chi_s : f \mapsto f(s)$ is a character of $\mathfrak{F}$ and $M_s = \ker \chi_s$ a (hyper) maximal ideal of $\mathfrak{F}$.*

PROOF. By choice of s, $\chi_s \neq 0$ and so χ_s is surjective and hence a character.

8.1.6. DEFINITION. A maximal ideal of the form M_s ($= \ker \chi_s = \{f \in \mathfrak{F} : f(s) = 0\}$) is called *fixed* or a fixed maximal ideal.

8.1.7. LEMMA. The ρ-norm $\|\cdot\|$ of a canonically ρ-normed algebra $\mathfrak{F} = \mathfrak{F}(S)$ satisfies the condition

$$\|f^2\| = \|f\|^2 \qquad (f \in \mathfrak{F}).$$

PROOF. $\|f^2\| = (\sup_s |f^2(s)|)^\rho = ((\sup_s |f(s)|)^\rho)^2 = \|f\|^2$.

8.1.8. LEMMA. *Let $\mathfrak{F} = \mathfrak{F}(S)$ be a function algebra with $\ker \mathfrak{F} = \emptyset$. Then:*

(i) *For $f \in \mathfrak{F}$, the range $\mathcal{R}(f) \subseteq \sigma'(f)$.*

(ii) *If $\mathfrak{F}$ is unital, $\mathcal{R}(f) \subseteq \sigma(f)$.*

(iii) *$\mathfrak{F}$ is h.s.s. and in particular q.s.s. and s.s..*

PROOF. (i) $f(s) = \chi_s(f) \in \sigma'(f)$, by 1.7.11.

(ii) This follows similarly from 1.7.24.

(iii) If $f \in \sqrt[h]{\mathfrak{F}}$ then $f(s) = \chi_s(f) = 0$ for all s and so $f = 0$. Thus $\sqrt[h]{\mathfrak{F}} = \{0\}$. Since, by 1.7.15,

$$\sqrt{\mathfrak{F}} \subseteq \sqrt[qn]{\mathfrak{F}} \subseteq \sqrt[h]{\mathfrak{F}} = \{0\},$$

$\mathfrak{F}$ is also q.s.s. and s.s. as required.

8.1.9. PROPOSITION. *Let $A = \mathfrak{F}(S)$ be a ρ-normed functionally continuous function algebra with its ρ-norm $\|\cdot\|$ sm. and*

$\ker A = \emptyset$. *Then every* $f \in A$ *is bounded and*

$$\|f\|_\infty \leqslant \|\hat{f}\|_\infty \leqslant \nu(f)^{\frac{1}{\rho}} \leqslant \|f\|^{\frac{1}{\rho}}.$$

In particular, the ρ *-norm* $\|\cdot\|$ *of* A *dominates the canonical* ρ *-norm* $\|\cdot\|_\infty^\rho$.

PROOF. Since A is functionally continuous and $\ker A = \emptyset$, for each $s \in S$,

$\chi_s \in \Delta = \Delta_c$, where χ_s is given by $\chi_s(f) = f(s) \quad (f \in A)$. By virtue of $(*)$ of 7.3.8 we have

$$\|f\|_\infty = \sup_s |f(s)| = \sup_s \chi_s(f) \leqslant \|\hat{f}\|_\infty \leqslant \nu(f)^{\frac{1}{\rho}} \leqslant \|f\|^{\frac{1}{\rho}},$$

completing the proof.

8.1.10. Let S be any set, $\mathbb{K}^S$ the set of all $\mathbb{K}$-valued functions on S. $\mathbb{K}^S$ is, under pointwise operations, a unital algebra over $\mathbb{K}$; the unity element of $\mathbb{K}^S$ is the constant function $1 : 1(s) = s \quad (\forall\ s \in S)$. $\mathbb{K}^S$ is a TA under the weak topology: a net $f_\alpha \to f$ in $\mathbb{K}^S$ iff $f_\alpha(s) \to f(s)$ for each $s \in S$. The subset $\mathsf{B}(S) = \mathsf{B}(S, \mathbb{K})$ of $\mathbb{K}^S$ comprising all bounded functions is a subalgebra which inherits the relative topology under which it is a TA. $\mathsf{B}(S)$ is also a Banach algebra under the sup norm $\|\cdot\|_\infty : \|f\|_\infty = \sup_s |f(s)|$. Since norm convergence clearly implies weak convergence, weak topology of $\mathsf{B}(S)$ is coarser than its norm topology.

If S is a topological space we have besides the algebra $\mathsf{B}(S)$ the algebra $\mathsf{C}(S)$ of all continuous functions on S. Further,

$$\mathsf{BC}(S) = \mathsf{B}(S) \bigcap \mathsf{C}(S)$$

is a subalgebra of both $\mathsf{B}(S)$ and $C(S)$.

Now let S be locally compact Hausdorff. A function $f \in \mathsf{C}(S)$ is said to ***vanish at*** ∞ if given $\epsilon > 0$ there is a compact set K in S such that $|f(s)| < \epsilon$ for all $s \in S \backslash K$. All functions in $\mathsf{C}(S)$ which vanish at ∞ form a subalgebra $\mathsf{C}_0(S)$ of $\mathsf{C}(S)$. Note that $\mathsf{C}_0(S) = \mathsf{C}(S)$ if S is compact.

8.1.11. PROPOSITION. $\mathsf{C}(S)$ *for compact Hausdorff* S *and*

$\mathsf{C}_0(S)$ *for locally compact Hausdorff* S *are Banach algebras under the sup norm. Moreover,* $\mathsf{C}(S)$, $C_0(S)$ *are s.s.*.

PROOF. The proof of the first statement being straightforward is omitted. For the second we note that for $\mathfrak{F} = \mathsf{C}(S)$ or $\mathsf{C}_0(S)$, $\ker \mathfrak{F} = \emptyset$, so that $\{M_s : s \in S\}$ are all maximal ideals, and clearly $\bigcap M_s = \{0\}$, whence $\mathcal{F}$ is s.s..

8.1.12. DEFINITION. Let $\mathfrak{F} = \mathfrak{F}(S)$ be a function algebra. A subset $\mathfrak{F}_0$ of $\mathfrak{F}$ is said to ***separate points*** (of S) if given any pair (s_1, s_2) of distinct points there is an $f \in \mathfrak{F}_0$ with $f(s_1) \neq f(s_2)$. $\mathfrak{F}_0$ is said to ***strongly separate points*** if it satisfies in addition the condition $\ker \mathfrak{F}_0 = \emptyset$. A family $\mathfrak{F}_0$ of functions separating (respy. strongly separating) points is also referred to as a separating (respy. strongly separating) family.

8.1.13. Remark. Let S be a locally compact Hausdorff space. Consider the Banach of algebra $\mathsf{C}_0(S)$.

Any strongly separating subalgebra A of $\mathsf{C}_0(S)$ which is s.a. (i.e. closed for conjugates) is dense by the extended Stone-Weirstrass theorem (see [26, pp.166-7, Theorems A,B]). Conversely, we have the result that any dense subalgebra A of $C_0(S)$ is strongly separating. To see this, observe first that S being locally compact Hausdorff is completely regular. So for any point $s_0 \in S$ we can choose $f \in \mathsf{C}_0(S)$ with $f(s_0) \neq 0$. By density of A we can choose $g \in A$ with $\|g - f\|_\infty < |f(s_0)|$. Then $g(s_0) \neq 0$. Again, if $s_1 \neq s_2$ are two points of S and $f_1 \in \mathsf{C}_0(S)$ is such that $f_1(s_1) \neq f_1(s_2)$, we can choose $g_1 \in A$ with $\|g_1 - f_1\|_\infty < \frac{1}{2}|f_1(s_1) - f_1(s_2)|$. Then $g_1(s_1) \neq g_1(s_2)$. Thus A is strongly separating.

8.1.14. LEMMA (Rickart). *Let* $\mathfrak{F} = \mathfrak{F}(S)$ *be a function algebra which strongly separates points of* S. *Then, given any finite set of points* $s_0, s_1, \cdots, s_n$ $(n \geqslant 1)$ *of* S, *we can find an* $f \in \mathfrak{F}$ *with*

$$f(s_0) \neq 0,\ f(s_j) = 0\ (j = 1, 2, \cdots, n).$$

PROOF. We first assume that $n = 1$ and write $s_0 = s$, $s_1 = t$ $(s \neq t)$. We shall shows that there is a function $f_s \in \mathfrak{F}$ with $f_s(s) \neq 0$, $f_s(t) = 0$.

By hypothesis there is an $f \in \mathfrak{F}$ with $f(s) \neq f(t)$. We have to consider three cases.

Case 1: $f(t) = 0$. We can take $f_s = f$.

Case 2: $f(s) \neq 0$, $f(t) \neq 0$. By replacing f by a suitable multiple we may assume that $f(t) = 1$ and then $f(s) \neq 1$. Set

$$f_s = f - f^2; \quad \text{then } f_s(t) = 0,\ f_s(s) \neq 0.$$

Case 3: $f(s) = 0$, $f(t) = 1$. Choose a $g \in \mathfrak{F}$ with $g(s) \neq 0$ (this is possible since $\ker \mathfrak{F} = \emptyset$). If $g(t) = 0$, we can take $f_s = g$, and if $g(t) \neq 0$ we take $f_s = f - g/g(t)$.

We next assume that $n \geqslant 2$. By the result for $n = 1$ just proved above we can choose $f_j \in \mathfrak{F}$ such that $f_j(s_0) \neq 0$, $f_j(s_j) = 0$ $(j = 1, \cdots, n)$. Set $f = f_1 \cdots f_n$ (product). Then $f(s_0) \neq 0$, while $f(s_j) = 0$ $(j = 1, \cdots n)$.

8.1.15. LEMMA. *Let $S = (S, \tau)$ be a compact Hausdorff space. Then the weak topology*† *τ_w on S induced by any separating family $\mathfrak{F}_0$ of $C(S)$ coincides with the initial topology τ.*

PROOF. τ_w is clearly coarser then τ. Moreover, τ_w is Hausdorff since $\mathfrak{F}_0$ is separating S. It follows that $\tau = \tau_w$ (since $\tau_w \subseteq \tau$, τ_w Hausdorff, τ compact).

8.1.16. COROLLARY. *If $S = (S, \tau)$ is locally compact Hausdorff and $\mathsf{C}_0(S)$ the algebra of $\mathbb{K}$-valued continuous functions on S which vanish at ∞, then the weak topology τ_w on S induced by any strongly separating family $\mathfrak{F}_0$ of $\mathsf{C}(S)$ coincides with τ.*

PROOF. Let $S_\infty = S \bigcup \{\infty\}$ be the 1-point compactification of S; S_∞ is compact Hausdorff. Extend the functions f in $\mathfrak{F}_0$ to S_∞ by defining $f(\infty) = 0$. Then the extended functions form a separating family of $\vec{C}(S_\infty)$. The required result now follows from 8.1.15.

8.1.17 PROPOSITION. *Let S be a compact Hausdorff space and A a s.a.*†† *inverse-closed subunital algebra of $\mathsf{C}(S)$. Then*

---

† A net $s_\alpha \to s$ in S under τ_w if $f(s_\alpha) \to f(s)$, $\forall\, f \in \mathfrak{F}_0$.

†† i.e., $f \in A \Rightarrow \overline{f} \in A$, where $f = f(s)$, $\overline{f}(s) = \overline{f(s)}$, bar denoting complex conjugate.

every maximal ideal M of A is fixed, i.e. of the form

$$M = M_s = \{f \in A : f(s) = 0\}, \text{ for some } s \in S.$$

In particular, every maximal ideal of $C(S)$ is fixed.

PROOF. Suppose that A has a maximal ideal M which is not fixed. Since $1 \in A$ $\ker A = \emptyset$, and by 8.1.5, for any $s \in S$, M_s is a (hyper) maximal ideal. So by our supposition $M \neq M_s$. By maximality of M, $M \not\subseteq M_s$, whence there is an $f_s \in M$ with $f_s(s) \neq 0$. By the continuity of f_s we have $f_s \neq 0$ on an open neighbourhood U_s of s. Since S is compact there is a finite number of points $s_1 \cdots, s_n$ in S with $\bigcup_{j=1}^{n} U_j = S$, where $U_j = U_{s_j}$. Set

$$f = f_1\bar{f}_1 + \cdots f_n\bar{f}_n, \text{ where } f_j = f_{s_j} \quad (j = 1, \cdots, n).$$

Then $f \in M$. Since

$$f = |f_1|^2 + \cdots + |f_n|^2 > 0,$$

f^{-1} exists in $\mathsf{C}(S)$. Since A is inverse-closed in $\mathsf{C}(S)$, $f^{-1} \in A$. Thus, M contains an invertible element f, which is impossible. So we must have $M = M_s$ for some s, completing the proof.

8.1.18. COROLLARY. *Let S be a locally compact Hausdorff space and $\mathsf{C}_0(S)$ the algebra of $\mathbb{K}$-valued continuous functions vanishing at ∞. Then every maximal ideal of $\mathsf{C}_0(S)$ is fixed.*

PROOF. Consider the 1-point compactification S_∞ of S. Then $\mathsf{C}(S_\infty)$ is† the unitization A_1 of $A = \mathsf{C}_0(S)$. The maximal ideals of $A_1 = \mathsf{C}(S_\infty)$ are, by 8.1.17, all fixed and they are $\{M_s^1 : s \in S_\infty\}$, where $M_s^1 = \{f_1 \in A_1 : f_1(s) = 0\}$. It follows that the maximal ideals of $\mathsf{C}_0(S)$ are

$$M_s^1 \bigcap A = M_s (s \in A),$$

where $M_s = \{f \in A : f(s) = 0\}$, and so are fixed.

8.1.19. Remark. In the proof of 8.1.17, the hypothesis that

† i.e. can be identified with.

A is s.a. was explicitty used. However, it is possible, for a subalgebra A of $\mathsf{C}(S)$ which is not s.a. to have the fixity property for the maximal ideals. For instance, the Banach algebra $\mathfrak{A}$ of example (ii) of 7.3.22 is a subalgebra of $\mathsf{C}(D)$, which is not s.a. ($\bar{f}_0 \notin \mathfrak{A}; f_0(z) = z$). $\mathfrak{A}$ has, nevertheless, the fixity property. In fact, if M is a maximal ideal of $\mathfrak{A}$ and $f_0(M) = z_0$ then $M = M_{z_0}$ (see [10, p.33]).

8.1.20. PROPOSITION. *Let the algebra $A = \mathsf{C}(S)$, where S is compact Hausdorff, be a TA under a topology τ which is finer than the weak topology τ_w. Then A is Gelfand and its spectrum $\Delta = \Delta_c$ is homeomorphic to S.*

PROOF. By 8.1.17, if $\chi \in \Delta = \Delta(A)$, then $\chi = \chi_s$ ($s \in S$). Since $\chi_s(f) = f(s)$, by definition of τ_w, χ_s is τ_w-continuous ($f_\alpha \to f$ if $\chi_s(f_\alpha) = f_\alpha(s) \to f(s) = \chi_s(f)$). Since $\tau \supseteq \tau_w, \chi_s$ is also τ-continuous. It follows that the maximal ideals $M = M_s (s \in S)$ are τ-closed, hypermaximal and so $A = (A, \tau)$ is Gelfand. Consider now the map $\Lambda : s \mapsto \chi_s$ of S to Δ. By 8.1.17, Λ is surjective. It is also injective since S is normal if $s_1 \neq s_2$, there is a $f \in \mathsf{C}(S)$ with $f(s_1) \neq f(s_2)$, so that $\chi_{s_1} \neq \chi_{s_2}$. Further, Λ is continuous, since if $s_\alpha \to s$ and f is continuous,

$$\chi_{s_\alpha}(f) = f(s_\alpha) \to f(s) = \chi_s(f).$$

Finally, since S is compact and Δ is Hausdorff (see 7.3.2) it follows that Λ is a homeomorphism.

8.1.21. COROLLARY. *The Banach algebra $(\mathsf{C}(S), \|\cdot\|_\infty)$, or more generally, the ρ-Banach algebra $(\mathsf{C}(S), \|\cdot\|_\infty^\rho)$ has S for its spectrum.*

PROOF. This follows from 8.1.20 by taking τ to be the norm topology.

8.1.22. THEOREM (Gelfand-Kolmogorov-Stone-Banach). *If S_1, S_2 are compact Hausdorff spaces such that the Banach algebras $\mathsf{C}(S_1)$ and $\mathsf{C}(S_2)$ are isomorphic (as algebras) then S_1 and S_2 are homeomorphic.*

PROOF. Since the spectrum Δ of the Banach algebra $A = \mathsf{C}(S)$ depends only on the algebraic structure of A, we have, by 8.1.21, $S_1 \simeq \Delta_1 \simeq \Delta_2 \simeq S_2$, where $\simeq$ stands for homeomorphism.

8.1.23. Remark. In the algebra $A = \mathsf{C}(\mathbb{R}) = \mathsf{C}(\mathbb{R}, \mathbb{K})$, *not* every maximal ideal is fixed. For, let

$$I = \{f \in A : f(t) = 0 \text{ for all } t > \text{ some } t_f\}.$$

Clearly I is an ideal and by Krull's lemma we can find a maximal ideal M containing I. It follows from the definition of I that M cannot be a fixed ideal.

8.1.24. Remark. Let S be a completely regular T_1-space and $A = \mathsf{BC}(S)$ the algebra of bounded continuous functions on S. Then A is a unital commutative Banach function algebra. If Δ is its spectrum then Δ is compact Hausdorff. It can be shown that Δ is the Stone-Čech compactification $\beta(S)$ of S (see [26, p.331] or [28, p.415]).

8.1.25. Theorem. *Let $A = (A, \|\cdot\|)$ be an ample, complex or strictly real, ρ-Banach algebra, with $\|\cdot\|$ sm.. Then A is t. isomorphic to a canonically Banach function algebra iff $\|\cdot\|$ satisfies the conditions*

$$C\|x\|^2 \leqslant \|x^2\| \text{ for all } x, \tag{$*$}$$

where C is some positive constant (cf. [28, p.409, Theorem. 4.8.]).

PROOF. Let $\|\cdot\|$ satisfy $(*)$. Then, by 7.4.15 or 7.4.19, A is commutative. Also, by 7.4.16, A is s.s.. By 7.4.12, the map $\mathcal{G} : x \mapsto \hat{x}$ is a t. isomorphism of A onto $\hat{A}$ – which is a canonically Banach function algebra.

Conversely, suppose that A is t. isomorphic to a canonically Banach function algebra $\mathfrak{F} = (\mathfrak{F}, \|\cdot\|^*)$. Then, by 8.1.7, we have

$$(\|y\|^*)^2 = \|y^2\|^* \quad (y \in \mathfrak{F}).$$

It follows from 7.4.11 that there is a constant C_2 with

$$C_2\|y\|^* \leqslant r(y)^\rho \quad (y \in \mathfrak{F}).$$

Let φ denote the t. isomorphism between A and $\mathfrak{F}$. Then we can transfer the norm of $\mathfrak{F}$ to A by setting for $x \in A$, $y = \varphi(x)$

$$\|x\|^* = \|\varphi(x)\|^* = \|y\|^*$$

then $\|\cdot\|^* \sim \|\cdot\|$. Since $r(y) = r(x)$, we get

$$C_2\|x\|^* \leqslant r(x)^\rho.$$

Since $\|\cdot\|^* \sim \|\cdot\|$ then is a constant $C_1 > 0$ such that $\|x\| \leqslant C_1\|x\|^*$. Therefore we get

$$\frac{C_2}{C_1}\|x\| \leqslant r(x)^\rho$$

whence, by 7.4.11, there is a $C > 0$ with

$$C\|x\|^2 \leqslant \|x^2\|$$

as required.

8.1.26. LEMMA. *Let A be a ρ-seminormed algebra. Then the Gelfand transform $\hat{A}$ is always a strongly separating subalgebra of $C_0(\Delta_c)$.*

PROOF. If $\chi \in \Delta_c$ and $a \in A\setminus \ker\chi$ then $\hat{a}(\chi) = \chi(a) \neq 0$. Again, if $\chi_1, \chi_2 \in \Delta_c$ and $\chi_1 \neq \chi_2$ there is an element $b \in A$ with $\chi_1(b) \neq \chi_2(b)$, and then $\hat{b}(\chi_1) \neq \hat{b}(\chi_2)$. Hence the lemma.

§ 2. Shilov Boundary

8.2.1. DEFINITION. Let S be a locally compact Hausdorff space, $\mathsf{C}_0(S) = \mathsf{C}_0(S, \mathbb{K})$ the algebra of all continuous $\mathbb{K}$-valued functions on S which vanish at ∞, and $A \subseteq \mathsf{C}_0(S)$ a subalgebra. A closed set $F \subseteq S$ is called a *pre-boundary* for A if for each $f \in A$ there is a point $s_f \in F$ such that $|f(s_f)| = \|f\|_\infty$.

8.2.2. Remark. The set S itself is a pre-boundary for A. For, if $f \in A, f \neq 0$, $f(s_0) \neq 0$ there is a compact set K such that $(*)|f(s)| < |f(s_0)|$ for $s \in S \backslash K$. Then $s_0 \in K$. Since K is compact there is a $s_1 \in K$ with $|f(s_1)| = \sup_{s \in K} |f(s)| = \sup_{s \in S} |f(s)|$ (by virtue of $(*)$).

8.2.3. DEFINITION. A pre-boundary (for A) is called a *Shilov boundary* or boundary if it is contained in every pre-boundary (for A). The Shilov boundary, whenever it exists, is clearly unique and is denoted by $\partial_A S$.

A pre-boundary F_0 is called a *minimal* pre-boundary if there is no other pre-boundary contained in F_0. Evidently $\partial_A S$ is a minimal pre-boundary.

8.2.4. LEMMA. *Every pre-boundary F_1 contains a minimal pre-boundary F_0.*

PROOF. The family $\mathcal{F}_1$ of pre-boundaries contained in F_1 is clearly a poset with respect to set-inclusion. By Zorn's lemma (or rather by the equivalent Hausdorff maximal principle[†]) there is a maximal chain $\mathfrak{F}_0$ in $\mathfrak{F}_1$ with $F_1 \in \mathfrak{F}_0$. Denote by F_0 the intersection of all sets F in $\mathfrak{F}_0$; then F_0 is closed. We claim F_0 is a pre-boundary. To see this, set for $f \in A$,

$$F_f = \{s \in S : |f(s)| = \|f\|_\infty\}.$$

If $f = 0, F_f = S$ and $F_0 \subseteq F_f$. If $f \neq 0$ then F_f is compact (since f vanishes at ∞). Since each $F \in \mathcal{F}_0$ is a pre-boundary, $F_f \bigcap F \neq \emptyset$ and further it is a closed subset of compact F_f. It follows that

$$F_f \bigcap F_0 = F_f \bigcap_{F \in \mathcal{F}_0} F) = \bigcap_{F \in \mathcal{F}_0} (F_f \bigcap F) \neq \emptyset.$$

Since f is an arbitary function in A the above relation shows that F_0 is a pre-boundary. Finally, by its construction F_0 is a minimal pre-boundary.

8.2.5. LEMMA. *Let F be a minimal pre-boundary for A. Then an element $s_0 \in S$ belongs to F iff for every open neighbourhood V of s_0 there exists an $f \in A$ such that*

[†] See [16, pp.31-36].

$$\sup_{s \in S \setminus V} |f(s)| < \sup_{s \in V} |f(s)| = \|f\|_\infty. \qquad (*)$$

PROOF. If $s_0 \in F$ then $F \setminus V$ is not a pre-boundary and so there is an f satisfying $(*)$. Conversely, suppose for $s_0 \in S$ and an arbitrary open neighbourhood V of s_0 the inequality $(*)$ holds for some f. Then since $\|f\|_\infty = \sup_{s \in F} |f(s)|$, $F \not\subseteq S \setminus V$, so that $V \bigcap F \neq \emptyset$. It follows that $s_0 \in \overline{F} = F$.

8.2.6. LEMMA. *If S is infinite and $A \subseteq \mathsf{C}_0(S)$ a strongly separating subalgebra, then every pre-boundary F (for A) is infinite.*

PROOF. If possible let $F = \{s_1, \cdots, s_n\}$ and s_0 any point of $S \setminus F$. By 8.1.14, we can choose $f \in A$ with

$$f(s_0) \neq 0, \quad f(s_j) = 0 \ (j = 1, \cdots, n).$$

Then $\|f\|_\infty > 0$ and $f = 0$ on F, contradicting that F is a pre-boundary. So F is infinite as required.

8.2.7. PROPOSITION. *The Shilov boundary $\partial_A S$ exists for every strongly separating subalgebra A of $\mathsf{C}_0(S)$, where S is a locally compact Hausdorff space.*

PROOF (cf. [23, pp.133-4]). First assume that S is finite, say, $S = \{s_1, \cdots, s_n\}$. Then S being compact and discrete, $\mathsf{C}_0(S) = \mathsf{C}(S) = \mathbb{K}^n$. By 8.1.14, there is a $f_j \in A (j = 1, \cdots, n)$ such that

$$f_j(s_j) \neq 0, \ f_j(s_k) = 0 \ (k \neq j).$$

Since $\|f_j\|_\infty$ is attained only at s_j, every pre-boundary F must contain all the s_j which means that $F = S$. Hence $\partial_A S$ exists and $\partial_A S = S$.

Next assume that S is infinite; by 8.2.6, every pre-boundary F is infinite. By 8.2.4, we can find a minimal pre-boundary F_0. Let F be any pre-boundary. We shall show that $F_0 \subseteq F$.

Take any point $s_0 \in F_0$ and let N be any open neighbourhood of s_0; since F_0 is infinite we can certainly select $t_0 \in F_0$ with $t_0 \neq s_0$. Set $N_0 = N \setminus \{t_0\}$. Then $N_0 \subseteq N$ and we have $\emptyset \neq F_0 \setminus N_0 \neq F_0$ (since $t_0 \in F_0 \setminus N_0$, $s_0 \in F_0, s_0 \notin F_0 \setminus N_0$).

Since N_0 is an open neighbourhood of s_0, and the topology of S is the same as the weak topology induced by A (see 8.1.16) we can find functions $f_j (j = 1, \cdots, n)$ in A and an $\epsilon > 0$ such that

$$N_0 \supseteq N_1 = \{t \in S : |f_j(t) - f_j(s_0)| < \epsilon, \ j = 1, \cdots, n\}. \tag{1}$$

We have also $\emptyset \neq F_0 \backslash N_1 \neq F_0$. The minimality property of F_0 implies that $F_0 \backslash N_1$ is not a pre-boundary. It follows that we can find an element $s_1 \in F_0 \bigcap N_1$ and a $g \in A$ such that

$$|g(s_1)| = \|g\|_\infty = c > \sup\{|g(t)| : t \in F_0 \backslash N_1\}.$$

Therefore $|c^{-1}g(s_1)| = 1$, $|c^{-1}g(t)| < 1$ for $t \in F_0 \backslash N_1$. By replacing g by $c^{-1}g$ we may assume $c = 1$, so that we have

$$|g(s_1)| = 1, \ \ |g(t)| < 1 \text{ for all } t \in F_0 \backslash N_1. \tag{2}$$

By taking a sufficiently large integer n and setting $g_1 = g^n$ we get

$$\|g_1\|_\infty = |g_1(s_1)| = 1,$$

$$|g_1(t)| < \frac{\epsilon}{2 \sup_j \|f_j\|_\infty} \text{ on } F_0 \backslash N_1. \tag{3}$$

Set $h_j = f_j g_1 - f_j(s_j) g_1$. Then we have for $t \in F_0 \backslash N_1$ (using(3))

$$|h_j(t)| = |f_j(t) - f_j(s_0)||g_1(t)| < \epsilon \tag{4}$$

and for $t \in N_1$.

$$|h_j(t)| < \epsilon \|g_1\|_\infty = \epsilon. \tag{5}$$

From (4),(5) we conclude, since F_0 is a pre-boundary, that

$$\|h_j\|_\infty < \epsilon. \tag{6}$$

On the other hand, since F is also a pre-boundary there is a $t_1 \in F$ such that

$$|g_1(t_1)| = \|g_1\|_\infty = 1.$$

It follows that

$$|f_j(t_1) - f_j(s_0)| = |f_j(t_1) - f_j(s_0)||g_1(t_1)| = |h_j(t_1)| \leqslant \|h_j\|_\infty < \epsilon,$$

where we have used (6). Therefore, by (1), $t_1 \in N_1$, so that $t_1 \in N_1 \bigcap F$, whence

$$N \bigcap F \supseteq N_0 \bigcap F \supseteq N_1 \bigcap F \neq \emptyset.$$

Since N is an arbitary neighbourhood of s_0 we conclude that

$$s_0 \in \bar{F} = F, \quad F_0 \subseteq F.$$

Hence $F_0 = \partial_A S$.

8.2.8. COROLLARY. *If* $A \subseteq \mathsf{C}_0(S)$ *is only separating then also* $\partial_A S$ *exists.*

PROOF. Since A is only separating (and not strongly separating) all $f \in A$ vanish at a points $s_0 \in S$; by separation hypothesis there cannot be another point in S at which all f in A vanish. Set

$$S_0 = S \backslash \{s_0\}, \ A_0 = \{f|S_0 : f \in A\}.$$

Then A_0 is clearly a strongly separating subalgebra of $\mathsf{C}_0(S_0)$. So, by 8.2.7, $\partial_{A_0} S_0$ exists. It is easy to see that

$$\overline{\partial_A S_0} = \partial_A S.$$

8.2.9. REMARK. The above corollary may fail if A is not separating. A counter-example is furnished by the following algebra A. Write $S = [-1, 1]$ and denote by $\mathsf{C}(S)$ the algebra of continuous $\mathbb{K}$-valued functions on S. Let A be the subalgebra of $\mathsf{C}(S)$ consisting of even functions $(f(s) = f(-s))$. Then $\partial_A S$ does not exist since [0,1] [-1,0] are both minimal pre-boundaries. Note that A does not separate s and $-s$ $(0 < s \leqslant 1)$.

8.2.10. PROPOSITION. For every dense subalgebra A of $\mathsf{C}_0(S)$, S itself is the Shilov boundary: $\partial_A(S) = S$.

PROOF. For $s_0 \in S$, choose any compact neighbourhood V_c of s_0. By local compactness of S there exists a $g \in \mathsf{C}_0(S)$ such that $g(s_0) = 1$ and $g = 0$ on $S \backslash V_c$. Since A is dense in $\mathsf{C}_0(S)$ there is a $f \in A$ such that

$$|f(s) - g(s)| < \frac{1}{2} \text{ for all } s \in S. \tag{1}$$

Since $g(s_0) = 1$, $|f(s_0)| > |g(s_0)| - |g(s_0) - f(s_0)| > 1 - \frac{1}{2} = \frac{1}{2}$, so that

$$\|f\|_\infty > \frac{1}{2}\,. \tag{2}$$

On the other hand, if $s \in S\backslash V_c$, $g(s) = 0$,

$$|f(s)| < \frac{1}{2}\,. \tag{3}$$

From (2),(3) we get

$$\sup_{S\backslash V_c} |f(s)| < \|f\|_\infty.$$

Since any arbitary neighbourhood V of s_0 contains some compact neighbourhood the above inequality implies that the inequality $(*)$ of 8.2.5 holds for V, whence $s_0 \in S_m$, where S_m is a minimal pre-boundary for A. Since s_0 is an arbitary point of S we get $S = S_m = \partial_A(S)$.

Note that in the above proof we have shown directly that every minimal pre-boundary is S, so that we have not made use of *a priori* the existence of $\hat{\partial}_A(S)$ (which of course is assured by 8.1.3, 8.2.7).

8.2.11. PROPOSITION. *Let $A = (A, p)$ with p sm., be a ρ-seminormed algebra having $\Delta_c \neq \emptyset$. Then:*

(i) *$\hat{A}$ is a strongly separating subalgebra of $C_0(\Delta_c)$.*

(ii) *The Shilov boundary $\partial_{\hat{A}}\Delta_c$ exists.*

PROOF. By 7.3.12, Δ_c is locally compact Hausdorff and by virtue of 7.3.27(i), $\hat{A}$ is a subalgebra. Also, by 8.1.26, $\hat{A}$ is strongly separating. Finally, (ii) follows from (i) (see 8.2.7).

8.2.12. COROLLARY. *$\partial_{\hat{A}}\Delta$ exists if A is a ρ-seminormed Gelfand algebra. In particular it exists if $A \neq \sqrt{A}$ and A is a complex, formally real or strictly real, commutative ρ-Banach algebra.*

PROOF. Since A is Gelfand we have $\Delta = \Delta_c \neq \emptyset$ whence the first assertion follows from 8.2.11. The second statement follows from the first taking into account the results of 7.2.17, 7.2.19.

8.2.13. DEFINITION. A ρ-seminormed algebra A with $\Delta_c \neq \emptyset$ is called *self-adjoint* or s.a. if to each $x \in A$ there is a y with

$$\hat{y} = \bar{\hat{x}}, \text{ i.e. } y(\chi) = \overline{x(\chi)} \text{ for every } \chi \in \Delta_c.$$

Note that if A is real it is automatically s.a. (we can take $y = x$).

8.2.14. LEMMA. *The unitization A_1 of a s.a. algebra is a s.a. algebra.*

PROOF. If $x_1 = \lambda e_1 + x$ $(x \in A)$, Choose $y \in A$ such that $\hat{y} = \bar{\hat{x}}$ in A. Set $y_1 = \bar{\lambda} e_1 + y$. Consider $\chi_1 \in \Delta_1 = \Delta(A_1), \chi_1 \neq \chi_0$ (the distinguished character, ker $\chi_0 = A$). Write $\chi = \chi_1|A$; then $\chi \in \Delta = \Delta(A)$. We have

$$\begin{aligned} \hat{y}_1(\chi_1) &= \chi_1(\bar{\lambda} e_1 + y) = \bar{\lambda} + \chi(y) = \bar{\lambda} + \bar{\hat{x}}(\chi) \\ &= \overline{(\lambda + \hat{x}(\chi))} = \overline{(\hat{x}_1(\chi))}. \end{aligned}$$

Also, $\hat{y}_1(\chi_0) = \hat{\bar{\lambda}} = \overline{\hat{x}_1(\chi_0)}$.

Thus, $\hat{y}_1 = \bar{\hat{x}}_1$, proving A_1 is s.a..

8.2.15. PROPOSITION. *The transform algebra $\hat{A}$ of a s.a. ρ-seminormed algebra A (with p sm.) is dense in $\mathsf{C}_0(\Delta_c)$. Hence $\partial_{\hat{A}} \Delta_c = \Delta_c$.*

PROOF. By 8.2.11, $\hat{A}$ is a strongly separating subalgebra of $\mathsf{C}_0(\Delta_c)$. Also, by s.a. property of A, $\hat{A}$ is closed for conjugates (i.e. $f \in \hat{A} \Rightarrow \bar{f} \in \hat{A}$). By the extended Stone-Weierstrass theorem (see [26, pp.166-7]) $\hat{A}$ is dense in $\mathsf{C}_0(\Delta_c)$. Finally, by 8.2.10, $\partial_{\hat{A}} \Delta_c = \Delta_c$.

8.2.16. Some Examples of Shilov Boundary.

(i) Let D be the closed unit disc in the complex plane and $A = \mathfrak{H}(D)$ the Banach algebra of all continuous complex functions on D which are holomorphic on the interior D^o of D. By the maximum modulus principle we have $\partial_A D =$ unit circle S^1; thus here the Shilov boundary coincides with the topological boundary ∂D of D.

(ii) Consider the bi-cylinder $D^2 = \{(z_1, z_2) \in \mathbb{C}^2 : |z_1|, |z_2| \leqslant 1\}$. Let $\mathsf{C}(D^2)$ denote the Banach algebra of all complex continuous functions on D^2. Let $A = \mathsf{P}(D)$ be the uniform closure in $C(D^2)$ of the subset consisting of restrictions on D^2 of polynomials in z_1, z_2.

We shall now show that

$$\begin{aligned} \partial_A D^2 &= \text{distinguished boundary } \partial_0 D^2 \\ &= \{(z_1, z_2) \in D^2 : |z_1| = |z_2| = 1\}. \end{aligned}$$

Note that $\partial_0 D^2$ is a proper part of the topological boundary $\partial D^2 = \{(z_1, z_2) \in D^2 : |z_1| \text{ or } |z_2| = 1\}$. It follows from the maximum modulus principle that $\partial_0 D^2$ is a pre-boundary, so that we have $\partial_A D^2 \subseteq \partial_0 D^2$. Further, for each point $p = (\vec{e}^{\,i\theta_1}, \vec{e}^{\,i\theta_1}) \in \partial_0 D^2$ the function $f(z_1, z_2) = (z_1 + \vec{e}^{\,i\theta_1})(z_2 + \vec{e}^{\,i\theta_2})$ on D^2. satisfies, as we shall see, the inequality $(*)$ of 8.2.5 with respect to any open neighbourhood of p of the form

$$V_\eta = \{(z_1, z_2) \in D^2 : |z_1 - e^{\,i\theta_1}| < \eta, |z_2 - e^{\,i\theta_2}| < \eta\}.$$

Since any neighbourhood V of p contains some V_η it means that the inequality $(*)$ is satisfied for all V, and so by 8.2.5. $p \in \partial_A D^2$, proving $\partial_A D^2 = \partial_0 D^2$.

It remains to show that the inequality $(*)$ of 8.2.5 holds for $V = V_\eta$. First observe that $|f|$ attains its maximum value at $z_1 = e^{\,i\theta_1}, z_2 = e^{\,i\theta_2}$, so that $\|f\|_\infty = 4$. If $(z_1, z_2) \in D^2 \backslash V$ we have

$$|z_1 - e^{\,i\theta_1}| \geqslant \eta \quad \text{or } |z_2 - e^{\,i\theta_2}| \geqslant \eta. \tag{$*$}$$

If $|z_1|$ or $|z_2| = 1$ then we must have by $(*)$ above that either $\operatorname{amp} z_1 \neq \operatorname{amp} e^{\,i\theta_1}$ or $\operatorname{amp} z_2 \neq \operatorname{amp} e^{\,i\theta_2}$. It follows that either $|z_1 + e^{\,i\theta_1}| < 1$ or $|z_2 + e^{\,i\theta_2}| < 2$, so that once again the inequality $(*)$ of 8.2.5 holds.

(iii) Let $\mathsf{C}(D)$ denote the algebra of all continuous $\mathbb{K}$-valued functions on the closed unit disc D in $\mathbb{K}$. Then $\mathsf{C}(D)$ is a Banach algebra under sup norm. Its subalgebra A comprising restrictions, of polynomials over $\mathbb{K}$, to D is a normed subalgebra. By Stone-Weierstrass, A is a dense subalgebra, so that, by 8.2.10, $\partial_A D = D$.

8.2.17. PROPOSITION (Shilov). *Let A' be either a complex or a strictly real, unital commutative ρ-Banach algebra and A a closed subunital algebra of A'. Let Δ, Δ' denote respectively the* *spectra of* A, A'. *Then every character* $\chi \in \partial_{\hat{A}}\Delta$ *has an extension to a character* χ' *of* A'. *In particular, whenever* $\hat{A}$ *is dense in* $\mathsf{C}(\Delta)$ *every* $\chi \in \Delta$ *has an extension to a* $\chi' \in \Delta'$.

PROOF. Consider the restriction map $\eta : \Delta' \to \Delta$ given by $\eta(\chi') =$ restriction $\chi'|A$, which is clearly continuous. Since Δ' is compact, $\eta(\Delta')$ is compact and so closed in Δ. We shall show that $\eta(\Delta')$ is a pre-boundary for A. If $x \in A \subseteq A'$ and $\hat{x}' = \hat{x}'(\chi')$ then $\|\hat{x}'\|_\infty = \nu(x)^{\frac{1}{\rho}} = \|\hat{x}\|_\infty$. By compactness of Δ' there is a $\chi_0' \in \Delta'$ such that

$$\|\hat{x}'\|_\infty = |\hat{x}'(\chi_0')| = |\hat{x}(\eta(\chi_0'))|,$$

proving that $\eta(\Delta')$ is a pre-boundary. But then $\partial_{\hat{A}}\Delta \subseteq \eta(\Delta')$, which clearly implies that every $\chi \in \partial_{\hat{A}}\Delta$ has an extension χ', proving the first assertion. When $\hat{A}$ is dense in $\mathsf{C}(\Delta)$, by 8.2.10, $\partial_{\hat{A}}\Delta = \Delta$, whence the second assertion.

8.2.18. COROLLARY. *Every maximal ideal* M *of* A, *whose associated character* χ *belongs to* $\partial_{\hat{A}}\Delta$, *is contained in a maximal ideal* M' *of* A'.

PROOF. By 8.2.17, χ extends to a character χ' of A'; then $M = \ker\chi \subseteq \ker\chi' = M'$.

8.2.19. Remark. In general not every maximal ideal of A is contained in some maximal ideal of A'. For example, if we denote by S^1 the unit circle in the plane $\mathbb{C}$ and consider the algebra $A' = \mathsf{C}(S^1) = \mathsf{C}(S^1, \mathbb{C})$ and its subalgebra A comprising the functions which can be extended to holomorphic functions on the interior of the circle S^1. The function $z = \vec{e}^{\,i\theta} \in A \subseteq A'$. Further z is invertible in A' with $z^{-1} = \bar{z}$. Since $\bar{z}$ is not holomorphic, $\bar{z} \notin A$ and z is not invertible in A. It follows that there is a maximal ideal M of A with $z \in M$. This ideal M is not contained in any maximal ideal of A' since z is invertible in A'.

8.2.20. PROPOSITION (Shilov). *Let* A *be a complex unital* ρ*-Banach algebra which is monogenic, with t. generator* a. *Then* $\partial_{\hat{A}}\Delta$ *can be identified with* (*the topological boundary*) $\partial\sigma(a)$.

PROOF. By 7.3.23, we identify Δ with $\sigma(a)$ and regard $\hat{A}$ as a function algebra on $\sigma(a)$. By virtue of the maximum modulus principle the topological boundary is a pre-boundary for $\hat{A}$, so that $\partial_{\hat{A}}\sigma(a) \subseteq \partial\sigma(a)$. For proving the reverse inequality take a point $\lambda_0 \in \partial\sigma(a)$ and an open neighbourhood V of λ_0 in $\sigma(a)$. Then

$$V \supseteq \{\lambda \in \mathbb{C} : |\lambda - \lambda_0| < \epsilon\} \bigcap \sigma(a), \text{ for some } \epsilon > 0.$$

Since $\lambda_0 \in \partial\sigma(a)$ there is a $\lambda_1 \in \rho(a)$ such that $|\lambda_0 - \lambda_1| < \frac{\epsilon}{3}$. Set $b = (a - \lambda_1 e)^{-1}$. Then $\hat{b}(\lambda) = 1/(\lambda - \lambda_1)$, and $|\hat{b}(\lambda_0)| > \frac{3}{\epsilon}$, $|\hat{b}(\lambda)| \leqslant \frac{3}{2\epsilon}$ for $\lambda \in \sigma(a)\backslash V$ (since $|\lambda - \lambda_0| \geqslant \epsilon$, $|\lambda - \lambda_1| \geqslant |\lambda - \lambda_0| - |\lambda_1 - \lambda_0| \geqslant \epsilon - \frac{\epsilon}{3}$).
Since $\|\hat{b}\|_\infty > \frac{3}{\epsilon} > |\hat{b}(\lambda)|$ $(\lambda \in \sigma(a)\backslash V)$, by 8.2.5, $\lambda_0 \in \partial_{\hat{A}}\sigma(a)$. This completes the proof.

§ 3. Hull-Kernel Topology

8.3.1. Let A be an algebra (or more generally a ring) with $A \neq \sqrt{A}$, by Denote by $\mathcal{P}$ the set of prime (bi-) ideals P of A with $P \neq A$, by Σ the set of maximal regular bi-ideals of A, and by Π the set of primitive ideals of A.

8.3.2. LEMMA *We have* $\Sigma \subseteq \Pi \subseteq \mathcal{P}$; $\mathcal{P}$, $\Pi \neq \emptyset$.

PROOF. The inclusion relations follow from 1.5.10, 1.5.9. Since $A \neq \sqrt{A}$, $\Pi \neq \emptyset$.

8.3.3. For any subset S of $\mathcal{P}$ we write

$$\mathsf{k}(S) = I_S = \bigcap_{P \in S} P$$

and call $\mathsf{k}(S)$ the kernel of S; $\mathsf{k}(S)$ is a bi-ideal of A. We have clearly:

$$\text{For } S_1 \subseteq S_2,\ \mathsf{k}(S_2) \subseteq \mathsf{k}(S_1). \tag{1}$$

If I is a bi-ideal we set $\mathsf{h}(I) = \{P \in \mathcal{P} : P \supseteq I\}$ and call $\mathsf{h}(I)$ the hull of I. We have obviously:

$$\text{For } I \subseteq J,\ \mathsf{h}(J) \subseteq \mathsf{h}(I). \tag{2}$$

The following inclusions are clear from the definitions.

$$S \subseteq \mathsf{hk}(S). \tag{3}$$

$$I \subseteq \mathsf{kh}(I). \tag{4}$$

By virtue of (1),(3) we obtain

$$\mathsf{k}(S) = \mathsf{khk}(S). \tag{5}$$

Similarly, by (2),(4) we get

$$\mathsf{h}(I) = \mathsf{hkh}(I). \tag{6}$$

(By applying k to (3) we get, by (1), $\mathsf{khk}(S) \subseteq \mathsf{k}(S)$; on the other hand, $\mathsf{khk}(S) = \mathsf{kh}(\mathsf{k}(S)) \supseteq \mathsf{k}(S)$. Combining the above two inclusions we get (5). Similarly, (6) can be proved.)

8.3.4. LEMMA. (i) $\bigcap_\alpha \mathsf{h}(I_\alpha) = \mathsf{h}(\sum_\alpha I_\alpha)$;
(ii) $\mathsf{h}(IJ) = \mathsf{h}(I) \bigcup \mathsf{h}(J) = \mathsf{h}(I \bigcap J)$;
(iii) $\mathsf{h}(A) = \emptyset$;
(iv) $\mathsf{h}(\{0\}) = \mathcal{P}$.

PROOF. The properties (i),(iii),(iv) are immediate from the definition of h. For (ii), we first observe that since $IJ \subseteq I, J$ we have $\mathsf{h}(I), \mathsf{h}(J) \subseteq \mathsf{h}(IJ)$, so that $\mathsf{h}(I) \bigcup \mathsf{h}(J) \subseteq \mathsf{h}(IJ)$. On the other hand, if $P \in \mathsf{h}(IJ)$ then $P \supseteq IJ$ and since P is prime, $P \supseteq I$ or J, so that $P \in \mathsf{h}(I)$ or $\mathsf{h}(J)$, whence $\mathsf{h}(IJ) \subseteq \mathsf{h}(I) \bigcup \mathsf{h}(J)$. Combining this with the reverse inclusion relation obtained above we get the first equality in (ii). Again, since $\mathsf{h}(I) \bigcup \mathsf{h}(J) \subseteq \mathsf{h}(I \bigcap J) \subseteq \mathsf{h}(IJ)$, the second equality follows from the first.

8.3.5. PROPOSITION. *The set $\mathcal{P}$ can be topologized by taking as its closed sets the family $\{\mathsf{h}(I) : I$ a bi-ideal of $A\}$. The resulting topology is called the hull-kernel or* hk *topology. We denote $\mathcal{P}$ with this topology by $\mathcal{P}_{\mathsf{hk}}$.*

PROOF. It follows from 8.3.4 that the family $\{\mathsf{h}(I)\}$ of subsets of $\mathcal{P}$ is closed for arbitrary intersections and finite unions, contains the empty set and the whole space $\mathcal{P}$. Hence the proposition.

8.3.6. LEMMA. *If S is a subset of $\mathcal{P}_{\mathsf{hk}}$ then* $\bar{S} = \mathsf{hk}(S)$, *where bar denotes closure in* $\mathcal{P}_{\mathsf{hk}}$.

PROOF. By definition of hk topology, $\mathsf{hk}(S)$ is a closed set and $\mathsf{hk}(S) \supseteq S$ (by(3) of 8.3.3) Further, if $\mathsf{h}(I) \supseteq S$ then $\mathsf{h}(I) = \mathsf{hkh}(I) \supseteq \mathsf{hk}(S)$. Therefore $\bar{S} = \mathsf{hk}(S)$.

8.3.7. COROLLARY. *For a closed subset* S, *among the bi-ideals* I *of* A *with* $\mathsf{h}(I) = S$ *the largest one is* $\mathsf{k}(S)$.

PROOF. First observe that $\mathsf{k}(S)$ is a bi-ideal and $\mathsf{h}(\mathsf{k}(S)) = \mathsf{hk}(S) = S$ since S is closed. Next, if $\mathsf{h}(I) = S$, then $I \subseteq \mathsf{kh}(I) = \mathsf{k}(S)$.

8.3.8. PROPOSITION. $\mathcal{P}_{\mathsf{hk}}$ *is a* T_0*-space. The subspace* Π *is also a* T_0*-space while the subspace* Σ *is* T_1.

PROOF. If $P_1, P_2 \in \mathcal{P}_{\mathsf{hk}}$ and $P_2 \not\supseteq P_1$ then $P_2 \notin \mathsf{h}(\{P_1\}) = \overline{\{P_1\}}$, proving $\mathcal{P}_{\mathsf{hk}}$ is T_0. By a similar argument Π is also T_0. Finally, if $M \in \Sigma$ then $\overline{\{M\}} = \mathsf{h}(\{M\}) = \{M\}$ (by maximality of M, whence Σ is T_1.)

8.3.9. DEFINITION. $\Pi = \Pi_A$ is called the *structure space* of A, $\Sigma = \Sigma_A$ the *strong structure* space of A.

8.3.10. PROPOSITION ([23, p.79]). *Let* J *be a bi-ideal of an algebra* A. *Then:*

(i) $\Sigma_A \backslash \mathsf{h}(J)$ *is homomorphic with* Σ_J *under the map* $\theta : M \to J \bigcap M$.

(ii) *If* $A^{\#} = A/J$ *then the hull* $\mathsf{h}(J)$ *of* J *in* Σ_A *is homomorphic with* $\Sigma_{A^{\#}}$ *under the canonical map* $\theta^{\#} : M \mapsto M/J$.

PROOF. By 1.4.3, the map $\theta : M \mapsto M' = J \bigcap M$ is a bijection between $\Sigma_A \backslash h(J)$ and Σ_J. To prove that θ is a homeomorphism it is enough to prove that

$$\theta(\mathsf{hk}(S)) = \mathsf{hk}(\theta(S)).$$

Clearly $\mathsf{k}(\theta(S)) = J \bigcap \mathsf{k}(S)$. If $M \in \mathsf{hk}(S)$ then $M \supseteq \mathsf{k}(S)$ and $\theta(M) \supseteq J \bigcap \mathsf{k}(S)$, so that $\theta(M) \in \mathsf{h}(J \bigcap \mathsf{k}(S))$. Conversely, suppose that

$$J \bigcap M \supseteq \mathsf{k}(\theta(S)) = J \bigcap \mathsf{k}(S).$$

Then $M \supseteq J \bigcap \mathsf{k}(S)$, and since $M \not\supseteq J$, by primality of M, $M \supseteq \mathsf{k}(S)$, so that $M \in \mathsf{hk}(S)$. This completes the proof of (i).

To prove (ii), we observe that if I is a bi-ideal of A which contains J, and its image under the canonical homeomorphism is denoted by $I^{\#}$, then the correspondence $I \mapsto I^{\#}$ is easily seen to be a bijection preserving the inclusion relation. Therefore maximal bi-ideals correspond to maximal bi-ideals. The required conclusion now readily follows.

8.3.11. PROPOSITION(cf. [23, p.79]). *If I is a regular bi-ideal of A then $\mathsf{h}(I), \mathsf{h}(I) \bigcap \Pi, \mathsf{h}(I) \bigcap \Sigma$ are compact under the* hk *-topology.*

PROOF. Let $\{F_\alpha\}$ be a family of closed sets in any one of these spaces such that

$$\bigcap F_\alpha = \emptyset. \qquad (*)$$

Write $\mathsf{k}(F_\alpha) = I_\alpha$ and $K = \sum_\alpha{}^{\dagger} I_\alpha$. Since $F_\alpha \supseteq \mathsf{h}(I)$ we have

$$I \subseteq \mathsf{hk}(I) \subseteq \mathsf{k}(F_\alpha) = I_\alpha.$$

So

$$K \supseteq I_\alpha \supseteq I,$$

and K is regular. Suppose that $K \neq A$. Then by 1.2.10 there is an $M \in \Sigma$ with $K \subseteq M$. But then

$$M \supseteq K \supseteq I_\alpha, \text{ so that } M \in \bigcap \mathsf{h}(I_\alpha) = \bigcap F_\alpha.$$

(since $\mathsf{h}(I_\alpha) = \mathsf{hk}(F_\alpha) = F_\alpha$). The last conclusion contradicts $(*)$. So we must have $K = \sum_\alpha I_\alpha = A$. Let u be a relative (bi-)unity for I. Since

$$u \in A = K = \sum_\alpha I_\alpha$$

there is a finite subfamily of ideals $I_1, \cdots, I_n$ $(I_j = I_{\alpha_j})$ such that

$$u \in J = I_1 + \cdots + I_n.$$

† $\sum_\alpha I_\alpha$ denotes the smallest bi-ideal containing the I_α's.

But since $J \supseteq I_j \supseteq I$, J is regular. Therefore, since $u \in J$, $J = A$. This implies that

$$\bigcap_{j=1}^{n} F_j = \bigcap_{j=1}^{n} \mathsf{hk}(F_j) = \bigcap_{j=1}^{n} \mathsf{h}(I_j) = \mathsf{h}(J) = \emptyset.$$

which proves the desired compactness.

8.3.12. Let A be a TA and $\mathcal{M}$ (respy. $\mathcal{M}_c$) the set hypermaximal (respy. closed hypermaximal) ideals of A. We have clearly the inclusion relative

$$\mathcal{M}_c \subseteq \mathcal{M} \subseteq \Sigma$$

so that $\mathcal{M}, \mathcal{M}_c$ inherit, by relativization, the hk-topology. This topology on $\mathcal{M}$ (respy. $\mathcal{M}_c$ can be transferred to Δ (respy. Δ_c) in view of the bijection.

$$\mathcal{M} \to \Delta \text{ (respy. } \mathcal{M}_c \to \Delta_c) \text{ (see7.3.11)}.$$

If E is a subset of Δ or Δ we define

$$\mathsf{k}(E) = \bigcap \ker \chi \quad (\chi \in E).$$

If I is an ideal of A, the hull $\mathsf{h}(I)$ with respect to Δ (respy. Δ_c) is defined by

$$\mathsf{h}(I) = \{\chi \in \Delta(\text{ respy. } \Delta_c) : \ker \chi \supseteq I\}.$$

Then a subset $E \subseteq \Delta$ (respy. Δ_c) is closed iff $E = \mathsf{hk}(E)$.

8.3.13. LEMMA. *The* hk*-topology on* Δ *(or* Δ_c*) can be described in terms of net convergence in the following way: a net* $\chi_\alpha \to \chi$ *in the* hk*-topology if it satisfies the condition: for* $x \in A$ *and any subnet* $(\chi_{\alpha'})$ *of* (χ_α), "$\chi_{\alpha'}(x) = 0$ *for all* $\chi_{\alpha'}$" $\Rightarrow$ $\chi(x) = 0$.

PROOF. It suffices to note that if $M_{\alpha'} = \ker \chi_{\alpha'}$, $M = \ker \chi$ then

$$M \supseteq \bigcap M_{\alpha'} \text{ iff } \hat{x}(M_{\alpha'}) = 0 \text{ for all } M_{\alpha'}, \Rightarrow \hat{x}(M) = 0 \quad (x \in A).$$

8.3.14. COROLLARY. *If* τ_w *denotes the (weak) topology of the spectrum* Δ *(or* Δ_c*) and* τ_{hk} *the* hk*-topology on it, then* τ_{hk} *is coarser than* $\tau_w : \tau_{\mathsf{hk}} \subseteq^{\dagger} \tau_w$.

PROOF. Suppose that a net $\chi_\alpha \to \chi$ in Δ (or Δ_c) under τ_w. Then $\chi_\alpha(x) \to \chi(x)$ for all $x \in A$. In particular, if $\chi_{\alpha'}(x) = 0$ for all $\chi_{\alpha'}$ then $\chi(x) = 0$, which means that $\chi_\alpha \to \chi$ under τ_{hk}. Therefore $\tau_{\mathsf{hk}} \subseteq \tau_w$, as required.

8.3.15. DEFINITION. Let A be a TA with $\Delta_c \neq \emptyset$. A is called *quasi-unital* if there is an element $u_0 \in A$ such that $\hat{u}_0(\chi) \neq 0, \forall \chi \in \Delta_c$. If A is unital (with unity e) it is also quasi-unital since $\hat{e}(\chi) = 1 \neq 0$ $(\chi \in \Delta_c)$.

8.3.16. LEMMA. *Let* A *be a s.a.* ρ*-seminormed algebra such that* Δ_c *is non-empty and compact. Then* A *is quasi-unital.*

PROOF. For each $\chi \in \Delta_c$ choose an element x_χ with $\hat{x}_\chi(\chi) = \chi(x_\chi) \neq 0$. By continuity of $\hat{x}_\chi$ and compactness of Δ_c we can find, as in the proof of 8.1.17, a finite subset $\{\chi_1, \cdots, \chi_n\}$ of Δ_c, elements $x_j = x_{\chi_j}$ and open neighbourhood U_j of χ_j such that $\hat{x}_j(\chi) \neq 0$ on U_j $(j = 1, \cdots, n)$ and $\bigcup_{j=1}^{n} U_j = \Delta_c$. Since A is s.a., we can find $y_j \in A$ with $\hat{y}_j = \bar{\hat{x}}_j$ $(j = 1, \cdots, n)$. Set $u_0 = \sum_{j=1}^{n} x_j y_j$. Then

$$\hat{u}_0(\chi) = \sum_{j=1}^{n} x_j(\chi) y_j(\chi) = \sum_{j=1}^{n} |x_j(\chi)|^2 > 0$$

(since if $\chi \in U_j, x_0(\chi) \neq 0$).

8.3.17. PROPOSITION. *Let* A *be a s.a. complex spectrally-Gelfand* ρ*-Banach algebra with compact spectrum* $\Delta (= \Delta_c)$*. Then* $\sqrt{A}$ *is a regular bi-ideal.*

PROOF. If $\sqrt{A} = A, \sqrt{A}$ is trivially regular. So we may assume that $A \neq \sqrt{A}$. Let A_1 be the unitization of A and let

---

† i.e. every τ_{hk}-open set is τ_w-open.

Δ_1, Δ denote respectively the spectra of A_1, A. It follows (see 7.3.2(a)) that we have

$$\Delta_1 = \{\chi_1 \in \Delta_1 : \chi_1|A \in \Delta\} \bigcup \{\chi_0\}, \tag{*}$$

where χ_0 is the distinguished character of A_1. Since, by hypothesis Δ is compact, so is Δ_1. By 8.3.16, there is an $u_0 \in A$ with $u_0(\chi) \neq 0$ for all $\chi \in \Delta$. Since A is spectrally Gelfand we have:

$$\sigma'(u_0) = \hat{u}_0(\Delta) \dot{\bigcup} \{0\}.$$

It follows that

$$\sigma_1(u_0) = \hat{u}_0(\Delta) \dot{\bigcup} \{\hat{\chi}_0(u_0)\},$$

where σ_1 denotes the spectrum with respect to A_1. Since Δ is compact, $\hat{u}_0(\Delta)$ is also compact and hence closed. It follows from 7.5.14 that there is an idempotents $u \in A_1$ with $\hat{u}(\chi_0) = 0$, $\hat{u}(\Delta) = 1$. But then $\chi_0(u) = 0$, $u \in A$. Further, if

$$x \in A \text{ and } \chi = \chi_1|A \quad (\chi_1 \neq \chi_0)$$

then
$$\widehat{ux}(\chi) = \hat{u}(\chi)\hat{x}(\chi) = 1 \cdot \hat{x}(\chi) = \hat{x}(\chi).$$

This relation holds for every $\chi \in \Delta$, whence we get

$$ux - x \in \bigcap \ker \chi = \sqrt[h]{A} = \sqrt{A} \quad \text{(by 7.2.12)},$$

whence $\sqrt{A}$ is regular.

8.3.18. COROLLARY. *In a commutative s.a. complex ρ-Banach algebra A with compact spectrum, $\sqrt{A}$ is regular.*

PROOF. We may assume that $A \neq \sqrt{A}$. Then, by 7.2.17, A is Gelfand and so spectrally Gelfand. The required conclusion now follows from 8.3.17.

8.3.19. COROLLARY. *If A is a strictly real commutative ρ-Banach algebra with Δ compact then $\sqrt{A}$ is regular.*

PROOF The complexification $\tilde{A}$ of A is s.a.:

$$\overline{\widehat{x+iy}(\chi)} = \overline{\chi(x) + i\chi(y)} = \chi(x) - i\chi(y) = \widehat{x-iy}(\chi).$$

Since, by 7.3.3, $\tilde{\Delta} = \Delta(\tilde{A})$ is homeomorphic to $\Delta = \Delta(A)$ and Δ is compact, $\tilde{\Delta}$ is compact. It follows from 8.3.17 that $\sqrt{\tilde{A}}$ is regular. By 1.9.17, $\sqrt{\tilde{A}}$ is self-conjugate and

$$\sqrt{A} = A \bigcap \sqrt{\tilde{A}}.$$

It follows that

$$\sqrt{\tilde{A}} = \sqrt{\tilde{A}} \bigcap A + i(\sqrt{\tilde{A}} \bigcap A) = \sqrt{A} + i\sqrt{A}.$$

Therefore (see 1.9.16(i)) $\sqrt{A} = \mathrm{Re}\sqrt{\tilde{A}}$ is regular.

8.3.20. PROPOSITION. *Let A be a commutative s.a., complex or strictly real, ρ-Banach algebra. Then the following two statements are equivalent:*

(i) $A/\sqrt{A}$ is unital .

(ii) $\Delta = \Delta_c$ is compact .

PROOF. (i) $\Rightarrow$ (ii) : Since $A/\sqrt{A}$ is unital, $\sqrt{A}$ is regular whence $\sqrt[ch]{A} \supseteq \sqrt{A}^{\dagger}$ is also regular. So, by 7.3.12, (ii) holds. (ii)$\Rightarrow$ (i): By 8.3.18 or 8.3.19, $\sqrt{A}$ is regular, and so (i) holds.

8.3.21. COROLLARY. *If A is s.s., then A is unital iff Δ is compact.*

8.3.22. Remark. (cf. [12, p.52, RMK 4.7]). The above corollary may not hold if the hypothesis "A is s.s" is dropped. For example if A_1 is commutative unital Banach algebra and $A_2 (\neq \{0\})$ a Banach algebra with trivial multiplication. Then $\Delta(A_1) = \emptyset$ and compact, while $\Delta(A_2) \neq \emptyset$. If $A = A_1 \times A_2$ then by 7.3.14, $\Delta(A) \simeq \Delta(A_1)$ is compact but A is not unital.

§4. Completely Regular Algebras

8.4.1. DEFINITION. Following Willcox an algebra A is called *completely regular*[††] if it satisfies the two conditions:

[†] See (*) of 2.2.18

[††] Some authors especially Russian use the term regular for completely regular, following the usage of the term by Shilov who first introduced these algebras in the commutative case. The nomenclature completely regular is due to Rickart.

(i) The strong structure space Σ is Hausdorff.

(ii) Each point $M \in \Sigma$ has an open neighbourhood V such that the kernel $\mathsf{k}(V)$ is a regular bi-ideal.

When A is unital, condition (ii) above can be dropped since it is automatically satisfied (every ideal then being regular).

8.4.2. Examples of Completely regular algebras.

(i) The Banach algebra $\mathsf{C}_0(X)$ (see 8.4.15).

(ii) The group algebra $\mathsf{L}^1(G)$ of a locally compact Hausdorff commutative group G (see [22, p.426]).

(iii) The group algebra $\mathsf{L}^1(G)$ of any compact Hausdorff group G (see [23, p.83]).

(iv) A von Neumann algebra (see [23, p.290]).

(v) The Wiener-Żelazko algebra W^ρ (see Appendix).

8.4.3. PROPOSITION. *Let A be a completely regular algebra. Then:*

(i) *Σ is locally compact Hausdorff.*

(ii) *Σ is compact Hausdorff if A is unital.*

PROOF. (i) Since $\mathsf{k}(V)$ is a regular bi-ideal (by (ii) of 8.4.1) it follows by 8.3.11 that $\mathsf{hk}(V)$ is a compact neighbourhood of M.

(ii) When A is unital, every ideal is regular and so in particular, $\{0\}$ is regular. Therefore, by 8.3.11, $h(\{0\}) = \Sigma$ is compact.

8.4.4. PROPOSITION. *If A is a completely regular algebra and J a bi-ideal of A then J and $A^\# = A/J$ are completely regular.*

PROOF. Let Σ_J denote the strong structure space of J. By 8.3.10 (i),

$$\theta : M \mapsto J \bigcap M \text{ is a homeomorphism of } \Sigma \backslash \mathsf{h}(J) \text{ on } \Sigma_J.$$

By complete regularity of A, Σ is Hausdorff and hence by the homeomorphism θ, Σ_J is also Hausdorff. Further, if $M \in \Sigma$ has V as an open neighbourhood with $\mathsf{k}(V)$ regular then $\theta(M)$ has $\theta(V)$

as an open neighbourhood and $\mathsf{k}(\theta(V)) = J \bigcap \mathsf{k}(V)$ is regular (cf. proof of 1.4.2 (v)). Therefore J is completely regular.

It remains to prove that $A^{\#}$ is also completely regular. By 8.3.10 (ii), $\Sigma_{A^{\#}}$ is Hausdorff. If $\varphi : A \to A^{\#}$ is the canonical homeomorphism and $M = \varphi^{-1}(M^{\#})$ $(M^{\#} \in \Sigma_{A^{\#}})$ then $M \in \Sigma_A$. Let V be an open neighbourhood of $M_0 = \varphi^{-1}(M_0^{\#})$, $M_0^{\#} \in \Sigma_{A^{\#}}$, such that $\mathsf{k}(V)$ is regular. The intersection I, of all $M \in \Sigma_A$ with $M \supseteq \mathsf{k}(V)$, $M \supseteq J$, is also a regular bi-ideal (since $I \supseteq \mathsf{k}(V)$). It follows that if $V^{\#} = \{M^{\#} : M \supseteq \mathsf{k}(V)\}$ then $V^{\#}$ is an open neighbourhood of $M_0^{\#}$ with $\mathsf{k}(V^{\#}) = \varphi(\mathsf{k}(V))$ regular. This completes the proof.

8.4.5. PROPOSITION. *The unitization A_1 of an algebra A is completely regular iff A is completely regular.*

PROOF (cf. [23, p.84]). Assume that A is completely regular. Since A_1 is unital, to prove that it is completely regular we have only to show that $\Sigma_1 = \Sigma_{A_1}$ is Hausdorff. Since A is a bi-ideal of A_1, by 8.3.10 (i), $\Sigma = \Sigma_A$ is homeomorphic to $\Sigma_1 \backslash \{A\}$. By complete regularity of A, Σ and consequently $\Sigma_1 \backslash \{A\}$ is Hausdorff. To complete the proof that Σ_1 is Hausdorff it remains to show that the point A can be separated by open sets from any other point M_1^0 in Σ_1. Write $M_1^0 = A \bigcap M_1^0$. By complete regularity of A there is an open neighbourhood V of M_0 with $I = \mathsf{k}(V)$ a regular bi-ideal, with a relative unity u (say). By 1.4.7. (iv), $I_1 = \hat{I} = I + A_1(e_1 - u)$ is a bi-ideal of A_1; $I_1 \nsubseteq A$ since $e_1 - u \in I_1$. It follows that $A \in \Sigma_1 \backslash \mathsf{h}_1(I_1) = U_1$ (say) were h_1 denotes the hull relative to A_1. For $M \in V_0$ consider $M_1 = \hat{M}$. Then $A \bigcap M_1 = M$. Since $M \supseteq I = \mathsf{k}(V)$, $M_1 \supseteq I_1$. It follows that $V_1 = \theta^{-1}(V) \subseteq \mathsf{h}_1(I_1)$, where θ is the homeomorphism $M_1 \mapsto A \bigcap M_1 = M$ (see 8.3.10) so that $U_1 \bigcap V_1 = \emptyset$. This complete the proof of the "if" part.

For the "only if" part assume that Σ_1 is Hausdorff. Then $\Sigma \simeq \Sigma_1 \backslash \{A\}$, is also Hausdorff. If $M \in \Sigma$ then $M = A \bigcap M_1$ $(M_1 \in \Sigma_1)$. Since Σ_1 is Hausdorff we can choose an open neighbourhood V_1 of M_1 such that $A \notin \mathsf{h}_1 k(V_1)$, i.e., $\mathsf{k}(V_1) = I_1 \nsubseteq A$. It follows (see 1.4.7 (ii)) that $I = A \bigcap I_1$ is regular. Since V_1 is open we may assume that

$$V_1 = \mathsf{h}_1'(J_1) \equiv \Sigma_1 \backslash \mathsf{h}_1(J_1).$$

Then it is easy to see that $V = \mathsf{h}_1'(J)$, where $J = A \bigcap J_1$, is an open neighbourhood of M. Since $k(V) = A \bigcap \mathsf{k}(V_1) = A \bigcap I_1 = I$ is regular, the proof of complete regularity of A is finished.

8.4.6. PROPOSITION. *Let A be completely regular and F any closed subset of Σ. Then $k(F)$ is regular iff F is compact.*

PROOF. If $k(F)$ is regular then by 3.3.11, $\mathsf{h}(\mathsf{k}(F)) = \mathsf{hk}(F) = F$ is compact, which proves the "only if" part. For the "if" part assume that F is compact. Then by using complete regularity of A, we can find a finite open covering $\{V_j\ (j = 1, \cdots, n)\}$, with each $\mathsf{k}(V_j)$ regular. By virtue of 1.2.15, $\mathsf{k}(V_1 \bigcup \cdots \bigcup V_n) = \mathsf{k}(V_1) \bigcap \cdots \mathsf{k}(V_n) = I$ (say) is regular. Since $\mathsf{k}(F)$ is also regular, completing the proof.

8.4.7. THEOREM. *Let A be a ρ-seminormed algebra. If A is completely regular then the hull-kernel topology τ_{hk} on Δ_c coincides with the weak topology τ_w. Conversely, when A is ρ-seminormed Gelfand, $\tau_{hk} = \tau_w$ on $\Delta_c (= \Delta)$ implies that A is completely regular.*

PROOF. Suppose hat A is completely regular. We have always, by 8.3.14, $\tau_{\mathsf{hk}} \subseteq \tau_w$. To prove the reverse inclusion, let F be a τ_w-closed set in Δ_c. Take $\chi_0 \in \Delta_c \backslash F$. By complete regularity of A we can choose a τ_{hk}-open neighbourhood V of χ_0 such that $\mathsf{k}(V)$ is regular. Let u be a relative (bi-) unity for $\mathsf{k}(V)$. Set

$$F_0 = \{\chi \in F : \chi(u) = 1\};$$

F_0 is τ_w-closed.

By 7.3.13, the set

$$F_1 = \{\chi \in \Delta_C : |\chi(u)| \geqslant 1\}$$

is τ_w-compact. Therefore, F_0 as a closed subset of F_1 is also τ_w-compact. Since $\tau_{\mathsf{hk}} \subseteq \tau_w$, F_0 is also τ_{hk}-compact and so τ_{hk}-closed (τ_{hk} being Hausdorff by complete regularity of A). Since by choice $\chi_0 \notin F_0$ and τ_{hk} is Hausdorff, each point $\chi \in F_0$ has a τ_{hk}-open neighbourhood U with $\chi_0 \notin \overline{U} = \mathsf{hk}(U)$. By τ_{hk}-compactness there is a finite covering of F by open neighbourhoods $U_1, \cdots, U_n$ such that $\chi_0 \notin \mathsf{hk}(U_j)$ $(j = 1, \cdots, n)$. It follows that there is an element $a_j \in \mathsf{k}(U_j)$ with $\chi_0(a_j) \neq 0$. Also, since $\Delta_c \backslash V$ is τ_{hk}-closed and $\chi_0 \notin \Delta_c \backslash V$, there exists $a_0 \in \mathsf{k}(\Delta_c \backslash V)$ with $\chi_0(a_0) \neq 0$. Set $a = a_0 a_1 \cdots a_n$. Then:

$$\chi_0(a) \neq 0, \tag{1}$$

$$a \in \ker(U_1 \bigcup \cdots \bigcup U_n \bigcup \Delta_c \backslash V) \subseteq \ker(F_0 \bigcup \Delta_c \backslash V). \tag{2}$$

If $\chi \in V$ then $\ker \chi \supseteq \mathsf{k}(V)$ and so $\ker \chi$ has also u has a relative unity. It follows that $\chi(u) = 1$ whence $V \bigcap F \subseteq F_0$; so that we have

$$F \subseteq F_0 \bigcup (\Delta_c \backslash V).$$

From (2),(3) we conclude that $a \in \ker F$. Also, by (1), $\chi_0(a) \neq 0$. Therefore $\chi_0 \notin \mathsf{hk}(F)$, proving $F = \mathsf{hk}(F)$, so that F is τ_{hk}-closed and $\tau_{\mathsf{hk}} = \tau_w$.

For the converse part, assume now that A is Gelfand and $\tau_w = \tau_{\mathsf{hk}}$. Since A is Gelfand, $\Sigma = \Delta = \Delta_c$, so that Σ is Hausdorff. Further, if $\chi_0 \in \Delta$ we can choose a $b \in A$ with $\chi_0(b) = 2$. Setting $V = \{\chi \in \Delta_c : |\chi(b)| > 1\}$, V is an open neighbourhood of χ_0 and

$$\overline{V} \subseteq \{\chi \in \Delta_c : |\chi(b)| \geqslant 1\}$$

is compact for $\tau_w = \tau_{\mathsf{hk}}$. Therefore, by 8.4.6, $\mathsf{k}(\bar{V})$ is a regular ideal and hence the ideal $\mathsf{k}(V) \supseteq \mathsf{k}(\bar{V})$ is also regular, completing the proof of the converse (and also of the theorem).

8.4.8. DEFINITION. Let S be a topological space and $\mathcal{F}_0$ a family of continuous $\mathbb{K}$-valued functions on S. Following Naimark (Neumark) $\mathcal{F}_0$ is called *completely regular* or a *completely regular family* if it satisfies the condition:

To each closed set F in S and to each point $s_0 \notin F$ there is an $f = f(s)$ in $\mathcal{F}_0$ such that

$$f(s) = 0 \text{ on } F \text{ and } f(s_0) \neq 0. \tag{$*$}$$

$\mathcal{F}_0$ is called *normal* if it satisfies the condition:

to each pair of disjoint closed sets F_1, F_2 in S there is an $f \in \mathcal{F}_0$ with

$$f(s) = 0 \text{ on } F_1 \text{ and } f(s) = 1 \text{ on } F_2. \tag{$**$}$$

8.4.9. LEMMA. (a) *If S is T_1 then every normal family is completely regular.*

(b) *A space S is completely regular iff the family $\mathsf{C}(S)$ is completely regular.*

PROOF. Clear.

8.4.10. Let $S = (S, \tau)$ be a completely regular Hausdorff space and $\mathsf{C}(S) = \mathsf{C}(S, \mathbb{K})$ the algebra of all continuous $\mathbb{K}$-valued functions on S. Let A be a *strongly separating* subalgebra of $\mathsf{C}(S)$. Denote by $\mathcal{M}$ the maximal ideal spectrum of A. For $s \in S$, set $M_s = M_s^A = \{f \in A : f(s) = 0\}$. Since $\ker A = \emptyset$, each $M_s \in \mathcal{M}$. Further, since A separates points of S, the map $\Omega : s \mapsto M_s$ is injective. Write $\mathcal{M}_0 = \{M_s : s \in S\}$. Since

$$\mathcal{M}_0 \subseteq \mathcal{M} \subseteq \Sigma$$

we have hk-topology τ_{hk} on $\mathcal{M}_0$ (got by relativization). Moreover, since Ω is a bijection between S and $\mathcal{M}_0, \tau_{\mathsf{hk}}$ can be transferred from $\mathcal{M}_0$ to S and the transferred topology we denote again by τ_{hk}. We also write for an ideal I of $\mathsf{C}(S)$

$$\mathsf{h}_0(I) = \{M \in \mathcal{M}_0 : M \supseteq I\} = \mathcal{M}_0 \bigcap \mathsf{h}(I).$$

8.4.11. PROPOSITION (cf. [19, p.57]). *$\tau_{\mathsf{hk}} \subseteq \tau$, and $\tau_{\mathsf{hk}} = \tau$ iff A is a completely regular family of functions.*

PROOF. Let $F \subseteq S$ be a τ_{hk}-closed set so that $\mathsf{h}_0\mathsf{k}(F^*) = F^*$, where $F^* = \Omega(F)$. If $f \in A$ then $f \in \mathsf{k}(F^*)$ iff $f = 0$ on F. Further, $M_s \in \mathsf{h}_0\mathsf{k}(F^*)$ iff $s \in \bigcap \ker f$, where f ranges in $\mathsf{k}(F^*)$ and $\ker f = \{s \in S : f(s) = 0\}$. It follows that $F = \bigcap \ker f$. Since f is continuous each $\ker f$ is τ-closed. So F is τ-closed, whence $\tau_{\mathsf{hk}} \subseteq \tau$.

Suppose now that A is a completely regular family and $F (\subseteq S)$ is τ-closed; then $I = \mathsf{k}(F^*) = \{f \in A : f = 0 \text{ on } F\}$. Clearly

$$M_s \in \mathsf{h}_0(I) = \mathsf{h}_0\mathsf{k}(F^*) \text{ iff “} f = 0 \text{ on } F \Rightarrow f(s) = 0\text{''}.$$

If $s \notin F$ there is, by the complete regularity of the family A, an $f \in A$ with $f = 0$ on F but $f(s) \neq 0$. This means that $M_s \notin \mathsf{h}_0\mathsf{k}(F^*)$, whence $\mathsf{h}_0\mathsf{k}(F^*) = F^*$, so that F^* and therefore F is τ_{hk}-closed, proving that $\tau = \tau_{\mathsf{hk}}$.

Conversely, if $\tau = \tau_{\mathsf{hk}}$ and $F \subseteq S$ is closed, then for each point $s \notin F = \mathsf{h}_0\mathsf{k}(F)$ there is an $f \in A$ with $f = 0$ on F and $f(s) \neq 0$.

But this is precisely the condition to be fulfilled for the family A to be completely regular.

8.4.12. COROLLARY. *$\tau_{\mathsf{hk}} = \tau$ on S for the algebra $\mathsf{C}(S)$.*

PROOF. Since S is a completely regular space, by its definition $\mathsf{C}(S)$ is a completely regular family and hence the result (by 8.4.11).

8.4.13. PROPOSITION. *Let $S = (S, \tau)$ be compact Hausdorff and A a separating subunital algebra of $\mathsf{C}(S)$ such that every maximal ideal of A is fixed. Then $\tau_{\mathsf{hk}} = \tau$ (on S) iff A is completely regular algebra.*

PROOF. If A is completely regular then τ_{hk} is Hausdorff. Since, by 8.4.11, $\tau_{\mathsf{hk}} \subseteq \tau$, and τ is compact Hausdorff (by hypothesis) it follows from a well-known result in topology that $\tau_{\mathsf{hk}} = \tau$ (on S). Conversely, if $\tau_{\mathsf{hk}} = \tau$ then τ_{hk} on $\mathcal{M}_0 = \Sigma$ (every maximal ideal of A being fixed) is also Hausdorff. Therefore, A being unital, is completely regular.

8.4.14. COROLLARY. *The algebra $\mathsf{C}(S)$ is completely regular.*

PROOF. This follows from 8.4.12, 8.4.13.

8.4.15. COROLLARY. *Let S be a locally compact Hausdorff space. Then the Banach algebra $\mathsf{C}_0(S)$ is completely regular.*

PROOF. Let S_∞ be the 1-point compactification of S. Then, by 8.4.14, $\mathsf{C}(S_\infty)$ is completely regular and so, by 8.4.5, $\mathsf{C}_0(S)$ is completely regular since $\mathsf{C}(S_\infty)$ is the unitization of $\mathsf{C}_0(S)$.

8.4.16. PROPOSITION. *Let $S = (S, \tau)$ be a locally compact Hausdorff space and A be a strongly separating subalgebra of $\mathsf{C}_0(S) = \mathsf{C}_0(S, \|\cdot\|_\infty)$. Then the following two statements are equivalent:*

(i) *A is a completely regular family of functions.*

(ii) *The hull-kernel topology τ_{hk} coincides with τ.*

If A is also Gelfand then (ii) *(or* (i)*) is equivalent to*

(iii) *A is completely regular algebra.*

PROOF. The equivalence of (i) and (ii) has already been demonstrated in 8.4.11. The equivalence of (ii) and (iii) follows from 8.4.7.

8.4.17. PROPOSITION. *If A is a completely regular ρ-seminormed algebra then $\hat{A}$ is a completely regular family of functions. Conversely, if A is a ρ-seminormed Gelfand algebra such that $\hat{A}$ is a completely regular family then A is a completely regular algebra.*

PROOF. Assume that A is completely regular. Then, by 8.4.7, $\tau_{\mathsf{hk}} = \tau_w$ on Δ_c – which is locally compact Hausdorff. By 8.1.25, $\hat{A}$ is a strongly separating subalgebra of $\mathsf{C}_0(\Delta_c)$. So by 8.4.16, $\hat{A}$ is a completely regular family. For the converse, assume that A is also Gelfand and that $\hat{A}$ is a completely regular family. By 8.4.16, $\tau_w = \tau_{\mathsf{hk}}$ on $\Delta_c(= \Delta = \Sigma)$ so that by 8.4.7., A is completely regular.

8.4.18. PROPOSITION. *Let A be a completely regular ρ-seminormed algebra. Then $\partial_{\hat{A}}\Delta_c = \Delta_c$.*

PROOF. Write $\partial_{\hat{A}}\Delta_c = F$; then F is a closed set in Δ_c. If $\Delta_c \backslash F \neq \emptyset$, take a $\chi_0 \in \Delta_c \backslash F$. Since, by 8.4.17, $\hat{A}$ is a completely regular family, there is an $a \in A$ with $\hat{a}(F) = 0$ and $\hat{a}(\chi_0) \neq 0$. Then $\|\hat{a}\|_\infty > 0$, but $\sup\limits_{\chi \in F}|\hat{a}(\chi)| = 0$, contradicting that F is the Shilov boundary. Hence $\Delta_c = F = \partial_{\hat{A}}\Delta_c$, as required.

8.4.19. PROPOSITION (cf. [20, p.236], [12, p.54]). *Let A be a completely regular ρ- seminormed algebra. Then:*

(i) *Any closed set $F \subseteq \Delta_c = \Delta_c(A)$ is homeomorphic to $\Delta_c(A/\mathsf{k}(F))$.*

(ii) *Any open set $G \subseteq \Delta_c(A)$ is homeomorphic to $\Delta_c(\mathsf{k}(\Delta_c \backslash G))$.*

PROOF. By complete regularity of A, $\tau_{\mathsf{hk}} = \tau_w$ on Δ_c (see 8.4.7). Therefore $F = \mathsf{hk}(F)$. It follows that $\chi \in F$ iff $\chi = 0$ on $\mathsf{k}(F)$. The homeomorphism in (i) is now clear.

Write $\Delta_c \backslash G = F$. Then $\chi \in G$ iff $\chi \notin F$ iff $\chi \neq 0$ on $\mathsf{k}(F)$ iff $\chi|\mathsf{k}(F) \in \Delta_c(\mathsf{k}(F))$. Hence the homeomorphism in (ii).

8.4.20. PROPOSITION (cf. [12, p.55]). *Let A be a completely regular, complex or strictly real, commutative ρ-Banach algebra $A(\neq \sqrt{A})$. Let I be an ideal of A, F_1 a closed subset of $\Delta(=\Delta_c)$, F_2 a compact subset Δ such that*

$$(\mathsf{h}(I) \bigcup F_1) \bigcap F_2 = \emptyset. \tag{$*$}$$

Then there exists an element $a_0 \in I \bigcap \mathsf{k}(F_1)$ such that $\hat{a}_0 = 0$ on F_1 and $\hat{a}_0 = 1$ on F_2.

PROOF. Write $I_j = \mathsf{k}(F_j)$ $(j = 1, 2)$. Then $F_j = \mathsf{h}(I_j)$, so that condition $(*)$ above becomes $(\mathsf{h}(I) \bigcup \mathsf{h}(I_1)) \bigcap \mathsf{h}(I_2) = \emptyset$, i.e.,

$$\mathsf{h}(I \bigcap I_1) \bigcap \mathsf{h}(I_2) = \emptyset \tag{1}$$

which means that no character of χ of A can vanish at the same time on both $I \bigcap I_1$ and I_2. Since I_2 is clearly the intersection of all hypermaximal ideals containing I_2 and A is Gelfand, it follows that $A^\# = A/I_2$ is s.s.. Since by 8.4.19, $\Delta(A^\#) \simeq F_2$, and F_2 is compact we conclude by 8.6.6 that $A^\#$ has unity $u^\# = u + I_2$ (say). Denote by φ the canonical homomorphism $A \to A^\# = A/I_2$. We assert that $\varphi(I \bigcap I_1) = A^\#$. At any rate $J^\# = \varphi(I \bigcap I_1)$ is an ideal of $A^\#$. If $J^\# \neq A^\#$ there is a hypermaximal ideal $M^\#$ with $\varphi(I \bigcap I_1) \subseteq M^\#$. Then $M = \varphi^{-1}(M^\#)$ is a hypermaximal ideal of A with $I \bigcap I_1, I_2 \subseteq M$. But this contradicts relation (1). Therefore we must have $\varphi(I \bigcap I_1) = A$, as asserted. It follows in particular that there is an element $a_0 \in I \bigcap I_1$ such that $a_0 + I_2 = \varphi(a_0) = u^\#$. If $\chi \in F_1$ then $\chi = 0$ on $\mathsf{k}(F_1) = I_1$ so that $\hat{a}_0(\chi) = \chi(a_0) = 0$. On the other hand if $\chi \in F_2$ then $\chi = 0$ on $\mathsf{k}(F_2) = I_2$, so that

$$\hat{a}_0(\chi) = \chi(a_0) = \chi^\#(a_0 + I_2) = \chi^\#(u^\#) = 1,$$

where $\chi^\#$ is the character of $A^\#$ induced by χ. This completes the proof.

8.4.21. COROLLARY. *A being as in* 8.4.20, *if F_1 is a closed subset and F_2 a compact subset of $\Delta(A)$ such that $F_1 \bigcap F_2 = \emptyset$, then there exists an element $a_0 \in \mathsf{k}(F_1)$ with $\hat{a}_0 = 0$ on F_1 and $= 1$*

on F_2.

PROOF. Suffixes to take in 8.4.20, $I = A$ (then $\mathsf{h}(I) = \emptyset$).

8.4.22. COROLLARY. *If I is an ideal of A and F a compact subset of $\Delta(A)$ such that $h(I) \bigcap F = \emptyset$, there is an element $a_0 \in I$ such that $\hat{a}_0 = 1$ on F.*

PROOF. Suffices to take in 8.4.20, $F_1 = \emptyset$ (then $\mathsf{k}(F_1) = A$) and $F_2 = F$.

8.4.23. COROLLARY. *If A is a completely regular, complex or strictly real, commutative unital ρ-Banach algebra then $\hat{A}$ is a normal family of functions on Δ.*

Proof. This follows from 8.4.21, since now Δ being compact every closed set in Δ is compact.

8.4.24. DEFINITION. Let A be a completely regular ρ-seminormed algebra and F a closed subset of $\mathcal{M}_c$ (the space of closed hypermaximal ideals). Denote by $J = j(F)$ the set of all $x \in A$ such that $\hat{x}$ has compact support[†] disjoint with F. If F is a single point $\{M\}$ we write $j(M)$ for $j(F)$. Further, if $F = \emptyset$ we write J_0 for $j(\emptyset)$.

8.4.25. LEMMA. *$J = j(F)$ is a bi-ideal of A such that* $\mathsf{h}(J) = F$.

PROOF. Clearly $x \in j(F)$ iff there is a compact set C_x such that $\hat{x} = 0$ outside C_x and $C_x \bigcap F = \emptyset$. From this result it is easy to see that $j(F)$ is a bi-ideal. Further, if $\chi \in F$ then $\chi(x) = 0$ for every $x \in j(F)$, whence $\mathsf{h}(j(F)) \supseteq F$. Now consider a $\chi_0 \in \Delta_c \backslash F$. Since Δ_c is locally compact Hausdorff and F is a closed set there is an open set U in Δ_c with $\chi_0 \in U$ and $\overline{U}$ compact and disjoint with F. By complete regularity of A there is an x in A with $\hat{x}(\chi_0) \neq 0$ and $\hat{x} = 0$ on $\Delta_c \backslash U$. Then $x \in j(F)$ and $\chi_0 \notin \mathsf{h}(j(F))$, which proves that $\mathsf{h}(j(F)) = F$.

8.4.26. PROPOSITION. *Let A be a commutative completely*

[†] For a function f on a topological space S, by support of f we mean the closure of the largest subset on which it is not zero.

regular s.s., complex or strictly real, ρ-Banach algebra. Let F be a closed subset of $\Delta_c = \Delta$. Then $J = j(F)$ is the smallest of the ideals I such that $\mathsf{h}(I) = F$.

PROOF. In view of 8.4.25, it is enough to prove that for any I with $\mathsf{h}(I) = F$ we have $J \subseteq I$. If $x \in J$ and C is the support of $\hat{x}$ then C is compact and $C \bigcap F = \emptyset$. By 8.4.22 there is a $y \in I$ with $\hat{y} = 1$ on C. Since $\hat{x} = 0$ outside C we have clearly $\hat{x}\hat{y} = \hat{x}$. But A being s.s., we get $xy = x$, whence $x \in I$, $J \subseteq I$, completing the proof.

8.4.27. COROLLARY. *The closure $\overline{J}$ is the smallest closed ideal I of A with $\mathsf{h}(I) = F$.*

PROOF. If $M \in \Delta$ the M is a closed ideal, so that $M \supseteq J \Rightarrow M \supseteq \overline{J}$. Therefore $h(\overline{J}) = h(J) = F$. Further, if I is a closed ideal with $h(I) = F$ then $J \subseteq I$ and so $\overline{J} \subseteq \overline{I} = I$, completing the proof.

8.4.28. COROLLARY. *For each maximal regular ideal M of $A, \overline{j(M)}$ (respy. $j(M)$) is the smallest primary (respy. closed primary) ideal contained in M.*

PROOF. Apply 8.4.26, 8.4.27 with $F = \{M\}$.

8.4.29. THEOREM (Abstract Wiener Tauberian Theorem). *Let A be a commutative s.s. regular, complex or strictly real, ρ-Banach algebra and J_0 the ideal of elements x such that $\hat{x}$ has compact support. If J_0 is dense in A then every closed ideal $I(\neq A)$ is contained in a maximal regular ideal M.*

PROOF. If I is not contained in any maximal regular ideal then $h(I)$ (in $\mathcal{M}$) $= \emptyset$. Since maximal ideals of A are closed and J_0 is dense in A it is clear that $h(J_0) = \emptyset$. By 8.4.26 we get $J_0 \subseteq I$. But then since I is closed and J_0 dense we must have $I = A$ – a contradiction.

8.4.30. Remark. For connection between above theorem and classical Wiener Tauberian theorem see [19, pp.147-9] or [22, pp.426-7]. See also [23, p.326] and reference cited therein for other related results.

§ 5. Holomorphic Functional Calculus for Several Commutative Algebra Elements

8.5.1. We begin by recalling the notion of polynomial convexity. Let K be a *bounded* subset of $\mathbb{K}^n$. Set

$$\mathsf{h}(K) = \{\vec{\lambda} = (\lambda_1, \cdots, \lambda_n) \in \mathbb{K}^n : |P(\vec{\lambda})| \leqslant \|P\|_K$$

for each polynomial P over $\mathbb{K}^n\}$, where $\|P\|_K = \sup\{|P(\vec{\mu})| : \vec{\mu} \in K\}$. Then clearly $\mathsf{h}(K) \supseteq K$ and $\mathsf{h}(K)$ is called the *polynomial convex hull* of K. If $\mathsf{h}(K) = K$ then K is called *polynomially convex* or *p-convex*.

If K is the unit circle in $\mathbb{C}$, $\mathsf{h}(K)$ is the closed unit disc, as can be seen using the maximum modulus theorem.

8.5.2. LEMMA. (i) $\mathsf{h}(K)$ *is compact, in particular a p-convex set* K *is compact.*

(ii) K *is p-convex iff for each* $\vec{\lambda}^0 \in \mathbb{K}^n \backslash K$, *there is a polynomial* P_0 *with* $|P_0(\vec{\lambda}^0)| > \|P_0\|_K$.

(iii) *If* K *is polynomially convex and* $C > 0$, *then for each* $\vec{\lambda}^0 \in \mathbb{K}^n \backslash K$ *we can find a polynomial* Q *with*

$$Q(\vec{\lambda}^0)| > C, \quad \|Q\|_K \leqslant C.$$

PROOF. (i) It is clear from its definition that $\mathsf{h}(K)$ is closed. Also, if Λ_j is the polynomial $\Lambda_j(\lambda_1, \cdots, \lambda_n) = \lambda_j$ then by taking $P = \Lambda_j$ in the definition of $\mathsf{h}(K)$ we get for $\vec{\lambda} \in \mathsf{h}(K)$, $|\lambda_j| \leqslant \sup\{|\mu_j| : \vec{\mu} \in K\} = C_j < \infty$. This means that $\mathsf{h}(K)$ is bounded and consequently it is compact.

(ii) Clear from the definition of p-convexity.

(iii) Given $\vec{\lambda}^0$, choose first P_0 as in (ii) and then set $Q(\vec{\lambda}) = CP_0(\vec{\lambda})/\|P_0\|_K$.

8.5.3. Remark. It is known that a compact set K in $\mathbb{C}$ is p-convex iff $\mathbb{C}\backslash K$ is connected (see [30, p.37]). Therefore the result 7.3.24 can be restated as: the spectrum $\sigma(a)$ of a t.generator a of a complex monogenic ρ-Banach algebra is p-convex. In this form the result is generalized in 8.5.15 (ii).

8.5.4. Recall that a polydisc $\mathbb{P}_{\vec{\delta}}$ with poly-radius $\vec{\delta} =$

$(\delta_1, \cdots, \delta_n)$ is given by

$$\mathbb{P}_{\vec{\delta}} = \{\vec{\lambda} \in \mathbb{K}^n : |\lambda_j| \leqslant \delta_j,\ j = 1, \cdots, n\}.$$

A subset Π of $\mathbb{P}_{\vec{\delta}}$ is called a *p*-polyhedron (in $\mathbb{P}_{\vec{\delta}}$) if there are polynomials $P_1, \cdots, P_m$ such that

$$\Pi = \{\vec{\lambda} \in \mathbb{P}_{\vec{\delta}} : \|P_1(\vec{\lambda})|, \cdots, |P_m(\vec{\lambda})| \leqslant 1\}.$$

8.5.5. LEMMA. (a) $\mathbb{P}_{\vec{\delta}}$ *itself is a p-polyhedron.*
(b) *Every p-polyhedron is p-convex.*

PROOF. (a) Take $m = n$, $P_j = \delta_j^{-1}\Lambda_j$, where Λ_j is the polynomial $\Lambda_j(\vec{\lambda}) = \lambda_j$.

(b) If $\vec{\lambda}^0 \notin \mathbb{K}^n \backslash \Pi$ then either some $|\lambda_j^0| > \delta_j$ or some $|P_k(\vec{\lambda}^0)| > 1$. In the first case $|\Lambda_j(\vec{\lambda}^0)| = |\lambda_j^0| > \delta_j \geqslant \|\Lambda_j\|_\Pi$ and in the second case we have $|P_k(\vec{\lambda}^0)| > 1 \geqslant \|P_k\|_\Pi$. Thus in either case $\vec{\lambda}^0 \notin \mathsf{h}(\Pi)$ which means that $\mathsf{h}(\Pi) = \Pi$ and Π is *p*-convex

8.5.6. LEMMA. *Let K be a p-convex set with $K \subseteq \mathbb{P}_{\vec{\delta}}$ and G an open set with $K \subseteq G \subseteq \mathbb{K}^n$. Then there exists a p-polyhedron Π with $K \subseteq \Pi \subseteq G$.*

PROOF. By 8.5.2 (iii), we can choose for each $\vec{\lambda}^0 \in \mathbb{K}^n \backslash K$ a polynomial $P_{\vec{\lambda}^0}$ with $P_{\vec{\lambda}_0}(\vec{\lambda}^{\,0}) > 1$, $\|P_{\vec{\lambda}}\|_K \leqslant 1$. By continuity of $P_{\vec{\lambda}^{\,0}}$ we have $|P_{\vec{\lambda}^0}(\vec{\lambda})| > 1$ for $\vec{\lambda}$ in some (open) neighbourhood $\mathcal{N}_{\vec{\lambda}^0}$ of $\vec{\lambda}^0$. Writing $\mathbb{P} = \mathbb{P}_{\vec{\delta}}$ and allowing $\vec{\lambda}_0$ to range in $\mathbb{P}\backslash G (\subseteq \mathbb{P}\backslash \mathbb{K})$ and using the compactness of $\mathbb{P}\backslash G$ we can find a finite family of neighbourhoods $\mathcal{N}_{\vec{\lambda}^j}$, corresponding to polynomials $P_{\vec{\lambda}^j}$, such that

$$\mathcal{N}_{\vec{\lambda}^1} \bigcup \cdots \bigcup \mathcal{N}_{\vec{\lambda}^n} \supseteq \mathbb{P}\backslash G.$$

Set $\Pi = \{\vec{\lambda} \in \mathbb{P} : |P_{\vec{\lambda}^j}(\vec{\lambda})| \leqslant \|P_{\vec{\lambda}^j}\|_K \leqslant 1\}$, so that $K \subseteq \Pi$. Suppose next $\vec{\lambda} \notin G$. If also $\vec{\lambda} \notin \mathbb{P}$ then of course $\vec{\lambda} \notin \Pi$ (since $\Pi \subseteq \mathbb{P}$). On the other hand if $\vec{\lambda} \in \mathbb{P}$ then $\vec{\lambda} \in \mathbb{P}\backslash G$ and so $\vec{\lambda} \in \mathcal{N}_{\vec{\lambda}^j}$ for some j. Hence $|P_{\vec{\lambda}^j}| > 1$, whence $\vec{\lambda} \in \Pi$. Therefore $\Pi \subseteq G$, as required.

8.5.7. The notion of spectrum of an element can be generalized to that of joint spectrum of a finite set of elements as delineated below.

Let A be a *commutative* algebra with unity e and write $\vec{a} = (a_1, \cdots, a_n)$ $(a_j \in A)$. If $\vec{\lambda} = (\lambda_1, \cdots, \lambda_n)$ $(\lambda_j \in \mathbb{K})$ then we write $I = I(\vec{\lambda}) = I(\vec{\lambda}, \vec{a}) =$ the ideal of A generated by $a_j - \lambda_j e$ $(j = 1, \cdots, n)$. Clearly we have

$$I = I(\vec{\lambda}) = \sum_{j=1}^{n} A(a_j - \lambda_j e). \qquad (*)$$

The *joint* (or simultaneous) *spectrum*

$$\begin{aligned} \sigma(\vec{a}) &= \sigma(a_1, \cdots, a_n) \text{ is defined by :} \\ \sigma(\vec{a}) &= \{\vec{\lambda} \in \mathbb{K}^n : I(\vec{\lambda}) \neq A\}. \end{aligned}$$

Not that when $n = 1$, $\vec{a} = a$ and $\sigma(\vec{a}) = \sigma(a)$.

8.5.8. LEMMA. (i) $\vec{\lambda} \in \mathbb{K}^n \backslash \sigma(\vec{a})$ *iff there are elements* $b_1, \cdots, b_n \in A$ *with* $\sum_{j=1}^{n}(a_j - \lambda_j e)b_j = e$.

(ii) $\vec{\lambda} \in \sigma(\vec{a})$ *iff for every* $b_1, \cdots, b_n \in A$ *we have* $\sum_{j=1}^{n}(a_j - \lambda_j e)b_j \notin G_i$, *where* G_i *denotes the group of invertible elements of* A.

(iii) $\vec{\lambda} \in \sigma(\vec{a})$ *iff there is a maximal ideal* M *of* A, *with* $a_j - \lambda_j e \in M$ $(j = 1, \cdots, n)$.

PROOF. (i) The stated condition is clearly equivalent to: $I(\vec{\lambda}) = A$.

(ii) The condition is clearly necessary and sufficient for $I(\vec{\lambda}) \neq A$.

(iii) If $\vec{\lambda} \in \sigma(\vec{a})$ then $I(\vec{\lambda}) \neq A$ and so by Krull, $I(\vec{\lambda}) \subseteq$ some maximal ideal M, whence $a_j - \lambda_j e \in I(\vec{\lambda}) \subseteq M$. Conversely, if $a_j - \lambda_j e.$ $(j = 1, \cdots, n)$ are contained in some maximal ideal M then $I(\vec{\lambda}) \subseteq M \neq A$ and $\vec{\lambda} \in \sigma(\vec{a})$.

8.5.9. COROLLARY. $\vec{\lambda} \in \sigma(\vec{a}) \Rightarrow \lambda_j \in \sigma(a_j)$ $(j = 1, \cdots, n)$, *so that*

$$\sigma(\vec{a}) \subseteq \sigma(a_1) \times \cdots \times \sigma(a_n).$$

PROOF. $\vec{\lambda} \in \sigma(\vec{a}) \Rightarrow a_j - \lambda_j e \in M \Rightarrow \lambda_j \in \sigma(a_j)$.

8.5.10. PROPOSITION. *Let A be a unital commutative Gelfand algebra – in particular, a unital, complex or strictly real, commutative ρ-Banach algebra. Then $\vec{\lambda} \in \sigma(\vec{a})$ iff there is a character $\chi \in \Delta$ with*

$$\chi(a_j) = \lambda_j \ (j = 1, \cdots, n), \text{ where } \ \vec{\lambda} = (\lambda_1, \cdots, \lambda_n).$$

PROOF. If $\vec{\lambda} \in \sigma(\vec{a})$, by 8.5.8 (iii) there is a maximal ideal M with $a_j - \lambda_j e \in M$. Since A is Gelfand M is hypermaximal and let χ be the character determined by it. Then $\chi(a_j) = \lambda_j$, $(\vec{\lambda} = (\lambda_j))$. Conversely, if $\chi \in \Delta$ satisfies the stated condition then $a_j - \lambda_j e \in M$ $(M = \ker \chi)$ and $\vec{\lambda} \in \sigma(\vec{a})$.

8.5.11. COROLLARY. *Let $A = (A, \|\cdot\|)$ ($\|\cdot\|$sm.) be a unital, complex or strictly real, ρ-Banach algebra. Then $\sigma(\vec{a})$ is nonempty compact with*

$$\sigma(\vec{a}) \subseteq \prod_{j=1}^{n} \sigma(a_j) \subseteq \mathbb{P}^n_{\vec{\delta}},$$

where $\vec{\delta} = (\delta_j)$, $\delta_j = \|a_j\|^{\frac{1}{\rho}}$.

PROOF. Since A is Gelfand, $\Delta = \Delta_c \neq \emptyset$. If $\chi \in \Delta$ and $\chi(a_j) = \lambda_j$ then by 8.5.10,

$$\vec{\lambda} = (\lambda_j) \in \sigma(\vec{a}), \text{ so that } \sigma(\vec{a}) \neq \emptyset.$$

By 7.3.12, $\Delta = \Delta_c$ is compact and since the map $\chi \mapsto (\chi(a_1), \cdots, \chi(a_n))$ is continuous, $\sigma(\vec{a})$ is compact. Finally, if $\mu_j \in \sigma(a_j)$ then $|\mu_j| \leqslant r(a_j) \leqslant \|a_j\|^{\frac{1}{\rho}}$ (see 7.3.27). This completes the proof.

8.5.12. COROLLARY. *Let A be as in 8.5.10, and $a_1, \cdots, a_n \in A$. Then*

(i) *For any polynomial P over $\mathbb{K}$,*

$$\sigma(P(a_1, \cdots, a_n)) = P(\sigma(a_1, \cdots a_n)).$$

(ii) *If* $\alpha_j \in \mathbb{K}$ $(j = 1, \cdots, n)$ *then*

$$\sigma(\alpha_1 a_1, \cdots, \alpha_n a_n) = \{(\alpha_1\lambda_1, \cdots, \alpha_n\lambda_n) : \vec{\lambda} \in \sigma(a_1, \cdots, a_n)\}.$$

PROOF. It suffices to observe that each $\chi \in \Delta$ being a homomorphism we have:

$$\chi(P(a_1, \cdots, a_n)) = P(\chi(a_1), \cdots, \chi(a_n))$$

$$\chi(\alpha_j a_j) = \alpha_j \chi(a_j).$$

8.5.13. Remark. Arens has shown that the conclusion in Proposition 8.5.10 holds also if A is a unital commutative complex locally sm. $\mathfrak{F}$ algebra (see [31, p.105]).

8.5.14. PROPOSITION. *Let A be a unital commutative ρ-Banach algebra which is either complex or strictly real, and which is t. generated by a finite set of elements $a_1, \cdots, a_n$. Then the spectrum $\Delta(A)$ is homeomorphic to the joint spectrum $\sigma(\vec{a})$ under the map*

$$\theta : \chi \mapsto (\chi(a_1), \cdots, \chi(a_n)).$$

PROOF. By 8.5.10, θ is surjective. The map θ is also injective. To see this, suppose that $\theta(\chi_1) = \theta(\chi_2)$. Then, for a polynomials $P = P(\lambda_1, \cdots, \lambda_n)$ over $\mathbb{K}$ we have

$$\begin{aligned} \chi_1(P(a_1, \cdots, a_n)) &= P(\chi_1(a_1), \cdots, \chi_1(a_n)) \\ &= P(\chi_2(a_1), \cdots, \chi_2(a_n)) = \chi_2(P(a_1, \cdots, a_n)). \end{aligned}$$

Since the elements of A of the form $P(a_1, \cdots, a_n)$ – where P is a polynomial over $\mathbb{K}$ in n variables – form a dense subalgebra A_0 of A (since $a_1, \cdots, a_n$ t. generate A) and $\chi_1 = \chi_2$ on A_0, we conclude from the continuity of χ_1, χ_2 that $\chi_1 = \chi_2$ on A, proving injectiveness of θ. Further, the continuity of θ follows from that of the χ's. By 7.3.12, $\Delta = \Delta(A)$ is compact Hausdorff. Therefore θ is a homeomorphism.

8.5.15. PROPOSITION. *Let A be a unital commutative, complex or strictly real, ρ-Banach algebra t. generated by $a_1, \cdots, a_n$. Then:*

(i) *For any $C > 0$, $\alpha \in \mathbb{K}$ and $\vec{\lambda}^0 \in \mathbb{K}^n \backslash \sigma(\vec{a})$ there is a polynomial P (in n variables over $\mathbb{K}$) such that*

$$P(\vec{\lambda}^0) = \alpha, \ \|P(\vec{a})\| < C$$

where $\vec{a} = (a_1, \cdots, a_n)$. In particular, there is a polynomial Q with $Q(\vec{\lambda}^0) = 1 > \|Q(\vec{a})\|$.

(ii) *$\sigma(\vec{a})$ is polynomially convex.*

PROOF. If $\vec{\lambda}^0 \in \mathbb{K}^n \backslash \sigma(\vec{a})$ then

$$I(\vec{\lambda}^0) = \sum_{j=1}^{n} A(a_j - \lambda_j^0 e) = \sum_{j=1}^{n} (a_j - \lambda_j^0 e) A = A,$$

where e is the unity of A. It follows that $\alpha e \in A$ has a representation

$$\sum_{j=1}^{n} (a_j - \lambda_j^0 e) b_j = \alpha e \tag{1}$$

where $b_j \in A$.

Since A is t. generated by $\vec{a}$, each b_j can be approximated by elements of the form $P(a_1, \cdots, a_n)$ where P is a polynomial. It follows from this and equation (1) that there are polynomials P_j $(j = 1, \cdots, n)$ such that

$$\|\alpha e - \sum_{j=1}^{n} (a_j - \lambda_j^0 e) P_j(\vec{a})\| < C. \tag{2}$$

Write

$$P(\vec{\lambda}) = \alpha - \sum_{j-1}^{n} (\lambda_j - \lambda_j^0) P_j(\vec{\lambda}). \tag{3}$$

Clearly P is a polynomial in $\lambda_1, \cdots, \lambda_n$. Further, it is clear from (3), (2) that we have

$$P(\vec{\lambda}^0) = \alpha$$
$$\|P(\vec{a})\| < C. \tag{4}$$

If we take $C = 1 = \alpha$ in (4) then the corresponding polynomial P which we denote by Q satisfies:

$$Q(\vec{\lambda}_0) = 1 > \|Q(\vec{a})\|.$$

This completes the proof of (i).

To prove (ii) take a $\vec{\lambda} \in \sigma(\vec{a})$. By 8.5.10 we have

$$\vec{\lambda} = \chi(\vec{a}) = (\chi(a_1), \cdots, \chi(a_n))$$

for some $\chi \in \Delta$. It follows that the polynomial Q chosen above satisfies:

$$\begin{aligned} |Q(\vec{\lambda}^0)| &= |Q(\chi(\vec{a}))| = |\chi(Q(\vec{a}))| \\ &\leqslant \|Q(\vec{a})\|^{\frac{1}{\rho}} \qquad (\text{since } \|\chi\| \leqslant 1) \\ &< 1. \end{aligned} \tag{5}$$

Since $K = \sigma(\vec{a})$ is compact, (5) implies that $\|Q\|_K < 1$. On the other hand, $Q(\vec{\lambda}^0) = 1$ (by construction). Therefore, by 8.5.2 (ii), $\sigma(\vec{a})$ is p-convex.

8.5.16. Remark. From 8.5.14, 8.5.15 (ii), 8.5.11, it follows that the spectrum $\Delta = \Delta(A)$, of a finitely t. generated complex commutative ρ-Banach algebra A, is homeomorphic to a polynomially convex compact subset of $\mathbb{C}^n$. Conversely, it can be shown that if K is a polynomially convex compact subset of $\mathbb{C}^n$ then there is a finitely t.generated (with n t. generators) complex commutative Banach algebra A such that $\Delta(A)$ is homeomorphic with K (see [10, p.45]).

8.5.17. Remark. The joint spectrum or even the spectrum of a single element, of a commutative Banach algebra which is not finitely t. generated, may not be p-convex. For example, in the Banach algebra B of bounded complex-valued functions on the unit interval $[0, 1]$ (under the sup-norm) the function $f(t) = \vec{e}^{\,2\pi i t}$ $(0 \leqslant t \leqslant 1)$ has for its spectrum its range which is the unit circle. The unit circle, however, is not p-convex (since its p-convex hull is the closed unit disc).

To establish the main theorem of this section, which asserts that holomorphic functions of several variables operate on the space of Gelfand transforms, we need a number of preliminary results which we proceed to consider.

8.5.18. LEMMA (Arens-Caulderon trick [3′, p.205]). *Let A be a unital commutative, complex or strictly real, ρ-Banach algebra.*

Let $a_1, \cdots, a_n \in A$ and $G \subseteq \mathbb{K}^n$ an open set with $\sigma(a_1, \cdots, a_n) \subseteq G$. Then there exists a finitely t. generated closed subunital algebra B of A, with $a_j \in B$ $(j = 1, \cdots, n)$ and $\sigma_B(a_1, \cdots, a_n) \subseteq G$, where σ_B denotes the spectrum with respect to the algebra B.

PROOF. We may assume that the norm $\|\cdot\|$ of A is sm. Write

$$K = \{\vec{\lambda} \in \mathbb{K}^n : |\lambda_j| \leqslant \|a_j\|^{\frac{1}{\rho}} \ (j = 1, \cdots, n)\}.$$

Then K is compact. By 8.5.8 (i), for each $\vec{\lambda} \in \mathbb{K}^n \backslash \sigma(a_1, \cdots, a_n)$ there are elements $b_1, \cdots b_n \in A$ such that

$$\sum_{j=1}^{n} (a_j - \lambda_j e) b_j = e. \tag{1}$$

Denote by $B(\vec{\lambda})$ the closed subalgebra of A generated by $e, a_1, \cdots, a_n,\ b_1, \cdots, b_n$. By considering the relation (1) in $B(\vec{\lambda})$, we get, by 8.5.8 (i), that

$$\vec{\lambda} \notin \sigma_{B(\vec{\lambda})}(a_1, \cdots, a_n) = \sigma_{B(\vec{\lambda})} \text{ (say)}. \tag{2}$$

Since $\sigma_{B(\vec{\lambda})}$ is closed there is an open neighbourhood $G(\vec{\lambda})$ of $\vec{\lambda}$ such that

$$\sigma_{B(\vec{\lambda})} \bigcap G(\vec{\lambda}) = \emptyset. \tag{3}$$

It follows, using the compactness of $K \backslash G$ and (3) that there are points $\vec{\lambda}^k \in K \backslash G$ $(k = 1, \cdots, n)$ such that

$$\bigcup_{k=1}^{n} G(\vec{\lambda}^k) \supseteq K \backslash G \tag{4}$$

and

$$\sigma_{B(\vec{\lambda}^k)} \bigcap G(\vec{\lambda}^k) = \emptyset \ (k = 1, \cdots, n). \tag{5}$$

Since each $B_k = B(\vec{\lambda}^k)$ is finitely t. generated it is clear that there is a finitely t. generated subunital algebra B of A with all $B_k \subseteq B$. By 1.7.19

$$\sigma_B(a_1, \cdots, a_n) \subseteq \sigma_{B_k}(a_1, \cdots, a_n). \tag{6}$$

From (4),(5),(6) we obtain

$$\sigma_B(a_1, \cdots, a_n) \bigcap (K \backslash G) = \emptyset.$$

On the other hand, by 8.5.11,

$$\sigma_B(a_1, \cdots, a_n) \subseteq \mathbb{P}^n_{\bar{\delta}} = K.$$

Therefore, $\sigma_B(a_1, \cdots, a_n) \subseteq G$, as required.

8.5.19. THEOREM (Oka's Extension Theorem.) *Let $m, r > 0$ be integers, $\epsilon_j > 0$ $(1 \leqslant j \leqslant m + r)$ be real numbers and*

$$\mathbb{P} = \{\vec{\lambda} \in \mathbb{C}^{m+r} : |\lambda_j| \leqslant \epsilon_j \ (1 \leqslant j \leqslant m + r)\}.$$

Let $P_j (j = 1, \cdots, r)$ be complex polynomials in m variables and θ the Oka map of $\mathbb{C}^m \to \mathbb{C}^{m+r}$ given by:

$$\theta : \vec{\lambda} = (\lambda_1, \cdots, \lambda_m) \mapsto (\lambda_1, \cdots, \lambda_m, P_1(\vec{\lambda}), \cdots, P_r(\vec{\lambda})).$$

Then, for any f holomorphic on an open neighbourhood of $\theta^{-1}(\mathbb{P})$ there corresponds an F holomorphic on an open neighbourhood of $\mathbb{P}$ such that

$$F(\theta(\vec{\lambda})) = f(\vec{\lambda}) \ (\vec{\lambda} \in \theta^{-1}(\mathbb{P})).$$

PROOF. See [4, p.103].

8.5.20. PROPOSITION. *Let $A = (A, \|\cdot\|)$ be a unital commutative complex ρ-Banach algebra, $\vec{a} = (a_1, \cdots, a_n)$, $a_j \in A$ $(j = 1, \cdots n)$ and $G \subseteq \mathbb{C}^n$ an open neighbourhood of $\sigma(\vec{a})$. Then we can find a finite number of elements $a_{n+1}, \cdots, a_N \in A$ such that given any function f holomorphic on G there is a function F holomorphic on the polydisc*

$$\{\vec{\lambda} \in \mathbb{C}^N : |\lambda_j| \leqslant 1 + 2\|a_j\|^{\frac{1}{\rho}} \ (j = 1, \cdots, N)\}$$

such that

$$f(\hat{a}_1(\chi), \cdots \hat{a}_n(\chi)) = F(\hat{a}_1(\chi), \cdots, \hat{a}_N(\chi)) \ (\chi \in \Delta).$$

PROOF. By 8.5.18, we can find a finitely t. generated subunital algebra B containing $a_1, \cdots a_n$ with

$$\sigma_B(\vec{a}) = \sigma_B(a_1, \cdots, a_n) \subseteq G. \tag{1}$$

We can take as t. generators ($\neq e$) of B to be $a_1, \cdots, a_m$, where $n \leqslant m$. Fix $C > 0$ and set $\alpha = 2C^{\frac{1}{\rho}} + 2$. For any $\vec{\lambda} \in \mathbb{C}^n \backslash \sigma_B(\vec{a})$, there is, by 8.5.15(i), a polynomial P such that

$$P(\vec{\lambda}) = 2C^{\frac{1}{\rho}} + 2, \ \|P(\vec{a})\| < C.$$

Then we have

$$|P(\vec{\lambda})| = 2C^{\frac{1}{\rho}} + 2 > 1 + 2C^{\frac{1}{\rho}} \geqslant 1 + 2\|P(\vec{a})\|^{\frac{1}{\rho}}. \tag{2}$$

It follows that there is an open neighbourhood $V_P(\vec{\lambda})$ of $\vec{\lambda}$ such that

$$|P(\vec{\mu})| > 1 + 2\|P(\vec{a})\|^{\frac{1}{\rho}}. \tag{3}$$

for all $\vec{\mu} \in V_P(\vec{\lambda})$. Denote by π the natural projection

$$(\lambda_1, \cdots, \lambda_m) \mapsto (\lambda_1, \cdots, \lambda_n) \qquad (n \leqslant m)$$

of $\mathbb{C}^m$ onto $\mathbb{C}^n$. Write

$$\mathbb{P}^m = \{\vec{\lambda} \in \mathbb{C}^m : |\lambda_j| \leqslant 1 + 2\|a_j\| \ (j = 1, \cdots, m)\}. \tag{4}$$

$$\text{Then } : K = \mathbb{P}^m \backslash \pi^{-1}(G) \text{ is compact.} \tag{5}$$

Also, $\pi^{-1}(G)$ contains $\sigma(\vec{b})$, where $\vec{b} = (a_1, \cdots, a_m)$ since

$$\pi(\sigma(\vec{b})) = \sigma(\vec{a}) \subseteq G.$$

It follows from (1),(2),(3) and (5) that there are polynomials $P_1, \cdots, P_r$ such that $V_{P_1}, \cdots, V_{P_r}$ cover K. Therefore, for any $\vec{\lambda} \in K$, we have

$$\|P_k(\vec{\lambda})\| > 1 + 2\|P_k(\vec{b})\|^{\frac{1}{\rho}} \tag{6}$$

for at least one k $(1 \leqslant k \leqslant r)$.

Set

$$N = m + r, \ a_{m+k} = P_k(\vec{b}) \ \ (k = 1, \cdots, r),$$

$$\mathbb{P}^N = \{\vec{\lambda} \in \mathbb{C}^N : |\lambda_j| \leqslant 1 + 2\|a_j\|^{\frac{1}{\rho}} \ \ (j = 1, \cdots, N)\}.$$

Let θ be the Oka map

$$\theta(\lambda_1, \cdots, \lambda_m) = (\lambda_1, \cdots, \lambda_m, P_1(\vec{\lambda}), \cdots, P_r(\vec{\lambda}))$$

where $\vec{\lambda} = (\lambda_1, \cdots, \lambda_m)$. Since $\theta(\vec{\lambda}) \in \mathbb{P}^N \Rightarrow \vec{\lambda} \in \mathbb{P}^m$ and

$$|P_k(\vec{\lambda})| \leqslant 1 + 2\|P_k(\vec{b})\|^{\frac{1}{\rho}} \quad (k = 1, \cdots, r),$$

using (4),(6) we get

$$\theta^{-1}(\mathbb{P}^N) \subseteq \pi^{-1}(G).$$

The function $f \circ \pi$ is holomorphic on the open neighbourhood $\pi^{-1}(G)$ of $\theta^{-1}(\mathbb{P}^N)$. By 8.5.19, there is a function F holomorphic on an open neighbourhood of $\theta^{-1}(\mathbb{P}^N)$ such that

$$F(\theta(\lambda^{-1})) = f \circ \pi(\vec{\lambda}) \; (\vec{\lambda} \in \theta^{-1}(\mathbb{P}^N)). \tag{7}$$

For $\chi \in \Delta$, set

$$\lambda_j = \chi(a_j) \; (j = 1, \cdots, m).$$

Then

$$P_k(\vec{\lambda}) = P_k(\lambda_1, \cdots, \lambda_m) = \chi P_k(\vec{b}) = \chi(a_{m+k}) \; (k = 1, \cdots, r)$$

and so

$$\theta(\vec{\lambda}) = (\chi(a_1), \cdots, \chi(a_N)).$$

Since

$$|\lambda_j| = |\chi(a_j)| \leqslant \|a_j\|^{\frac{1}{\rho}} \quad (j = 1, \cdots, m)$$

and

$$|P_k(\vec{\lambda})| = \chi(a_{m+k}) \leqslant \|a_{m+k}\|^{\frac{1}{\rho}}$$

we conclude that $\theta(\vec{\lambda}) \in \mathbb{P}^N$, $\vec{\lambda} \in \theta^{-1}(\mathbb{P}^N)$. Finally we obtain, by (7)

$$f(\chi(a_1), \cdots, \chi(a_n)) = f \circ \pi(\vec{\lambda}) = F(\theta(\vec{\lambda})) = F(\chi(a_1), \cdots, \chi(a_n)),$$

which completes the proof.

The next lemma is needed for proving the main theorem which follows

8.5.21. LEMMA. *If* $\beta_n \in \mathbb{C}$, $\limsup\limits_{n\to\infty} |\beta_n| = l$ *and* $0 < \rho \leqslant 1$, *then* $\limsup\limits_{n\to\infty} |\beta_n|^\rho = l^\rho$.

PROOF. Given $\epsilon > 0$, choose $\eta > 0$ such that $\eta^\rho = \epsilon$. Since $\limsup\limits_{n\to\infty}|\beta_n| = l$ there is a N such that

$$|\beta_n| < l + \eta \text{ for } n > N. \tag{1}$$

Also, for any m, there is an $n > m$ with

$$|\beta_n| > l - \eta. \tag{2}$$

From (1) we get

$$|\beta_n|^\rho < (l+\eta)^\rho \leqslant l^\rho + \eta^\rho \leqslant l^\rho + \epsilon \quad (n \geqslant N) \tag{3}$$

and from(2) we obtain

$$|\beta_n|^\rho + \epsilon = |\beta_n|^\rho + \eta^\rho \geqslant (|\beta_n| + \eta)^\rho > l^\rho. \tag{4}$$

The inequalities (3),(4) show that $\limsup_{n\to\infty} |\beta_n|^\rho = l^\rho$.

8.5.22. THEOREM (Shilov-Arens-Caulderon)†. *Let A be a unital commutative, complex or strictly real, ρ-Banach algebra. Let $a_1, \cdots, a_n$ be elements of A, $G \subseteq \mathbb{K}^n$, an open set such that $\sigma(a_1, \cdots, a_n) \subseteq G$ and f a function holomorphic*†† *on G. Then there exists an element $b \in A$ such that*

$$\hat{b}(\chi) = \chi(b) = f(\hat{a}_1(\chi), \cdots, \hat{a}_n(\chi)) \quad (\chi \in \Delta). \tag{*}$$

The element b is uniquely determined iff A is s.s..

PROOF. First assume that A is complex. Then, by 8.5.20, there are elements $a_{n+1}, \cdots, a_N$ in A and a function F holomorphic on an open neighbourhood G^* of the polydisc

$$\mathbb{P} = \mathbb{P}^N = \{\vec{\lambda} \in \mathbb{C}^N : |\lambda_j| \leqslant 1 + 2\|a_j\|^{\frac{1}{\rho}} \ (j = 1, \cdots, N)\}$$

such that

$$F(\chi(a_1), \cdots, \chi(a_N)) = f(\chi(a_1), \cdots, \chi(a_n)).$$

Since $\mathbb{P}$ is compact and $\mathbb{P} \subseteq G^*$ the distance from $\mathbb{P}$ to the disjoint closed $\mathbb{C}^n \backslash G^*$ is positive. So we can find an open polydisc $\mathbb{P}_1^o$ with

$$\mathbb{P} \subset \mathbb{P}_1^o \subseteq G^*.$$

† These authors considered only the case of complex Banach algebras.

†† holomorphic in the real case means real analytic.

Let the Taylor series representation of F on $\mathbb{P}_1^o$ be

$$F(\vec{\lambda}) = \sum_{\vec{k}\in\mathbb{N}^N} \alpha_{\vec{k}}\lambda_1^{k_1}\cdots\lambda_N^{k_N}$$

where $\mathbb{N} = \{0, 1, 2, \cdots\}$. By Abel's lemma, the (multiple) series for F converges absolutely on $\mathbb{P}$. In particular, the series converges at $\vec{\mu}$ with $\mu_j = 2\|a_j\|^{\frac{1}{\rho}} (j = 1, \cdots, N)$. By root test† for convergence we have

$$\limsup_{\|\vec{k}\|\to\infty}(|\alpha_{\vec{k}}|2^{k_1}\|a_1\|^{\frac{k_1}{\rho}}\cdots 2^{k_N}\|a_N\|^{\frac{k_N}{\rho}})^{\frac{1}{\|\vec{k}\|}} \leqslant 1 \qquad (*)$$

where $\|\vec{k}\| = k_1 + \cdots + k_N$.

The inequality $(*)$ reduces to

$$\limsup_{\|\vec{k}\|\to\infty}(|\alpha_{\vec{k}}|\|a_1\|^{\frac{k_1}{\rho}}\cdots\|a_N\|^{\frac{k_N}{\rho}})^{\frac{1}{\|\vec{k}\|}} \leqslant \frac{1}{2}.$$

By applying 8.5.21, we obtain

$$\limsup_{\|\vec{k}\|\to\infty}(|\alpha_{\vec{k}}|^{\rho}\|a_1\|^{k_1}\cdots\|a_N\|^{k_N})^{\frac{1}{\|\vec{k}\|}} \leqslant \frac{1}{2^{\rho}} < 1.$$

Therefore, by the root test, the multiple series

$$\sum|\alpha_{\vec{k}}|^{\rho}\|a_1\|^{k_1}\cdots\|a_N\|^{k_N}$$

converges, which means that the series

$$\sum\alpha_{\vec{k}}a_1^{k_1}\cdots a_N^{k_N}$$

converges absolutely and consequently the series converges to an element $b \in A$, i.e.,

$$\sum\alpha_{\vec{k}}a_1^{k_1}\cdots a_N^{k_N} = b.$$

---

† See [2, p.364].

If $\chi \in \Delta$, then χ being a continuous homomorphism we have

$$\begin{aligned}\chi(b) &= \sum \alpha_{\vec{k}} \chi(a_1)^{k_1} \cdots \chi(a_N)^{k_N} \\ &= F(\chi(a_1), \cdots, \chi(a_N)) = f(\chi(a_1), \cdots, \chi(a_N)),\end{aligned}$$

so that we have

$$\hat{b}(\chi) = f(\hat{a}_1(\chi), \cdots, \hat{a}_n(\chi)) \quad (\chi \in \Delta).$$

This proves the theorem when A is complex.

Next, let A be strictly real and $\tilde{A}$ its complexification. Here we have

$$\sigma(a_1, \cdots, a_n) \subseteq G \subseteq \mathbb{R}^n$$

and f holomorphic on G. Then it is known† that there is a (complex) holomorphic function F on an open set $\tilde{G} \subseteq \mathbb{C}^n$ with $G \subseteq \tilde{G}$ and $F|G = f$. Since A is strictly real we have

$$\tilde{\sigma}(a_1, \cdots, a_n) = \sigma(a_1, \cdots, a_n) \subseteq G \subseteq \tilde{G},$$

where $\tilde{\sigma}$ denotes the joint spectrum with reference to $\tilde{A}$. Since $\tilde{A}$ is a complex algebra, by applying the theorem for this case (just proved) we can find a $\tilde{b} \in \tilde{A}$ such that

$$\tilde{\chi}(\tilde{b}) = F(\tilde{\chi}(a_1), \cdots, \tilde{\chi}(a_n)) = F(\chi(a_1), \cdots, \chi(a_n))$$

where $\tilde{\chi} \in \tilde{\Delta}$, and $\chi = \tilde{\chi}|A$; by 1.9.18, 1.6.14, χ is real-valued. Since $\tilde{b} \in \tilde{A}$, $\tilde{b} = b + ic$ $(b, c \in A)$.

Therefore we have

$$\begin{aligned}\chi(b) + i\chi(c) &= \tilde{\chi}(b + ic) = \tilde{\chi}(\tilde{b}) \\ &= F(\chi(a_1), \cdots, \chi(a_n)) = f(\chi(a_1), \cdots, \chi(a_n)).\end{aligned}$$

Since the RHS is real we must have

$$\chi(b) = f(\chi(a_1), \cdots, \chi(a_n))$$

which completes the proof of the existence of b for the strictly real case.

† See [3, p.134].

For the uniqueness of b result (in both the cases) we note that if b_1 is also an element such that $(*)$ holds when b is replaced by b_1 then clearly we have $b - b_1 \in \bigcap \ker \chi = \sqrt[h]{A} = \sqrt{A}$ (since A is a Gelfand algebra). So $b_1 = b$ iff $\sqrt{A} = \{0\}$ iff A is s.s..

8.5.23. PROPOSITION. *Let A be a unital commutative, complex or strictly real, pseudo-Michael algebra such that in its projective limit decomposition $A = \varprojlim \hat{A}_\alpha$, each $\hat{A}_\alpha$ is s.s.. Let Δ_c be the set of continuous characters of A. For $a_1, \cdots, a_n \in A$ write*

$$\sigma^*(a_1, \cdots, a_n) = \{\chi(a_1), \cdots, \chi(a_n) : \chi \in \Delta_c\}.$$

If f is a holomorphic function on an open set $G \supseteq \sigma^*(a_1, \cdots, a_n)$ then there is a unique element $b \in A$ such that

$$\hat{b}(\chi) = f(\hat{a}_1(\chi), \cdots, \hat{a}_n(\chi)) \quad (\chi \in \Delta_c).$$

PROOF. By 7.3.19 we have $\Delta_c = \bigcup \Delta_\alpha, \Delta_\alpha \simeq \hat{\Delta}_\alpha$, where $\hat{\Delta}_\alpha =$ the space of all characters of $\hat{A}_\alpha =$ the space of all continuous characters of $\hat{A}_\alpha$ (by 7.1.6(c), 7.1.9). Since each $\hat{A}_\alpha$ is a Gelfand algebra we have (see 8.5.10) the relation

$$\sigma^*(a_1, \cdots, a_n) = \bigcup_\alpha \sigma(a_{1\alpha}, \cdots, a_{n\alpha}),$$

where $a_{j\alpha} = \varphi_\alpha(a_j)$, and φ_α the canonical map $A \to A_\alpha$. Since $G \supseteq \sigma(a_{1\alpha}, \cdots, a_{n\alpha})$, by 8.5.22 there is an element $b_\alpha \in \hat{A}_\alpha$ such that

$$\hat{\chi}_\alpha(b_\alpha) = f(\hat{\chi}_\alpha(a_{1\alpha}), \cdots, \hat{\chi}_\alpha(a_{n\alpha})) \quad (\hat{\chi}_\alpha \in \hat{\Delta}_\alpha).$$

Since $b_{j\beta} = \varphi_{\alpha\beta}(a_{j\beta})$ we have

$$\begin{aligned} \hat{\chi}_\alpha(b_\alpha) &= f(\hat{\chi}_\alpha \circ \varphi_{\alpha\beta}(a_{1\beta}), \cdots, \hat{\chi}_\alpha \circ \varphi_{\alpha\beta}(a_{n\beta})) \\ &= \hat{\chi}_\alpha \circ \varphi_{\alpha\beta}(b_\beta) = \hat{\chi}_\alpha(\varphi_{\alpha\beta}(b_\beta)). \end{aligned}$$

Since $\hat{A}_\alpha$ is s.s. we have $\varphi_{\alpha\beta}(b_\beta) = b_\alpha$. It follows that $\varprojlim b_\alpha = b$ exists and we have clearly

$$\hat{b}(\chi) = f(\hat{a}_1(\chi), \cdots, \hat{a}_n(\chi)) \quad (\chi \in \Delta_c).$$

To prove the uniqueness of b it is enough to show that $\sqrt[ch]{A} = \{0\}$. Suppose that $x \in A$, $x \neq 0$. Then for some α, $x_\alpha \neq 0$. Since $\hat{A}_\alpha$ is s.s. Gelfand there is $\hat{\chi}_\alpha$ with $\hat{\chi}_\alpha(x_\alpha) \neq 0$. Then $\chi_\alpha = \hat{\chi}_\alpha \circ \varphi_\alpha \in \Delta_c$ and $\chi_\alpha(x) = \hat{\chi}_\alpha(x_\alpha) \neq 0$. This shows that $\sqrt[ch]{A} = \{0\}$,† which completes the proof.

8.5.24. Remark. The holomorphic functional calculus embodied in 8.5.22, for t. finitely generated unital commutative complex Banach algebras, using *A*. Weil's integral formula for holomorphic functions of several variables, was obtained by Shilov. This was extended by Arens and Caulderon for arbitrary unital commutative complex Banach algebras. An approach to holomorphic functional calculus based on exterior differential forms had earlier been given by L. Waelbroeck (see [29, pp.41-50], [12′]). Waelbroeck's method enabled him to show that the correspondence

$$f \mapsto b = f(a_1, \cdots, a_n) = f(\vec{a}).$$

is a homomorphism – a result not obtained by either Shilov or Arens. Further, Waelbroeck developed the functional calculus for any complex commutative locally convex complete Hausdorff algebra. For an introduction to Waelbroeck's method (for commutative Banach algebras) see [12, pp.41-50].

The above proof of theorem 8.5.22, based on Oka's extension theorem instead of any Cauchy integral formula is modelled after the proof found in Bonsall and Duncan's book [4, §20].

§ 6. Shilov Idempotent Theorem

8.6.1. LEMMA. *Let R be a ring with unity e, which is a direct sum $R = I_1 + I_2$ of two l. ideals $I_j \neq \{0\}$ $(j = 1, 2)$. Then there are idempotents $e_j \neq 0$ $(j = 1, 2)$ in R such that*

$$e = e_1 + e_2,\ e_1 e_2 = e_2 e_1 = 0,\ I_j = Re_j\ (j = 1, 2).$$

(*A similar result for r. ideal decomposition.*)

PROOF. Since $R = I_1 + I_2$, we have for e the decomposition

† It follows from this that the algebra A is s.s. (since $\sqrt{A} \subseteq \sqrt[ch]{A} = \{0\}$).

$$e = e_1 + e_2 \ (e_1 \in I_1, e_2 \in I_2).$$

Therefore, for $x \in R$, we have

$$x = xe = xe_1 + xe_2.$$

If $x \in I_1$, then $x - xe_1 \in I_1$, $xe_2 \in I_2$ and since $I_1 \bigcap I_2 = \{0\}$, we get

$$x = xe_1, \ xe_2 = 0.$$

Therefore, $I_1 = Re_1$, $e_1^2 = e_1$, $e_1e_2 = 0$. Similarly, $I_2 = Re_2$, $e_2^2 = e_2$, $e_2e_1 = 0$.

8.6.2. PROPOSITION. *Let A be a unital t. spectrally Gelfand algebra – in particular a Gelfand algebra. If A has an idempotent* $u \neq 0, e$ *then* Δ_c *is disconnected.*

PROOF. By 1.7.9, $\sigma(u) = \{0,1\}$. Since A is t. spectrally Gelfand there are $\chi_1, \chi_2 \in \Delta_c$ with $\chi_1(u) = 0$, $\chi_2(u) = 1$. Further, for any $\chi \in \Delta_c$, $\chi(u) = 0$ or 1 (since $\chi(u) = \chi(u^2) = \chi(u)^2$). It follows that if we set

$$F_1 = \{\chi \in \Delta_c : \chi(u) = 0\}; \quad F_2 = \{\chi \in \Delta_c : \chi(u) = 1\}$$

then the F_j $(j = 1,2)$ are non-empty (since $\chi_j \in F_j$) closed sets in Δ_c such that

$$F_1 \dot{\bigcup} F_2 = \Delta_c.$$

So Δ_c is disconnected.

8.6.3. COROLLARY. *If a unital t. spectrally Gelfand algebra A is a decomposable l. (or r.) A-module then* Δ_c *is disconnected.*

PROOF. By 8.6.1, there are idempotents $e_1 \neq 0$, $e_2 \neq 0$ with $e_1 + e_2 = e$, when e is the unity of A. Clearly e_1 (or e_2) $\neq 0, e$. Therefore, by 8.6.2, Δ_c is disconnected.

8.6.4. THEOREM (Shilov† idempotent theorem). *Let A be a unital commutative, complex or strictly real, ρ-Banach algebra whose spectrum* $\Delta = \Delta_c$ *is disconnected, so that we have a*

† He obtained the result for the case of complex Banach algebras.

decomposition $\Delta = \Delta_1 \dot{\bigcup} \Delta_0$ *with* Δ_1, Δ_0 *clopen.*† *Then there exists a unique*†† *idempotent* u *in* A *such that* $\hat{u} = 1$ *on* Δ_1 *and* $\hat{u} = 0$ *on* Δ_0.

PROOF. Since $\Delta_1 \bigcap \Delta_0 = \emptyset$, for each $\chi_1 \in \Delta_1$ and $\chi_0 \in \Delta_0$ there is an element $a_{\chi_1\chi_0} \in A$ with $\chi_1(a_{\chi_1,\chi_0}) \neq \chi_0(a_{\chi_1,\chi_0})$. It follows that we can choose open neighbourhoods $V(\chi_1)$ of χ_1 and $W(\chi_0)$ of χ_0 in Δ such that

$$\chi_1'(a_{\chi_1\chi_0}) \neq \chi_0'(a_{\chi_1,\chi_0}) \text{ for } \chi_1' \in V(\chi_1), \chi_0' \in W(\chi_0).$$

The open sets $V(\chi_1) \times W(\chi_0)$ cover the compact set $\Delta_1 \times \Delta_0$ and consequently there is a finite subcovering $V(\chi_{1j}) \times W(\chi_{0j})$ $(j = 1, \cdots, n)$. Rename the associated elements $a_{\chi_{1j}\chi_{0j}}$ as $a_1, \cdots, a_n$ and denote by $\hat{a}$ the mapping

$$\hat{a} : \chi \mapsto (\hat{a}_1(\chi), \cdots, \hat{a}_n(\chi)).$$

Then $\hat{a}(\Delta_1) \bigcap \hat{a}(\Delta_0) = \emptyset$, since if $\chi_1 \in \Delta_1$, $\chi_0 \in \Delta_0$ then (χ_1, χ_0) belongs to some $V(\chi_{1j}) \times W(\chi_{0j})$, so that

$$\chi_1(a_{\chi_{1i}\chi_{0j}}) \neq \chi_0(a_{\chi_{1j}\chi_{0j}}).$$

Since Δ is compact Hausdorff there exists disjoint open sets G_1, G_0 in K^n with $\hat{a}(\Delta_j) \subseteq G_j$ $(j = 1, 0)$. Write $G = G_1 \dot{\bigcup} G_0$ and define f on G by setting $f = 1$ on G_1 and $f = 0$ on G_0. Then f is holomorphic on G, and by 8.5.22, there is a $b \in A$ such that

$$\hat{b} = f(\hat{a}_1, \cdots \hat{a}_n) \text{ on } \Delta.$$

Note that $\hat{b}(\chi) = 1$ or 0 according as $\chi \in \Delta_1$ or Δ_0. Hence, since $\Delta_1, \Delta_0 \neq \emptyset$, $\sigma(b) = \{0, 1\}$. By 5.3.8, there is a unique element $d \in \sqrt{A}$ such that $u = b + d$ is idempotent. Then $\hat{u} = \hat{b} = 1$ on Δ_1 and $\hat{u} = \hat{b} = 0$ on Δ_0, completing the proof.

8.6.5. PROPOSITION. *Let* A *be a unital commutative, complex or strictly real, pseudo-Michael algebra. If its spectrum* Δ_c *is disconnected, with* $\Delta_c = \Delta_1 \dot{\bigcup} \Delta_0$, Δ_j $(j = 1, 0)$ *non-empty clopen then there is a unique idempotent* u *in* A *with*
$\hat{u} = 1$ *on* Δ_1 *and* $\hat{u} = 0$ *on* Δ_0.

---

† clopen=closed and open.

†† The uniqueness follows from the monomorphism property of ω proved on pp.397-8.

PROOF. By 4.5.1, A has a projective limit decomposition $A = \varprojlim \hat{A}_\alpha$, where $\hat{A}_\alpha$ are commutative pseudo-Banach algebras. In the notation of 7.3.19, we have

$$\Delta_c = \bigcup_\alpha \Delta_\alpha; \ \Delta_\alpha \simeq \hat{\Delta}_\alpha$$

where $\Delta_c, \hat{\Delta}_\alpha$ are the spectra of A, $\hat{A}_\alpha$ respy. and $\Delta_\alpha = \{\chi \in \Delta_c :$ χ is $p_\alpha-$continuous $\}$.

Set

$$\Delta_{\alpha j} = \Delta_j \bigcap \Delta_\alpha \ (j = 1, 0).$$

If for an α we have $\Delta_{\alpha 1}$, $\Delta_{\alpha 0} \neq \emptyset$, then by applying 8.6.4. to $\hat{A}_\alpha$ we can determine a unique element $u_\alpha \in \hat{A}_\alpha$ with $u_\alpha(\hat{\chi}_\alpha) = 1$ or 0 (where $\hat{\chi}_\alpha = \chi_0 \varphi_\alpha$) according as $\chi \in \Delta_{\alpha 1}$ or $\Delta_{\alpha 0}$. If for an α, $\Delta_{\alpha 1} = \emptyset$, then choose $u_\alpha = 0_\alpha$ and if $\Delta_{\alpha 0} = \emptyset$, choose $u_\alpha = e_\alpha$, where $0_\alpha, e_\alpha$ are the zero and unity elements of $\hat{A}_\alpha$.

Suppose now that $\alpha \prec \beta$, $\chi_\alpha \in \Delta_{\alpha j}$ then $\chi_\alpha \circ \hat{\varphi}_{\alpha\beta} \in \Delta_{\beta j}$ and

$$\chi_\alpha(\hat{\varphi}_{\alpha\beta}(u_\beta)) = \chi_\alpha \circ \hat{\varphi}_{\alpha\beta}(u_\beta) = 1 \text{ or } 0$$

according as $\chi_\alpha \in \Delta_{\alpha 1}$ or $\Delta_{\alpha 0}$. Since $\hat{\varphi}_{\alpha\beta}(u_\beta)$ is an idempotent the uniqueness of the choice of the u_α implies that we have

$$\hat{\varphi}_{\alpha\beta}(u_\beta) = u_\alpha.$$

But this means that

$$u = (u_\alpha) \in \varprojlim A_\alpha = A,$$

and $\chi(u) = 1$ or 0 according as $\chi \in \Delta_1$ or Δ_0. Finally, by 4.5.6 (iii), u is an idempotent of A and the proof is complete.

8.6.6. COROLLARY. *Let A be a commutative, complex or strictly real, pseudo-Michael (in particular ρ-Banach algebra) such that its spectrum Δ_c is compact and non-empty. Then $\sqrt{A}$ is a regular ideal. In particular, if A is s.s., then A is unital* (*cf.* 8.3.17, 8.3.18).

PROOF. Let A_1 be the unitization of A. Then $\Delta_1 = \Delta \bigcup \{\chi_0\}$ where χ_0 is the distinguished character of A_1. By 7.3.2(c), $\Delta_{1c}(=$

$(\Delta_1)_c$) and Δ_c are Hausdorff. Also, by our hypothesis, Δ_c is compact, whence it follows that Δ_c is clopen in $(\Delta_1)_c$. So, by 8.6.5, there is an idempotent $u \in A_1$ such that

$$\hat{u} = 1 \text{ on } \Delta_c \text{ and } \hat{u}(\chi_0) = 0.$$

Since $\hat{u}(\chi_0) = 0$ it follows that $u \in A$. Further, for $x \in A$ and $\chi \in \Delta_c$

$$\chi(xu - x) = \chi(x)\chi(u) - \chi(x) = \chi(x) \cdot 1 - \chi(x) = 0$$

(since $\chi(u) = \hat{u}(\chi) = 1$). It follows that $xu - x \in \sqrt[h]{A} = \sqrt{A}$, whence $\sqrt{A}$ is regular with u as relative unity. When A is s.s., $\{0\} = \sqrt{A}$ is regular which means that $xu - x = 0$ $(x \in A)$, i.e. u is unity of A.

8.6.7. A partially ordered set or poset S, with ordering relation $\prec$, is called a *laltice* if any two elements a, b in S has a least upper bound or laltice sum $a \vee b$ and a greatest lower bound or laltice product $a \wedge b$. The element $a \vee b$ of S is characterized by the properties: $a, b \prec a \vee b$ and if $a, b \prec x \in S$ then $a \vee b \prec x$. Similarly $a \wedge b$ has the properties: $a \wedge b \prec a, b$ and $x \prec a, b \Rightarrow x \prec a \wedge b$.

The least element 0 and the greatest element 1 of S – whenever they exist – are characterized by the properties: $0 \prec x \prec 1$ for every $x \in S$.

A laltice L is called *distributive* if it satisfies the distributive law

$$a \wedge (b \vee c) = (a \wedge b) \vee (a \wedge c) \text{ for every } a, b, c \in L.$$

Also, a lattice L with 0,1 is called *complemented* if each $a \in L$ has a complement a' satisfying

$$a \wedge a' = 0, \ a \vee a' = 1.$$

A complemented distributive laltice is called a *Boolean algebra.*

8.6.8. PROPOSITION(Stone.) *The set $\mathcal{B}_i$ of idempotents of any unital commutative ring R is a Boolean algebra.*

PROOF. Define an order "$\prec$" in $\mathcal{B}_i$ by: for $u, v \in \mathcal{B}_i$, $u \prec v$ if $uv = u$. It is easily checked that "$\prec$" is a partial order.

Moreover, it is straightforward to verify that $\mathcal{B}_i$ is a laltice with laltice operations $\vee, \wedge$ given by:

$$u \vee v = u + v - uv; \ u \wedge v = uv.$$

$\mathcal{B}_i$ is further complemented with $u' = e - u$ as complement of u, where e is the unity of R. Finally, the distributivity of the ring operations imply the distributivity of the laltice operations. Thus, $\mathcal{B}_i$ is a complemented distributive laltice or a Boolean Algebra.

8.6.9. LEMMA. *Let X be a topological space. Then the clopen sets in X form a Boolean algebra $\mathcal{B}_{co}(X)$, with set-union, set-intersection and set-complementation as the Boolean operations.*

PROOF. Clear.

8.6.10. LEMMA. *Let A be a TA and $\Delta_c \neq \emptyset$. If $u \in A$ is an idempotent and*

$$G_u = \{\chi \in \Delta : \chi(u) = 1\}, \quad G_{uc} = \{\chi \in \Delta_c : \chi(u) = 1\}$$

then G_u, G_{uc} are clopen sets of Δ, Δ_c respectively.

PROOF. Since u is an idempotent, for any $\chi \in \Delta$,

$$\chi(u)^2 = \chi(u^2) = \chi(u), \text{ so that } \chi(u) = 0 \text{ or } 1.$$

Write

$$G_u^0 = \{\chi \in \Delta : \chi(u) = 0\}.$$

Then clearly $\Delta = G_u \dot{\bigcup} G_u^0$. By definition of the (weak) topology of Δ it is clear that G_u, G_u^0 are closed. So G_u is clopen in Δ. Similarly it can be seen that G_{uc} is clopen in Δ_c.

8.6.11. PROPOSITION. *Let A be a unital commutative TA. Then the map $\omega : u \mapsto G_u$ is a homomorphism between the Boolean algebras $\mathcal{B}_i = \mathcal{B}_i(A)$ and $\mathcal{B}_{co}(\Delta)$. Similarly, $\omega_c : u \mapsto G_{uc}$ is a homomorphism between $\mathcal{B}_i$ and $\mathcal{B}_{co}(\Delta_c)$. If A is also spectrally Gelfand (respy. t. spectrally Gelfand) then the map ω (respy. ω_c) is a monomorphism.*

PROOF. Since

$$\chi(u \vee v) = \chi(u) + \chi(v) - \chi(u)\chi(v) = 1 \text{ if } \chi(u) \text{ or } \chi(v) = 1, \quad (1)$$

we get
$$G_{u\vee v} = G_u \bigcup G_v.$$
Again, since $\chi(u \wedge v) = \chi(u)\chi(v) = 1$ iff $\chi(u) = \chi(v) = 1$, we obtain
$$G_{u\wedge v} = G_u \bigcap G_v. \tag{2}$$
Finally, if u' denotes the complement of u then, $\chi(u') = \chi(e-u) = 1 - \chi(u) = 1$ iff $\chi(u) = 0$, whence
$$G_{u'} = \Delta \backslash G_u. \tag{3}$$
It follows from (1),(2),(3) that ω is a (Boolean algebra) homomorphism. Similarly ω_c is a homomorphism.

Suppose now that A is spectrally Gelfand and $G_{u_1} = G_{u_2}$; then $\chi(u_1) = 1$ iff $\chi(u_2) = 1$. Since u_1, u_2 are idempotents, $\chi(u_1), \chi(u_2) = 1$ or 0. Therefore, $\chi(u_1 - u_2) = 0$, for all $\chi \in \Delta$, whence
$$u_1 - u_2 \in \sqrt[h]{A} = \sqrt{A}.$$
Since u_1, u_2 are idempotents, $u_1 - u_1u_2$ is also an idempotent (as easily checked). But
$$u_1 - u_1u_2 = u_1(u_1 - u_2) \in \sqrt{A},$$
whence by 1.2.24(d), $u_1 - u_1u_2 = 0$ or $u_1 = u_1u_2$. Similarly, $u_2 = u_2u_1 = u_1u_2$. So $u_1 = u_2$. This completes the proof that ω is a monomorphism. Similarly it can be shown that ω_c is a monomorphism (when A is t. spectrally Gelfand).

8.6.12. DEFINITION. A unital commutative TA is called a *Shilov algebra* if the map $u \mapsto G_{uc}$ of $\mathcal{B}_i(A) \to \mathcal{B}_{co}(\Delta_c)$ is an isomorphism.

8.6.13. THEOREM. *Any unital commutative, complex or strictly real, pseudo-Michael algebra A is a Shilov algebra. In particular, a unital commutative, complex or strictly real, pseudo-Banach algebra is a Shilov algebra.*

PROOF. By 7.2.21, 8.6.11, ω_c is a monomorphism. Also, by 8.6.5, ω_c is an epimorphism. Thus ω_c is an isomorphism and A is a Shilov algebra.

8.6.14. DEFINITION. A unital algebra A (over $\mathbb{K}$) is called *spectrally connected* if the spectrum $\sigma(x)$ of every element $x \in A$ is a connected subset of $\mathbb{K}$.

8.6.15. THEOREM. *A unital commutative, complex or strictly real, pseudo-Michael (in particular, pseudo-Banach) algebra A is spectrally connected iff its spectrum Δ_c is connected.*

PROOF. If Δ_c is disconnected then by 8.6.4, there is an idempotent $u \neq 0, e$ in A, so that $\sigma(u) = \{0,1\}$ is disconnected, whence A is not spectrally connected. On the other hand, if Δ_c is connected then $\chi \mapsto \chi(a)$ being a surjective continuous map of Δ_c onto $\sigma(a), \sigma(a)$ is connected for every $a \in A$, so that A is spectrally connected.

8.6.16. COROLLARY. *A is spectrally disconnected iff A has an idempotent $u \neq 0$, e.*

PROOF. If $u \neq 0$, e is an idempotent then by 1.7.9, $\sigma(u) = \{0,1\}$, so that $\sigma(u)$ is disconnected, which proves the "if" part.

To prove the "only if" part assume that A has an element a with $\sigma(a)$ disconnected. Since $\chi \mapsto \chi(a)$ is a continuous map of Δ_c onto $\sigma(a)$ it follows that Δ_c is also disconnected. But then, by 8.6.5, there is an idempotent $u \neq 0, 1$ in A, completing the proof.

CHAPTER IX

NORM UNIQUENESS THEOREMS

§ 1. Norm- uniqueness Theorem of Gelfand

9.1.1. LEMMA. *Let* $|\cdot|_j$ $(j = 1,2)$ *be two complete* (F) *norms on a LS* X *(over* $\mathbb{K}$*) such that*

$$|x|_1 \leqslant C|x|_2 \text{ for all } x \in X \text{ and some } C > 0. \qquad (*)$$

Then $|\cdot|_1 \sim |\cdot|_2$.

PROOF. The condition $(*)$ implies that the identity map

$$I : (X, |\cdot|_1) \to (X, |\cdot|_2)$$

is continuous. By the open mapping theorem (3.1.15) for (F) spaces, I is open and consequently I is a homeomorphism and hence $\|\cdot\|_1 \sim \|\cdot\|_2$.

9.1.2. LEMMA. *Let* $|\cdot|_j$ $(j = 1,2)$ *be complete* (F) *norms on a LS* X*. Then* $|\cdot|_1 \sim |\cdot|_2$ *iff* $|\cdot| = |\cdot|_1 + |\cdot|_2$ *is a complete* (F) *norm.*

PROOF. First note that $|\cdot|$ (as defined above) is always an (F) norm. If $|\cdot|$ is complete then since $|\cdot|_j \leqslant |\cdot|$, by 9.1.1, $|\cdot|_j \sim |\cdot|$ $(j = 1,2)$. Hence $|\cdot|_1 \sim |\cdot|_2$.

Conversely, assume that $|\cdot|_1 \sim |\cdot|_2$. Suppose that $|x_n - x_m| \to 0$. Then $|x_n - x_m|_j \to 0$ $(j = 1,2)$. By completeness of $|\cdot|_1$ there is an element x with $|x_n - x|_1 \to 0$. Since $|\cdot|_2 \sim |\cdot|_1$ we have also $|x_n - x|_2 \to 0$. So $|x_n - x| = |x_n - x|_1 + |x_n - x|_2 \to 0$ and $|\cdot|$ is a complete (F) norm.

9.1.3. COROLLARY. *If* $|\cdot|_j$ $(j = 1,2)$ *are two complete* (F) *norms and there is a* (F) *norm* $|\cdot|_0$ *with* $|x|_0 < |x|_1, |x|_2$ $(x \in A)$ *then* $|\cdot|_1 \sim |\cdot|_2$.

PROOF. Suppose that $|\cdot|_0 = |\cdot|_1 + |\cdot|_2$ and $|x_n - x_m| \to 0$. Then $|x_n - x_m|_j \to 0$ $(j = 1,2)$. By the completeness of the

(F) norms $|\cdot|$ there are elements $x, y \in A$ with $|x_n - x|_1 \to 0, |x_n - y|_2 \to 0$. Since $|\cdot|_0 \leqslant |\cdot|_1, |\cdot|_2$ we obtain $|x_n - x|_0 \to 0$, $|x_n - y|_0 \to 0$. By uniqueness of limits property, we get $x = y$. Therefore $|x_n - x| = |x_n - x|_1 + |x_n - x|_2 \to 0$, proving $|\cdot|$ is complete. Now we can apply 9.1.2 to conclude that $|\cdot|_1 \sim |\cdot|_2$, completing the proof.

9.1.4. THEOREM (Gelfand†). *Let A be a commutative s.s. algebra which is a ρ-Banach algebra under either the ρ-norms $\|\cdot\|_j$ $(j = 1, 2)$. Then $\|\cdot\|_1 \sim \|\cdot\|_2$.*

PROOF. First assume that A is complex. We can also assume that the $\|\cdot\|_j$ are sm.. Then, by 7.4.6 $(A, \|\cdot\|_j)$ is a GB algebra, so that we have

$$r(x) = \nu_1(x)^{\frac{1}{\rho}} = \nu_2(x)^{\frac{1}{\rho}}, \text{ where } \nu_j = \nu_j(x) \quad (j = 1, 2)$$

are topological spectral radii functions with respect to the ρ-norms $\|\cdot\|_j$. Writing

$$\nu(x) = \nu_1(x) = \nu_2(x),$$

by virtue 4.8.4, 7.4.5 and the hypothesis that A is s.s. we obtain that ν is a ρ-norm. Since $\nu(x) \leqslant \|x\|_j$ $(x \in A)$, we can apply 9.1.3 to conclude that $\|\cdot\|_1 \sim \|\cdot\|_2$.

It remains to consider the case where the algebra A is real. If $\tilde{A}$ is the complexification of A, then by 4.1.16, $\|\cdot\|_j$ can be extended to $\|\tilde{\cdot}\|_j$ over $\tilde{A}$ such that $(\tilde{A}, \|\tilde{\cdot}\|_j)$ $(j = 1, 2)$ are commutative complex ρ-Banach algebras. Furthermore, by 1.6.16, $\tilde{A}$ is s.s.. So applying the result just proved for complex algebra we get $\|\tilde{\cdot}\|_1 \sim \|\tilde{\cdot}\|_2$, whence (by restriction) $\|\cdot\|_1 \sim \|\cdot\|_2$.

§ 2. Rickart Separating Function

9.2.1. DEFINITION. Let A be a LS over $\mathbb{K}$ and $|\cdot|_j$ $(j = 1, 2)$ be two (F) norms on A. Set

---

† Gelfand considered only the complex Banach algebra case of the theorem.

$$\Delta(a) = \Delta_{12}(a) = \inf_x\{|x|_1 + |a - x|_2\} \quad (a, x \in A).$$

The function Δ is called the ***Rickart separating function,*** or just separating function, for $|\cdot|_j$.

9.2.2. LEMMA. (i) $\Delta_{12}(a) = \Delta_{21}(a) \quad (a \in A)$.
(ii) $\Delta(a) \leqslant \min\{|a|_1, |a|_2\}$.
(iii) $\Delta(0) = 0$.
(iv) $\Delta(-a) = \Delta(a)$.
(v) $\Delta(a + b) \leqslant \Delta(a) + \Delta(b) \quad (a, b \in A)$.
(vi) If $|\cdot|_j = \|\cdot\|_j$ are ρ-norms then

$$\Delta(\lambda a) = |\lambda|^\rho \Delta(a) \quad (a \in A, \lambda \in \mathbb{K}).$$

PROOF. (i) This follows by relacing x by $a - x$ in the definition of $\Delta_{12}(a)$.

(ii) By taking $x = a$ in the definitions of $\Delta_{12}(a), \Delta_{21}(a)$.

(iii) This follows from (ii) since $|0|_1 = 0$ $(= |0|_2)$.

(iv) This follows from the definition of Δ, using the relations $|-x|_j = |x|_j$ $(j = 1, 2)$.

(v) Given $\epsilon > 0$, by definition of Δ, there are $x_1, x_2 \in A$ with

$$\begin{aligned} |x_1|_1 + |a - x_1|_2 &\leqslant \Delta(a) + \epsilon \\ |x_2|_1 + |b - x_2|_2 &\leqslant \Delta(b) + \epsilon. \end{aligned}$$

Therefore

$$\begin{aligned} \Delta(a + b) &\leqslant |x_1 + x_2|_1 + |a + b - x_1 - x_2|_2 \\ &\leqslant |x_1|_1 + |x_2|_1 + |a - x_1|_2 + |b - x_2|_2 \\ &\leqslant \Delta(a) + \Delta(b) + 2\epsilon. \end{aligned}$$

From the arbitrariness of ϵ, the inequality (v) follows.

(vi) For $\lambda = 0$ the equality clearly holds (since both sides are zero). Assume next that $\lambda \neq 0$.

Then

$$\begin{aligned} \Delta(\lambda a) &\leqslant \|\lambda x\|_1 + \|\lambda a - \lambda x\|_2 \leqslant |\lambda|^\rho \ \{\|x\|_1 + \|a - x\|_2\} \\ &\leqslant |\lambda|^\rho \Delta(a) \ \ (\text{ by taking inf over } x). \end{aligned}$$

On the other hand

$$|\lambda|^\rho \Delta(a) = |\lambda|^\rho \Delta(\lambda^{-1}\lambda a) \leqslant |\lambda|^\rho |\lambda^{-1}|^\rho \Delta(\lambda a) = \Delta(\lambda a).$$

Combining the two inequalities we get the equality (vi).

9.2.3. LEMMA. *If A is an algebra and $|\cdot|_j$ $(j = 1, 2)$ sm. then*

$$\Delta(a, b) \leqslant \begin{cases} \Delta(a)(|b|_1 + |b|_2) \\ (|a|_1 + |a|_2)\Delta(b) \end{cases} \qquad (a, b \in A).$$

PROOF.

$$\begin{aligned} \Delta(ab) &\leqslant |ab - ax|_1 + |ax|_2 \leqslant |a|_1|b - x|_1 + |a|_2|x|_2 \\ &\leqslant (|a|_1 + |a|_2)(|b - x|_1 + |x|_2) \\ &\leqslant (|a|_1 + |a|_2)\Delta(b). \end{aligned}$$

Similarly

$$\begin{aligned} \Delta(ab) &\leqslant |ab - xb|_1 + |xb|_2 \leqslant |a - x|_1|b|_1 + |x|_2|b|_2 \\ &\leqslant \Delta(a)(|b|_1 + |b|_2), \end{aligned}$$

completing the proof of the lemma.

9.2.4. Let Δ be the separating function for the (F) norms $|\cdot|_j$ $(j = 1, 2)$. If $\Delta(a) = 0$, a is called a *separating* element for the $|\cdot|_j$. Write

$$\mathfrak{S} = \{a \in A : \Delta(a) = 0\}.$$

9.2.5. PROPOSITION. (a) *$a \in \mathfrak{S}$ iff there is a sequence (x_n) in A with $|x_n|_1 \to 0, |a - x_n|_2 \to 0$.*

(b) *$\mathfrak{S}$ is a subspace of A which is closed with respect to both $|\cdot|_j$ $(j = 1, 2)$.*

(c) *If A is an algebra then $\mathfrak{S}$ is a bi-ideal.*

PROOF. (a) Clear from the definition of Δ.

(b) That $\mathfrak{S}$ is a subspace follows from 9.2.2((iii)-(v)), result (a) above, and continuity of scalar multiplication with respect to either (F) norm. That $\mathfrak{S}$ is closed can be easily shown, using (a).

(c) This follows from (b) and 9.2.3 .

9.2.6. PROPOSITION. *Two (F) norms* $|\cdot|_j$ $(j=1,2)$ *making a LS X into (F) spaces are equivalent iff* $\mathfrak{S}=\{0\}$.

PROOF. First suppose that $\mathfrak{S}=\{0\}$. Then, by 9.2.5.(a), for any sequence (x_n) in A, $a \in A$,

$$|x_n|_1 \to 0, |a-x_n|_2 \to 0 \Rightarrow a=0.$$

But this means that the identity map $(A,|\cdot|_1) \to (A,|\cdot|_2)$ is closed and so continuous, by the closed graph theorem (3.1.16) and consequently it is a homeomorphism by the open mapping theorem. Thus, $\mathfrak{S}=\{0\}$ implies that $|\cdot|_1 \sim |\cdot|_2$. On the other hand, if $|\cdot|_1 \sim |\cdot|_2$ and $|x_n|_1 \to 0$, $|a-x_n|_2 \to 0$ then also $|x_n|_2 \to 0$ (since $|\cdot|_1 \sim |\cdot|_2$). Thus, $x_n \to a, 0$ under $|\cdot|_2$. By the uniqueness of limit property, $a=0$, $\mathfrak{S}=\{0\}$.

9.2.7. Let $|\cdot|_j$ $(j=1,2)$ be (F) norms on an algebra A such that $(A,|\cdot|_j)$ are (F) algebras. Let I be a bi-ideal of A closed with respect to both (F) norm topologies and

$$\pi : A \to A/I = A^{\#} \text{ the canonical homomorphism .}$$

Let $|\cdot|_j^{\#}$ be the corresponding quotient (F) norms. Let $\mathfrak{S}, \mathfrak{S}^{\#}$ be the sets of separating elements for $|\cdot|_j, |\cdot|_j^{\#}$ respectively. Then we have

9.2.8. LEMMA. $\pi(\mathfrak{S}) \subseteq \mathfrak{S}^{\#}$. *Hence* $\mathfrak{S}^{\#}=\{0\} \Rightarrow \mathfrak{S} \subseteq I$.

PROOF. If $a \in \mathfrak{S}$ then $|x_n|_1 \to 0, |x_n-a|_2 \to 0$ for some sequence (x_n) in A. Since $|\pi(x_n)|_1^{\#} \leqslant |x_n|_1 \to 0$ and $|\pi(x_n) - \pi(a)|_2^{\#} = |\pi(x_n-a)|_2^{\#} \leqslant |x_n-a|_2 \to 0$ it follows that $\pi(a) \in \mathfrak{S}^{\#}$, whence $\pi(\mathfrak{S}) \subseteq \mathfrak{S}^{\#}$. If $\mathfrak{S}^{\#}=\{0\}$ then $\pi(\mathfrak{S})=\{0\}$, so that $\mathfrak{S} \subseteq I$.

9.2.9. DEFINITION. An (F) algebra $A=(A,|\cdot|)$ is said to have ***norm uniqueness property*** if for any (F) norm $|\cdot|'$ on A making A into a (F) algebra we have $|\cdot|' \sim |\cdot|$. Similarly, a ρ-Banach algebra $A=(A,\|\cdot\|)$ is said to have norm uniqueness property if for any ρ-norm $\|\cdot\|'$ making A into a ρ-Banach algebra we have $\|\cdot\|' \sim \|\cdot\|$.

9.2.10. THEOREM. *Let $A = (A, |\cdot|_1)$ be a t. spectrally Gelfand functionally continuous s.s. (F) algebra. If $|\cdot|_2$ is an (F) norm on A such that $(A, |\cdot|_2)$ is a t. spectrally Gelfand (F) algebra then $|\cdot|_1 \sim |\cdot|_2$.*

PROOF. Let $\{M_\alpha\}$ be the family of $|\cdot|_2$-closed hypermaximal ideals of A. Since A is s.s., by 7.2.12,

$$\bigcap M_\alpha = \sqrt[ch]{A} = \sqrt{A} = \{0\}.$$

Since $(A, |\cdot|_1)$ is functionally continuous, each M_α is also $|\cdot|_1$-closed. Since $A^\# = A/M_\alpha \simeq \mathbb{K}$ it follows that $|\cdot|_1^\# \sim |\cdot|_2^\#$. So, by 9.2.6, $\mathfrak{S}^\# = \{0\}$, whence by 9.2.8, $\mathfrak{S} \subseteq M_\alpha$. Therefore $\mathfrak{S} \subseteq \bigcap M_\alpha = \{0\}$, so $|\cdot|_1 \sim |\cdot|_2$.

9.2.11. COROLLARY. (cf. Michael [20,p59]). *Let $A = (A, \mathcal{P})$ be a commutative functionally continuous locally sm. pseudo-$\mathcal{F}$ algebra which is s.s.. If $\mathcal{Q} = \{q_j\}$ be a family of locally sm. pseudo-seminorms on A making it into a locally sm. pseudo-$\mathcal{F}$ algebra. Then $\mathcal{Q} \sim \mathcal{P}$.*

PROOF. Assume first that A is a complex algebra. Then by 7.2.21, A is t. spectrally Gelfand. The result for complex algebras now follows from 9.2.10. The result for real algebras A is obtained by considering the complexification $\tilde{A}$ of A and applying the result for complex algebras, and deducing from it the result for A (using the fact that a pseudo-seminorm of A is the restriction of its extension to $\tilde{A}$).

§ 3. Topological Modules

9.3.1. Definition. Let A be an algebra (over $\mathbb{K}$) and X a TLS. X is called a *topological* A-module or a t. A-module if:

(i) X is an A-module;

(ii) The map $m^{\#} : (a, x) \mapsto ax \quad (a \in A,\ x \in X)$ is continuous.

9.3.2. LEMMA. *Let A be a ρ-normed algebra and X a ρ-normed linear space which is an A-module. Then X is a t. A-module iff there is a constant $C > 0$ such that*

$$\|ax\|^* \leqslant C\|a\|\ \|a\|^* \text{ for all } a \in A,\ x \in X \tag{$*$}$$

where $\|\cdot\|$, $\|\cdot\|^*$ are respectively the ρ-norms of A, X.

PROOF. If $(*)$ is satisfied then clearly the map $m^{\#}$ is continuous and X is a t. A-module.

Conversely, suppose that X is a t. A-module. Then from the continuity of the map $m^{\#}$ at $(0,0)$ we get for a given $\epsilon > 0$, a $\delta > 0$ such that

$$\|ax\|^* \leqslant \epsilon \text{ if } \|a\|,\ \|x\|^* \leqslant \delta. \tag{1}$$

For any $a \neq 0$ in A and $x \neq 0$ in X, write

$$a_1 = \delta^{\frac{1}{\rho}} a\|a\|^{-\frac{1}{\rho}},\ x_1 = \delta^{\frac{1}{\rho}} x(\|x\|^*)^{-\frac{1}{\rho}}.$$

Then $\|a_1\| = \delta$, $\|x_1\| = \delta$ so that by (1), we get $\|a_1x_1\|^* \leqslant \epsilon$ which reduces to

$$\|ax\| \leqslant C\|a\|\ \|x\|^*\ (C = \epsilon/\delta^2). \tag{2}$$

The inequality (2) is trivially satisfied if a or x is zero. Thus (2) holds for all a, x, completing the proof.

9.3.3. DEFINITION. If A, X are as in 9.3.2 and condition $(*)$ is satisfied then we call X a *ρ-normed* A-module. If besides, A and X are complete with respect to their ρ- norms we call X a *ρ-Banach* A-module.

9.3.4. PROPOSITION. *Let $A = (A, \|\cdot\|)$ be a ρ-Banach algebra (with $\|\cdot\|$ sm.) and I a closed regular l. ideal of A. Then the quotient module $A^{\#} = A/I$ is canonically a cyclic ρ-Banach module. Further, $A^{\#}$ is irreducible iff I is maximal.*

PROOF. In view of 1.5.16 it is enough to prove that $A^{\#}$ is a ρ- Banach A-module. Set

$$\|x + I\|^{\#} = \inf_t \{\|x + t\| : t \in I\}.$$

It is straightforward to check that $\|\cdot\|^*$ is a ρ-norm on $A^\#$ ($\|\cdot\|$ is called the canonical ρ-norm induced by $\|\cdot\|$ on $A^\#$). Further, $A^\#$ is complete with respect to $\|\cdot\|^\#$ (cf. proof of 3.4.14). Finally, we have

$$\begin{aligned} \|a(x+I)\|^\# &= \|ax+I\|^\# \leqslant \inf\{\|ax+t\| : t \in I\} \\ &\leqslant \inf\{\|ax+at\| : t \in I\} \\ &\leqslant \inf\{\|a\|\,\|x+t\| : t \in I\} \\ &\leqslant \|a\|\,\|x+I\|^\#. \end{aligned}$$

9.3.5. LEMMA. *Let $A = (A, \|\cdot\|)$ be a ρ-Banach algebra and $X = (X, \|\cdot\|^*)$ a ρ-Banach cyclic A-module, with generator x_0. Then $I = \ker x_0$ is closed in A and $\Phi : a + I \mapsto ax_0$ is a t. isomorphism of $A^\# = A/I$ onto X.*

PROOF. Since the map $a \mapsto ax_0$ is continuous, $I = \ker x_0$ is closed. Also, by 1.5.14 (i), I is regular. Therefore, by 9.3.4, $A^\#$ is a ρ-Banach cyclic A-module and Φ is clearly an isomorphism. Further

$$\|\Phi(a+I)\|^* = \|ax_0\|^* \leqslant C\|a\|\,\|x_0\|^*.$$

If $a_1 \in a + I$ then $ax_0 = a_1x_0$, so that

$$\|\Phi(a+I)\|^* = \|ax_0\|^* = \|a_1x_0\|^* \leqslant C\|a_1\|\,\|x_0\|^*.$$

By taking the minimum over all $a_1 \in a + I$, we get

$$\|\Phi(a+I)\|^* \leqslant C\|a+I\|^\#\|x_0\|^*.$$

It follows that Φ is continuous and consequently by the open mapping theorem[†] Φ is a t. isomorphism, as required.

9.3.6. PROPOSITION. *Let A be a ρ-Banach algebra, X a cyclic ρ-Banach A-module and D_c the set of all continuous A-endomorphisms of X. Then:*

(i) *D_c is a ρ-normed algebra.*

(ii) *D_c is also a division algebra if X is an irreducible module.*

[†] The open mapping theorem is available since ρ-Banach spaces are (F)-spaces.

PROOF. Let x_0 be a generator of X, $I = \ker x_0$ and Φ the t. isomorphism $A^{\#} = A/I \to X$ (see 9.3.5). For $T \in D_c$ write

$$y = \Phi^{-1}Tx_0,\ ax_0 = \Phi(a+I).$$

Then

$$\Phi^{-1}T\Phi(a+I) = \Phi^{-1}Tax_0 = a\Phi^{-1}Tx_0 = ay.$$

It follows that

$$\|\Phi^{-1}T\Phi(a+I)\|^* = \|ay\|^* = \|a_1y\|^* \leqslant \|a_1\|\ \|y\|^*$$

for any a_1 with $a_1 + I = a + I$. By taking the infimum over all such a_1 we get

$$\|\Phi^{-1}T\Phi(a+I)\|^* \leqslant \inf\{\|a_1\|\ \|y\|^* : a_1\} = \|a+I\|^{\#}\|y\|^*.$$

Thus, $T^{\#} = \Phi^{-1}T\Phi$ is a bounded linear operator of $A^{\#}$. Set $\|T\|^{\#} = \|T^{\#}\|$. Then $\|\cdot\|^{\#}$ is clearly a sm. ρ-norm on D_c and D_c is a normed algebra, proving (i).

For proving (ii), consider the algebra D of all A-endomorphisms of X. When X is irreducible, by 1.5.19, D is a division algebra. Clearly, D_c is a subalgebra of D. Further, if $T \in D_c$, $T \neq 0$ then $T^{-1} \in D$ (D being a division algebra). But, by the open mapping theorem applied to X we have $T^{-1} \in D_c$, proving (ii).

9.3.7. COROLLARY. *For an irreducible ρ-Banach A-module X, D_c is t. isomorphic to $\mathbb{C}$ if A is complex and t. isomorphic to $\mathbb{R}, \mathbb{C}$ or $\mathbb{H}$ if A is real.*

PROOF. This follows from 9.3.6, 6.5.6, 6.5.12.

§ 4. Norm-uniqueness Theorem for Non-commutative † Algebras

9.4.1 LEMMA (Bonsall-Duncan). *Let A be a ρ-Banach algebra and X an irreducible ρ-Banach A-module. Then, for any*

† non-commutative = not necessarily commutative

non-zero element x_0 *of* X, *if* $M = \ker x_0$ *and* I *a closed l. ideal with* $I \not\subseteq M$ *then we can find an* $\eta > 0$ *such that*

$$\eta \overline{B}_1 x_0 \subseteq (I \bigcap \overline{B}_1) x_0$$

where $\overline{B}_1$ *denotes the closed unit ball of* A.

PROOF. By 1.5.14, M is a maximal regular l. ideal which by 9.3.5, is closed. Further, by the choice of I and maximality of M, we have: $I + M = A$. It follows that the map

$$\pi_0 : a \mapsto a + M \quad (a \in I)$$

of the ρ-Banach space I to the ρ-Banach space $A^{\#} = A/M$ is surjective. By the open mapping theorem, $\pi_0(I \bigcap B_1)$ † is open, so that we can find an $\eta > 0$ such that for every $a + M$ with $\|a + M\| \leqslant \eta^{\rho}$ we have $a + M \in \pi_0(I \bigcap \overline{B}_1)$. This means that there is an element $b \in I \bigcap \overline{B}_1$ with $b + M = a + M, bx_0 = ax_0$. If $a \in \eta \overline{B}_1$ then $\|a + M\| \leqslant \|a\| \leqslant \eta^{\rho}$, so that $ax_0 = bx_0$ with $b \in I \bigcap \overline{B}_1$, proving the lemma.

9.4.2. THEOREM (Johnson). *Let* A *be a primitive* ρ*-Banach algebra and* π *a faithful irreducible representation of* A *on a* ρ*-normed linear space* X *such that* $\pi(a) \in \mathbf{B}(X)$ *for all* a *in* A. *Then the map*

$$\pi : A \to \mathbf{B}(X)$$

is continuous, where $\mathbf{B}(X)$ *denotes the algebra of all bounded linear operators of* X.

PROOF. If X is finite-dimensional so is $\mathbf{B}(X)$ and then also A (since π is $1-1$). Thus, in this case π, being a linear map between two finite-dimensional Hausdorff TLS's, is automatically continuous. ††

Assume now that X is infinite-dimensional. Let r_x denote the map

$$a \mapsto ax = r_x(a) \qquad (a \in A, x \in X).$$

Write

† B_1 denotes the open unit ball of A.

†† See [5,14, Cor. 2]

$$X_0 = \{x \in X : r_x \in \mathbf{B}(A, X)\},$$

where $\mathbf{B}(A, X)$ denotes the space of all bounded linear transformations of A into X. Then X_0 is a submodule of $X = (X, \pi)$, since

$$\|r_{bx}(a)\| = \|abx\| = \|r_x(ab)\| \leqslant \|r_x\|\|ab\| \leqslant \|r_x\|\|a\|\|b\|.^{\dagger}$$

Since X is irreducible, $X_0 = X$ or $\{0\}$.

If $X_0 = X$, and $\overline{B}_1$ denotes the closed unit ball of X, for $x \in \overline{B}_1$ and $a \in X$ we have

$$\|r_x(a)\| = \|ax\| = \|\pi(a)x\| \leqslant \|\pi(a)\|\|x\| \leqslant \|\pi(a)\|.$$

By Banach-Steinhauss theorem (see 3.2.16) there is a constant $C > 0$ with $\|r_x\| \leqslant C \quad (x \in \overline{B}_1)$. It follows that

$$\|\pi(a)x\| = \|r_x(a)\| \leqslant C\|a\| \qquad (x \in \overline{B}_1, a \in A)$$

whence $\|\pi(a)\| \leqslant C(\|a\|)$, so that by 3.5.5, π is continuous.

To complete the proof it is enough to rule out the other alternative $X_0 = \{0\}$. Suppose therefore that $X_0 = \{0\}$. Then the definition of X_0 implies that for each $x \neq 0$ in X the operator r_x is not bounded, so that (by 3.5.2):

the set $\{\|ax\| : \|a\| \leq 1\}$ is not bounded. (1)

Now by 9.3.7, D_c is a division algebra over $\mathbb{K}$ with $\dim D_c \leqslant 4$. Since X is infinite-dimensional over $\mathbb{K}$ it is also infinite dimensional over D_c. It follows that X contains an infinite sequence (x_n) of D_c-independent vectors. By replacing x_n if necessary by a suitable scalar multiple we may assume that $\|x_n\| = 1$ for all n. Set

$$M_n = \ker x_n, \; L_n = M_1 \bigcap \cdots \bigcap M_{n-1} \;\; (n = 2, 3, \cdots).$$

By density theorem (1.5.21) we can find an $a \in A$ such that

$$ax_j = 0 \; (1 \leqslant j \leqslant n-1), \quad ax_n \neq 0 \;\; (n \geqslant 2).$$

† We have assumed that the norm $\|\cdot\|$ of A is sm..

Then $a \in L_n, a \notin M_n$, so that $L_n \not\subseteq M_n$. Since by (1), $\overline{B}_1 x_n$ is unbounded, by 9.4.1

$$(L_n \bigcap \overline{B}_1)x_n \text{ is unbounded.} \tag{2}$$

We now inductively choose a sequence (a_n) with

$$a_n \in L_n,\ \|a_n\| \leqslant 2^{-n}. \tag{3}$$

and

$$\|a_n x_n\| > n + \|(a_1 + \cdots + a_{n-1})x_n\|. \tag{4}$$

To justify the inductive step, suppose we have chosen $a_1, \cdots, a_{n-1}$. Then, by using(2), we can select a $b_n \in L_n \bigcap \overline{B}_1$ such that

$$\|b_n x_n\| > 2^{n+1}(n + \|(a_1 + \cdots + a_{n-1})x_n\|). \tag{5}$$

Set $a_n = 2^{-(n+1)/\rho} b_n$. Then

$$\|a_n\| = 2^{-(n+1)}\|b_n\| \leqslant 2^{-(n+1)} < 2^{-n} \ \ (\text{ using } b_n \in \overline{B}_1),$$

and using (5),

$$\|a_n x_n\| = 2^{-(n+1)}\|b_n x_n\| > n + \|(a_1 + \cdots + a_{n-1})x_n\|.$$

Next write

$$c = \sum_{k=1}^{\infty} a_k, \quad c_n = \sum_{k=n+1}^{\infty} a_k.$$

These elements are defined since the series representing them converge absolutely (by virtue of the inequality in (3)) and hence also in A. Since $a_k \in M_n$ $(k > n)$ and M_n is closed, we conclude that $c_n \in M_n$ and consequently, $c_n x_n = 0$.

It follows that we have

$$cx_n = \left(\sum_{k=1}^{n} a_k + c_n\right) x_n = \left(\sum_{k=1}^{n} a_k\right) x_n,$$

so that by inequality (4) we get

$$\|cx_n\| \geqslant \|a_n x_n\| - \|(a_1 + \cdots + a_{n-1})x_n\| \geqslant n.$$

This means that $\pi(c) = l_c$ is not a bounded operator contradicting the hypothesis on π. Therefore, $X_0 \neq \{0\}$, completing the proof.

9.4.3. THEOREM (Johnson). Every primitive ρ-Banach algebra A has ρ-norm uniqueness property.[†]

PROOF. Let $A = (A, \|\cdot\|_1)$ and $\|\cdot\|_2$ a complete ρ-norm on A such that $(A, \|\cdot\|_2)$ is also a ρ-Banach algebra. We have to show that $\|\cdot\|_2 \sim \|\cdot\|_1$.

Since A is primitive there is a maximal regular l. ideal M such that the left regular representation $\mathcal{L}^{\#}$ on $A^{\#} = A/M$ is faithful. By 7.1.9, M is closed with respect to either of the norms $\|\cdot\|_j$ $(j = 1, 2)$. Let $\|\cdot\|_j^{\#}$ be the canonical ρ-norm on $A^{\#}$ induced by $\|\cdot\|_j$. Then $(A^{\#}, \|\cdot\|_j^{\#})$ are ρ-Banach spaces (see 3.1.22, 3.4.16). It follows that π is a faithful representation of the ρ-Banach algebra $A = (A, \|\cdot\|_1)$ on the ρ-Banach space $A^{\#} = (A^{\#}, \|\cdot\|_2^{\#})$. By 9.4.2, π is continuous and consequently we have

$$\|\pi(a)\|_2^{\#} \leqslant C\|a\|_1 \quad (a \in A)$$

for some $C > 0$. This implies that

$$\begin{aligned} \|\pi(a)(x+M)\|_2^{\#} &\leqslant \|\pi(a)\|_2^{\#}\, \|x+M\|_2^{\#} \\ &\leqslant C\|a\|_1\, \|x+M\|_2^{\#}. \end{aligned} \tag{1}$$

Let u be the relative r. unity for M. Then

$$\pi(a)(u+M) = a(u+M) = a+M,$$

so that

$$\|a+M\|_2^{\#} = \|\pi(a)(u+M)\|_2^{\#} \leqslant C\|a\|_1\, \|u+M\|_2^{\#},$$

where we have used the inequality (i) with $x = u$. If

$$a + M = b + M \qquad (2),$$

then

$$\|a+M\|_2^{\#} = \|b+M\|_2^{\#} \leqslant C\|b\|_1 \|u+M\|_2^{\#}. \tag{3}$$

[†] Johnson obtained the theorem for Banach algebras (i.e. for the case $\rho = 1$).

Taking the infimum in (3) over all b satisfying (2) we obtain

$$\|a+M\|_2^\# \leqslant C\|a+M\|_1^\# \|u+M\|_2^\# . \tag{4}$$

The inequality (4) together with the open mapping theorem implies that $\|\cdot\|_1^\#$ and $\|\cdot\|_2^\#$ are equivalent. Suppose now that a in A is a separating element with respect to the norms $\|\cdot\|_j$ $(j=1,2)$, so that there is a sequence (x_n) in A with

$$\|x_n\|_1 \to 0, \ \|a-x_n\|_2 \to 0.$$

Since $\|x+M\|_j^\# \leqslant \|x_j\|$ it follows that $a+M$ is a separating element of $A^\#$ with respect to $\|\cdot\|_j^\#$. Since $\|\cdot\|_1^\# \sim \|\cdot\|_2^\#$ we must have

$$a+M=M, \ \ a\in M, \ \ \mathfrak{S}\subseteq M,$$

where $\mathfrak{S}$ is the separating set for the $\|\cdot\|_j$. Since by 9.2.5(a), $\mathfrak{S}$ is a bi-ideal we have, by 1.5.3,

$$\mathfrak{S}\subseteq (M:A)=\{0\} \quad (\text{ since } \mathcal{L}^\# \text{ is faithful }).$$

So, by 9.2.6, $\|\cdot\|_1 \sim \|\cdot\|_2$, as we wished to show.

9.4.4. THEOREM. *Every s.s.* (*real or complex*) ρ*-Banach algebra* A *has* ρ*-norm uniqueness property.*

PROOF. If P is a primitive ideal of A then A/P is a primitive ρ-Banach algebra and so has, by 9.4.3, ρ-norm uniqueness property. Let $\mathfrak{S}$ be the separating ideal for the ρ-norms $\|\cdot\|_j$ on A, where $A=(A,\|\cdot\|_1)$ and $\|\cdot\|_2$ is a second ρ-norm on A making it again into a ρ-Banach algebra. Then we have $\mathcal{S}\subseteq P$ (the proof is similar to that above showing $\mathcal{S}\subseteq M$). It follows that

$$\mathcal{S}\subseteq \bigcap P=\sqrt{A}=\{0\} \quad (\text{ since } A \text{ is s.s. })$$

whence A has ρ-norm uniqueness property.

9.4.5. Remark. Theorem 9.4.4 admits a strengthening to the following form: If $\|\cdot\|_j$ $(j=1,2)$ are ρ_j-norms $(0<\rho_j\leqslant 1)$ on a s.s. algebra A such that $(A,\|\cdot\|_j)$ are ρ_j-Banach algebras, then $\|\cdot\|_1 \sim \|\cdot\|_2$.

To obtain the above result we first remark that we can assume that $\rho_1 \leqslant \rho_2$ and $\|\cdot\|_j$ are sm. If we set $\|\cdot\|_1^* = \|\cdot\|_2^{\frac{\rho_1}{\rho_2}}$, then by 3.2.9. $\|\cdot\|_1^*$ is a ρ_1-norm which from its form is clearly (topologically) equivalent to $\|\cdot\|_2$. Further it is sm.:

$$\|xy\|_1^* = (\|xy\|_2)^{\frac{\rho_1}{\rho_2}} \leqslant (\|x\|_2\|y\|_2)^{\frac{\rho_1}{\rho_2}} = \|x\|_1^*\|y\|_1^*.$$

It follows that $(A, \|\cdot\|_1^*)$ is a ρ_1 Banach algebra. By 9.4.4, $\|\cdot\|_1 \sim \|\cdot\|_1^*$ and so $\|\cdot\|_1 \sim \|\cdot\|_2$ (since $\|\cdot\|_1^* \sim \|\cdot\|_2$) as claimed.

9.4.6. PROPOSITION. *Let A, A^* be sm. (F) algebras and $\varphi : A \to A^*$ an epimorphism. If A^* is s.s. and has norm-uniqueness property then φ is continuous.*

PROOF. Write $I = \ker\varphi$. By 3.6.23(a), 7.1.17, I is closed and therefore $A^\# = A/I$ is a sm. (F) algebra with its (F) norm $|\cdot|^\#$ satisfying $|\cdot|^\# \leqslant |\cdot|$ (see 3.1.22, 3.4.15). Since $A^\#$ is isomorphic to A^* we can shift the (F)-norm $|\cdot|^\#$ to A^*. Then

$$\varphi : A \to A^* = (A^*, |\cdot|^\#)$$

is continuous. By hypothesis A^* has norm-uniqueness property. It follows that $|\cdot|^*$ (of A^*) $\sim |\cdot|^\#$, whence

$$\varphi : A \to A^* = (A^*, |\cdot|^*) \text{ is continuous .}$$

9.4.7. COROLLARY. *If A, A^* are ρ-Banach algebras; A^* s.s. and $\varphi : A \to A^*$ an epimorphism then φ is continuous.*

PROOF. By 9.4.4, A^* has norm-uniqueness property and so the corollary follows from 9.4.6.

9.4.8. COROLLARY. *Any automorphism φ of a s.s. sm. (F) algebra A with norm-uniqueness property is a t. automorphism. In particular, every automorphism of a s.s. ρ-Banach algebra is a t. automorphism.*

PROOF. By 9.4.6 applied to φ, φ^{-1} we conclude that they are continuous, so that φ is a t. automorphism. The second statement follows from the first in view of 9.4.4.

APPENDIX

LEMMA A. *The character space* $\Delta = \Delta(W^\rho)$ *is homeomorphic to the unit circle* S^1 *in* $\mathbb{C}$.

PROOF. We have already seen in 7.5.16 that the characters χ of W^ρ are precisely the maps

$$\chi_{t_1} : f \mapsto f(t_1) \quad (f \in W^\rho, t_1 \in \mathbb{R}).$$

Since $\chi_{t_1}(e^{it}) = e^{it_1}$ determines χ_{t_1} it follows that $\chi_{t_1} = \chi_{t_2}$ iff $e^{it_1} = e^{it_2}$. Thus

$$\chi_{t_1} \to e^{it_1} = \chi_{t_1}(e^{it})$$

is a bijection of Δ on S^1. Moreover it is continuous (by definition of the (weak) topology of Δ). Since Δ is compact (by 7.3.12) and S^1 is Hausdorff, the map $\chi \mapsto \chi(e^t)$ is a homeomorphism.

LEMMA B. *Let* F, C *be respectively a closed set and a compact set in* $\mathbb{R}^n$ *such that* $F \cap C = \emptyset$. *Then there exists a* C^∞ *function* g *on* $\mathbb{R}^n$ *such that* $g = 0$ *on* F *and* $g = 1$ *on* C.

PROOF. See [H, p.3].

LEMMA C. *Let* $f = f(t)$ *be a* C^∞ 2π*-periodic complex function on* $\mathbb{R}$ *and* $f^{(j)}$ *denote the* j*th derivative of* f. *If* $\hat{f}(n), \hat{f^{(j)}}(n)$ $(n \in \mathbb{Z})$ *denote the* n*th Fourier coefficients of* f *and* $f^{(j)}$ *respectively then we have, for* $n \neq 0$,

$$|\hat{f}(n)| \leqslant |n|^{-j} \|f^{(j)}\|_{L^1},$$

where $\|f^{(j)}\|_{L^1} = \frac{1}{2\pi} \int_0^{2\pi} |f^{(j)}(t)| dt$.

PROOF. See [K, p.24, 4.4].

THEOREM. W^ρ *is completely regular.*

PROOF. Since by Lemma A, Δ can be identified with S^1, for proving the theorem it is enough (in view of 8.14.7) to show that given a closed subset F of S^1 and a point $e^{it_0} \in S^1 \backslash F$, there is a $f \in W^\rho = \hat{W}^\rho$ such that $f = 0$ on F and $f(e^{it_0}) \neq 0$. Now by Lemma B we can find a C^∞ function $g = g(x, y)$ on $\mathbb{R}^2$ with $g = 0$ on F and $g(e^{it_0}) \neq 0$. If we set

$$f(t) = g(e^{it}) = g(\cos t, \sin t)$$

then f is a 2π-periodic C^∞ function on $\mathbb{R}$ with $f(t) = 0$ if $e^{it} \in F$ and $f(t_0) \neq 0$. Choose integer j such that $j > \frac{1}{\rho}$. Then, by Lemma C,

$$\sum_{n\in\mathbb{Z}} |\hat{f}(n)|^\rho \leqslant |\hat{f}(0)|^\rho + 2\sum_{n=1}^{\infty} \frac{C}{|n|^{j\rho}} < \infty,$$

where $C = \|f^{(j)}\|_{L^1}^\rho$. It follows that $f \in W^\rho$, completing the proof.

References

[H] H. HELGASON: Differential Geometry and Symmetric Spaces, Academic Press, New York, 1962.

[K] Y. KATZNELSON: An Introduction to Harmonic Analysis, Douer Publications, New York, 2nd edition, 1976.

TOPOLOGICAL ALGEBRA

TYPE CONNECTION CHART

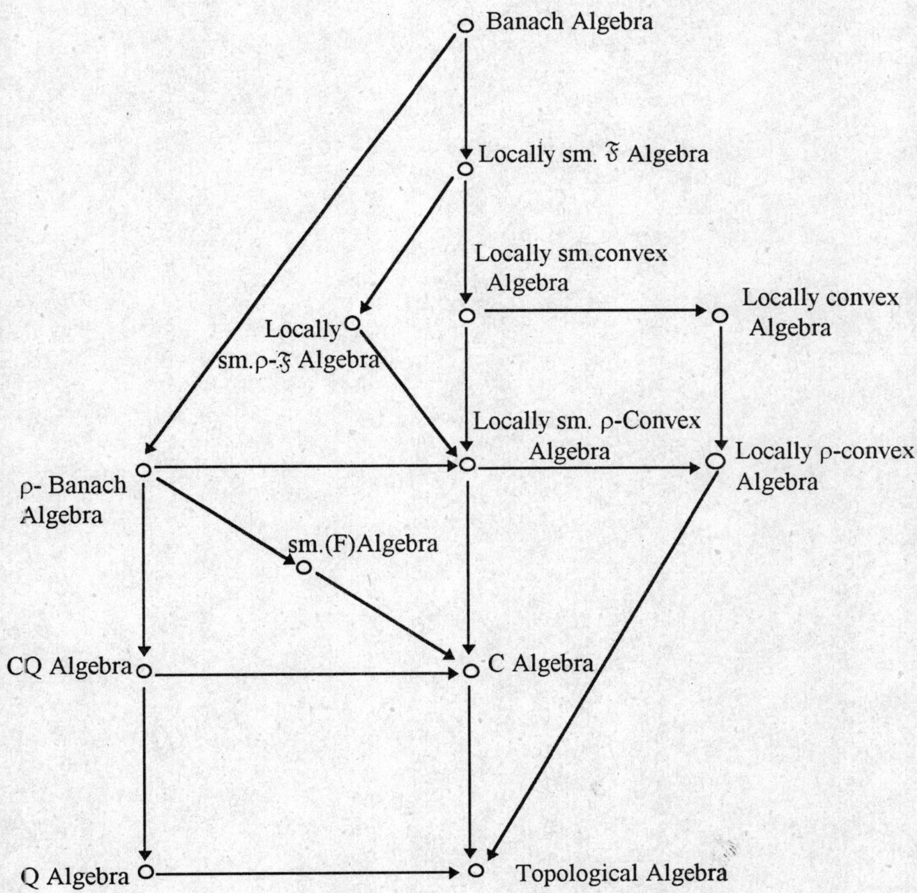

The Arrow indicates that the algebra at the tail of the arrow has the properties of the algebra at its tip.

BIBLIOGRAPHY

Books Referred

[1] L.V. AHLFORS: Complex Analysis, McGraw-Hill, International Student Edition, 1979.

[2] T.M. APOSTOL: Mathematical Analysis, Addision Wesley, World Student Edition, 1963.

[3] S. BOCHNER and W.T. MARTIN: Several Complex Variables, Princeton Univ. Press, Princeton, 1948.

[4] F.F. BONSALL and J. DUNCAN: Complete Normed Algebras, Springer Verlag, New York, 1973.

[5] N. BOURBAKI: Topological Vector Spaces, Chapters 1-5, Springer Verlag, New York, 1987.

[6] N. BOURBAKI: Théories Spectrales, Chapters 1-2, Hermann, Paris, 1967.

[7] G. CHOQUET: Lectures on Analysis, Vol. 1, W.A. Benjamen, New York, 1969.

[8] J.B. CONWAY: Functions of One Complex Variable, Springer International Student Edition, Narosa, New Delhi, 2nd revised edit., 1980.

[9] J. DUGUNDJI: Topology, Printice-Hall of India, New Delhi, 1975.

[10] I.M. GELFAND, D.A. RAIKOV and G.E. Shilov commutative Normed Rings, Chelsea, New York, 1964.

[11] L. GILLMAN and M. GERRISON: Rings of Continuous Functions, Springer Verlag, New York, 1976.

[12] A. GUICHARDT: Special Topics in Topological Algebras, Gordon and Breach, New York, 1968.

[13] N. JACOBSON: Lectures in Abstract Algebras V.I., East-West Student Edition, Affiliated East-West Press, New Delhi, 1964.

[14] E. HILLE and R.S. PHILLIPS: Amer. Math. Soc. Coll. Publ. 31, Providence, R.I., 1957.

[15] R.V. KADISON and J.R. RINGROSE: Fundamentals of the Theory of Operator Algebras, V.I., Academic Press, New York, 1983.

[16] J.L. KELLEY: General Topology, Van Nostrand, Princeton, 1955.

[17] J.L. KELLEY and I. NAMIOKA: Linear Topological Spaces, East-West Student Edition, Affiliated East-West Press, New Delhi, 1968.

[18] G. KÖTHE: Topologische Lineare Raüme, Springer Verlag, Berlin, 1960.

[19] L.H. LOOMIS: An Introduction to Abstract Harmonic Analysis, Van Nostrand, New York, 1953.

[20] E.A. MICHAEL: Locally Multiplicatively-Convex Topological Algebras, Memoirs Amer. Math. Soc., No. 11, Providence, R.I., 1952.

[21] D. MONTGOMERY and L. ZIPPIN: Topological Transformation Groups, Interscience Publishers, New York, 1955.

[22] M.A. NEUMARK: Normierten Algebren, VEB Deutscher Verlag Der Wissenshaftin, Berlin, 1959.

[23] G.E. RICKART: General Theory of Banach Algebras, Van Nostrand, 1960, reprinted with by R.E. Kreiger Publishing Co., New York, 1974.

[24] W. RUDIN: Functional Analysis, Tata McGraw-Hill, New Delhi, 1974.

[25] W. RUDIN: Real and Complex Analysis, Tata McGraw-Hill, Bombay - New Delhi, 1966.

[26] G.F. SIMMONS: Topology and Modern Analysis, McGraw-Hill, New York, 1963.

[27] A.M. SINCLAIR: Automatic Continuity of Linear Operators, L.N. Ser. 21, London Math. Soc., CUP, 1976.

[28] A.E. TAYLOR and D.C. LAY: Introduction to Functional Analysis, John Wiley, New York, 2nd edit., 1980.

[29] L. WAELBROECK: Topological Vector Spaces and Algebras, LNM No. 230, Springer Verlag, New York, 1971.

[30] J. WERMER: Banach Algebras and Several complex Variables, GTM. No. 35, Springer Verlag, New York, 1976.

[31] W. ŻELAZKO: Selected Topics in Topological Algebras, LN Ser.No.31, Mathematisk Institut, Aarhus Universitet, 1971.

Papers Referred

[1′] R. ARENS: The space L^w and convex topological rings, Bull. Amer. Math. Soc. 52 (1946) 931 - 935.

[2′] R. ARENS: Linear topological division algebras, Bull. Amer. Math. Soc. 52 (1974) 623-630.

[3′] R. ARENS and A.P. CALDRON: Analytic functions of several Banach algebra elements, Ann. of Math (2) 62 (1955) 204 - 216.

[4′] G. BERGMAN: A ring primitive on the right but not on the left, Proc. Amer. Math. Soc. 15 (1964) 473 - 475.

[5′] F.T. BIRTELL: Singly-generated Liouville F algebras, Michigen Math. J. 11 (1964) 89-94.

[6′] V.G. IYER: On the spaces of integral functions I, J. Ind. Math. Soc. 12 (1948) 13-30.

[7′] I.M. GELFAND, D.A. RAIKOV and G.E. SILOV: Commutative normed rings, Uspehi Matem Nauk I (1946) 48-146; Amer. Math. Soc. Transl. (2) 5 (1957) 115 - 220.

[8′] I. KAPLANSKY: Normed algebras, Duke Math. J. 16 (1949) 399-418.

[9′] M. LANDSBERG: Lineare topologische Raüme, die nicht lokalkonvercen sind, Math. Z. 65 (1956) 104-112.

[10′] B. MITIAGIN, S. ROLEWICZ, W. ŻELAZKO: Entire functions in B_0-algebras, Studia Math. 21 (1962) 291 - 306.

[11′] M. NAGUMO: Einige alalytische Untersuchungen in Linearen metrischen Ringen, Jap. J. Math. 13 (1936) 61-80.

[12′] S.R. ROLEWICZ: On a certain class of linear metric spaces, Bull. Acad. Polon. Sci., cl.III 5 (1957) 471-473.

[13′] L. WAELBROECK: Le calcul symbolique dans les algebras commutative, J. de Math. P. et App. 33 (1954) 147-186.

[14′] J.H. WILLIAMSON: On topologizing the field $\mathbb{C}(t)$: Proc. Amer. Math. Soc. 5 (1954) 729-734.

INDEX

The bold face figures after the subsection numbers below indicate the corresponding pages in which these numbers occur.

List of Special Symbols

The numbers after the symbols indicate the subsections in which they are introduced

$\sqrt{A}$, 1.2.24-25, $\sqrt[h]{A}$, 1.3.7, $\sqrt[ch]{A}$, 2.218,

$\sqrt[s]{A}$, 1.5.11, A^{qn}, 1.7.13

$\mathbb{C}$, 1.6.1

Δ, 1.3.1 , Δ_c 2.2.18, $\partial_A S$, 8.23

$E(x)$, 5.2.1; $E_x(\lambda)$, 5.2.10; $E_q(x)$, 5.2.6

$\vec{E}^{\lambda}$, 5.2.1; $\vec{e}^{\lambda}$, 5.2.1; $\vec{e}$, 5.2.1

G_δ, 3.6.16

G_i, 1.1.4 ; G_i^l, 3.7.1; G_i^r, 3.7.1

G_q, 1.1.17; G_q^l, 3.7.1; G_q^r, 3.7.1

G_{ie}, 5.2.7; G_{q0}, 5.2.7

$\mathbb{K}$, 1.6.1

l_a, 2.2.2; l_a^0, 3.6.1

$\nu_p(x)$, 3.3.5 ; $\nu(x)$ 3.3.7

π_A, 8.3.1; $\sum_A$, 8.3.1

$\mathbb{R}$, 1.6.1

r_a, 2.2.2; r_a°, 3.6.1

$r(x)$, 1.8.10; $r_1(x)$, 6.6.1; $r_2(x)$, 6.6.1; $r_3(x)$, 6.6.1

$\rho(x)$, 1.7.1; $\rho'(x)$, 1.7.5; $\rho^p(x)$, 6.3.1

$\sigma(x)$, 1.7.1; $\sigma'(x)$, 1.7.5

S^l, 3.7.1; S^r, 3.7.1; S^{bi}, 3.7.1

x_λ, 6.2.1; x'_λ, 6.3.1

$\mathfrak{Z}^{lt}$, 3.7.3; $\mathfrak{Z}^{rt}$, 3.7.3; $\mathfrak{Z}^{bit}$, 3.7.3

$a \leftrightarrow b =$ a commutes with $b : ab = ba$, 1.1.7

$\emptyset =$ empty set

List of Special Abbreviations

GB (algebra)	Gelfand Beurling
LS	linear space
STG	semi-topological group
TA	topological algebra
TG	topological group
TID	topological integral domain
TLS	topological linear space
TSG	topological group
l. ideal	left ideal
r. ideal	right ideal
bi-ideal	both l.r. ideal
q.i	quasi-inverse
l.q.i.	left quasi-inverse
r.q.i	right quasi-inverse
q.	quasi in q. invertible
l.q.	left quasi in l.q. invertible
r.q.	right quasi in r.q. invertible
q. nilpotent	quasi-nilpotent
q.sq.r	quasi-square root
s.q.r.	square root
ext.	extended in ext. q. nilpotent
t.	topological in t. automorphism, t. isomorphism
t.z.d.	topological zero divisor
l.t.z.d.	left topological zero divisor
r.t.z.d.	right topological zero divisor
z.d.	zero divisor
l.z.d.	left zero divisor,
r.z.d.	right zero divisor